RELATIVISTIC, QUANTUM ELECTRODYNAMIC, AND WEAK INTERACTION EFFECTS IN ATOMS

AIP
CONFERENCE
PROCEEDINGS 189

RITA G. LERNER
SERIES EDITOR

RELATIVISTIC, QUANTUM
ELECTRODYNAMIC,
AND WEAK INTERACTION
EFFECTS IN ATOMS

SANTA BARBARA, CA 1988

EDITORS:

WALTER JOHNSON
UNIVERSITY OF NOTRE DAME

PETER MOHR
NATIONAL INSTITUTE OF
STANDARDS AND TECHNOLOGY

JOSEPH SUCHER
UNIVERSITY OF MARYLAND

American Institute of Physics   **New York**

L.C. Catalog Card No. 89-084431
ISBN 0-88318-389-7
DOE CONF 8807162

Printed in the United States of America.

CONTENTS

*The asterisk symbol indicates the person who attended the conference and presented the lecture.

FOREWORD

We were privileged to participate in the first month of the program *Relativistic, Quantum Electrodynamic, and Weak Interaction Effects in Atoms* held in January–July of 1988 at the Institute of Theoretical Physics in Santa Barbara. In fact, this was the first atomic physics program to be held at the ITP, Santa Barbara.

Perhaps the "*Weak Interaction*" in the title of the program applied to us; we were accepted into the program only because of credentials of having done atomic theory long ago, one of us, very long ago. Nonetheless, we greatly enjoyed hearing of the considerable systematization that had been made in theoretical atomic physics, especially in the many-body aspects, and in modern developments in obtaining interactions from quantum electrodynamics. These lectures should serve both as a text for students wishing to learn the subject and for the research worker who wants to follow modern developments.

We went to the first lectures of Ingvar Lindgren, who covered many-body theory in an elegant and systematic fashion. Great progress has been made in evaluating many-body effects in phenomena such as the hyperfine structure in atomic physics.

Joe Sucher treated the relativistic problems carefully, with special reference to effects from virtual pairs. Peter Mohr set up the formalism for systematic calculations of higher-order effects.

Because our visit was short, we could only get the flavor of the meeting. But it is very enjoyable to see the many high-quality research papers collected here.

We were pleased to see the great vitality of the atomic physics program and are sure that these proceedings will be of considerable use.

Hans Bethe   Gerry Brown
Ithaca, New York   Stony Brook, New York

                                                                                      November 1988

PREFACE

This volume is a product of the program on *Relativistic, Quantum Electrodynamic, and Weak Interaction Effects in Atoms* which took place at the Institute for Theoretical Physics at Santa Barbara during the period of January 1, 1988 to June 30, 1988. The aim of this program was to address the problem of bringing quantum electrodynamics, the quantum field theory of electromagnetic interactions, to bear on practical calculations of the structure of atoms with more than one electron. R. P. Feynman, in the closing section of *Theory of Fundamental Processes*, said:

> You are now able to calculate many of the problems of physics by yourself. You still cannot do everything: example—the many-electron atom. The answers are contained in these rules but you will have to learn to use them in their nonrelativistic form, where they correspond to the Schrödinger equation.*

Since 1958, when these remarks were made, sophisticated and accurate relativistic theories have been developed and applied to many-electron atoms, but in most cases these theories are not fully based on the underlying "fundamental processes." An improvement in the methodology is necessary to take into account quantum electrodynamic effects that are not presently included and to provide a framework for dealing in a convincing manner with the calculation of parity and time-reversal violating effects in atomic systems.

The program brought together people with varied backgrounds and approaches to such problems. A series of lectures was given by the participants on various current issues in this field. It was felt that a written record of these lectures would be useful not only to the atomic physics community but also, in view of the interdisciplinary character of much of the subject matter, to a wider audience.

The lectures were pedagogical reviews or reports on current research; in many cases the written version is an expansion of the original material. For papers with more than one author, the name of the person who attended the program and presented the lecture is indicated by an asterisk (*) in the Table of Contents. We present the lectures not in the chronological order in which they were given, but grouped together into several categories determined roughly by subject.

We gratefully acknowledge the hospitality provided by the Institute for Theoretical Physics, and we thank Robert Schrieffer, Ivan Muzinich, and Daniel Hone for their part in organizing the program. In addition, we would like to express our appreciation for the excellent support provided by the secretarial staff at the Institute. We also acknowledge the support for the Institute provided by the National Science Foundation under Grant No. PHY82-17853 with supplemental funds provided by the National Aeronautics and Space Administration.

Walter Johnson   Peter Mohr   Joseph Sucher
South Bend, Indiana   Gaithersburg, Maryland   College Park, Maryland

February 1989

*From R. P. Feynman, *Theory of Fundamental Processes* (Benjamin, New York, 1961), p. 168. Reprinted by permission of Addison-Wesley Publishing Co., Inc., Reading, Mass.

INTRODUCTORY REVIEWS

3

Many-Body Theory

Ingvar Lindgren

Department of Physics, Chalmers University of Technology/
University of Gothenburg, Göteborg, S 412 96 Sweden

In these lectures I shall give a review of the non-relativistic many-body perturbation theory (MBPT). In the first section a brief summary will be given of the Rayleigh-Schrödinger perturbation theory for non-degenerate and degenerate systems, and in the second section the second quantization and the graphical representation are introduced. In the third section the diagrammatic expansions are developed for closed- and open-shell systems, and in the fourth section, finally, the procedure is generalized to the all-order and coupled-cluster approaches.

I. Perturbation theory

I.A. Non-degenerate case. Closed-shell systems.

We are interested in solving the Schrödinger equation

$$H \Psi = E \Psi \tag{I.1}$$

The hamiltonian, H, is partitioned into a *zeroth-order* or *model* hamiltonian, H_0, and a *perturbation*, V,

$$H = H_0 + V \tag{I.2}$$

and we assume that we know the solutions of the zeroth-order equation

$$H_0 \Phi_i = E_0^{(i)} \Phi_i \tag{I.3}$$

One of the solutions, Φ,

$$H_0 \Phi = E_0 \Phi \tag{I.4}$$

is assumed to be an approximation to the solution of (I.1).

The **projection operator** for the unperturbed state, Φ, is

$$P = |\Phi\rangle\langle\Phi| \tag{I.5a}$$

and for the *complementary space*

$$Q = 1 - P \tag{I.5b}$$

Using the **intermediate normalization,**

$$\langle\Phi|\Psi\rangle = \langle\Phi|\Phi\rangle = 1 \tag{I.6}$$

we have

$$P\Psi = |\Phi\rangle\langle\Phi|\Psi\rangle = |\Phi\rangle \tag{I.7}$$

and the full wave function becomes

$$\Psi = P\,\Psi + Q\,\Psi = \Phi + Q\,\Psi \tag{I.8}$$

With intermediate normalization we also have

$$E = \langle \Phi | H | \Psi \rangle = E_0 + \Delta E$$

$$\Delta E = \langle \Phi | V | \Psi \rangle \tag{I.9}$$

The Schrödinger equation (I.1) can be written in the form

$$(E_0 - H_0)\,\Psi = (V - \Delta E)\,\Psi \tag{I.10}$$

and the equation for the perturbed part, $Q\,\Psi$,

$$(E_0 - H_0)\,Q\,\Psi = Q\,(V\,\Psi - \Psi\,\Delta E) = Q\,(V\,\Psi - \Psi\,\langle \Phi | V | \Psi \rangle) \tag{I.11}$$

This can be solved formally by means of the *resolvent operator*

$$R = \frac{Q}{E_0 - H_0} \tag{I.12}$$

$$Q\,\Psi = R\,(V\,\Psi - \Psi\,\langle \Phi | V | \Psi \rangle) \tag{I.13}$$

This leads to the *Rayleigh-Schrödinger perturbation expansion*

$$\Psi = \Phi + \Psi^{(1)} + \Psi^{(2)} + \dots$$

$$E = E_0 + E^{(1)} + E^{(2)} + \dots$$

$$\Psi^{(1)} = R\,V\,\Phi = \frac{Q}{E_0 - H_0}\,V\,\Phi \tag{I.14}$$

$$\Psi^{(2)} = (R\,V\,R\,V - R^2\,V\,P\,V)\,\Phi =$$

$$= \left(\frac{Q}{E_0 - H_0}\,V\,\frac{Q}{E_0 - H_0}\,V - \left(\frac{Q}{E_0 - H_0} \right)^2 V\,P\,V \right)\Phi$$

etc.

$$E^{(n)} = \langle \Phi | V | \Psi^{(n-1)} \rangle$$

I. B. Degenerate case. Open-shell system.

We consider now the more general case, where the unperturbed state (I.4) is degenerate. We then have to consider all the corresponding exact states simultaneously, say,

$$H\,\Psi^{(a)} = E^{(a)}\,\Psi^{(a)} \qquad (a=1,2,\dots d) \tag{I.15}$$

which form a *"target space"*. For each of the target functions we assume that there exists a corresponding *model function*, $\Psi_0^{(a)}$, which is an eigenfunction of the approximate Schrödinger equation

$$H_0\,\Psi_0^{(a)} = E_0\,\Psi_0^{(a)} \qquad (a=1,2,\dots d) \tag{I.16}$$

These functions form the **model space** (P). Note that these model functions are generally *linear combinations of the basis functions* (I.3) and not known *ab initio.*

It is also possible to work with an *extended model space*, containing the eigenfunctions corresponding *several* eigenvalues of H_0. Then the model functions are no longer eigenfunctions of H_0. (In this formalism, all degenerate eigenstates must be treated in the same way, i.e. they have to be either inside or outside P.)

A ***wave operator***, Ω, transforms all the model functions into the corresponding target functions

$$\Psi^{(a)} = \Omega \, \Psi_o^{(a)} \tag{I.17}$$

The ***effective hamiltonian***, H_{eff}, operating on any of the model functions generates the corresponding *exact* energy,

$$H_{eff} \, \Psi_o^{(a)} = E^{(a)} \, \Psi_o^{(a)} \tag{I.18}$$

Operating on this equation with Ω from the left, gives

$$\Omega \, H_{eff} \, \Psi_o^{(a)} = E^{(a)} \, \Psi^{(a)} = H \, \Omega \, \Psi_o^{(a)} \tag{I.19}$$

Since this holds for all a, it can be written as an *operator equation*

$$\Omega \, H_{eff} \, P = H \, \Omega \, P \tag{I.20}$$

referred to as the ***Bloch equation.*** By partitioning the effective hamiltonian in analogy with (I.2)

$$H_{eff} \, P = (H_0 + V_{eff}) \, P \tag{I.21}$$

the Bloch equation can be written in the form of the commutator relation

$$[\Omega, H_0] \, P = (V \, \Omega - \Omega \, V_{eff}) \, P \tag{I.22}$$

V_{eff} represents the *effective Coulomb interaction.*

The equations above hold regardless of the choice of the normalization. The *intermediate normalization* (I.6) implies in this case that the model functions are the *projection of the corresponding target functions onto the model space*

$$\Psi_o^{(a)} = P \, \Psi^{(a)} \tag{I.23}$$

which leads to

$$P \, \Omega \, P = P \tag{I.24}$$

The effective hamiltonian and the effective interaction are in this scheme

$$H_{eff} \, P = P \, H \, \Omega \, P \; ; \quad V_{eff} \, P = P \, V \, \Omega \, P \tag{I.25}$$

and the Bloch equation (I.22) becomes

$$[\Omega, H_0] \, P = (V \, \Omega - \Omega \, P \, V \, \Omega) \, P \tag{I.26}$$

In the open-shell case we can not get an expansion of the wave functions directly. Instead we *expand the wave operator* order by order

$$\Omega = 1 + \Omega^{(1)} + \Omega^{(2)} + \ldots \tag{I.27}$$

and insert into (I.26), which yields the *general Rayleigh-Schrödinger perturbation expansion*

$$[\Omega^{(1)}, H_0] P = (V - P V) P = Q V P$$

$$[\Omega^{(2)}, H_0] P = (Q V \Omega^{(1)} - \Omega^{(1)} P V) P \tag{I.28}$$

etc.

and for the effective hamiltonian (I.25)

$$H_{eff}^{(n)} = P V \Omega^{(n-1)} \tag{I.29}$$

(This can be compared with the corresponding expansion in the non-degenerate case (I.14).)

According to (I.18), *the eigenvalues of the effective hamiltonian within the model space yield the energies of the target functions and eigenvectors the corresponding model functions*. Note that *the effective hamiltonian is generally non-hermitian with intermediate normalization*. By another choice of normalization it is possible to formulate the theory so that this operator becomes hermitian.

II. Second Quantization. Graphical representation.

II.A. Orbital classification

We assume now that the hamiltonian is of the form [*)]

$$H = \sum_{n=1}^{N} (- 1/2 \, \nabla^2 - Z/r)_i + \sum_{n<m=1}^{N} 1/r_{ij} \tag{II.1}$$

with the zeroth-order hamiltonian and the perturbation given by

$$H_0 = \sum_{n=1}^{N} h_0(i) ; \quad h_0 = - 1/2 \, \nabla^2 - Z/r + U \tag{II.2a}$$

$$V = - \sum_{n=1}^{N} U(i) + \sum_{n<m=1}^{N} 1/r_{ij} \tag{II.2b}$$

[*)] *Hartree atomic units are used, i.e.,*

$$e = m = h/2\pi = 4 \, \pi \, \epsilon_0 = 1$$

which gives the velocity of light $c = 1/\alpha \approx 137$, the unit of length equal to the Bohr radius and the **unit of energy**

$$\mathbf{1 \ H = 2 \ h \ c \ Ry = \alpha^2 \, m \, c^2 \approx 27.2 \ eV}$$

N is the number of electrons, and U is a single-electron potential, which is optional. The eigenfunctions of h_0

$$h_0\,\phi_i = \varepsilon_i\,\phi_i \qquad\qquad (II.3)$$

form the *single-electron basis functions* or *(spin-)orbitals*. The eigenfunctions of H_0 (I.3)

$$H_0\,\Phi_i = E_0^{(i)}\,\Phi_i \qquad\qquad (II.4)$$

are assumed to be antisymmetrized products (Slater determinants) of the orbitals.

The model space (P) is spanned by a number of determinants, corresponding to one or several eigenvalues of H_0, and the orbitals are separated into the following three categories (see Fig. 1):

a) **core orbitals,** which are occupied in all determinants in P;

b) **valence orbitals,** which are occupied in some, unoccupied in some determinants in P;

c) **virtual orbitals,** which are unoccupied in all determinants in P.

The model space is said to be complete, if it contains all the determinants that can be formed by distributing the valence electrons in all possible ways among the valence orbitals.

With the intermediate normalization the **linked-diagram theorem,** *to be discussed in sect. III, is valid only for complete model spaces.* It is possible, however, to choose the normalization in such a way that the theorem is valid for arbitrary, incomplete *model spaces.*

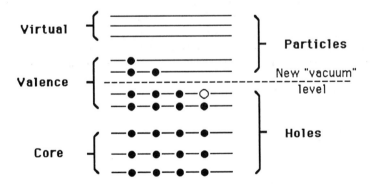

Figure 1. *Orbital classification in the particle-hole representation.*

8

II.B. Particle-hole representation

Using the second quantization, the zeroth-order hamiltonian and the perturbation (II.2) can be expressed

$$H_0 = \sum_i a_i^{\dagger} a_i \varepsilon_i \tag{II.5a}$$

$$V = \sum_{ij} a_i^{\dagger} a_j \langle i| -U | j \rangle + 1/2 \sum_{ijkl} a_i^{\dagger} a_j^{\dagger} a_l a_k \langle ij | 1/r_{12} | kl \rangle \tag{II.5b}$$

where $a_i^{\dagger}$ (a_j) are single-electron creation (destruction) operators, satisfying the ordinary anti-commutation rules.

It follows from the orbital classification above that

$$a_{\text{virtual}} P = a^{\dagger}_{\text{core}} P = 0 \tag{II.6}$$

while $a_{\text{valence}} P$ or $a^{\dagger}_{\text{valence}} P$ need not be zero.

The operators in (II.5) are in *normal form with respect to the true vacuum*, i.e., with the creation operators to the left of the destruction operators. In many-body theory it is often more convenient to use the **particle-hole representation**, *i.e., the normal form with respect to some closed-shell state*, forming a **new vacuum level** (see Fig. 1). Particle (hole) states are unoccupied (occupied) in that state. In the normal form *the particle-hole creation operators are placed to the left of the particle-hole destruction operators.* (Hole creation/destruction = destruction/creation of an electron in a hole state).

In the p-h representation the valence orbitals can be of two types, *valence particles and valence holes*, while the core orbitals are assumed to be hole states and the virtual orbitals to be particle states. The orbitals can be represented as shown in Fig. 2a. *The electron creation (destruction) operators are represented by lines directed out from (towards) the inter-action vertex.* This has the consequence that **the lines above (below) the interaction line represent p-h creation (destruction) operators** (see Fig. 2b).

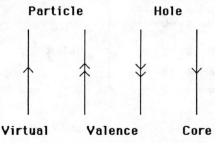

Figure 2a. *Graphical representation of the orbitals.*

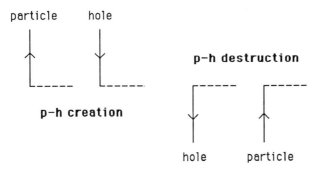

Figure 2b. *Particle-hole* **creation** *is represented by orbital lines* **above** *the interaction line and corresponding* **destruction** *by orbital lines* **below** *the that line.*

The idea is now to express the perturbation and the wave operator in the p-h representation and insert them into the Rayleigh-Schrödinger expansion (I.28). By means of Wick's theorem the rhs can be transformed to normal form and identifications be made with the commutator on the lhs. This will lead to the *linked-diagram (Brueckner-Goldstone) expansion* for complete model spaces. The same procedure applied to the Bloch eqn (I.26) directly will lead to the corresponding *iterative (all-order) procedure* to be discussed in section IV.

The perturbation (II.5b) becomes in the p-h representation

$$V = V_0 + V_1 + V_2$$

$$V_0 = \sum_a \langle a|-U|a\rangle + 1/2 \sum_{ab} \langle ab| \widetilde{1/r_{12}} |ab\rangle$$

$$V_1 = \sum_{ij} \{a_i{}^\dagger a_j\} \langle i|v|j\rangle \tag{II.7a}$$

$$V_2 = 1/2 \sum_{ijkl} \{a_i{}^\dagger a_j{}^\dagger a_l a_k\} \langle ij| 1/r_{12}|kl\rangle$$

$$\langle i|v|j\rangle = \langle i|-U|j\rangle + \sum_a \langle ia| \widetilde{1/r_{12}} |ja\rangle \tag{II.7b}$$

where the curly brackets represent normal order in the p-h representation and the tilde indicates that exchange is included. In this representation the contractions within the perturbation V are done once and for all. The one-

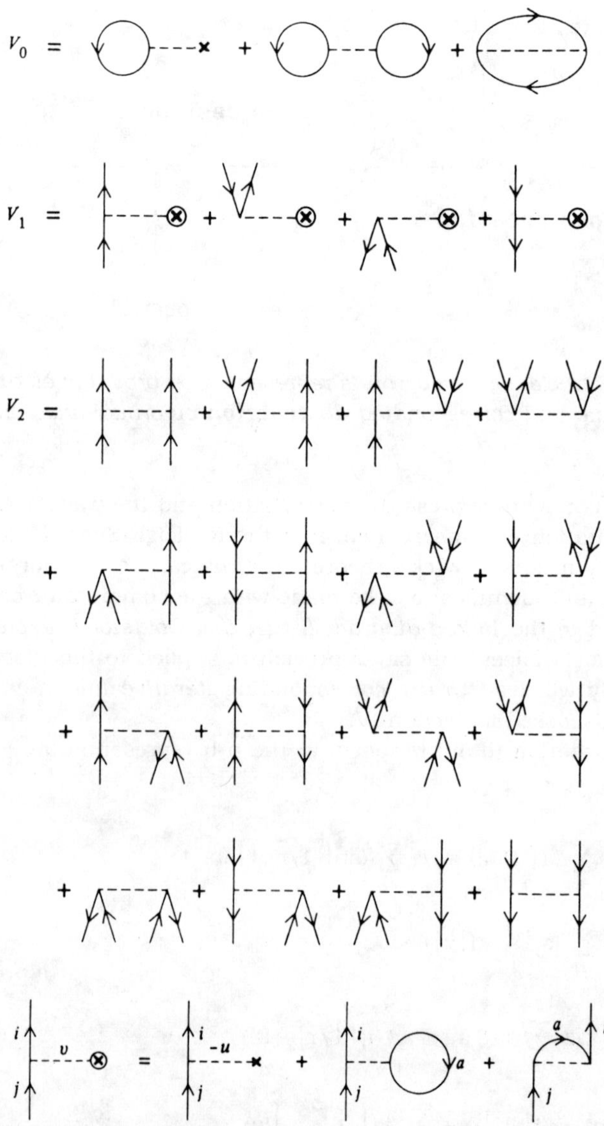

Figure 3. Graphical representation of the perturbation (II.7a) and the effective potential (II.7b) in the particle-hole representation. The closed loops are summed over the hole states, i.e., orbitals occupied in the new vacuum state.

body part, V_1, vanishes when U equals the Hartree-Fock potential of the new vacuum state. The graphical representation of the normal-ordered perturbation is shown in Fig. 3.

Similarly, the wave operator can be separated into n-body parts

$$\Omega = 1 + \Omega_1 + \Omega_2 + \ldots$$

$$\Omega_1 = \sum_{ij} \{a_i{}^\dagger a_j\} \, x_j^i \tag{II.8}$$

$$\Omega_2 = 1/2 \sum_{ijkl} \{a_i{}^\dagger a_j{}^\dagger a_l a_k\} \, x_{kl}^{ij}$$

The wave operator operates to the right always on the model space, P, and we consider only terms which here give non-zero contributions. Then there can be no a_{virtual} or $a^\dagger{}_{\text{core}}$ operators in Ω, while a_{valence} as well as $a^\dagger{}_{\text{valence}}$ may appear. Therefore, in the graphical representation of Ω there can be no other free lines at the bottom than valence lines. With valence orbitals only of *particle* type we get the representation shown in Fig. 4.

Ω :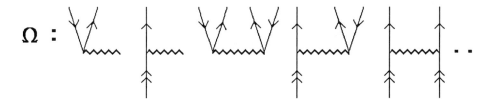

Figure 4. *Graphical representation of the wave operator, assuming valence particle only (no valence holes).*

Wick's theorem for a *product of normal-ordered operators* can be expressed

$$A \, B = \{A \, B\} + \{\overset{\frown}{A \, B}\} \tag{II.9}$$

where $\{A \, B\}$ represents the product without any contractions and $\{\overset{\frown}{A \, B}\}$ represents all possible single, double, . . contractions between p-h destruction operators in A and p-h creation operators in B. There are no contractions within A or B. Graphically, this relation is represented by **all diagrams that can be formed by joining zero, one, two . . . lines at the bottom of A with the lines at the top of B** (with the arrows being continuous), as illustrated in Fig. 5.

12

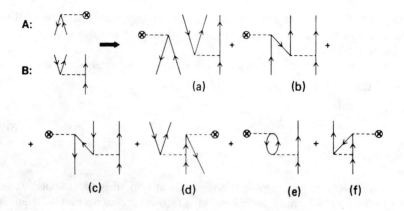

Figure 5. *Illustration of Wick's theorem in graphical form. Diagram (a) represents the product without any contraction, diagrams (b-d) represent single and (e,f) double contractions.*

III. Diagrammatic expansion

III.A. Closed-shell case.

We now apply the technique described above to the Rayleigh-Schrödinger expansion (I.28), and we shall express both sides in normal form by using Wick's theorem and then make the identifications. The commutators on the lhs are easily found using the second-quantized form (II.8) of the wave operator, and we then get for the one- and two-body parts, respectively.

$$[\Omega_1, H_0] = \sum_{ij} \{a_i{}^\dagger a_j\} (\epsilon_i - \epsilon_j) x_j^i \qquad (III.1)$$

$$[\Omega_2, H_0] = 1/2 \sum_{ijkl} \{a_i{}^\dagger a_j{}^\dagger a_l a_k\} (\epsilon_k + \epsilon_l - \epsilon_i - \epsilon_j) x_{kl}^{ij}$$

Thus, the commutator simply supplies a factor equal to the excitation energy in each term. The rhs is obtained by means of the normal form of the perturbation (Fig. 3) and Wick's theorem (Fig. 5).

In the closed-shell case there are *no valence orbitals*, so the (single-electron) hole states are identical to the core orbitals and the (single-electron) particle states are identical to the virtual orbitals (See Fig. 1). The model space consists of a single (many-electron) function, which is also the nwe vacuum state (Φ) for the normal-ordering. In this state all core orbitals (hole states) are filled and all virtual orbitals (particle states) empty (II.6)

$$a^\dagger{}_{core} \Phi = a_{virtual} \Phi = 0 \quad \text{or} \quad a^\dagger{}_{core} P = a_{virtual} P = 0 \qquad (III.2)$$

This means that an operator in normal form operating on P gives zero if it contains any $a^\dagger{}_{core}$ or $a_{virtual}$ operator, since the operators can be permuted within the normal product without contractions. Graphically, this implies that *no free lines are allowed at the bottom when operating on the model space.*

First order

In the first-order RS eqn (I.28)

$$[\Omega^{(1)}, H_0] P = Q V P \qquad (III.3)$$

the rhs is given by the part of the normal-ordered perturbation (Fig. 3), which does not have any lines at the bottom and which is not completely closed (because of the Q operator). This gives the following diagrams

14

$$\mathcal{Q}VP: \qquad = a_r{}^\dagger a_a \langle r|v|a\rangle \qquad\qquad \text{(III.4a)}$$

$$= a_r{}^\dagger a_s{}^\dagger a_b a_a \langle rs|1/r_{12}|ab\rangle \qquad\qquad \text{(III.4b)}$$

a, b, . . represent core orbitals and r, s, . . virtual orbitals. The operators are here automatically in normal order. The first-order wave operator, $\Omega^{(1)}$, can be represented by the same diagrams, if we include an ***energy denominator equal to the excitation energy*** according to (III.1), i.e.,

$$\Omega^{(1)}: \qquad = a_r{}^\dagger a_a \langle r|v|a\rangle / (\varepsilon_a - \varepsilon_r) \qquad\qquad \text{(III.5a)}$$

$$= a_r{}^\dagger a_s{}^\dagger a_b a_a \langle rs|1/r_{12}|ab\rangle / (\varepsilon_a + \varepsilon_b - \varepsilon_r - \varepsilon_s) \qquad \text{(III.5b)}$$

The corresponding energy is obtained by means of (I.29)

$$H_{\text{eff}}{}^{(n)} = P\, V\, \Omega^{(n-1)} \quad \text{or} \quad E^{(n)} = \langle \Phi|V\,\Omega^{(n-1)}|\Phi\rangle \qquad\qquad \text{(III.6)}$$

i.e., by "closing" the wave-operator diagrams by the perturbation. This gives the second-order energy diagrams

$$E^{(2)}: \qquad = \sum_{a,r} \frac{\langle a|v|r\rangle\langle r|v|a\rangle}{\varepsilon_a - \varepsilon_r} \qquad\qquad \text{(III.7a)}$$

$$= \frac{1}{2}\sum_{abrs} \frac{\langle ab|\frac{1}{r_{12}}|rs\rangle\langle rs|\frac{1}{r_{12}}|ab\rangle}{\varepsilon_a + \varepsilon_b - \varepsilon_r - \varepsilon_s} \qquad\qquad \text{(III.7b)}$$

$$= -\frac{1}{2}\sum_{abrs} \frac{\langle ab|\frac{1}{r_{12}}|sr\rangle\langle rs|\frac{1}{r_{12}}|ab\rangle}{\varepsilon_a + \varepsilon_b - \varepsilon_r - \varepsilon_s} \qquad\qquad \text{(III.7c)}$$

Second order

The second-order RS expression is

$$[\Omega^{(2)}, H_o] P = (\varrho V \Omega^{(1)} - \Omega^{(1)} P V) P \qquad (III.8)$$

Note that in PVP no operators or free lines at all are allowed, since V operates on P in both directions. This leaves only the closed part, V_o, which is a C number. That cancels the V_o part of the first term, and we can write the second-order expression as

$$[\Omega^{(2)}, H_o] P = \varrho (V_1 + V_2) \Omega^{(1)} P \qquad (III.9)$$

The rhs is obtained graphically by operating on $\Omega^{(1)}$ above with V_1 and V_2 from Fig. 3, using Wick's theorem. Again, no free lines are allowed at the bottom, and the diagrams must not be entirely closed (at least one free line at the top). The V_2 part of the rhs is shown in Fig. 6. Note that the diagrams of V_1 and $\Omega^{(1)}$ need not be connected to each other. The diagrams of $\Omega^{(2)}$ can be represented by the same diagrams as the rhs by including the energy denominator corresponding to the last excitation.

The corresponding third-order energy diagrams, shown in Fig. 7, are obtained by closing the $\Omega^{(2)}$ diagrams in analogy with (III.8). Note that the wave-operator diagrams with more than two free lines cannot be closed by the perturbation.

Third order

The third-order RS expression is

$$[\Omega^{(3)}, H_o] P = \varrho (V_1 + V_2) \Omega^{(2)} P - \Omega^{(1)} P (V_1 + V_2) \Omega^{(1)} P \qquad (III.10)$$

eliminating the V_o part as before. When V operates on a *disconnected* diagram of $\Omega^{(2)}$ (Fig. 6), **unlinked diagrams** can be formed, i.e., diagrams **with a closed, disconnected part,** as illustrated in Fig. 8. These diagrams are **cancelled** by the last term, provided the **exclusion principle is ignored** in the first term. (The corresponding **linked** diagrams are shown in Fig. 9.) This kind of cancellation occurs in each order, which is the **linked-diagram** or **linked-cluster theorem.** The expansion can then - in the closed-shell case - be expressed

$$[\Omega^{(n)}, H_o] P = (\varrho V \Omega^{(n-1)} P)_{linked} \qquad (III.11)$$

Since the model space contains only one function (Φ) in this case, we can also get an explicit expansion for the full wave function

$$\Psi = \sum_{n=0}^{\infty} \left(\frac{\varrho}{E_o - H_o} V \right)^n_{linked} \qquad (III.12)$$

using the resolvent operator (I.12). The corresponding energy expression is

16

$$E = E_0 + \sum_{n=0}^{\infty} \langle \Phi | V \left(\frac{Q}{E_0 - H_0} V \right)_{\text{linked}}^{n} | \Phi \rangle \tag{III.13}$$

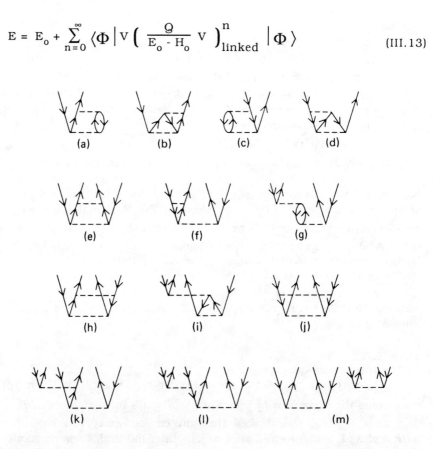

Figure 6. *Second-order wave-operator diagrams for closed-shell systems. The effective-potential diagrams are left out.*

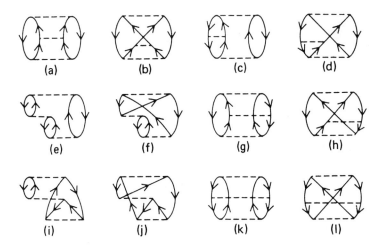

Figure 7. *Third-order energy diagrams for closed-shell systems. (Potential diagrams left out).*

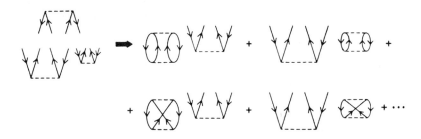

Figure 8. *By operating with the perturbation on a disconnected wave-operator diagram, unlinked diagrams can be formed. These are cancelled by the last term in the RS expansion, provided the exclusion principle is ignored. The corresponding linked diagrams are shown in Fig. 9.*

18

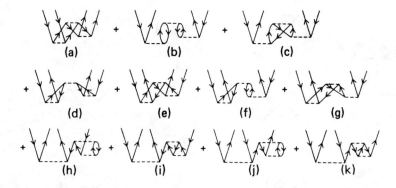

Figure 9. *Linked two-body diagrams formed by operating with the perturbation on disconnected second-order wave operator diagrams. (C.f. the
coupled-cluster diagrams in Fig. 14).*

III. B. Open-shell case

In the open-shell case there are - in addition to the core and virtual
orbitals - also *valence orbitals*, which are partially filled in the model space
(see Fig. 1). The relations (III.2) still hold, but there are no similar relations
for the valence orbitals, since these can be created as well as absorbed,
when operating on the model space,

$$a^{\dagger}_{valence} \, P \neq 0 \quad \text{and} \quad a_{valence} \, P \neq 0 \qquad (III.14)$$

This has the important consequence that *free valence lines are allowed at
the bottom when operating on the model space.* The diagrams without
valence lines are constructed as before. In addition, there will be diagrams
with one, two, . . valence lines at the bottom, depending on the number of
valence electrons in the system.

First-order wave-operator

The first-order wave operator is obtained by means of the first-order RS
eqn as before (I.28, III.3)

$$[\Omega^{(1)}, H_0] \, P = Q \, V \, P \qquad (III.15)$$

with the difference that also free lines are now allowed at the bottom, provided they are valence lines. Assuming that all valence lines are of particle
type, we get the graphical representation shown in Fig. 10. It should be
noted that the outgoing lines may be valence lines, but **_all_ in- and outgoing
lines must not be valence lines** in order to reach the Q space. The denomi-

nator of the diagrams are generally given by the rule

$$D = \Sigma \left(\varepsilon_{in} - \varepsilon_{out} \right) \qquad (III.16)$$

where ε_{in} (ε_{out}) represent the orbital energies of lines coming in to (going out from) the interaction line.

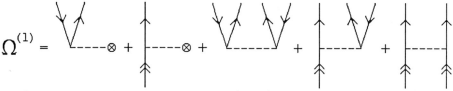

Figure 10. *Graphical representation of the first-order wave operator for open-shell systems.*

Effective hamiltonian

The effective hamiltonian is given by (I.29, III.6), i.e., generally by "closing" the diagrams of the wave operator by the perturbation. *A closed diagram* operates only within the model space, which means that *all free lines are valence lines.* Diagrams without free lines also belong to the effective hamiltonian. They represent a C-number equal to the energy of the vacuum state.

The first-order effective hamiltonian is simply represented by the part of the perturbation in Fig. 3, which can operate within the model space. In addition to the core diagrams, represented by V_0, there are only two such diagrams (with valence particles only), shown in Fig. 11.

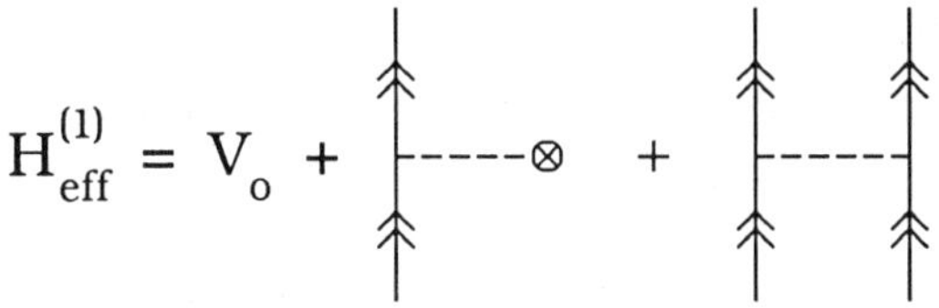

$$H^{(1)}_{eff} = V_o + \quad\dots\quad + \quad\dots$$

Figure 11. *Representation of the first-order effective-operator for open-shell systems, assuming valence particles only.*

The second-order effective hamiltonian is obtained by closing the diagrams of the first-order wave operator in Fig. 10. This leads to a considerable number of diagrams, of which the one- and two-body parts (with one and two pairs of free lines, respectively) are shown in Fig. 12.

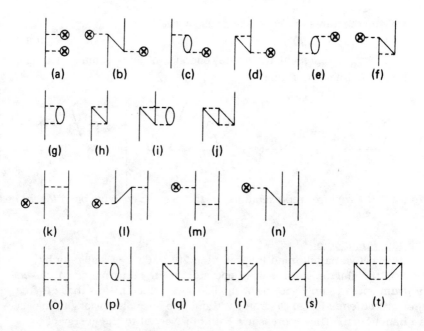

Figure 12. *The second-order effective hamiltonian. Only the one- and two-body parts are shown. All free lines are valence lines.*

<u>*Second-order wave-operator. Folded diagrams*</u>

The second-order wave-operator diagrams are constructed as before by means of (II.16)

$$[\Omega^{(2)}, H_0] P = (Q V \Omega^{(1)} - \Omega^{(1)} P V) P \tag{III.17}$$

The V_0 part of the perturbation can still be eliminated, but there are additional contributions from the last term, not occurring in the closed-shell case

$$[\Omega^{(2)}, H_0] P = Q (V_1 + V_2) \Omega^{(1)} P - \Omega^{(1)} P (V_1 + V_2) P \tag{III.18}$$

As in the closed-shell case, the first term can give rise to **unlinked diagrams,** now defined as diagrams with a **disconnected, closed part** (with no other free lines than valence lines) as illustrated in Fig. 13. The last term is essentially the first-order wave operator operating on the first-order effective hamiltonian. This gives rise to connected and disconnected diagrams, of which the former are unlinked and cancel the unlinked diagrams from the first term, again provided the exclusion principle is ignored. The connected diagrams from the last term, on the other hand, do not cancel, and they represent a new type of diagrams, called **folded diagrams,** not present for closed-shell systems. (Note that the connecting line must be a valence line). Traditionally these diagrams are drawn in a folded way, the reason being that the standard Goldstone rules can then be used for their evaluation (in the general case by considering all possible time-orderings and then using the factorization theorem). With the procedure described here, however, starting from the Bloch eqn, the evaluation is straightforward and there is no need for drawing them folded.

It can be shown that the cancellation demonstrated in second order occurs in each order for a *complete model space* (see p. 5) , and we can then express the **linked-diagram theorem for open-shell systems** in analogy with (II.19) as

$$[\Omega^{(n)}, H_0] P = (V \Omega - \Omega P V \Omega)^{(n)}_{\text{linked}} \qquad \text{(III.19)}$$

The last term represents here the folded diagrams. For closed-shell systems there are no folded diagrams, and the expression goes over into (III.11). By summing to all orders, we can express the theorem in terms of a modified "Bloch eqn"

$$[\Omega, H_0] P = (V \Omega - \Omega \ P V \Omega)_{\text{linked}} P \qquad \text{(III.20)}$$

$P V \Omega$ appearing in the eqns above represents the effective interaction - or together with H_0 the effective hamiltonian. The corresponding diagrams are obtained, as before, by closing the wave-operator diagrams by the perturbation, which implies that no other free lines than valence lines are allowed. Graphically, the effective-interaction diagrams are quite similar to the wave-operator diagrams, the difference being that all outgoing lines are valence lines and that there are no folded diagrams.

22

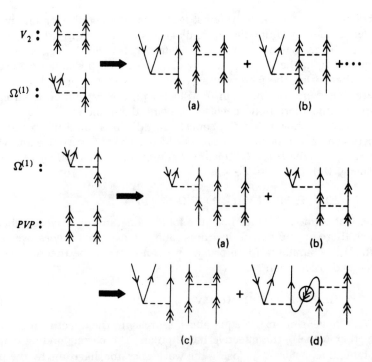

Figure 13. *Illustration of the cancellation of unlinked diagrams and the appearance of folded diagrams (d) in the open-shell MBPT.*

IV. All-order procedure. Coupled-cluster approach.

IV.A. All-order procedure

Instead of performing a perturbation expansion order by order, it is often convenient to start directly with the n-body expansion (II.8) of the wave operator

$$\Omega = \Omega_1 + \Omega_2 + \ldots \tag{IV.1}$$

$$\Omega_1 = \sum_{ij} \{a_i{}^\dagger a_j\}\, x_j^i \tag{IV.2a}$$

$$\Omega_2 = 1/2 \sum_{ijkl} \{a_i{}^\dagger a_j{}^\dagger a_l a_k\}\, x_{kl}^{ij} \tag{IV.2a}$$

with the graphical representation given in Fig. 4,

$$\tag{IV.3a}$$

$$\tag{IV.3b}$$

By inserting this into the all-order Bloch eqn (II.27), a set of coupled n-body equations is obtained

$$[\Omega_n, H_0]\, P = (\, V\,\Omega - \Omega\, P\, V\,\Omega)_{n\text{-body, linked}}\; P \tag{IV.4}$$

The rhs is transformed to normal form, using Wick's theorem in algebraic or graphical form, and identifications are then made with the lhs, as before. Solving these eqns self-consistently is equivalent to evaluating the corresponding effects to **all orders of perturbation theory.**

For closed-shell systems the energy is obtained, as before, by "closing" the wave-operator diagrams. This means that the full correlation energy is represented by the same diagrams as in second order, if the first Coulomb interaction is replaced by the all-order wave operator. This gives the following simple expression for the **correlation energy to all orders** (leaving out the potential contributions)

$$= \frac{1}{2} \sum_{abrs} \langle ab | \frac{1}{r_{12}} | rs \rangle x_{ab}^{rs} \qquad (IV.5a)$$

$$= -\frac{1}{2} \sum_{abrs} \langle ab | \frac{1}{r_{12}} | sr \rangle x_{ab}^{rs} \qquad (IV.5b)$$

For open-shell systems the diagrams of the effective hamiltonian (or effective interaction) are obtained in a similar way by closing the diagrams of the wave operator so that no other free lines than valence lines appear. These diagrams are quite similar to the second-order diagrams of the effective hamiltonian shown in Fig. 12.

III.B. Coupled-cluster approach

A related but more powerful procedure is the **coupled-cluster approach,** where *the wave operator is expressed as an exponential*

$$\Omega = \exp S = 1 + S + 1/2 \, S^2 + ... \qquad (IV.6)$$

S is referred to as the **cluster operator.** For general open-shell systems it is most convenient to use a *normal-ordered exponential*

$$\Omega = \{\exp S\} = 1 + S + 1/2 \, \{S^2\} + ... \qquad (IV.7)$$

which implies that there are no contractions between the cluster operators. The idea is then to use an expansion for S, similar to (IV.1,2)

$$S = S_1 + S_2 + . . . \qquad (IV.8)$$

$$\Omega_1 = \sum_{ij} \{a_i{}^\dagger a_j\} \, s_j^i \qquad (IV.9a)$$

$$\Omega_2 = 1/2 \sum_{ijkl} \{a_i{}^\dagger a_j{}^\dagger a_l a_k\} \, s_{kl}^{ij} \qquad (IV.9b)$$

which can be represented graphically

$$\mathbf{S}_1 = \text{[diagram]} + \text{[diagram]} \tag{IV.10a}$$

$$\mathbf{S}_2 = \text{[diagram]} + \text{[diagram]} + \text{[diagram]} \tag{IV.10b}$$

and to insert that into the Bloch eqn. It can then be shown that for a complete model space this will lead to a **cluster operator which is fully connected** - in contrast to the wave operator which may have disconnected, open parts (see Fig. 6m). The result can then be expressed

$$[S_n, H_0] = (V\Omega - \Omega PV\Omega)_{\text{n-body, connected}} \tag{IV.11}$$

Summing to all orders, this can also be written in analogy with (III.20) as

$$[S, H_0] = (V\Omega - \Omega PV\Omega)_{\text{connected}} \tag{IV.12}$$

In constructing the rhs of the cluster eqn the expansion (IV.7) of the wave operator is used. The diagrams containing more than one cluster operator are the particular **coupled-cluster diagrams** appearing in this formalism. There are no disconnected diagrams, in contrast to the corresponding eqns based on the all-order eqn (IV.4). The graphical representation of the one- and two-body coupled-cluster equations are shown in Fig. 14.

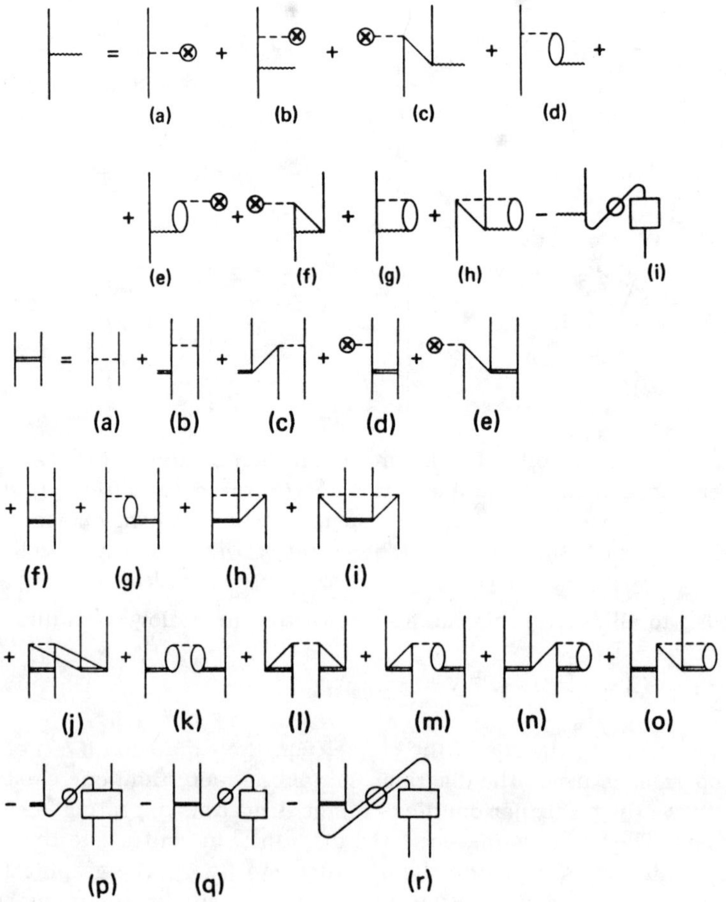

26

Figure 14. Graphical representation of the one- and two-particle coupled-cluster eqns. The diagrams (j-o) in the two-particle eqn are the coupled-cluster diagrams originating from operating with V on {S²} (the remaining ones left out). Diagram (i) in the one-particle eqn and (p-r) in the two-particle eqn are the folded diagrams. The box represents the effective interaction PVΩ. The incoming lines can be core or valence lines. In the former case there are no folded diagrams (p-r).

The procedure described above leads to a set of eqns for the x or s coefficients in the wave operator (IV.2) or cluster operator (IV.9), respectively. This procedure requires that a basis set is used to evaluate the matrix elements appearing in the eqns. Alternatively, **one-, two-,.. particle functions** can be defined by summing the one-, two-, . . body contributions over the excited (virtual) states, keeping the orbitals being excited fixed,

$$\rho_a = \sum_r s_a^r \phi_r \qquad \text{(IV.14a)}$$

$$\rho_{ab} = \sum_{rs} s_{ab}^{rs} \phi_r \phi_s \qquad \text{(IV.14b)}$$

etc., where ϕ_r and ϕ_s are single-electron basis functions. The functions (IV.14) satisfy **inhomogeneous differential equations,** which are obtained by summing the rhs of the eqns above in the corresponding way. By using the closure property, the infinite sums can be eliminated, and practically useful equations are obtained. By means of these functions, for instance, the all-order correlation energy (IV.5) becomes

$$E_{corr} = \frac{1}{2} \sum_{ab} [\, \langle ab | \frac{1}{r_{12}} | \rho_{ab} \rangle - \langle ab | \frac{1}{r_{12}} | \rho_{ba} \rangle \,] \qquad \text{(IV.15)}$$

General reference

I.Lindgren and J.Morrison, *Atomic Many-Body Theory*, second ed., Springer Series in Atoms and Plasmas, Vol 3 (1986); see also the lecture notes by Ann-Marie Mårtensson-Pendrill in this volume.

BOUND STATE QED

J. Sucher

Department of Physics, University of Maryland, College Park, MD 20742

ABSTRACT

A review is given of the foundations of the relativistic theory of atomic systems within the framework of QED. The emphasis is on an approach in which in first approximation the system is described by a configuration-space Hamiltonian which takes into account the degree of freedom associated with virtual electron-positron pairs.

I. Introduction

The theme of this workshop is the study of relativistic, quantum electrodynamic and weak interaction effects in atoms and the aim of these initial "pedagogical" lectures is to review relevant language and techniques to provide a common background for further discussion and later lectures by the participants. Ingvar Lindgren has set a deplorably high standard for such lectures with his review of many-body techniques in the context of ordinary quantum mechanics.[1]

Although weak interaction effects can always be treated by perturbation theory this is not true for relativistic and QED effects. One often needs a starting point which incorporates at least a major part of these effects and this is one of the topics I will review.

Before getting into this and to set the stage for later lectures by others, let's consider what the phrase "bound-state QED" encompasses. For simplicity let us confine our attention to physical systems which can be usefully regarded as composed of N electrons ($N \geq 1$), together with a single positively charged constituent C_+ of mass m_+, such as a positron, muon, proton, or nucleus. We can then distinguish between two cases: (i) $m_+ \sim m_e$ and (ii) $m_+ >> m_e$. For $N = 1$, we have in the first case only one known system, positronium with $m_+ = m_e$, while in the second case we have muonium, hydrogen, and the sequence of hydrogenic ions. For $N \geq 2$, we have the usual atoms, $He, Li \ldots$, and the associated ions, as well as the negative H-ion.

The accurate treatment of the $N = 1$ cases leads to the study of the relativistic two-body problem in the context of field theory and to the topic of "relativistic two-body wave equations." However, except for positronium, it is an excellent first approximation to regard C_+ as infinitely heavy, serving only as the source of an external electromagnetic field. In the context of this "external field approximation" (EFA) one is led to the study of radiative corrections, (defined in the present context as level shifts arising from the emission and subsequent absorption of virtual photons by the electron or from closed loops) in their simplest setting. Nevertheless, if one then wishes to take into account the finite mass of C_+ *i.e.* include recoil corrections, one is again led back to the relativistic two-body problem. There is a long history and a highly-developed technology for dealing with these problems; some of this will be reviewed in later lectures, especially those of Adkins, of Yennie and of Grotch.[2]

Let us now turn to the cases where $N \geq 2$ and C_+ is a nucleus. This is the domain of "many-electron atoms," a phrase which, by a common abuse of language, also includes many-electron ions. It is then always a good first approximation to treat the nucleus as the source of an external field; the inclusion of nuclear spin in order to treat hyperfine structure causes no essential complications. We are then dealing with the two-or-more electron problem in the EFA. With Ze the nuclear charge, this domain of problems can again be divided into two categories: (i) $Z >> N$ and (ii) $Z \sim N$.

In the first category one can, as in the $N = 1$ case, use the Furry representation for the Dirac field associated with the electrons and thereby take into account, at least formally, the interaction with the external field exactly. Working in the radiation gauge, one can then treat not only the interaction with the quantized transverse radiation field as a perturbation but also the self-interaction of the Dirac field, corresponding to the Coulomb energy of a charge distribution. Alternatively, one can use the Feynman gauge with the associated Gupta-Bleuler quantization of the electromagnetic four-potential, involving an indefinite metric. Either way, one has as a starting point wave functions which are antisymmetrized products of Dirac orbitals and which represent exact eigenstates of the zero-order Hamiltonian. A further simplification is that one can then use a symmetrized version of the Gell-Mann–Low level shift formula to compute shifts arising from photon exchange between electrons and from radiative corrections, in a manner which is formally covariant. Such an approach was advocated long ago, but has been exploited only in recent years. This topic will also be covered in later lectures, by Mohr, by Saperstein and by Johnson.[3]

With regard to the second category, we can distinguish first the case $N = 2$ and $Z = 2$. Here one can still hope to do highly accurate calculations, with control on the errors being made, primarily because one can compute the eigenfunctions of the nonrelativistic two-electron Hamiltonion, which includes the Coulomb interaction, with great numerical precision. The relativistic and QED effects can then be treated as perturbations. This is also a calculational arena of long standing, which in recent years has become of renewed interest, especially because of measurements of the Rydberg levels of helium by Lundeen and collaborators and because of its connection with the quantum theory of long-range forces. The lectures of Drake, of Morgan, and of Au will deal with various aspects of this topic.[4]

30

However, already in this case one encounters a peculiarity which must be taken into account in any systematic treatment. Let h_D denote the Dirac Hamiltonian for a free electron ($\hbar = c = 1$)

$$h_D = \boldsymbol{\alpha} \cdot \boldsymbol{p}_{op} + \beta m \tag{1.1}$$

and $h_{D;ext}$ the Dirac Hamiltonian for an electron moving in an external field:

$$h_{D;ext} = h_D + V \tag{1.2}$$

On intuitive grounds one might expect that a decent starting point for the study of relativisitc effects in a two-electron atom would be provided by the Dirac-Coulomb Hamiltonian, defined by

$$H_{DC}(1,2) = H_D^{(o)}(1,2) + V_C(1,2) \tag{1.3}$$

where

$$H_D^{(o)}(1,2) = h_{D;ext}(1) + h_{D;ext}(2) \tag{1.4}$$

and $V_C(1,2)$ is the electron-electron Coulomb potential:

$$V_C(1,2) = e^2/4\pi r . \tag{1.5}$$

However, it has been known for a long time that $H_{DC}(1,2)$ has no normalizable eigenfunctions.[5] This phenomenon is a consequence of the fact that V_C mixes bound-state product eigenfunctions of $H_D^{(o)}$ with product eigenfunctions in which, say, "1" and "2" are in continuum states with positive- and negative energy respectively. It is therefore natural to refer to it as "continuum dissolution" (CD).[6,7] This fact seems to be now generally accepted, so in the spirit of Thomas Huxley, who urged that one not slay the slain more than once, I will not belabor it. (Huxley was referring to a verse from Dryden's "Alexander's Feast": "The King grew vain; Fought all his battles over again; And thrice he routed all his foes and thrice he slew the slain.")

As we will see, CD is a pseudo-problem, arising from the failure of Dirac's "one-electron theory" to generalize smoothly to more than one electron. When one considers the description of many-electron atoms within the framework of QED rather than a follow-your-nose extension of Dirac's one-electron theory one arrives in a quite straightforward way at Hamiltonians which do not suffer from CD: The terms involving interaction between the electrons are found to contain positive-energy projection operators which forbid the mixing between single-particle positive- and negative-energy wave functions.

The key to a sensible description of atoms in terms of relativistic configuration-space Hamiltonians is the recognition that processes involving virtual electron-positron pairs can in most cases be treated perturbatively (as can those involving the exchange of virtual transverse photons). The magnitude of virtual-pair processes can already be studied within the framework of the ordinary external-field

Dirac Hamiltonian (1.2); we will therefore do this first, in the next section, before turning our attention to the many-electron problem. A more detailed discussion of much of this material and references to related work may be found in some earlier reviews.[8]

I. External-field Dirac Equation and Virtual Pairs

A. c-number theory.

1. Free electron

The Dirac equation for a free electron is

$$\frac{i\partial\psi}{\partial t} = h\,\psi \tag{2.1}$$

where h_D is defined by (1.1). With $E(\boldsymbol{p})$ defined by $E(\boldsymbol{p}) = (m^2 + \boldsymbol{p}^2)^{1/2}$ and a corresponding operator E_{op} defined by

$$E_{op} = E(\boldsymbol{p}_{op}) \tag{2.2}$$

we can introduce positive- and negative-energy projection operators via

$$\Lambda_{\pm}^{op} = (E_{op} \pm h_D)\,/2E_{op} \tag{2.3}$$

and write the wave function ψ as the sum of a positive-energy part ψ_+ and a negative-energy part ψ_-,

$$\psi = \psi_+ + \psi_- \tag{2.4}$$

where

$$\psi_{\pm} = \Lambda_{\pm}^{op}\psi \,. \tag{2.5}$$

Note that in momentum space these operators just reduce to the usual Casimir projection operators $\Lambda_{\pm}(\boldsymbol{p})$.

It is now easy to see that the Dirac equation (2.1) is equivalent to two *uncoupled* systems of equations for ψ_+ and ψ_-:

$$i\partial\psi_+/\partial t = E_{op}\psi_+ \,, \quad \Lambda_+^{op}\psi_+ = \psi_+ \tag{2.6}$$

and

$$i\partial\psi_-/\partial t = -E_{op}\psi_- \,, \quad \Lambda_-^{op}\psi_+ = \psi_- \,, \tag{2.7}$$

Moreover, each of these systems is separately invariant under the Lorentz group, with the same linear transformation law, viz. $\psi'(x') = S\psi(\Lambda x)$. One can therefore interpret a ψ_+ wave packet as describing a freely propagating electron.

2. Inclusion of a static external field

In the presence of an external electromagnetic field, with four-potential $A^\mu(x)$, (2.1) takes the form

$$(i\partial/\partial t)\psi(x) = h_{D;ext}\psi(x) , \qquad (2.8a)$$

$$h_{D;ext} \equiv h_D + V \qquad (2.8b)$$

where, for a particle of charge $q = -e$, $V = -e(A^\circ - \boldsymbol{\alpha} \cdot \boldsymbol{A})$. For a static field there are stationary state solutions of the form $\psi(x) = e^{-iEt}\psi(\boldsymbol{x})$ with

$$E\psi(\boldsymbol{x}) = h_{D;ext}\psi(\boldsymbol{x}) . \qquad (2.9)$$

The most familiar and important case is that of a pure Coulomb field. I want to discuss it briefly in order to comment on a feature which is worth noting and which is relevant for the discussion of the many-electron problem. With $A^\circ = Ze/4\pi r$ and $\boldsymbol{A} = 0$, $V = -Z\alpha/r$ where $\alpha = e^2/4\pi$ is the fine structure constant. The discrete eigenvalues of (2.8) are then given by

$$E_{n,j} = m\left\{1 + \lambda/\left[a + (b^2 - \lambda)^{1/2}\right]^2\right\}^{-1/2} \qquad (2.10)$$

where $a = n - (j + 1/2)$, $b = j + 1/2$ and $\lambda \equiv (Z\alpha)^2$. The reason for writing down this formula is to emphasize the fact that E is not only an analytic function of $Z\alpha$ for small enough values of this quantity, which one might expect, but E is actually an analytic function of λ, the square of $Z\alpha$. Why this should be so is not at all obvious from the structure of the coupled partial differential equations which determine the eigenvalue. In view of the importance of this equation for relativistic atomic physics it would be nice to understand this on general grounds, i.e. without actually solving the equation. One way to gain some insight is to consider the reduced form obtained by writing

$$\psi = \psi^{(+)} + \psi^{(-)} , \quad \psi^\pm \equiv \beta^\pm \psi \qquad (2.11)$$

with

$$\beta^\pm = (1 \pm \beta)/2 . \qquad (2.12)$$

In the standard rep for the Dirac matrices this is just a decomposition into upper and lower components. On eliminating $\psi^{(-)}$ and introducing scaled lengths and energies via $\rho' = Zr/a_o$ (with $a_o = (\alpha m)^{-1} =$ Bohr radius, in our units) and $\epsilon = (E - m)/(Z\alpha)^2 m$, one finds that $\psi^{(+)}$ satisfies

$$\epsilon\psi^{(+)} = \left\{\boldsymbol{\sigma}_D \cdot \boldsymbol{p}'_{op}\left[2 + \lambda\left(\epsilon + \rho^{-1}\right)^{-1}\right]\boldsymbol{\sigma}_D \cdot \boldsymbol{p}'_{op} - \rho^{-1}\right\}\psi^{(+)} \qquad (2.13)$$

where $p'_{op} = -i\partial/\partial\rho$ and $\sigma_D = \alpha \times \alpha/2i$. Only the square λ of $Z\alpha$ appears explicitly in (2.13), so the fact that ϵ and hence E can be expanded in powers of λ is not implausible. It remains to be seen whether this observation can be used as a basis for constructing a proof. The difficulty is that the expansion of the square bracket in powers of λ breaks down for small enough values of ρ.

Be that as it may, the form (2.10) shows if the nucleus is treated as a point charge only even powers of $Z\alpha$ enter in the corrections to the nonrelativistic energy levels arising from use of the external-field Dirac equation. I will come back to this point shortly.

3. Positive-energy equation

It is instructive to separate the eigenfunctions ψ into positive- and negative-energy parts ψ_+ and ψ_- as in the free case. On multiplying (2.8) by Λ_+^{op} and Λ_-^{op} in turn one now gets two coupled equations, from which ψ_- can be eliminated to yield

$$E\psi_+ = h_+\psi_+ \tag{2.14a}$$

$$\Lambda_+^{op}\psi_+ = \psi_+ \tag{2.14b}$$

where

$$h_+ = h_+^{(o)} + V^{\text{pair}} \tag{2.15}$$

with

$$h_+^{(o)} = E_{op} + \Lambda_+^{op} V \Lambda_+^{op} \tag{2.16a}$$

and

$$V^{\text{pair}} = \Lambda_+^{op} V \frac{\Lambda_-^{op}}{E + E_{op} - \Lambda_-^{op} V \Lambda_-^{op}} V \Lambda_+^{p} . \tag{2.16b}$$

(a) Physical interpretation

While Eq. (2.14) looks complicated and corresponds to a nonlinear eigenvalue problem (as does the more familiar Eq. (2.13), it has the virtue of admitting a clear-cut physical interpretation. To see this it is convenient to consider the scattering rather than the bound state problem, *i.e.* to assume that $E > m$. Then, as will be seen later from the viewpoint of field theory, $h_+^{(o)}$ takes into account all virtual scattering processes involving only single-electron intermediate states and the inclusion of V^{pair} generates the contribution to the scattering amplitude arising from the creation and subsequent annihilation of one or more electron-positron pairs.

(b) Reduction to Pauli form

Equation (2.14) may be reduced to Pauli form without any approximation. This is because the constraint (2.14b) determines the relation between the upper and lower components $\psi_+^{(+)}$ and $\psi_+^{(-)}$ of ψ_+, in a way which is independent of the dynamics. It is convenient to introduce instead of $\psi_+^{(+)}$ a new function ϕ which in $\boldsymbol{p}$-space differs from $\psi_+^{(+)}$ only by a slowly varying multiplicative factor,

$$\phi = A_{op}^{-1}\psi_+^{(+)} , \quad A_{op} = \left[(E_{op} + m)/2E_{op}\right]^{1/2} \tag{2.17}$$

and to write the relation between ψ_+ and ϕ in the form[6]

$$\psi_+^{(+)} = U\phi \tag{2.18}$$

where

$$U = (1 + R) A_{op} \beta^{(+)} , \quad R \equiv \boldsymbol{\alpha} \cdot \boldsymbol{p}_{op}/(m + E_{op}) . \tag{2.19}$$

The inclusion of the factors A and $\beta^{(+)}$ has the virtue that U is pseudo-unitary, satisfying the relations

$$U^\dagger U = \beta^{(+)} , \quad U U^\dagger = \Lambda_+^{op} . \tag{2.20}$$

The inverse of the relation (2.18) is

$$\phi = U^\dagger \psi . \tag{2.21}$$

On multiplying (2.14a) on the left by $U^\dagger$ one then finds that

$$E\phi = h_{\text{red}}\phi \tag{2.22a}$$

with

$$h_{\text{red}} = \text{``}U^\dagger h_+ U\text{''} = E_{op} + V_{\text{red}} + V_{\text{red}}^{\text{pair}} . \tag{2.22b}$$

Here the quotes mean that after elimination of the $\boldsymbol{\alpha}$ matrices any remaining β matrices are replaced by unity; this is allowed because $\beta^{(+)}\phi = \phi$.

(c) Application to fine structure

As an example, with V even in Dirac matrices, e.g. with $\boldsymbol{A} = 0$ in the case of interaction with an electromagnetic field,

$$V_{\text{red}} = A_{op} \left[V + \hat{R}V\hat{R}\right] A_{op} , \quad \hat{R} \equiv \boldsymbol{\sigma}_D \cdot \boldsymbol{p}_{op}/(m + E_{op}) . \tag{2.23}$$

In the standard rep ϕ has only upper components, and $\boldsymbol{\sigma}_D$ may be replaced by the Pauli spin matrix $\boldsymbol{\sigma}$. Expansion of V_{red} in powers of $\boldsymbol{p}_{op}/m$ yields, as leading corrections to V for $V = V(r)$, the spin-orbit interaction

$$V_{s.o.} = V'(r)\boldsymbol{\sigma} \cdot \boldsymbol{\ell}_{op}/4m^2 r \tag{2.24}$$

and a spin-independent operator

$$V_{s.i.} = \frac{-1}{8m^2}\left(p_{op}^2 V + V p_{op}^2 - 2\boldsymbol{p}_{op}\cdot V\boldsymbol{p}_{op}\right) = \frac{-1}{8m^2}\left[\boldsymbol{p}_{op}\cdot,[\boldsymbol{p}_{op},V]\right] = \boldsymbol{\nabla}^2 V/8m^2 \ .$$

$$(2.25a)$$

For a pure Coulomb potential, $V(r) = -Z\alpha/r$, this reduces to

$$V_{s.i.} = Z\alpha\pi\delta(\boldsymbol{r})/2m \ ,$$

$$(2.25b)$$

the so-called S-state interaction. Expansion of E_{op} in powers of $\boldsymbol{p}_{op}/m$ yields directly the operator which corrects the nonrelativistic kinetic energy operator, viz,

$$-\boldsymbol{p}_{op}^2/8m^3 \ .$$

$$(2.25c)$$

Thus *all* the familiar operators which give level shifts corrections of order $(Z\alpha)^4 m$ in one-electron atoms are contained in $h_+^{(o)}$. Moreover these operators emerge directly, with much greater transparency than in the usual treatments, which require the use of the nonrelativistic Schrödinger equation to get the operators in their final form or the use of the Foldy-Wouthuysen transformation (see, *e.g.* the messy calculation on pp. 49-50 of the book by Bjorken and Drell). Furthermore, the derivation shows that *none* of these level-shift corrections involve virtual pairs. That this is so is certainly not obvious for the contact term, which has no classical counterpart. Note also that although the kinetic energy and spin-orbit corrections do have such counterparts, the diamagnetic term $e^2 A^2/2m$ provides an example of a "classical term" which nevertheless has its origin in processes involving virtual pairs, from the viewpoint of field theory.

4. Magnitude of pair effects.

To see the order of magnitude of the operator V^{pair} consider its expectation value δE_+ in a bound state described in first approximation by an eigenfunction $\psi_+^{(o)}$ of $h_+^{(o)}$,

$$h_+^{(o)}\psi_+^{(o)} = E_+\psi_+^{(o)}$$

$$(2.26)$$

On replacing E by m and neglecting V in the denominator of V^{pair} one finds that

$$\delta E_+ \approx -\langle\psi_+^{(o)}|\tilde{V}^{pair}|\psi_+^{(o)}\rangle$$

$$(2.27a)$$

where

$$\tilde{V}^{pair} = \Lambda_+^{op}\frac{V\Lambda_-^{op}}{2m}V\Lambda_+^{op} \ .$$

$$(2.27b)$$

In the weak binding limit and for $\ell \neq 0$ states further expansion in powers of $\boldsymbol{p}/m$ then gives, with $V = -eA^\circ$ and $\boldsymbol{E} \equiv -\boldsymbol{\nabla} A^\circ$,

$$\delta E_+ \approx \langle \phi_{nr} | \frac{-1}{2} \alpha_E^{pair} \boldsymbol{E}^2 | \phi_{nr} \rangle \; . \tag{2.27c}$$

Here ϕ_{nr} is the nonrelativistic wave function and α_E^{pair} is an electric polarizability associated with virtual pairs, defined by

$$\alpha_E^{pair} = -\alpha/4m^3 = -\left(\alpha^4/4\right) a_o^3 \; . \tag{2.28}$$

With $V = -Z\alpha/r$, this yields

$$\Delta E \sim (Z\alpha)^6 \, m \; . \tag{2.29}$$

For $\ell = 0$, $\langle 1/r^4 \rangle$ diverges and a more careful analysis yields

$$\Delta E \sim (Z\alpha)^5 \, m \; . \tag{2.30}$$

However the eigenvalue $E_+^{(o)}$ then also has a $(Z\alpha)^5 m$ term, in an expansion in powers of $Z\alpha$, with a coefficient which is equal in magnitude but opposite in sign from the one found for (2.30). This is in agreement with the fact that for a pure Coulomb field only even powers of $Z\alpha$ occur in the expansion of the eigenvalues of the Dirac Hamiltonian, as discussed earlier.

A corollary to all this is that for such a potential the eigenvalues of the no-pair Hamiltonian h_+ coincide with those of $h_{D;ext}$ through terms of order $(Z\alpha)^4 m$. Moreover, an energy-independent Hamiltonian whose eigenvalues are still accurate through terms of order $(Z\alpha)^6 m$ is given by

$$h'_+ = h_+^{(o)} + \tilde{V}^{pair} \tag{2.31}$$

where $\tilde{V}^{pair}$ is defined by (2.27b).

B. q-number theory

1. Free Dirac field

As described in many texts, the free Dirac field $\psi_D(x)$ satisfies (2.1), now regarded as an equation of motion in the Heisenberg picture, and anticommutation relations such as

$$\left\{ \psi_{D,\alpha}(\boldsymbol{x}, t), \psi_{D,\beta}^\dagger(\boldsymbol{x}'; t) \right\} = \delta_{\alpha\beta} \delta(\boldsymbol{x} - \boldsymbol{x}') \; . \tag{2.32}$$

These requirements are satisfied by the familiar expansion in terms of plane wave solutions of the corresponding c-number equation, viz.

$$\psi_D(x) = (2\pi)^{-3/2} \int dp \sum_r \left(a_r(\boldsymbol{p}) u_r(\boldsymbol{p}) e^{-ip\cdot x} + b_r^\dagger(\boldsymbol{p}) v_r(-\boldsymbol{p}) e^{ip\cdot x} \right) \qquad (2.33a)$$

with

$$\left\{ a_r(\boldsymbol{p}) , a_s^\dagger(\boldsymbol{p}') \right\} = \left\{ b_r(\boldsymbol{p}) , b_s^\dagger(\boldsymbol{p}') \right\} = \delta_{rs} \delta(\boldsymbol{p} - \boldsymbol{p}') \qquad (2.33b)$$

and

$$\{ a_r(\boldsymbol{p}) , b_s(\boldsymbol{p}') \} = 0 , \quad \{ a_r(\boldsymbol{p}), b_s^\dagger(\boldsymbol{p}') \} = 0 . \qquad (2.33c)$$

The framework is completed by (i) the requirement that there exist a state $|\text{vac}\rangle$ which is annihilated by all the a's and b's and which can be used to build up the whole Hilbert space by repeated application at the $a^\dagger$'s and $b^\dagger$'s and (ii) the form of the energy operator, viz.

$$H_D = \int d\boldsymbol{x} \, \psi_D^\dagger(x) h_D \psi(x) + \text{const.} \qquad (2.34)$$

I write all this down only to show explicitly how already in this simplest of circumstances projection operators make a natural appearance. For consider a generic one-electron state (the adjective is justified by the later introduction of interactions with the e.m. field), given by

$$\Psi = \int dp \sum_r f_r(\boldsymbol{p}) a_r^\dagger(\boldsymbol{p}) |\text{vac}\rangle . \qquad (2.35)$$

The amplitude $f_r(\boldsymbol{p})$ is just the Fock-space wave function of this state, a two-component object if we think of r as an "index" rather than a "variable" like $\boldsymbol{p}$. To get back to something like a Dirac wave function associated with this state it is natural to introduce a four-component spinor $\tilde{\psi}(\boldsymbol{p})$ defined by

$$\tilde{\psi}(\boldsymbol{p}) = \sum_r f_r(\boldsymbol{p}) u_r(\boldsymbol{p}) , \qquad (2.36a)$$

which then satisfies, by construction, the condition

$$\Lambda_+(\boldsymbol{p}) \tilde{\psi}(\boldsymbol{p}) = \tilde{\psi}(\boldsymbol{p}) . \qquad (2.36b)$$

If we introduce a Schrödinger picture at $t = 0$, then the time evolution of the state yields a factor $\exp(-iE(p)t)$ for $f_r(\boldsymbol{p})$ and hence the associated spinor $\tilde{\psi}(\boldsymbol{p}, t)$ satisfies

$$(i\partial/\partial t) \, \tilde{\psi}(\boldsymbol{p}, t) = (\boldsymbol{\alpha} \cdot \boldsymbol{p} + \beta m) \, \tilde{\psi}(\boldsymbol{p}, t) . \qquad (2.36c)$$

The Fourier transform $\psi(\boldsymbol{x}, t)$ of $\tilde{\psi}(\boldsymbol{p}, t)$ therefore satisfies the free Dirac equation together with the positive-energy constraint of (2.6).

2. Interaction with an external field: Free, Furry and Fuzzy S-pictures

For simplicity let us consider only a static external field. In the S-picture this can be included by taking the Hamiltonian as

$$H_{D;ext} = \int d\boldsymbol{x} \psi_D(\boldsymbol{x})\, h_{D;ext} \psi_D(\boldsymbol{x}) + \text{const.}\,, \tag{2.37}$$

with $h_{D;ext}$ given by (2.7b) in the case of an external electromagnetic field. There are now a variety of ways to proceed, depending on how one chooses to expand the field operator $\psi_D(\boldsymbol{x})$. (i) One may stay with the expansion in terms of plane wave solutions — I will call this the use of the "free S-picture". (ii) For V such that the spectrum of $h_{D;ext}$ splits into a continuum part with $E < -m$ and a part for which $E > 0$ (discrete in the interval $0 < E < m$ and continuous for $E > m$) it is very useful to use an expansion in positive- and negative-energy eigenfunctions $u_n(\boldsymbol{x})$ and $v_m(\boldsymbol{x})$ of $h_{D;ext}$; the operator coefficients of $\psi_D(\boldsymbol{x})$ are then still called electron creation and positron annihilation operators. Such an expansion was first introduced by Furry. I will refer to it as the use of the external-field or Furry S-picture. (iii) Finally one can expand in terms of eigenfunctions of $h_D + U$ where U is some potential intermediate between zero (the free S-picture) and the full V (the Furry S-picture). One might call this the intermediate or fuzzy S-picture.

As an application of the free S-picture, consider the scattering of an electron in the presence of V. The operator $H_{D;ext}$ may be written in the form

$$H_{D;ext} = H_D + H_1 \tag{2.38}$$

where

$$H_1 = \int d\boldsymbol{x} \; : \psi_D^\dagger V \psi_D : \,, \tag{2.39a}$$

On substitution of the expansion (2.33a), H_1 takes the form

$$H_1 = H_{sc}\left(e^-, e^-\right) + H_{sc}\left(e^+, e^+\right) + H_{cr}(e^-, e^+) + H_{ann}(e^-, e^+)\,. \tag{2.39b}$$

Here the first two terms involve products of the form $a'^\dagger a$ and $b'^\dagger b$ and so represent scattering of electrons and positrons in the external field whereas the last two terms involve products of the form $a^\dagger b^\dagger$ and ab and so represent creation and annihilation of pairs by the external field.

The transition amplitude for scattering from a state $|i\rangle = a_r^\dagger(\boldsymbol{p})|\text{vac}\rangle$ to a state $|f\rangle = a_{r'}^\dagger(\boldsymbol{p}')|\text{vac}\rangle$ is given by

$$T = T^{(1)} + T^{(2)} + \dots \tag{2.40}$$

where
$$T^{(1)} = \langle f|H_1|i\rangle, \quad T^{(2)} = \langle f|H_1(E - H_{D;ext})^{-1}H_1|i\rangle, \ldots \quad (2.41)$$

with $E = E(\boldsymbol{p}) = E(\boldsymbol{p}')$. One can see immediately that

$$T^{(1)} = \langle \boldsymbol{p}'|V|\boldsymbol{p}\rangle, \quad (2.42)$$

where I suppress the spin labels r and r'. The second-order term gets a contribution $T_a^{(2)}$ from one-particle intermediate states coming from the $H_{sc}(e^-, e^-)$ part of H_1, which may be written in the form

$$T_a^{(2)} = \langle \boldsymbol{p}'|V\frac{\Lambda_+^{op}}{E - h_D}V|f\rangle \quad (2.43)$$

and which coincides with the contribution from positive-energy states in the one-electron Dirac theory. There is also a contribution $T_b^{(2)}$ from 3-particle intermediate states, involving the initial and final electrons together with a positron, arising from the action of H_{cr} followed by H_{ann}. For $\boldsymbol{p}' \neq \boldsymbol{p}$ this is given by

$$T_b^{(2)} = -\int d\boldsymbol{q} \sum_s \frac{\left[u_{r'}^\dagger(\boldsymbol{p}')\tilde{V}(\boldsymbol{p}'+\boldsymbol{q})v_s(-\boldsymbol{q})\right]\left[v_s^\dagger(-\boldsymbol{q})\tilde{V}(-\boldsymbol{q}-\boldsymbol{p})u_r(\boldsymbol{p})\right]}{E - E_{int}} \quad (2.44)$$

where $E_{int} = 2E + E(\boldsymbol{q})$. This may be rewritten in the form

$$T_b^{(2)} = \langle \boldsymbol{p}'|V\frac{\Lambda_-^{op}}{E - h_D}V|\boldsymbol{p}\rangle. \quad (2.45)$$

Thus the contribution from these states is the same as that from negative-energy states in the c-number Dirac theory. Note the vital role played by the factor -1 in (2.44), which arises from the ACR (2.33c), in obtaining this agreement. One can show by explicit calculation that this agreement holds to all orders in perturbation theory.[9]

As an application of the Furry picture we can show that this must be so, without any detailed calculation. For in the Furry S-picture the generic form of a "one-electron" state is

$$\psi = \sum_n f(n)\, a^\dagger(n)|Vac\rangle \quad (2.46a)$$

and the associated spinor amplitude defined by

$$\psi(\boldsymbol{x}, t) = \sum_n f(n)\exp(-iE_n t)\, u_n(\boldsymbol{x}) \quad (2.46b)$$

satisfies

$$(i\partial/\partial t)\,\psi\,(\boldsymbol{x},t) = h_{D;ext}\psi(x,t) \tag{2.46c}$$

which is just the external-field Dirac equation. Note that ψ also satisfies

$$L_+\psi = \psi \tag{2.47}$$

where L_+ is the projection operator for the positive-energy part of the spectrum, is the external-field counterpart of Λ_+^{op}. For a stationary state of positive energy E this constraint is automatically satisfied so we just get back to the formalism of the one-electron Dirac theory. Another way to see this is to note that in the Furry picture the Hamiltonian is diagonal in creation and annihilation operators,

$$H = \sum_n E_n a^\dagger(n)a(n) + \sum_m E_m b^\dagger(m)b(m) \tag{2.48}$$

so that all the effects of interaction are already contained in the single-particle wave functions.

3. Use of a "no-pair" Furry picture

As a final example, let us consider an intermediate S-picture obtained in the following way. Since $\Lambda_+^{op} + \Lambda_-^{op} = 1$ we may write V in the form

$$V = V_{++} + V_{--} + V_{+-} + V_{-+} \tag{2.49a}$$

where

$$V_{\pm\pm} = \Lambda_\mp^{op}V\Lambda_\pm^{op} \,. \tag{2.49b}$$

Consider now the hermitian operator $\tilde{h}$ defined by

$$\tilde{h} = h_D + V_{++} + V_{--} \,. \tag{2.50}$$

Note that $\tilde{h}$ commutes with both Λ_+^{op} and Λ_-^{op}. In the subspace of ψ's which satisfy $\Lambda_+^{op}\psi = \psi$, $\tilde{h}$ is equivalent to the operator $h_+^{(o)}$ defined by (2.15) and in the subspace of ψ's which satisfy $\Lambda_-^{op}\psi = \psi$, $\tilde{h}$ is equivalent to an operator $h_-^{(o)}$ defined by

$$h_-^{(o)} = -\left(E_{op} - V_{--}\right) \,. \tag{2.51}$$

With V attractive (e.g. $V = -Z\alpha/r$) and not too strong, the spectrum of $\tilde{h}$ will be similar in character to that of h_D and we may expand the field operator ψ in eigenfunctions of $\tilde{h}$, with eigenvalues $\tilde{E}$.

The Hamiltonian may then be written in the form

$$H_{D;ext} = \widetilde{H}_D + \widetilde{H}_1 \tag{2.52}$$

where

$$\widetilde{H}_D = \sum \widetilde{E}_n \tilde{a}^\dagger(n)a(n) + \sum \widetilde{E}_m \tilde{b}^\dagger(m)b(m) \tag{2.53a}$$

and

$$\widetilde{H}_1 = \widetilde{H}_{cr}(e^-,e^+) + \widetilde{H}_{ann}(e^-,e^+) \tag{2.53b}$$

with $\widetilde{H}_{ann} = \widetilde{H}_{cr}^\dagger$ and

$$\widetilde{H}_{cr} = \sum_{n,m} \tilde{a}^+(n)\tilde{b}^\dagger(m)\langle \tilde{u}_n|V|\tilde{v}_m\rangle \ . \tag{2.54}$$

The tilde indicates that the wave functions and energies as well as the creation and annihilation operators are associated with the operator $\tilde{h}$. Note that with this breakup the virtual scatterings of the free electrons and positrons are included in the zero-order wave functions and only pair creation and annihilation needs to be studied perturbatively.

We can now consider the level shift arising from $\tilde{H}_1$ in a bound state defined by $\Psi = \tilde{a}^\dagger(n)|Vac\rangle$. To second order in $\widetilde{H}_1$ we then get, after a calculation analogous to that done for $T^{(2)}$,

$$\Delta E = \langle \Psi|\widetilde{H}_1 \left(E_n - H_D\right)^{-1} \widetilde{H}_1|\Psi\rangle \tag{2.55a}$$

or

$$\Delta E = \langle n|V\Lambda_-^{op} \left(E_n - \tilde{h}\right)^{-1} V|n\rangle \ . \tag{2.55b}$$

This coincides with the lowest-order level shift arising from V^{pair} [Eq. (2.16b)] provided we replace E by E_n in V^{pair}. We have thus justified from first principles the physical interpretation given earlier of this shift. One might call the picture introduced here the "no-pair Furry picture". It differs from the Furry picture precisely by splitting off the effects of virtual pairs, thus making them "visible", and it does so in a way which is symmetric between electron and positrons.

III. External-field QED

A. Preliminaries

To include interaction between the electrons and to allow for the possibility of radiation one must, in the Coulomb gauge version of QED, add three terms to the Hamiltonian $H_{D;ext}$. One of these is the operator H_C which corresponds to the self-interaction of the charge density $j^o(\boldsymbol{x})$, viz.

$$H_C = (1/8\pi) \int\int d\boldsymbol{x}\, d\boldsymbol{x}'\, j^o(\boldsymbol{x})j^o(\boldsymbol{x}')/[\boldsymbol{x} - \boldsymbol{x}'] \ . \tag{2.56}$$

42

The second is the free Hamiltonian H_{rad} associated with the quantized transverse radiation field $A_T(x)$,

$$H_{rad} = \frac{1}{2} \int dx : E_T^2 + H_T^2 : \qquad (2.57)$$

where

$$A_T(x) = (2\pi)^{-3/2} \int \frac{dk}{\sqrt{2w}} \sum_\lambda (A_\lambda(k)e^{ik \cdot x} + H.c.) \qquad (2.58)$$

with the usual C.R. for the A's and $A^\dagger$'s, and the third is the interaction H_T of the three-current density $j(x)$ with A_T:

$$H_T = - \int dx\, j(x) \cdot A_T(x). \qquad (2.59)$$

Here $j^\mu(x) = -e : \bar\psi_D(x)\, \gamma^\mu \psi_D(x): (\mu = 0,1,2,3)$ and the S-picture is being used. The total Hamiltonian is then

$$H = H_{D;ext} + H_{rad} + H_C + H_T + c.t.. \qquad (2.60)$$

where $c.t.$ denotes renormalization counter terms.

B. One electron; two-electrons, with $Z >> 2$

For these cases an efficient approach is to use the Furry picture and to treat all of $H_C + H_T$ as a perturbation. To calculate level shifts one can then use a modification of a formula of Gell-Mann and Low in which the level shift is expressed not in terms of the Moller wave operator $U(0, -\infty; \epsilon)$ but in terms of the adiabatically switched-on S-matrix S_ϵ, calculated for charge λe, viz.[10]

$$\Delta E = \lim_{\epsilon \to 0} \frac{i\epsilon}{2} \frac{\partial}{\partial \lambda} \log\langle \Psi | S_\epsilon[\lambda] | \Psi \rangle_{conn}|_{\lambda=1}. \qquad (2.61)$$

Here the subscript "conn" indicates that only connected diagrams are kept in the matrix element. As emphasized ages ago, one virtue of this formula is that one can switch to the Feynman gauge and use Feynman-like diagrams rather than time-ordered diagrams in writing down higher-order contributions. This will be discussed in some detail in the lectures of Mohr.

C. Two electrons, with $Z \sim 2$

In this case one cannot treat all of $H_C + H_T$ as perturbation. As is evident from the nonrelativistic limit, at least a part of H_C must be included in the zero-order Hamiltonian. The study of this case was initiated by Breit in the late 20's and has a somewhat dark history (pun intended). As already mentioned, the naive Hamiltonian (1.3) suffers from CD. One approach, which is based on field theory but involves four-dimensional machinery, is based on the external-field analogue of the two-body Bethe-Salpeter equation.[11,12] On a reduction to equal times, one finds that a suitable starting point for the application of perturbation theory is given by[12]

$$h_{++}\phi = E\phi \tag{2.62a}$$

and

$$L_+(i)\phi = \phi \tag{2.62b}$$

where h_{++} is the no-pair Coulomb-ladder Hamiltonian defined by

$$h_{++} = h_{D;ext}(1) + h_{D;ext}(2) + L_{++}U_C(1,2)L_{++} \tag{2.63}$$

with $U_C(1,2) = e^2/4\pi\, r_{12}$, the electron-electron Coulomb potential. Here L_{++} is the product of positive-energy external-field projection operators, $L_{++} = L_+(1)L_+(2)$, with

$$L_+(i) = \sum_n |u_n(i)\rangle\langle u_n(i)|. \tag{2.64}$$

One can derive the same equation more simply, and with an obvious field-theoretic interpretation for the wave function, by working in Fock space and using the fact that virtual-pair effects can be treated perturbatively. In particular, if one uses the Furry picture and takes as the zero-order Hamiltonian the operator

$$H_D^{np} = H_{D;ext} + H_{rad} + H_C^{np} \tag{2.65}$$

where H_C^{np} is the no-pair part of H_C, the number operators associated with this picture commute with H_D^{np}. The equation

$$H_D\Psi = E\Psi \tag{2.66}$$

is then precisely equivalent to a set of Fock-space equations which do not couple the different sectors of Hilbert space. To be specific, the generic two-electron state now has the form

$$\Psi = \sum_{n_1,n_2} f(n_1, n_2)\, a^\dagger(n_1)a^\dagger(n_2)|Vac\rangle . \tag{2.67}$$

If we define a spinor wave function by

$$\phi = \sum_{n_1,n_2} f(n_1, n_2)\, u_{n1}(\boldsymbol{r}_1)u_{n_1}(\boldsymbol{r}_2) \tag{2.68}$$

then, as is readily shown, the equation determining the Fock space amplitude $f(n_1, n_2)$ is equivalent to (2.62) for ϕ.

D. Truly many electrons: $N \geq 3$

The splitting of the Hamiltonian into a no-pair part and a remainder, as described above, can of course be used to obtain equations similar to (2.62) for any number of electrons. The specific form depends only on the choice of picture. To take into account all choices at once write

$$V = U + (V - U) , \tag{2.69}$$

define

$$h_{D;U} = h_D + U ,$$

and use the eigenfunctions of h_{DU} for the expansion of $\psi_D(x)$. The zero-order no-pair configuration space equation for determining the energy levels of an N-electron atom then takes the form

$$\left(\sum_i h_{D;U}(i) + L_+^{tot} \left(V^{tot} - U^{tot} + U_C^{tot} \right) L_+^{tot} \right) \phi = E\phi \tag{2.70}$$

where L_+^{tot} is the product of projection operators $L_{+;U}(i)$ defined by the positive-energy eigenfunctions of $h_{D;U}$ and U_C^{tot} is the sum of all electron-electron Coulomb potentials,

$$U_C^{tot} = \sum_{i \epsilon j} U_C(i,j) . \tag{2.71}$$

The wave function ϕ also satisfies

$$L_{+;U}(i)\phi = \phi \quad (i = 1, 2, \ldots N) . \tag{2.72}$$

For $U = 0$, we are dealing with the free S-picture and (2.70) gives the many-electron generalization of (2.62), viz.

$$\left(\sum_i h_D(i) + \Lambda_+^{tot} \left(V^{tot} + U_C^{tot} \right) \Lambda_+^{tot} \right) \phi = E\phi , \tag{2.73a}$$

and

$$\Lambda_+^{op}(i)\phi = \phi . \tag{2.73b}$$

For $U = V$, *i.e.* use of the Furry S-picture, we get the obvious generalization of the no-pair Coulomb ladder equation (2.62)

$$\left(\sum h_{D;ext}(i) + L_{+;ext}^{tot} U_C^{tot} L_{+;ext}^{tot} \right) \phi = \phi . \tag{2.74}$$

If one chooses U to be, *e.g.* some kind of mean potential such as a Dirac-Hartree-Fock or Thomas-Fermi potential, one can use (2.70) to give an *a posteriori* interpretation of the so-called relativistic configuration interaction (RCI) calculations in which the wave function is expanded in a fixed basis.[13] The main points are that in each case:

(i) The starting equation has a definite field-theoretic origin and the quantities calculated from it are susceptible to systematic improvement. For example, to take into account the leading effects of transverse-photon exchange one may add the Breit potential $U_B(i,j)$ to $U_C(i,j)$ in any of these equations; the old *caveat* about using this potential only in lowest order does not hold for equations with projection operators (and was never really cogent to begin with).[14]

(ii) Both variational and many-body techniques may be applied to find approximate solutions to such equations, in complete analogy to the nonrelativistic case. We will hear about some of this in later lectures.

The projection-operator constraint is easy to handle in the free S-picture. In that case it can be readily used to eliminate, say in the standard representation, the components of a multi-Dirac spinor $\psi_{\alpha_1, \ldots, \alpha_N}$ in which any index α is greater than two in favor of those in which every index is less than or equal to two, so that one is dealing in effect with a Pauli wave function, before any further approximations are made. This approach has been used by Hess to study several many-electron atoms.[15]

If one imagines having in hand the positive-energy eigenfunctions of the operator $\tilde{h}$ defined by (2.50) and chooses U accordingly, *i.e.*

$$U = \Lambda_+^{op} \left(V + V \frac{\Lambda_-^{op}}{2m} V \right) \Lambda_+^{op} , \qquad (2.75)$$

the resulting Hamiltonian will have eigenvalues which are correct, apart from radiative corrections, through terms of order $(Z\alpha)^6 m$, as far as powers of $Z\alpha$ are concerned. Such a choice may prove to be useful in the context of many-electron atoms.[16]

As a final remark, as discussed in Ref. 8, the inclusion of radiative corrections within the framework described here introduces no new questions of principle. It of course remains as a challenging computational problem.

References

1. I. Lindgren, these proceedings.

2. G.S. Adkins; D.R. Yennie; H. Grotch, these proceedings.

3. P.J. Mohr; J. Saperstein; W.R. Johnson, these proceedings.

4. G.W.F. Drake; J.D. Morgan III; C.-K. Au, these proceedings.

5. G.E. Brown and D.G. Ravenhall, *Proc. Roy. Soc. London*, Sec. A**208**, 552 (1951).

6. J. Sucher, *Phys. Rev. A* **30**, 703 (1980).

7. For a solvable model which exhibits CD, see J. Sucher, *Phys. Rev. Lett.* **55**, 1023 (1985).

8. (a) J. Sucher, *Physica Scripta* **36**, 271 (1987); (b) J. Sucher, in *NATO Advanced Study Institute on Atoms in Unusual Conditions*, Cargèse (1985), (Edited by J.-P. Briand), Gordon and Breach, New York (1986).

9. S. Weiskop and J. Sucher, unpublished U. of MD report (1983).

10. J. Sucher, *Phys. Rev.* **107**, 1448 (1957).

11. H. Araki, *Prog. Theoret. Phys.* **17**, 1 (1957).

12. J. Sucher, Columbia University Ph.D. dissertation (1958, unpublished) and *Phys. Rev.* **109**, 1010 (1958).

13. J. Sucher, *Int. J. Quantum Chem.* **25**, 1 (1984).

14. For a discussion of this point, see J. Sucher, in *Proceedings of the Argonne Workshop on the Foundation of Relativistic Theory of Atomic Structure*, (Edited by H.G. Berry, K.T. Cheng, W.R. Johnson and Y.-K. Kim); ANL-80-116 (Argonne National Laboratory, 1980).

15. B. Hess, *Phys. Rev. A* **32**, 756 (1985); *ibid* **33**, 3742 (1986).

16. Other approaches to the many-electron problem are discussed by B. Zygelman (these proceedings) and T. Fulton (these proceedings); see also K. Dietz, *Physica Scripta* **37**, 651 (1988).

QUANTUM ELECTRODYNAMICS PERTURBATION THEORY

Peter J. Mohr
National Institute of Standards and Technology
Gaithersburg, MD 20899

ABSTRACT

A review is given of quantum electrodynamics (QED) with a view toward applications to multielectron atoms. An attempt is made to provide a basis for incorporating QED corrections into a many-body framework.

INTRODUCTION

The objective of these lectures is to describe bound-state QED within a framework that can easily be carried over to a many-body formulation of the few- or many-electron problem. The standard bound interaction picture is a suitable starting point for such an approach.

Much of the material to be discussed is "known", but an attempt will be made to make it self-contained with emphasis on the aspects of the theory relevant to the problem at hand that are probably not familiar.

Some examples of the non-standard methodology needed are: Special treatment of Wick's theorem to accommodate vacuum polarization, general expressions for QED operators in terms of electron creation and annihilation operators, cancellation of singularities in the Gell-Mann Low and Sucher expression for the level shift, and "retardation" terms in the level shift formula.

Also needed for a useful calculation are practical methods to regulate and renormalize the operators and numerical methods to get the result. These topics are no less important, but are somewhat more specialized and will be deferred to other lectures.

FURRY PICTURE

The Furry picture can be formulated as a transformation from the equations of free-particle QED with an external binding field to the bound interaction picture in which the creation and annihilation operators correspond to bound states.[1] On the other hand, bound state creation and annihilation operators are as physically motivated as free particle creation and annihilation operators, and no less fundamental, so that will be the starting point here.

In this picture, the zeroth-order wave functions are solutions of the Dirac equation in an external potential V, which for high-Z one- and two-electron atoms can be taken as the Coulomb potential (units are chosen such that $\hbar = c = m_e = 1$)

$$\left[-i\vec{\alpha} \cdot \vec{\nabla} + V(\vec{x}) + \beta - E_n\right] \phi_n(\vec{x}) = 0 \qquad (1)$$

The corresponding time-dependent solutions are given by

$$\phi_n(x) = \phi_n(\vec{x})e^{-iE_n t} \tag{2}$$

As an alternative procedure, for a many-electron atom, one might begin with eigenfunctions in an external field other than the Coulomb field to account for average effects of the electrons, and then correct for this approximation term by term in perturbation theory. This approach is covered in other lectures.

In terms of the Coulomb eigenfunctions, the electron-positron field operator is

$$\psi(x) = \sum_{E_n > 0} a_n \phi_n(x) + \sum_{E_m < 0} b_m^\dagger \phi_m(x) \tag{3}$$

where a_n is the annihilation operator for an electron in the positive energy state n, and $b_m^\dagger$ is the creation operator for a positron in the negative energy state m. The electron and positron creation and annihilation operators obey the standard Fermion anticommutation relations

$$\{a_n, a_m^\dagger\} = \delta_{nm} ; \qquad \{a_n, a_m\} = 0 ; \qquad etc. \tag{4}$$

The interaction between the electron-positron field and the photon field is represented by the interaction Hamiltonian

$$H_I(x) = j^\mu(x)A_\mu(x) - \delta M(x) \tag{5}$$

where $j^\mu(x)$ is the current operator

$$j^\mu(x) = -\frac{e}{2}\left[\bar{\psi}(x)\gamma^\mu, \psi(x)\right] \tag{6}$$

$\delta M(x)$ is the mass renormalization operator

$$\delta M(x) = \frac{\delta m}{2}\left[\bar{\psi}(x), \psi(x)\right] \tag{7}$$

and $A_\mu(x)$ is the photon field operator.

LEVEL SHIFTS

A basic requirement to predict energy levels is to have a useful expression to evaluate. Any formula will have to be treated with some caution, because the relation between the level shift expression and the actual observations made in experiments will require further analysis at some level of precision.

A feature of the bound-state problem that requires some attention is the fact that the perturbation expansion for the S matrix, which can provide the starting point for level shift evaluations, contains infinite terms corresponding to intermediate states degenerate with the state under consideration. These singularities ultimately cancel other terms in the S-matrix expansion, but their existence makes it desirable to employ a systematic method of regularizing the expressions so that they are finite at intermediate stages of the calculation. The standard many-body formulation deals with this problem and provides final formulas that are free of infinities. However the derivation takes advantage of the fact that the interaction is instantaneous, and the generalization to retarded photon interactions is not obvious. The generalization is not likely to be completely trivial, because there are terms in the complete calculation that do not exist for instantaneous interactions.

One method that appears to provide a systematic regularization in the intermediate stages of calculation is the Gell-Mann Low level shift formula in the symmetric form given by Sucher.[2,3] The formula is based on the adiabatically damped Hamiltonian

$$H_I^\epsilon(t) = e^{-\epsilon|t|} H_I(t) \tag{8}$$

where

$$H_I(t) = \int d\vec{x}\; H_I(x) \tag{9}$$

The corresponding time development operator is defined by the equation

$$i\,\frac{\partial}{\partial t}\, U_{\epsilon,\lambda}(t,t_0) = \lambda H_I^\epsilon(t) U_{\epsilon,\lambda}(t,t_0) \tag{10}$$

with the well-known solution

$$U_{\epsilon,\lambda}(t,t_0) = \sum_{j=0}^{\infty} (-i\lambda)^j \int_{t_0}^{t} dt_j \int_{t_0}^{t_j} dt_{j-1} \ldots \int_{t_0}^{t_2} dt_1 \left[H_I^\epsilon(t_j) \ldots H_I^\epsilon(t_1) \right] \tag{11}$$

or

$$U_{\epsilon,\lambda}(t,t_0) = \sum_{j=0}^{\infty} \frac{(-i\lambda)^j}{j!} \int_{t_0}^{t} dt_j \int_{t_0}^{t} dt_{j-1} \ldots \int_{t_0}^{t} dt_1 T\left[H_I^\epsilon(t_j) \ldots H_I^\epsilon(t_1) \right] \tag{12}$$

where T denotes the time-ordering symbol. The S matrix is then given by

$$S_{\epsilon,\lambda} = U_{\epsilon,\lambda}(\infty,-\infty) \tag{13}$$

or

$$S_{\epsilon,\lambda} = 1 + \sum_{j=1}^{\infty} S_{\epsilon,\lambda}^{(j)} \tag{14}$$

with

$$S_{\epsilon,\lambda}^{(j)} = \frac{(-i\lambda)^j}{j!} \int d^4x_j \ldots \int d^4x_1\; e^{-\epsilon|t_j|} \ldots e^{-\epsilon|t_1|} T\left[H_I(x_j) \ldots H_I(x_1) \right] \tag{15}$$

A reasonable definition for the energy level shift is

$$\Delta E_n = \lim_{\substack{\epsilon \to 0 \\ \lambda \to 1}} \frac{\langle n|U_{\epsilon,\lambda}(\infty,0)[H_0 + \lambda H_I^\epsilon(0) - E_n]U_{\epsilon,\lambda}(0,-\infty)|n\rangle}{\langle n|U_{\epsilon,\lambda}(\infty,-\infty)|n\rangle} \qquad (16)$$

where $|n\rangle$ is the unperturbed state. With a few pages of algebra one can show that[2,3]

$$\langle n|U_{\epsilon,\lambda}(\infty,0)[H_0 + \lambda H_I^\epsilon(0) - E_n]U_{\epsilon,\lambda}(0,-\infty)|n\rangle = \frac{i\epsilon\lambda}{2}\frac{\partial}{\partial\lambda}\langle n|S_{\epsilon,\lambda}|n\rangle \qquad (17)$$

so the level shift is given by

$$\Delta E_n = \lim_{\substack{\epsilon \to 0 \\ \lambda \to 1}} \frac{i\epsilon\lambda}{2}\frac{\frac{\partial}{\partial\lambda}\langle n|S_{\epsilon,\lambda}|n\rangle}{\langle n|S_{\epsilon,\lambda}|n\rangle} \qquad (18)$$

To eliminate disconnected graphs from consideration, one notes that

$$\langle n|S_{\epsilon,\lambda}|n\rangle = \langle n|S_{\epsilon,\lambda}|n\rangle_c \langle 0|S_{\epsilon,\lambda}|0\rangle \qquad (19)$$

Substitution of this expression into the level shift formula yields

$$\Delta E_n = \lim_{\substack{\epsilon \to 0 \\ \lambda \to 1}} \frac{i\epsilon\lambda}{2}\frac{\left[\frac{\partial}{\partial\lambda}\langle n|S_{\epsilon,\lambda}|n\rangle_c\right]\langle 0|S_{\epsilon,\lambda}|0\rangle + \langle n|S_{\epsilon,\lambda}|n\rangle_c\left[\frac{\partial}{\partial\lambda}\langle 0|S_{\epsilon,\lambda}|0\rangle\right]}{\langle n|S_{\epsilon,\lambda}|n\rangle_c\langle 0|S_{\epsilon,\lambda}|0\rangle} \qquad (20)$$

which provides the simpler level shift formula

$$\Delta E_n = \lim_{\substack{\epsilon \to 0 \\ \lambda \to 1}} \frac{i\epsilon}{2}\frac{\frac{\partial}{\partial\lambda}\langle n|S_{\epsilon,\lambda}|n\rangle_c}{\langle n|S_{\epsilon,\lambda}|n\rangle_c} + \text{constant} \qquad (21)$$

The additive constant is the same for all states, so it does not contribute to level differences.

The motivation for this particular formula for the level shift is primarily its usefulness in setting up calculations. There are other ways of arriving at level shifts, such as looking for poles in the full propagation function. These methods are in agreement with the expressions that are produced by (21) for the level shifts, at least in the case of hydrogen, to high precision. Of course, the complete interpretation of the results of such calculations depends on the experiment in question. No measurement looks directly at the level shift formula or the poles of a propagator, but usually involves a variation on the scattering process shown in the Feynman diagram in Fig. 1. That figure depicts excitation of an atom by an external potential, propagation of the electron influenced by the virtual radiation field (triple line), and subsequent photon emission. At sufficiently high precision, the resonance center and line shape depend on corrections that cannot be encompassed in a general formula, but depend on a detailed analysis. Fortunately, as shown by Low,[4] the pole of the propagator gives the resonance

Figure 1: Scattering process

center to high accuracy, and as shown by Sucher,[3] the Gell-Mann Low and Sucher formula agrees with the pole term expression to high precision.

An apparent advantage of this formulation is the fact that the singularity cancellation can be carried out in the general expressions for the corrections no matter how many particles are present. This is particularly advantageous in applications to multielectron atoms.

PERTURBATION THEORY

Evaluation of the expression in the Gell-Mann Low formula is carried out with the aid of Wick's theorem.[5] The statement of the theorem useful for this purpose is

$$T[ABCD\ldots] = :ABCD\ldots: + :AB\,CD\ldots:$$
$$+ :AB\,CD\ldots: + \text{ all possible contractions} \tag{22}$$

In this expression, the expansion includes all possible numbers and combinations of contractions. If the vacuum expectation value of both sides is taken, only the term with all contracted operators survives on the right hand side. It is useful not to do this, but rather keep all terms in order to arrive at general expressions for the QED corrections. The operators in this expression are not restricted to be either creation or annihilation type. In general they will be electron-positron field operators which contain both types of operators. The case of two Fermion operators with equal times requires special attention in the bound-state problem. In particular, the Dirac current contains two electron-positron fields with the same coordinates. In the free QED case, the fields in the current are normal ordered and the corresponding rule is to omit contractions between equal-time operators. In the bound-state case, contractions between equal-time fields in the currents produce vacuum polarization corrections that must be included. This case can be accommodated in Wick's theorem by the following two observations. First, the current can be simplified by taking advantage of the identity

$$T\left[AB\tfrac{1}{2}(CD - DC)EF\ldots\right] = T[ABCDEF\ldots] \tag{23}$$

valid for Fermion operators. Second, the time-ordering operator is defined for equal-time operators by writing

52

$$T[A(t)B(t)] = \tfrac{1}{2}A(t)B(t) - \tfrac{1}{2}B(t)A(t) \tag{24}$$

This determines the equal-time contraction symbol via the definition

$$\underset{\sqcup}{AB} = \langle 0|T[AB]|0\rangle \tag{25}$$

which leads to the proper expression for the vacuum polarization corrections.

Application of Wick's theorem then produces the correct vacuum polarization terms. The rule for dealing with the current operator is to write it as a single product according to the first observation above, and then apply Wick's theorem as stated above with contractions between all operators, including equal-time operators.

The contraction symbols from Wick's theorem lead to the following propagation functions. For the electron-positron field we have

$$S_F(x_2, x_1) = \langle 0|T[\psi(x_2)\bar\psi(x_1)]|0\rangle$$

$$= \begin{cases} \displaystyle\sum_{E_n > 0} \phi_n(x_2)\bar\phi_n(x_1) & t_2 > t_1 \\[2em] \displaystyle -\sum_{E_n < 0} \phi_n(x_2)\bar\phi_n(x_1) & t_2 < t_1 \end{cases} \tag{26}$$

$$= \frac{1}{2\pi i}\int_{-\infty}^{\infty} dz \sum_n \frac{\phi_n(\vec x_2)\bar\phi_n(\vec x_1)}{E_n - z(1+i\delta)}\, e^{-iz(t_2-t_1)}$$

$$= \frac{1}{2\pi i}\int_{-\infty}^{\infty} dz\ G\big(\vec x_2, \vec x_1, z(1+i\delta)\big)\gamma^0 e^{-iz(t_2-t_1)}$$

where the Green's function G is the solution of the equation

$$\left[-i\vec\alpha \cdot \vec\nabla_2 + V(\vec x_2) + \beta - z\right] G(\vec x_2, \vec x_1, z) = \delta(\vec x_2 - \vec x_1) \tag{27}$$

The Green's function is an analytic function of z except for the bound state poles in the complex z plane with branch points at $z^2 = 1$ and cuts corresponding to the condition $\mathrm{Re}\big((1-z^2)^{1/2}\big) > 0$ as depicted in Fig. 2.

The photon propagation function D_F defined by

$$\langle 0|T[A_{\mu_2}(x_2)A_{\mu_1}(x_1)]|0\rangle = g_{\mu_2\mu_1} D_F(x_2 - x_1) \tag{28}$$

is

$$D_F(x_2 - x_1) = -\frac{i}{(2\pi)^4}\int d^4q\, \frac{e^{-iq\cdot(x_2-x_1)}}{q^2 + i\delta}$$

$$= \frac{1}{2\pi i}\int_{-\infty}^{\infty} dq_0\ H(\vec x_2 - \vec x_1, q_0)e^{-iq_0(t_2-t_1)} \tag{29}$$

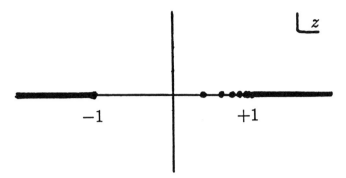

Figure 2: Singularities of the Dirac Green's function

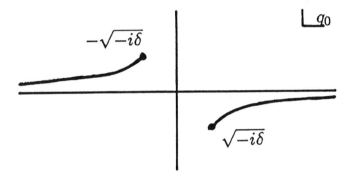

Figure 3: Singularities of the photon Green's function

The function H, which is obtained by carrying out the integration over the space components of q in the above expression, is

$$H(\vec{x}_2 - \vec{x}_1, q_0) = -\frac{e^{-bx_{21}}}{4\pi x_{21}}$$

$$x_{21} = |\vec{x}_2 - \vec{x}_1| \; ; \; b = -i(q_0^2 + i\delta)^{1/2} \; , \; \mathrm{Re}(b) > 0$$

(30)

The function H is analytic in the complex q_0 plane except for branch points and cuts corresponding to the condition $\mathrm{Re}(b) > 0$ as shown in Fig. 3.

APPLICATIONS

The level-shift formula has the explicit perturbation expansion

$$\frac{\lambda \frac{\partial}{\partial \lambda} \langle S_{\epsilon,\lambda} \rangle_c}{\langle S_{\epsilon,\lambda} \rangle_c} \Bigg|_{\lambda=1} = \frac{\langle S_\epsilon^{(1)} \rangle_c + 2 \langle S_\epsilon^{(2)} \rangle_c + 3 \langle S_\epsilon^{(3)} \rangle_c + \cdots}{1 + \langle S_\epsilon^{(1)} \rangle_c + \langle S_\epsilon^{(2)} \rangle_c + \langle S_\epsilon^{(3)} \rangle_c + \cdots}$$

$$= \langle S_\epsilon^{(1)} \rangle_c + 2 \langle S_\epsilon^{(2)} \rangle_c - \langle S_\epsilon^{(1)} \rangle_c^2$$

$$+ 3 \langle S_\epsilon^{(3)} \rangle_c - 3 \langle S_\epsilon^{(1)} \rangle_c \langle S_\epsilon^{(2)} \rangle_c + \langle S_\epsilon^{(1)} \rangle_c^3 \tag{31}$$

$$+ 4 \langle S_\epsilon^{(4)} \rangle_c - 4 \langle S_\epsilon^{(1)} \rangle_c \langle S_\epsilon^{(3)} \rangle_c - 2 \langle S_\epsilon^{(2)} \rangle_c^2$$

$$+ 4 \langle S_\epsilon^{(1)} \rangle_c^2 \langle S_\epsilon^{(2)} \rangle_c - \langle S_\epsilon^{(1)} \rangle_c^4 + \cdots$$

where

$$\langle S_{\epsilon,\lambda} \rangle_c = \langle n | S_{\epsilon,\lambda} | n \rangle_c$$
$$\langle S_\epsilon^{(j)} \rangle_c = \langle n | S_{\epsilon,1}^{(j)} | n \rangle_c \tag{32}$$

The denominator will be seen to play a critical role in the cancellation of singularities.

As a warm-up exercise, consider the simple example of an atom in a perturbing potential in first-order perturbation theory. This case is trivial, so it is a useful way to illustrate some general features of the formulation with relatively simple equations. The external field case is implemented by substitution of the external field vector potential for the photon field operator in the expression for the current and proceeding as indicated above.

The vector potential is

$$A_\mu^e(x) = \delta_{\mu 0} \Phi(\vec{x}) \tag{33}$$

where Φ is the electrostatic potential, and the interaction Hamiltonian density is

$$H_I(x) = -\frac{e}{2} \left[\bar{\psi}(x) \gamma^0, \psi(x) \right] \Phi(\vec{x}) = \tfrac{1}{2} V(\vec{x}) \left[\bar{\psi}(x) \gamma^0, \psi(x) \right] \tag{34}$$

The first-order S matrix is

$$S_\epsilon^{(1)} = -i \int d^4x \; e^{-\epsilon |t|} T \left[H_I(x) \right] \tag{35}$$

Application of Wick's theorem leads to either no contraction or one contraction

$$T \left[H_I(x) \right] = V(\vec{x}) \, T \left[\bar{\psi}(x) \gamma^0 \psi(x) \right]$$

$$= V(\vec{x}) \left[: \bar{\psi}(x) \gamma^0 \psi(x) : + \overline{\bar{\psi}(x) \gamma^0 \psi(x)} \right] \tag{36}$$

Figurè 4: External-field Feynman graphs

with the corresponding operator Feynman diagrams shown in Fig. 4. The contracted graph on the right is disconnected and discarded, so we are left with

$$\langle S_\epsilon^{(1)} \rangle_c = -i \int d^4x \; e^{-\epsilon|t|} V(\vec{x}) \left\langle \sum_{nm} a_n^\dagger \phi_n^\dagger(x) a_m \phi_m(x) \right\rangle$$

$$= -i \int d^4x \; e^{-\epsilon|t|} \sum_{nm} e^{i(E_n - E_m)t} \phi_n^\dagger(\vec{x}) V(\vec{x}) \phi_m(\vec{x}) \langle a_n^\dagger a_m \rangle$$

$$= -i \sum_{nm} \frac{2\epsilon}{(E_n - E_m)^2 + \epsilon^2} \int d\vec{x} \; \phi_n^\dagger(\vec{x}) V(\vec{x}) \phi_m(\vec{x}) \langle a_n^\dagger a_m \rangle$$

$$= -\frac{2i}{\epsilon} \sum_{nm} \delta(E_n, E_m) V_{nm} \langle a_n^\dagger a_m \rangle + \mathcal{O}(\epsilon) \tag{37}$$

where

$$\delta(E_n, E_m) = \begin{cases} 1 & \text{if} \quad E_n = E_m \\ 0 & \text{if} \quad E_n \neq E_m \end{cases} \tag{38}$$

In (37), it is assumed that the zeroth-order state contains only electrons and no positrons. The corresponding energy level shift is

$$E^{(1)} = \lim_{\epsilon \to 0} \tfrac{1}{2} \, i\epsilon \langle S_\epsilon^{(1)} \rangle_c = \sum_{nm} \delta(E_n, E_m) V_{nm} \langle a_n^\dagger a_m \rangle \tag{39}$$

For the case of a one-electron atom in the 1S state, one has

$$|1S\rangle = a_{1S}^\dagger |0\rangle \tag{40}$$

so

$$\langle a_n^\dagger a_m \rangle = \delta_{n,1S} \delta_{m,1S} \tag{41}$$

and

$$E^{(1)} = V_{1S,1S} \tag{42}$$

56

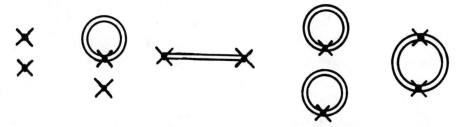

Figure 5: Feynman diagrams for the second-order correction

as expected.

CANCELLATION OF SINGULARITIES

In higher orders of perturbation theory, higher inverse powers of ϵ appear in the intermediate stages of the calculation as described above. They ultimately cancel each other when all relevant terms are combined to give a finite result. The cancellation can be made while the operators are still written in a general form, so the temporary appearance of the singular terms is not as disruptive as it might be. It is instructive to see how these terms actually cancel, so a simple example in which they appear is examined here.

Consider a perturbation as above in second order, with the additional assumption that the perturbing potential is central as would be the case for the finite nuclear radius correction. The second-order level shift is

$$E^{(2)} = \lim_{\epsilon \to 0} \frac{i\epsilon}{2} \left[2\langle S_\epsilon^{(2)} \rangle_c - \langle S_\epsilon^{(1)} \rangle_c^2 + \ldots \right] \tag{43}$$

with the second-order S matrix given by

$$S_\epsilon^{(2)} = -\tfrac{1}{2} \int d^4 x_2 \int d^4 x_1 \; e^{-\epsilon|t_2|} e^{-\epsilon|t_1|} T\left[H_I(x_2) H_I(x_1) \right] \tag{44}$$

where

$$T\left[H_I(x_2) H_I(x_1) \right] = V(\vec{x}_2) V(\vec{x}_1) \; T\left[\bar{\psi}(x_2) \gamma^0 \psi(x_2) \bar{\psi}(x_1) \gamma^0 \psi(x_1) \right] \tag{45}$$

The operator Feynman diagrams corresponding to the various terms in the Wick expansion are shown in Fig. 5. Only the first and third diagrams are connected, so the connected S-matrix element is

$$\langle S_\epsilon^{(2)}\rangle_c = -\tfrac{1}{2}\int d^4x_2 \int d^4x_1 e^{-\epsilon|t_2|}e^{-\epsilon|t_1|}V(\vec{x}_2)V(\vec{x}_1)$$

$$\times \langle : \bar{\psi}(x_2)\gamma^0\psi(x_2)\bar{\psi}(x_1)\gamma^0\psi(x_1) :$$

$$+ : \bar{\psi}(x_2)\gamma^0\psi(x_2)\bar{\psi}(x_1)\gamma^0\psi(x_1) : \qquad (46)$$

$$+ : \bar{\psi}(x_2)\gamma^0\,\psi(x_2)\bar{\psi}(x_1)\gamma^0\psi(x_1) :\rangle$$

or

$$\langle S_\epsilon^{(2)}\rangle_c = -\tfrac{1}{2}\int d^4x_2 \int d^4x_1\; e^{-\epsilon|t_2|}e^{-\epsilon|t_1|}V(\vec{x}_2)V(\vec{x}_1)$$

$$\times \left\{ \sum_{nm} \bar{\phi}_{n_2}(x_2)\gamma^0\phi_{m_2}(x_2)\bar{\phi}_{n_1}(x_1)\gamma^0\phi_{m_1}(x_1)\langle a_{n_2}^\dagger a_{n_1}^\dagger a_{m_1} a_{m_2}\rangle \right.$$

$$\left. + 2\sum_{nm} \bar{\phi}_n(x_2)\gamma^0 S_F(x_2,x_1)\gamma^0\phi_m(x_1)\langle a_n^\dagger a_m\rangle \right\}$$

$$(47)$$

Integration over time in the first term yields

$$I_1 = \int dt_2 \int dt_1 e^{-\epsilon|t_2|}e^{-\epsilon|t_1|}e^{i(E_{n_2}-E_{m_2})t_2}e^{i(E_{n_1}-E_{m_1})t_1}$$

$$= \frac{2\epsilon}{(E_{n_2}-E_{m_2})^2+\epsilon^2}\frac{2\epsilon}{(E_{n_1}-E_{m_1})^2+\epsilon^2} \qquad (48)$$

$$= \frac{4}{\epsilon^2}\,\delta(E_{n_2},E_{m_2})\,\delta(E_{n_1},E_{m_1})+\mathcal{O}(\epsilon^0)$$

The second term gives

$$I_2 = \int dt_2 \int dt_1 e^{-\epsilon|t_2|}e^{-\epsilon|t_1|}e^{iE_n t_2}\frac{1}{2\pi i}\int dz\, \frac{e^{-iz(t_2-t_1)}}{E_i - z(1+i\delta)}e^{-iE_m t_1}$$

$$= \frac{1}{2\pi i}\int dz\, \frac{1}{E_i - z(1+i\delta)}\frac{2\epsilon}{(E_n-z)^2+\epsilon^2}\frac{2\epsilon}{(z-E_m)^2+\epsilon^2} \qquad (49)$$

This integral is (at least) of order ϵ^0 unless $E_n = E_m$, so it is only necessary to consider the simpler integral

$$I_2 = -\frac{1}{2\pi i}\,\delta(E_n, E_m)\int dz\,\frac{1}{z - E_i(1 - i\delta)}\,\frac{(2\epsilon)^2}{[(z - E_n)^2 + \epsilon^2]^2} + \mathcal{O}(1)$$

$$= -i\delta(E_n, E_m)\,\epsilon\frac{\partial}{\partial\epsilon}\frac{1}{\epsilon}\,\frac{1}{E_n - E_i + i\epsilon\frac{E_i}{|E_i|}} + \mathcal{O}(1)$$

$$= \begin{cases} \dfrac{2}{\epsilon^2}\,\delta(E_n, E_m) + \mathcal{O}(1), & \text{if } E_i = E_n \\[3mm] \dfrac{i}{\epsilon}\,\delta(E_n, E_m)\,\dfrac{1}{E_n - E_i} + \mathcal{O}(1), & \text{if } E_i \neq E_n \end{cases}$$

(50)

The order in ϵ of the result depends on whether the intermediate state energy is degenerate with the external line energy. In the case where they are degenerate, the intermediate state energy is positive by the assumption that external lines are only electrons and not positrons.

The complete second-order expression is given by

$$2\langle S_\epsilon^{(2)}\rangle_c - \langle S_\epsilon^{(1)}\rangle_c^2$$

$$= -\frac{4}{\epsilon^2}\sum_{nm}\delta(E_{n_2}, E_{m_2})\,\delta(E_{n_1}, E_{m_1})V_{n_2, m_2}V_{n_1, m_1}\langle a_{n_2}^\dagger a_{n_1}^\dagger a_{m_1} a_{m_2}\rangle$$

$$-\frac{2i}{\epsilon}\sum_{nm}\sum_{E_i \neq E_n}\delta(E_n, E_m)V_{n,i}\frac{1}{E_n - E_i}V_{i,m}\langle a_n^\dagger a_m\rangle$$

$$-\frac{4}{\epsilon^2}\sum_{nm}\delta(E_n, E_i)\,\delta(E_i, E_m)V_{n,i}V_{i,m}\langle a_n^\dagger a_m\rangle$$

$$+\frac{4}{\epsilon^2}\sum_{nm}\delta(E_{n_2}, E_{m_2})\,\delta(E_{n_1}, E_{m_1})V_{n_2, m_2}V_{n_1, m_1}\langle a_{n_2}^\dagger a_{m_2}\rangle\langle a_{n_1}^\dagger a_{m_1}\rangle$$

(51)

In order to get a finite result in the limit $\epsilon \to 0$, it is necessary that the terms of order ϵ^{-2} add up to zero. In fact, they do, but not in an obvious way. To see how, we employ the identity

$$a_{n_2}^\dagger a_{n_1}^\dagger a_{m_1} a_{m_2} = -a_{n_2}^\dagger a_{n_1}^\dagger a_{m_2} a_{m_1}$$

$$= -\delta_{n_1, m_2} a_{n_2}^\dagger a_{m_1} + a_{n_2}^\dagger a_{m_2} a_{n_1}^\dagger a_{m_1}$$

(52)

and note that when it is substituted in the first line of the complete expression, the first term cancels the next-to-last term of the complete expression. The remaining singular terms are then proportional to

$$\sum_{nm}\delta(E_{n_2}, E_{m_2})\,\delta(E_{n_1}, E_{m_1})V_{n_2, m_2}V_{n_1, m_1}\left[\langle a_{n_2}^\dagger a_{m_2} a_{n_1}^\dagger a_{m_1}\rangle\right.$$

$$\left. - \langle a_{n_2}^\dagger a_{m_2}\rangle\langle a_{n_1}^\dagger a_{m_1}\rangle\right]$$

(53)

$$= \langle Q^2\rangle - \langle Q\rangle^2$$

where

$$Q = \sum_{nm} \delta(E_n, E_m) V_{n,m} \, a_n^\dagger a_m \tag{54}$$

The singular term vanishes if the operator Q has no off-diagonal matrix elements. In fact, this is the case for the problem at hand. In particular, we consider an even parity spherically symmetric external potential for the binding and a perturbation that also has even parity and spherical symmetry. Then because of the energy conservation factor in (54), Q will be diagonal in the basis states to which attention is restricted here, i.e., eigenfunctions of parity, angular momentum, and energy. If there are degenerate states with the same symmetry, then Q will not necessarily be diagonal and further analysis is needed. This is the degenerate perturbation theory case.

The final result for the second-order correction is

$$E^{(2)} = \sum_{nm} \sum_{E_i \neq E_n} \delta(E_n, E_m) V_{n,i} \, \frac{1}{E_n - E_i} \, V_{i,m} \langle a_n^\dagger a_m \rangle \tag{55}$$

which for the case of a hydrogenlike atom in the state a reduces to

$$E_a^{(2)} = \sum_{E_i \neq E_a} V_{a,i} \, \frac{1}{E_a - E_i} \, V_{i,a} \tag{56}$$

which is the expected result.

The above discussion derives the standard perturbation theory results generalized to the many-body form. These expressions could have been easily derived with the aid of standard many-body methods. However, the present approach will also deal with retarded interactions which are not encompassed in the standard many-body theory. Such applications will be illustrated in the remainder of these notes.

PHOTON EXCHANGE

The exchanged-photon operator arises from a time-dependent perturbation, namely the interaction with the quantized photon field. This time dependence manifests itself in the fact that the energy of the external line appears in the operator. This is also true for the self-energy correction, but attention is confined to the photon exchange correction here. The second-order time-ordered product is

$$T\left[j^{\mu_2}(x_2) A_{\mu_2}(x_2) j^{\mu_1}(x_1) A_{\mu_1}(x_1)\right]$$
$$= e^2 T\left[\bar{\psi}(x_2) \gamma^{\mu_2} \psi(x_2) \bar{\psi}(x_1) \gamma^{\mu_1} \psi(x_1)\right] g_{\mu_2 \mu_1} D_F(x_2 - x_1) \tag{57}$$

where the photon vacuum expectation value is understood. The exchanged-photon correction comes from the no-contraction term in the Wick expansion

$$\langle S_\epsilon^{(2)} \rangle_{PE} = -\frac{e^2}{2} \int d^4x_2 \int d^4x_1 \; e^{-\epsilon|t_2|} e^{-\epsilon|t_1|} D_F(x_2 - x_1)$$

$$\times \sum_{nm} \bar{\phi}_{n_2}(x_2)\gamma_\mu \phi_{m_2}(x_2)\bar{\phi}_{n_1}(x_1)\gamma^\mu \phi_{m_1}(x_1)\langle a_{n_2}^\dagger a_{n_1}^\dagger a_{m_1} a_{m_2}\rangle \tag{58}$$

The time integration is

$$I = \int dt_2 \int dt_1 e^{-\epsilon|t_2|} e^{-\epsilon|t_1|} \frac{1}{2\pi i} \int dq_0 H(\vec{x}_2 - \vec{x}_1, q_0)e^{-iq_0(t_2-t_1)}$$

$$\times e^{i(E_{n_2}-E_{m_2})t_2} e^{i(E_{n_1}-E_{m_1})t_1}$$

$$= \frac{1}{2\pi i} \int dq_0 H(\vec{x}_2 - \vec{x}_1, q_0) \frac{2\epsilon}{(E_{n_2} - E_{m_2} - q_0)^2 + \epsilon^2} \frac{2\epsilon}{(q_0 + E_{n_1} - E_{m_1})^2 + \epsilon^2} \tag{59}$$

The leading term in ϵ corresponds to the conditions

$$E_{n_2} - E_{m_2} - q_0 = 0$$
$$q_0 + E_{n_1} - E_{m_1} = 0 \tag{60}$$

or

$$E_{n_2} + E_{n_1} = E_{m_2} + E_{m_1}$$
$$q_0 = E_{n_2} - E_{m_2} \tag{61}$$

Evaluation of the integral over q_0 is facilitated by expanding the function H in a power series about the dominant value of q_0 and keeping only the leading term

$$I = \frac{1}{2\pi i} H(\vec{x}_2 - \vec{x}_1, E_{n_2} - E_{m_2}) \, \delta(E_{n_2} + E_{n_1}, E_{m_2} + E_{m_1})$$

$$\times \int dq_0 \frac{(2\epsilon)^2}{[(q_0 - E_{n_2} + E_{m_2})^2 + \epsilon^2]^2} + \mathcal{O}(1) \tag{62}$$

which gives

$$I = \frac{1}{i\epsilon} H(\vec{x}_2 - \vec{x}_1, E_{n_2} - E_{m_2}) \, \delta(E_{n_2} + E_{n_1}, E_{m_2} + E_{m_1}) + \mathcal{O}(1) \tag{63}$$

with the result

$$E_{PE}^{(2)} = -2\pi\alpha \int d\vec{x}_2 \int d\vec{x}_1 \sum_{nm} \phi_{n_2}^\dagger(\vec{x}_2)\alpha_\mu \phi_{m_2}(\vec{x}_2)\phi_{n_1}^\dagger(\vec{x}_1)\alpha^\mu \phi_{m_1}(\vec{x}_1)$$

$$\times H(\vec{x}_2 - \vec{x}_1, E_{n_2} - E_{m_2}) \, \delta(E_{n_2} + E_{n_1}, E_{m_2} + E_{m_1})\langle a_{n_2}^\dagger a_{n_1}^\dagger a_{m_1} a_{m_2}\rangle \tag{64}$$

Figure 6: The one-photon operator Feynman diagram

for the one-photon correction. The corresponding operator Feynman diagram is shown in Fig. 6.

The final formula contains the energy of the external lines in the function H. This has the consequence that higher-order perturbation theory will generate derivatives of H with respect to the energy, as is easily seen by considering an additional perturbation such as the finite nuclear radius correction. If the one-photon expression were calculated with the exact external potential, then the size correction would be contained in the external line energy as well as in the wave functions. Then an expansion to arrive at the terms linear in the size correction would involve the derivative of the function H corresponding to the correction in the energy. The source of such terms in this formalism will be indicated in the next section. More detail on the second-order QED corrections along the lines described here is given elsewhere.[6]

DERIVATIVE TERMS

In this section, the origin of the derivative terms will be sketched in the context of the combined effects of an external potential and one-photon exchange, each of which is considered separately above. The complete calculation would be cumbersome to present, but the main features can be placed in the framework of the calculations described above. Incidentally, the external potential corrections to the one-photon exchange correction have the same structure as the mass renormalization corrections that accompany the radiative corrections to the one exchanged photon correction.

Consider an interaction term that contains both an external potential and the quantized radiation field potential. The relevant terms are third order overall, with second-order terms in the radiation field and first-order terms in the external potential. The relevant part of the level shift expression is given by

$$E^{(3)} = \lim_{\epsilon \to 0} \frac{i\epsilon}{2} \left[3\langle S_\epsilon^{(3)} \rangle_c - 3\langle S_\epsilon^{(1)} \rangle_c \langle S_\epsilon^{(2)} \rangle_c \dots \right] \tag{65}$$

The operator Feynman diagrams are shown in Fig. 7.

As in the earlier calculation, there will be cancellations between singular terms in ϵ between the various diagrams listed. In the diagrams with the electron-positron propagator, the singular term arises from states degenerate in energy with the external line attached to the external potential. In this case, the degenerate-state term has a correction term of one higher power in ϵ which makes a nonvanishing contribution to the level shift. In particular, if the function H is expanded about the dominant point as

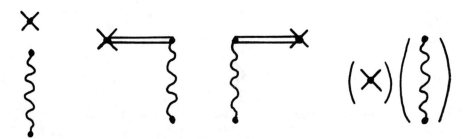

Figure 7: Operator Feynman diagrams for the combined
external potential and one-photon corrections

$$H(\vec{x}, q_0) = H(\vec{x}, E_0) + (q_0 - E_0)H'(\vec{x}, E_0) + \ldots \qquad (66)$$

then the leading term is the singular term that cancels other terms, and the
first correction gives the derivative contribution. Higher-derivative terms may
be neglected in this case.

The result obtained this way agrees with the result obtained by substituting the perturbed energy and wave functions in the exchanged-photon expression
derived above and retaining terms linear in the external potential.

The derivative terms are clearly necessary, so that any alternative formulation of the theory that eliminates degenerate intermediate states at an earlier
stage of the calculation would require an alternative scheme to generate such
terms through, for example, a resummation of infinite subsets of diagrams.

REFERENCES

1. W. H. Furry, Phys. Rev. **81**, 115 (1951).
2. M. Gell-Mann and F. Low, Phys. Rev. **84**, 350 (1951).
3. J. Sucher, Phys. Rev. **107**, 1448 (1957).
4. F. Low, Phys. Rev. **88**, 53 (1952).
5. G. C. Wick, Phys. Rev. **80**, 268 (1950).
6. P. J. Mohr, Phys. Rev. A **32**, 1949 (1985).

THE TWO-BODY PROBLEM IN QED

APPLICATION OF THE BOUND STATE FORMALISM
TO POSITRONIUM

Gregory S. Adkins

Franklin and Marshall College, Lancaster, PA   17604

ABSTRACT

This report contains a discussion of a bound state formalism for QED and its application to positronium. The formalism is based on the Bethe-Salpeter equation for the two-to-two Green's function. The energy levels of positronium are found as poles of this Green's function. A perturbation scheme for the energy levels is set up by comparing the Bethe-Salpeter equation with a soluble reference equation. A useful reference problem is examined, and the reference solution is obtained by comparison with the non-relativistic Schrödinger-Coulomb Green's function. Renormalization effects are discussed. As an example of the use of this formalism, an outline is given of the calculation of the ground state hyperfine splitting of positronium at the $O(m\alpha^4)$ and $O(m\alpha^5)$ levels.

I. INTRODUCTION TO POSITRONIUM

Positronium is the e^+e^- bound state.[1] The states of positronium are labeled by principal quantum number n ($n = 1, 2, 3, \ldots$), total spin s ($s = 0$: parapositronium, $s = 1$: orthopositronium), orbital angular momentum L ($L = S, P, D, \ldots$), and total angular momentum j ($j = 0, 1, 2, \ldots$). State labels are written as $n^{2s+1}L_j{}^{PC}$, where $P = (-1)^{l+1}$ is the space parity and $C = (-1)^{l+s}$ is the charge parity. The gross structure of the positronium energy spectrum can be obtained from the non-relativistic (NR) Schrödinger equation with a Coulomb potential. The NR energy levels are[2]

$$E_n = -\frac{1}{2}\left(\frac{m}{2}\right)\frac{\alpha^2}{n^2} \, . \tag{1}$$

The NR energy levels are similar to those of hydrogen, but are compressed by a factor of approximately two because of the different electron reduced mass.

Degeneracy under s, L, and j is broken in positronium at $O(m\alpha^4)$ by relativistic, spin-orbit, magnetic spin-spin, and annihilation effects (Pirenne,[3] Berestetski,[4] Ferrell[5]). For example, the $O(m\alpha^4)$ contributions to S state energies are[6]

$$\frac{m\alpha^4}{n^3}\left[\frac{11}{64n} - \frac{1}{2} + \frac{7}{12}\delta_{1,s}\right] \, . \tag{2}$$

An energy level diagram for the $n = 1$ and $n = 2$ states is given in Fig. 1. At $O(m\alpha^5)$ radiative corrections appear including Lamb shift-like effects (although the Lamb shift does not split degenerate states in positronium as in hydrogen), recoil and retardation, and two-photon annihilation contributions (Karplus and

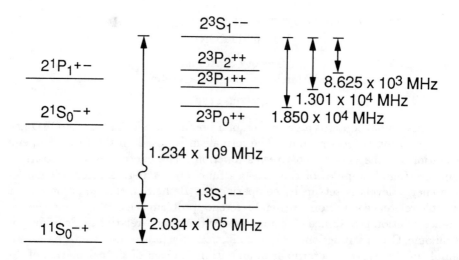

Fig. 1. Energy level diagram for the $n = 1$ and $n = 2$ states of positronium. Theoretical values of transition frequencies (Ref. 7 and 8) for the experimentally interesting transitions are given.

Klein,[7] Fulton and Martin[8]). For example, the $O(m\alpha^5)$ contributions to S state energies are[6]

$$\frac{m\alpha^5}{\pi n^3}\left[\frac{31}{24} + \frac{7}{12}\ln 2 - \frac{3}{4}\ln\alpha - \frac{2}{3}\ln\left(k_0(n,0)\right) - \left(\frac{8}{9} + \frac{1}{2}\ln 2\right)\delta_{1,s}\right]. \qquad (3)$$

Here $\ln\left(k_0(n,l)\right)$ is the Bethe logarithm.[9] Contributions of $O(m\alpha^6\ln\alpha)$ and $O(m\alpha^6)$ have been found by many workers.[10,11] The complete $O(m\alpha^6\ln\alpha)$ contribution to the ground state hyperfine splitting $\left(E(1^3S_1) - E(1^1S_0)\right)$ is [10]

$$\frac{5}{24}m\alpha^6\ln\left(\frac{1}{\alpha}\right). \qquad (4)$$

Some but not all of the pure $O(m\alpha^6)$ contributions have been calculated.[11]

Experimentally, several of the positronium $n = 1$ and $n = 2$ energy intervals have been measured. The most accurate measurement is of the ground state hyperfine splitting. It was first measured by Deutsch and Dulit in 1951.[12] Several other determinations were made over the years,[13] the latest being[14]

$$203{,}389.10 \pm 0.74 \text{ MHz}. \qquad (5)$$

The theoretical result for this splitting is

$$\Delta E = m\alpha^4\left[\frac{7}{12} - \frac{\alpha}{\pi}\left(\frac{8}{9} + \frac{1}{2}\ln 2\right) + \frac{5}{24}\alpha^2\ln\left(\frac{1}{\alpha}\right) + O(\alpha^2)\right]$$
$$= 203{,}400 \text{ MHz}. \qquad (6)$$

The uncalculated term of $O(m\alpha^6)$ would contribute 18.7 MHz if its coefficient were unity.

Positronium decays to photons. The charge parity of a positronium state is $(-1)^{l+s}$, and that of an n photon state is $(-1)^n$, so positronium decays to an even number of photons when $\ell + s$ is even and to an odd number of photons when $\ell+s$ is odd. Decays to a single photon are forbidden by energy-momentum conservation. In particular, the 1^1S_0 parapositronium state decays to two (and four, and six, ...) photons. The decay rate is[15]

$$\Gamma_{LO}(1^1S_0) = \frac{1}{2}m\alpha^5 \cong 8(\text{ns})^{-1} . \tag{7}$$

The 1^3S_1 orthopositronium state decays to three (and five, and seven, ...) photons, with rate[16]

$$\Gamma_{LO}(1^3S_1) = \frac{2}{9\pi}(\pi^2 - 9)m\alpha^6 \cong 7\,(\mu s)^{-1} . \tag{8}$$

It is interesting that a recent measurement[17] of this orthopositronium decay rate is apparently not in accord with theory. The measured value is

$$\Gamma_{\exp}(1^3S_1) = 7.0516(13)\ \ (\mu s)^{-1} , \tag{9}$$

while the theoretical result is[18-20]

$$\Gamma_{th}(1^3S_1) = \Gamma_{LO}(1^3S_1)\left[1 - 10.282(3)\left(\frac{\alpha}{\pi}\right) - \frac{\alpha^2}{3}\ln\left(\frac{1}{\alpha}\right) + A\left(\frac{\alpha}{\pi}\right)^2 + O(\alpha^3)\right]$$

$$= 7.03830(7)\,(\mu s)^{-1} + \Gamma_{LO}(1^3S_1)\left[A\left(\frac{\alpha}{\pi}\right)^2 + O(\alpha^3)\right] . \tag{10}$$

Comparison suggests that $A \cong 300$, which seems excessively large.

II. GRAPHICAL DERIVATION OF THE BOUND STATE EQUATION

In this section I will describe a graphical derivation of the Bethe-Salpeter equation for the electron-positron Green's function. I will begin by listing the Feynman rules of QED, and then focus on the Feynman graphs involved in electron-positron propagation.

Processes in QED can be represented by Feynman diagrams.[21] These diagrams contain straight lines representing fermion propagators, wiggly lines representing photon propagators, and vertices representing photon-fermion interactions. The mathematical expression for a diagram is found by multiplying the various factors contained in the diagram and integrating over closed loops.

68

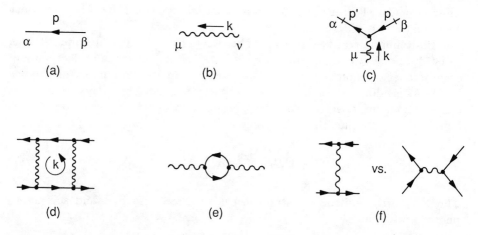

(a)   (b)   (c)

(d)   (e)   vs.   (f)

Fig. 2. Feynman diagrams in QED. Fig. 2(a) shows a fermion propagator, 2(b) shows a photon propagator, and 2(c) shows a fermion-photon vertex. The barred external legs in Fig. 2(c) are not part of the vertex, they are pictured only to show where the photon and fermion lines attach. Fig. 2(d) depicts a closed loop with loop momentum k, which must be integrated over. Fig. 2(e) contains a fermion loop which involves an extra minus sign. Fig. 2(f) shows two graphs having a relative minus sign because they differ only by the interchange of two (upper left and lower right) fermion legs.

The three basic diagram elements are pictured in Fig. 2(a,b,c). The fermion propagation factor of Fig. 2(a) is

$$iS_{\alpha\beta}(p) = \left(\frac{i}{\gamma p - m}\right)_{\alpha\beta} = \frac{i(\gamma p + m)_{\alpha\beta}}{(p^2 - m^2)} , \qquad (11)$$

where α and β are Dirac indices and p is the momentum flow in the direction of the line orientation. The photon propagation factor of Fig. 2(b) is $iD^{\mu v}(k)$, where μ and v are Lorentz indices and k is the momentum of the photon. In the general convariant gauge the photon propagator has the form

$$D^{\mu v}(k) = \frac{-1}{k^2}\left(g^{\mu v} + \xi \frac{k^\mu k^v}{k^2}\right). \qquad (12)$$

Here ξ is a gauge parameter, with $\xi = 0$ corresponding to Feynman gauge, $\xi = -1$ to Landau gauge, and $\xi = 2$ to Fried-Yennie gauge. Another useful gauge is Coulomb gauge, in which

$$D_C^{\mu v}(k) = \begin{pmatrix} \frac{1}{\vec{k}^2} & 0 \\ 0 & \frac{1}{k^2}\left(\delta_{ij} - \frac{k_i k_j}{\vec{k}^2}\right) \end{pmatrix}. \qquad (13)$$

Physical results in QED are gauge independent, so any convenient gauge can be used for the photon propagator. Coulomb gauge and Fried-Yennie gauge are particularly useful, since they are infrared (IR) safe. Renormalization can be carried out on mass shell in these gauges without the need for IR regularization.[22-24] The vertex factor Fig. 2(c) is

$$-ie(\gamma^\mu)_{\alpha\beta} , \qquad (14)$$

where e is the fermion charge. The "barred" lines shown in Fig. 2(c) are not included in the vertex. They are pictured only to show where the external photon and fermion lines attach. Momentum conservation is required at each vertex, so $p' = p + k$. Processes with closed loops, such as that shown in Fig. 2(d), can occur with arbitrary loop momentum. These loop momenta must be integrated with weight $(dk)' = d^4k/(2\pi)^4$. Finally, there are two sign rules: graphs with closed fermion loops [such as Fig. 2(e)] have an extra minus sign, and two graphs differing only by the interchange of two fermion lines [such as the graphs shown in Fig. 2(f)] have a relative minus sign.

As an example of the application of these rules, I will write down the expression $A^{CL}_{\alpha\beta,\sigma\tau}(P;p,p')$ for the "crossed ladder" graph shown in Fig. 3. This amputated, two-particle irreducible graph contributes to the Bethe-Salpeter kernel that will be defined later. The expression for the crossed ladder graph is

$$A^{CL}_{\alpha\beta,\sigma\tau}(P;p,p') = \int (dk)'[(-ie\gamma^\mu)\,iS(\tfrac{1}{2}P + p' + k)(-ie\gamma^\lambda)]_{\alpha\sigma}$$

$$\times [(-ie\gamma^\kappa)\,iS(-\tfrac{1}{2}P + p - k)(-ie\gamma^v)]_{\tau\beta}\,iD_{v\lambda}(k)\,iD_{\mu\kappa}(k + p' - p) . \qquad (15)$$

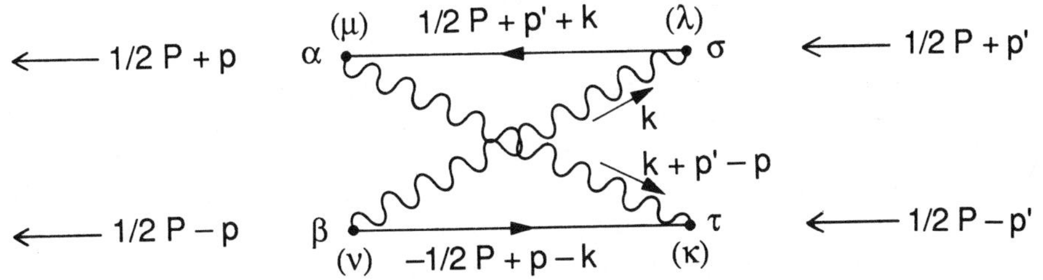

Fig. 3. The crossed ladder graph $A^{CL}_{\alpha\beta,\sigma\tau}(P;p,p')$.

In a notation where Dirac indices are implicit, this is

$$A^{CL}(P;p,p') = \int (dk)'[(-ie\gamma^\mu)\,iS(\tfrac{1}{2}P + p' + k)(-ie\gamma^\lambda)]^{(1)}$$

$$\times [(-ie\gamma^\kappa)\,iS(-\tfrac{1}{2}P + p - k)(-ie\gamma^v)]^{(2)T}\,iD_{v\lambda}(k)\,iD_{\mu\kappa}(k + p' - p) , \qquad (16)$$

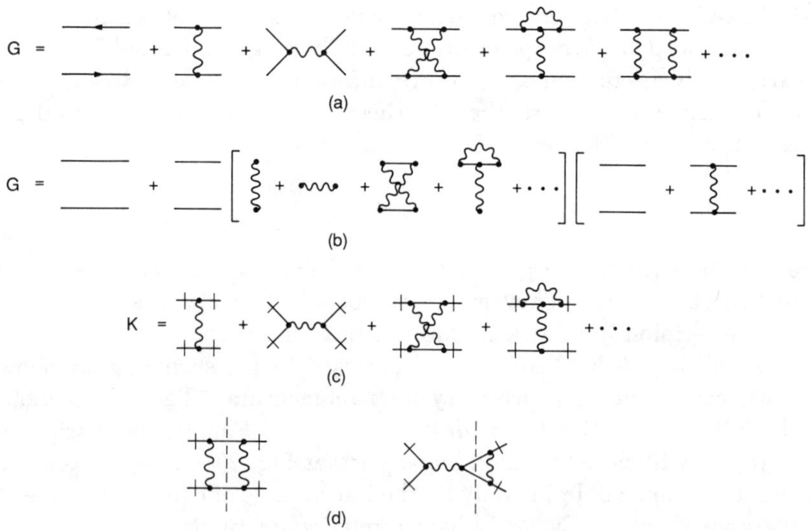

Fig. 4. The Bethe-Salpeter equation for the two-to-two Green's function G of Fig. 4(a) is shown in Fig. 4(b). The kernel K is shown in Fig. 4(c), where barred legs are to be amputated. Fig. 4(d) shows two two-particle reducible graphs which are not part of the kernel.

where (1) indicates the electron (upper) line, (2) indicates the positron (lower) line, and T stands for matrix transposition.

The Bethe-Salpeter equation[25] for the two-to-two Green's function is easy to derive graphically. The electron-positron to electron-positron Green's function G is pictured in Fig. 4(a).[26] An alternate representation of this infinite set of graphs is shown in Fig. 4(b), where the second factor in the second term on the right hand side is the kernal K, and the last factor is G itself. The kernel K is shown again in Fig. 4(c), where again the barred external lines are not to be included. The kernel contains all connected graphs of G that are "two-particle irreducible," i.e., that cannot be cut in two by breaking just the two fermion lines. The graphs shown in Fig. 4(d) are examples of two-particle reducible graphs that are not included in K. A symbolic rendering of the inhomogeneous Bethe-Salpeter equation pictured in Fig. 4(b) is

$$G = S + SKG, \tag{17}$$

where momentum variables and spin indices are implicit. In (17) the quantity S represents the free propagation of the two fermions

$$S_{\alpha\beta,\sigma\tau}(P; p, p') = (2\pi)^4 \delta(p - p') S_{\alpha\beta,\sigma\tau}(P; p), \tag{18a}$$

$$S_{\alpha\beta,\sigma\tau}(P; p) = i S_{\alpha\sigma}(\tfrac{1}{2}P + p) i S_{\tau\beta}(-\tfrac{1}{2}P + p). \tag{18b}$$

The four dimensional delta function in (18a) reflects the fact that the electron and positron do not exchange momentum. With all indices explicit, the BS equation can be written as

$$G_{\alpha\beta,\sigma\tau}(P;p,p') = S_{\alpha\beta,\sigma\tau}(P;p,p')$$
$$+ S_{\alpha\beta,\gamma\delta}(P;p) \int (dk)' K_{\gamma\delta,\lambda\kappa}(P;p,k) G_{\lambda\kappa,\sigma\tau}(P;k,p') . \tag{19}$$

III. BOUND STATE PERTURBATION THEORY

We want to find the energies of the bound states of G, which satisfies the BS equation

$$G = S + SKG . \tag{20}$$

In practice, this equation is impossible to solve. What we can do is solve a "reference problem"

$$G_0 = S + SK_0G_0 , \tag{21}$$

where K_0 is "close" to K and is chosen so that (21) can actually be solved. The idea is to relate G to G_0, and then develop a perturbative expansion for the energies of the bound states described by G in terms of the known reference quantities.

The Green's function G can be related to the known G_0 by eliminating S between (20) and (21). From (21) one finds that

$$S = (1 + G_0K_0)^{-1}G_0 . \tag{22}$$

Then from (20) one has

$$G = (1 - SK)^{-1}S = (1 - G_0\delta K)^{-1}G_0$$
$$= G_0 + G_0\delta K G_0 + G_0\delta K G_0\delta K G_0 + \dots , \tag{23}$$

where $\delta K = K - K_0$.

We can now derive a perturbative expansion for the bound state energies E_n of G. Bound states appear in the Green's function as poles.[27] Assuming that G and G_0 have corresponding bound states, we can write

$$G \to i\frac{\Psi_n \bar{\Psi}_n}{(E - E_n)} \qquad \text{as } E \to E_n , \tag{24a}$$

$$G_0 \to i\frac{\Psi_n^0 \bar{\Psi}_n^0}{(E - E_n^0)} \qquad \text{as } E \to E_n^0 . \tag{24b}$$

Let us define C_n as a counterclockwise closed contour in the complex E plane that circles E_n and E_n^0 but no other poles of G or G_0. Let L and R be two arbitrary quantities dependent on two spin indices and a relative momentum. Then using Cauchy's theorem, we have

$$\oint_{C_n} \frac{dE}{2\pi i} E \, LGR = E_n i(L\Psi_n)(\bar{\Psi}_n R) \,, \tag{25a}$$

$$\oint_{C_n} \frac{dE}{2\pi i} LGR = \quad i(L\Psi_n)(\bar{\Psi}_n R) \,, \tag{25b}$$

so that

$$E_n = \frac{\oint_{C_n} \frac{dE}{2\pi i} (E_n^0 + (E - E_n^0)) LGR}{\oint_{C_n} \frac{dE}{2\pi i} LGR}$$

$$= E_n^0 + \frac{\oint_{C_n} \frac{dE}{2\pi i} (E - E_n^0) LGR}{\oint_{C_n} \frac{dE}{2\pi i} LGR} \,. \tag{26}$$

We want to express the right hand side of (26) in terms of known quantities like Ψ_n^0, $\bar{\Psi}_n^0$, E_n^0, and δK, and elminate L, R, and G. We can eliminate G by using (23). We will need to expand the numerator and denominator in Laurent series. Let us write

$$G_0 = \hat{G}_0 + i \frac{\Psi_n^0 \bar{\Psi}_n^0}{(E - E_n^0)} \tag{27}$$

so that the "subtracted" Green's function $\hat{G}_0$ has no pole at E_n^0. Then we have

$$G = \hat{G}_0 + i \frac{\Psi_n^0 \bar{\Psi}_n^0}{(E - E_n^0)} + \hat{G}_0 \delta K \hat{G}_0$$

$$+ \frac{i}{(E - E_n^0)} [\hat{G}_0 \delta K \Psi_n^0 \bar{\Psi}_n^0 + \Psi_n^0 \bar{\Psi}_n^0 \delta K \hat{G}_0]$$

$$+ \frac{i^2}{(E - E_n^0)^2} \Psi_n^0 (\bar{\Psi}_n^0 \delta K \Psi_n^0) \bar{\Psi}_n^0 + O(\delta K)^2 \,. \tag{28}$$

The contour integrals in Eq. (26) can be done using Cauchy's theorem, with results

$$\text{Num} = i^2 (L\Psi_n^0)(\bar{\Psi}_n^0 \delta K \Psi_n^0)(\bar{\Psi}_n^0 R) + O(\delta K)^2 \,, \tag{29a}$$

$$\text{Den} = i(L\Psi_n^0)(\bar{\Psi}_n^0 R) + O(\delta K) \,. \tag{29b}$$

In the ratio (Num)/(Den) all dependence on L and R vanishes. The series through terms of $O(\delta K)^3$ is[28-31]

$$
\begin{aligned}
E_n = {}& E_n^0 + (\delta K) + (\delta K \hat{G}_0 \delta K) + (\delta K)(\delta K)' \\
& + (\delta K \hat{G}_0 \delta K \hat{G}_0 \delta K) + (\delta K)(\delta K \hat{G}_0 \delta K)' \\
& + (\delta K)'(\delta K \hat{G}_0 \delta K) + (\delta K)((\delta K)')^2 \\
& + \frac{1}{2}(\delta K)^2 (\delta K)'' + O(\delta K)^4 \,,
\end{aligned}
\tag{30}
$$

where

$$
(X) \equiv i\bar{\Psi}_n^0 X \Psi_n^0
\tag{31a}
$$
$$
(X)' \equiv i\bar{\Psi}_n^0 [dX/dE|_{E=E_n^0}] \Psi_n^0 \,,
\tag{31b}
$$

etc. In practice, the orders of the various terms calculated in renormalized Coulomb or Fried-Yennie gauge are

$$
(\delta K) = O(m\alpha^4) \,,
\tag{32a}
$$
$$
(\delta K \hat{G}_0 \delta K) = O(m\alpha^5) \,,
\tag{32b}
$$
$$
(\delta K \hat{G}_0 \delta K \hat{G}_0 \delta K) = O(m\alpha^6) \,,
\tag{32c}
$$
$$
(\delta K)' = O(\alpha^2) \,,
\tag{32d}
$$

etc., where only the lowest order contained in a particular term is indicated.

IV. THE SCHRÖDINGER-COULOMB GREEN'S FUNCTION

The Coulomb force is responsible for the binding of the electron and positron in positronium. Consequently, any reasonable reference kernel K_0 must describe the exchange of a Coulomb photon in some approximate way. We will obtain our solution of the reference problem (21) by grafting relativity and spin onto the NR solution for a particle moving in a Coulomb field. In this section we will find a useful form of the NR Schrödinger-Coulomb Green's function.[32]

We need a solution to the inhomogenous Schrödinger-Coulomb equation

$$
\int (d^3q)' [(E - \frac{\vec{p}^{\,2}}{2\mu})(2\pi)^3 \delta(\vec{p}-\vec{q}) - V(\vec{p}-\vec{q})] g(E; \vec{q}, \vec{p}\,') = (2\pi)^3 \delta(\vec{p}-\vec{p}\,')
\tag{33}
$$

where $(d^3q)' = d^3q/(2\pi)^3$ and

$$
V(\vec{p}-\vec{q}) = \frac{-4\pi Z\alpha}{(\vec{p}-\vec{q})^2} \,.
\tag{34}
$$

This equation describes the motion of a particle of mass μ in a potential $-Z\alpha/r$. Suppose we define the free NR Green's function by

$$s(E; \vec{p}, \vec{p}\,') = (2\pi)^3 \delta(\vec{p} - \vec{p}\,') s(E; \vec{p}) \ . \tag{35a}$$

$$s(E; \vec{p}) = \frac{-2\mu}{(\vec{p}^{\,2} + \gamma^2)} \tag{35b}$$

where $\gamma^2 = -2\mu E$. Then we can rewrite (33) as

$$g = s + sVg \ , \tag{36}$$

where again momentum variables and integrations are implicit. Equation (36) has the series solution

$$g = (1 - sV)^{-1}s = s + sVs + sVsVs + sVsVsVs + \ldots \ . \tag{37}$$

The first interesting term in (37) is $sVsVs$. After performing the integrals, one finds that

$$[VsV](E; \vec{p}, \vec{p}\,') = \frac{(-2\mu)4\pi(Z\alpha)^2}{|\vec{p} - \vec{p}\,'|\sqrt{Q}} \tan^{-1}\left(\frac{\sqrt{Q}}{\gamma|\vec{p} - \vec{p}\,'|}\right) \tag{38}$$

where

$$Q = (\vec{p}^{\,2} + \gamma^2)(\vec{p}^{\,\prime 2} + \gamma^2) - \gamma^2(\vec{p} - \vec{p}\,')^2 \ . \tag{39}$$

The result (38) can be reexpressed in a one-parameter form

$$[VsV](E; \vec{p}, \vec{p}\,') = (-2\mu)4\pi(Z\alpha)^2 \int_0^1 dx \, \frac{2\gamma}{H(E; \vec{p}, \vec{p}\,')} \tag{40}$$

where

$$H(E; \vec{p}, \vec{p}\,') = 4x\gamma^2(\vec{p} - \vec{p}\,')^2 + (1 - x)^2(\vec{p}^{\,2} + \gamma^2)(\vec{p}^{\,\prime 2} + \gamma^2) \ . \tag{41}$$

The nice feature of the one-parameter form is that it can be generalized according to

$$[(Vs)^n VsV](E; \vec{p}, \vec{p}\,') =$$
$$(-2\mu)4\pi(Z\alpha)^2 \int_0^1 dx \left(\frac{-\mu Z\alpha}{\gamma}\right)^n \frac{\ln^n x}{n!} \frac{2\gamma}{H(E; \vec{p}, \vec{p}\,')} \ . \tag{42}$$

Thus the series for g can be summed:

$$g(E; \vec{p}, \vec{p}\,') = s(E; \vec{p}, \vec{p}\,') + s(E; \vec{p})V(\vec{p} - \vec{p}\,')s(E; \vec{p}\,')$$
$$+ (-2\mu)4\pi(Z\alpha)^2 s(E; \vec{p}) \int_0^1 dx \, x^{-\xi} \frac{2\gamma}{H(E; \vec{p}, \vec{p}\,')} s(E; \vec{p}\,') \tag{43}$$

where $\xi = (\mu Z \alpha)/\gamma$, since

$$\sum_{n=0}^{\infty}(-\xi)^n \frac{\ln^n x}{n!} = x^{-\xi} . \tag{44}$$

The Green's function g has the simple form $0P + 1P + MP$, where the zero-potential $(0P)$ and one-potential $(1P)$ terms are separated out and the many-potential (MP) term contains the contributions of two, three, ... actions of the potential.

The Schrödinger-Coulomb Green's function contains complete information about the NR energy levels and wave functions. The bound states are contained in the MP term of g, and can be seen if $1/H$ is expanded in a Taylor series:

$$\begin{aligned}
\int_0^1 dx\, x^{-\xi} \frac{1}{H} &= \int_0^1 dx\, x^{-\xi} \sum_{n=1}^{\infty} \frac{x^{n-1}}{(n-1)!}\left(\frac{d}{dx}\right)^{n-1}\frac{1}{H}\bigg|_{x=0} \\
&= \sum_{n=1}^{\infty}\frac{1}{(n-1)!}\left(\frac{d}{dx}\right)^{n-1}\frac{1}{H}\bigg|_{x=0}\frac{1}{(n-\xi)} .
\end{aligned} \tag{45}$$

The bound state energies are given by $(n - \xi) = 0$, which is equivalent to $\gamma = (\mu Z \alpha)/n$ and to

$$E_n = -\frac{\mu}{2}\frac{(Z\alpha)^2}{n^2} . \tag{46}$$

It is sometimes useful to express the Green's function as

$$g(E;\vec{p},\vec{p}\,') = \sum_{n\ell m}\frac{\psi_{n\ell m}(\vec{p})\psi_{n\ell m}^*(\vec{p}\,')}{(E - E_n)} + \int_0^{\infty} dk \sum_{\ell m}\frac{\psi_{\ell m}(k;\vec{p})\psi_{\ell m}^*(k;\vec{p}\,')}{(E - k^2/2\mu)} , \tag{47}$$

which is a sum over bound states plus continuum states. The wave functions have simple forms in momentum space. For example, the ground state wave function is

$$\psi_{100}(\vec{p}) = \left(\frac{\gamma^3}{\pi}\right)^{1/2}\frac{8\pi\gamma}{(\vec{p}^{\,2} + \gamma^2)^2} \tag{48}$$

where here $\gamma = \mu Z \alpha$.

V. A SOLUBLE REFERENCE BOUND STATE EQUATION

The reference bound state equation has the form

$$\begin{aligned}
G_0 &= S + SK_0 G_0 \\
&= S + SK_0 S + SK_0 SK_0 S + \cdots ,
\end{aligned} \tag{49}$$

where S is the product of two free Dirac propagators and K_0 is an approximation to K. Barbieri and Remiddi[33] have given an explicit form for K_0 for

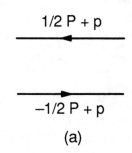

$$\text{1/2 P + p}$$

$$-\text{1/2 P + p}$$

(a)

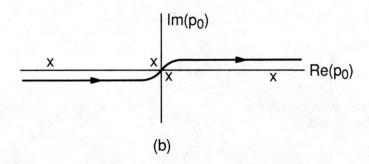

(b)

Fig. 5. Fig. 5(a) shows the noninteracting two particle propagator S. Fig. 5(b) shows the pole structure of S.

which (49) can be solved. This K_0 differs from the kernel for the exchange of a Coulomb photon only in its small components and relativistic behavior.

The rationale for Barbieri and Remiddi's choice of K_0 can be made clear by looking at the propagator S, which is pictured in Fig. 5(a). Using (18) we can write

$$
\begin{aligned}
S(P;p) &= \left[\frac{i}{\gamma(\frac{1}{2}P + p) - m + i\epsilon} \right]^{(1)} \left[\frac{i}{\gamma(-\frac{1}{2}P + p) - m + i\epsilon} \right]^{(2)T} \\
&= \left[\frac{i\Lambda_+(\vec{p})\gamma^0}{W + p_0 - \omega_p + i\epsilon} + \frac{i\Lambda_-(\vec{p})\gamma^0}{W + p_0 + \omega_p - i\epsilon} \right]^{(1)} \\
&\quad \times \left[\frac{i\Lambda_+(\vec{p})\gamma^0}{-W + p_0 - \omega_p + i\epsilon} + \frac{i\Lambda_-(\vec{p})\gamma^0}{-W + p_0 + \omega_p - i\epsilon} \right]^{(2)T}
\end{aligned}
\tag{50}
$$

in the center-of-mass frame where $P = (2W, \vec{0})$. The projection operators $\Lambda_\pm(\vec{p})$

are defined by

$$\Lambda_\pm(\vec{p}) = \frac{1}{2\omega_p}[\omega_p \pm (m - \vec{\gamma}\cdot\vec{p})\gamma^0] \tag{51}$$

and satisfy the relations

$$\Lambda_\pm(\vec{p})\Lambda_\pm(\vec{p}) = \Lambda_\pm(\vec{p}) \tag{52a}$$
$$\Lambda_\pm(\vec{p})\Lambda_\mp(\vec{p}) = 0\,, \tag{52b}$$
$$\Lambda_+(\vec{p}) + \Lambda_-(\vec{p}) = 1\,, \tag{52c}$$

where $\omega_p = (\vec{p}^{\,2} + m^2)^{1/2}$. To lowest approximation positronium is just a hydrogenic atom with reduced mass $\mu = m/2$, so we know that $\omega_p = m + O(m\alpha^2)$ and $W = m + O(m\alpha^2)$. Consequently $(\omega_p - W)$ is small and the poles of S at $p_0 = \pm(\omega_p - W) \mp i\epsilon$ dominate the p_0 integral [see Fig 5(b)], which can be considered a contour integral closed in either the upper or lower half plane. Also, $\Lambda_+(\vec{p})$ has large upper components, corresponding to a particle on line 1, and $\Lambda_-(\vec{p})$ has large lower components, corresponding to an antiparticle on line 2. The dominant term in S is

$$\left[\frac{i\Lambda_+(\vec{p})\gamma^0}{p_0 - \omega_p + W + i\epsilon}\right]^{(1)}\left[\frac{i\Lambda_-(\vec{p})\gamma^0}{p_0 + \omega_p - W - i\epsilon}\right]^{(2)T}$$
$$= \frac{-i\pi\Delta(p_0)}{(\omega_p - W)}[\Lambda_+(\vec{p})\gamma^0]^{(1)}[\Lambda_-(\vec{p})(-\gamma^0)]^{(2)T} \tag{53}$$

where

$$\Delta(p_0) = \frac{(\omega_p - W)}{(-i\pi)(p_0^2 - (\omega_p - W - i\epsilon)^2)}\,. \tag{54}$$

The quantity $\Delta(p_0)$ acts like $\delta(p_0)$ since $\Delta(p_0)$ is small when p_0 is not close to zero and

$$\int_{-\infty}^{\infty} dp_0\, \Delta(p_0) = 1\,. \tag{55}$$

Barbieri and Remiddi define a kernel K_0 with the projection operators required to destroy all but the dominant term of S. Their kernel is

$$K_0(P; p, p') = [\gamma^0\Lambda_+(\vec{p})\tfrac{1}{2}(1 + \gamma^0)\Lambda_+(\vec{p}\,')]^{(1)}$$
$$\times [(-\gamma^0)\Lambda_-(\vec{p}\,')\tfrac{1}{2}(1 - \gamma^0)\Lambda_-(\vec{p})]^{(2)T} \tag{56}$$
$$\times \left(\frac{2\omega_p}{\omega_p + m}\right)\left(\frac{2m}{\omega_p + W}\right)^{1/2}(-i)V(\vec{p}-\vec{p}\,')\left(\frac{2\omega_{p'}}{\omega_{p'} + m}\right)\left(\frac{2m}{\omega_{p'} + W}\right)^{1/2},$$

where $Z = 1$ in positronium. In the NR limit this reduces to

$$K_0 \to [\tfrac{1}{2}(1 + \gamma^0)]^{(1)}[\tfrac{1}{2}(1 - \gamma^0)]^{(2)T}(-i)V(\vec{p}-\vec{p}\,')\,, \tag{57}$$

which for particles on line 1 and antiparticles on line 2 is just the kernel for a Coulomb interaction. Consider the product

$$[K_0 S K_0](P; p, p') = \int (d\ell)' \, K_0(P; p, \ell) S(P; \ell) K_0(P; \ell, p') \tag{58}$$

inside of which

$$\int \frac{d\ell_0}{2\pi} \, S(P; \ell) \rightarrow \frac{(-i/2)}{(\omega_\ell - W)} [\Lambda_+(\vec{\ell}) \gamma^0]^{(1)} [\Lambda_-(\vec{\ell})(-\gamma^0)]^{(2)T} . \tag{59}$$

Then, using

$$\frac{1}{2}(1 \pm \gamma^0) \Lambda_\pm(\vec{p}) \frac{1}{2}(1 \pm \gamma^0) = \frac{1}{2}(1 \pm \gamma^0) \left(\frac{\omega_p + m}{2\omega_p} \right) \tag{60}$$

one has

$$[K_0 S K_0](P; p, p') =$$

$$\int (d^3\ell)' \left(\frac{2\omega_p}{\omega_p + m} \right) \left(\frac{2m}{\omega_p + W} \right)^{1/2} \left(\frac{2\omega_\ell}{\omega_\ell + m} \right)^2 \left(\frac{2\omega_{p'}}{\omega_{p'} + m} \right) \left(\frac{2m}{\omega_{p'} + W} \right)^{1/2}$$

$$\times [\gamma^0 \Lambda_+(\vec{p}) \frac{1}{2}(1 + \gamma^0) \Lambda_+(\vec{\ell}) \Lambda_+(\vec{\ell}) \gamma^0 \gamma^0 \Lambda_+(\vec{\ell}) \frac{1}{2}(1 + \gamma^0) \Lambda_+(\vec{p}\,')]^{(1)}$$

$$\times [(-\gamma^0) \Lambda_-(\vec{p}\,') \frac{1}{2}(1 - \gamma^0) \Lambda_-(\vec{\ell}) \Lambda_-(\vec{\ell})(-\gamma^0)(-\gamma^0) \Lambda_-(\vec{\ell}) \frac{1}{2}(1 - \gamma^0) \Lambda_-(\vec{p})]^{(2)T}$$

$$\times (-i) V(\vec{p} - \vec{\ell}) \frac{(-i/2)}{(\omega_\ell - W)} \left(\frac{2m}{\omega_\ell + W} \right) (-i) V(\vec{\ell} - \vec{p}\,') \tag{61}$$

$$= \left(\frac{2\omega_p}{\omega_p + m} \right) \left(\frac{2m}{\omega_p + W} \right)^{1/2} [\gamma^0 \Lambda_+(\vec{p}) \frac{1}{2}(1 + \gamma^0) \Lambda_+(\vec{p}\,')]^{(1)}$$

$$\times [(-\gamma^0) \Lambda_-(\vec{p}\,') \frac{1}{2}(1 - \gamma^0) \Lambda_-(\vec{p})]^{(2)T} \left(\frac{2\omega_{p'}}{\omega_{p'} + m} \right) \left(\frac{2m}{\omega_{p'} + W} \right)^{1/2}$$

$$\times (-i) \int (d^3\ell)' \, V(\vec{p} - \vec{\ell}) s(\vec{\ell}) V(\vec{\ell} - \vec{p}\,')$$

where $\gamma^2 = m^2 - W^2$ in $s(\vec{\ell})$. The crucial fact here is that the intermediate momentum ℓ has disappeared from the spin factors and only appears in the NR form $[V s V]$, which is known from (40). The particular $\vec{p}$ and $\vec{p}\,'$ dependent scalar factors in (56) were chosen for this purpose. Equation (61) can be generalized: the term with $N + 1$ actions of K_0 is

$$[K_0 S K_0 \ldots S K_0](P; p, p') = \left(\frac{2\omega_p}{\omega_p + m} \right) \left(\frac{2m}{\omega_p + W} \right)^{1/2}$$

$$\times [\gamma^0 \Lambda_+(\vec{p}) \frac{1}{2}(1 + \gamma^0) \Lambda_+(\vec{p}\,')]^{(1)} \tag{62}$$

$$\times [(-\gamma^0) \Lambda_-(\vec{p}\,') \frac{1}{2}(1 - \gamma^0) \Lambda_-(\vec{p})]^{(2)T} \left(\frac{2\omega_{p'}}{\omega_{p'} + m} \right) \left(\frac{2m}{\omega_{p'} + W} \right)^{1/2}$$

$$\times (-i) \int (d^3\ell_1)' \ldots (d^3\ell_N)' V(\vec{p} - \ell_1) s(\vec{\ell}_1) V(\vec{\ell}_1 - \vec{\ell}_2) \ldots s(\vec{\ell}_N) V(\vec{\ell}_N - \vec{p}\,') .$$

Equation (49) for the Green's function G_0 can now be summed. The result is

$$G_0(P;p,p') = S(P;p,p') + S(P;p)K_0(P;p,p')S(P;p')$$
$$+ S(P;p)\left(\frac{2\omega_p}{\omega_p+m}\right)\left(\frac{2m}{\omega_p+W}\right)^{1/2}[\gamma^0\Lambda_+(\vec{p})\frac{1}{2}(1+\gamma^0)\Lambda_+(\vec{p}\,')]^{(1)} \qquad (63)$$
$$\times[(-\gamma^0)\Lambda_-(\vec{p}\,')\frac{1}{2}(1-\gamma^0)\Lambda_-(\vec{p})]^{(2)T}\left(\frac{2\omega_{p'}}{\omega_{p'}+m}\right)\left(\frac{2m}{\omega_{p'}+W}\right)^{1/2}S(P;p')$$
$$\times 4\pi i m\alpha^2\int_0^1 dx\, x^{-\xi}\frac{2\gamma}{H(E;\vec{p},\vec{p}\,')}$$

where $\gamma^2 = m^2 - W^2$ and $E = -\gamma^2/m$.

The MP term of G_0 contains the bound state poles at energies fixed by $\gamma = m\alpha/(2n)$. One finds that $W = m(1 - \alpha^2/(4n^2))^{1/2}$ and

$$E_n^0 = 2W = 2m - \frac{m\alpha^2}{4n^2} - \frac{m\alpha^4}{64n^4} + O(m\alpha^6)\,. \qquad (64)$$

The $n=1$ wave functions are

$$\Psi_{sm}^0(p) = 2\pi\Delta(p_0)\left(\frac{\omega_p+W}{4W}\right)^{1/2}\left(\frac{2\omega_p}{\omega_p+m}\right)[\Lambda_+(\vec{p})\Gamma_{sm}\Lambda_-(\vec{p})(-\gamma^0)]\phi(\vec{p})$$
$$= 2\pi\Delta(p_0)\left(\frac{\omega_p+W}{4W}\right)^{1/2}\left(\frac{\omega_p+m}{2\omega_p}\right)$$
$$\times\begin{pmatrix}\chi_{sm}\vec{\sigma}\cdot\tilde{\vec{p}} & \chi_{sm} \\ \vec{\sigma}\cdot\tilde{\vec{p}}\chi_{sm}\vec{\sigma}\cdot\tilde{\vec{p}} & \vec{\sigma}\cdot\tilde{\vec{p}}\chi_{sm}\end{pmatrix}\phi(\vec{p})\,, \qquad (65a)$$

$$\bar{\Psi}_{sm}^0(p) = 2\pi\Delta(p_0)\left(\frac{\omega_p+W}{4W}\right)^{1/2}\left(\frac{2\omega_p}{\omega_p+m}\right)[\Lambda_-(\vec{p})\Gamma_{sm}^\dagger\Lambda_+(\vec{p})(\gamma^0)]^T\phi(\vec{p})$$
$$= 2\pi\Delta(p_0)\left(\frac{\omega_p+W}{4W}\right)^{1/2}\left(\frac{\omega_p+m}{2\omega_p}\right)$$
$$\times\begin{pmatrix}-\vec{\sigma}\cdot\tilde{\vec{p}}\chi_{sm}^\dagger & \vec{\sigma}\cdot\tilde{\vec{p}}\chi_{sm}^\dagger\vec{\sigma}\cdot\tilde{\vec{p}} \\ \chi_{sm}^\dagger & -\chi_{sm}^\dagger\vec{\sigma}\cdot\tilde{\vec{p}}\end{pmatrix}^T\phi(\vec{p})\,. \qquad (65b)$$

where $\gamma = m\alpha/2$, $W = m(1-\alpha^2/4)^{1/2}$, and

$$\phi(\vec{p}) = \left(\frac{\gamma^3}{\pi}\right)^{1/2}\frac{8\pi\gamma}{(\vec{p}^{\,2}+\gamma^2)^2}\,, \qquad (66a)$$

$$\Gamma_{sm} = \begin{pmatrix} 0 & \chi_{sm} \\ 0 & 0 \end{pmatrix} , \tag{66b}$$

$$\chi_{00} = 1 \quad ; \quad \chi_{1m} = \vec{\sigma} \cdot \hat{\epsilon}_m , \tag{66c}$$

$$\hat{\epsilon}_0 = (0,0,1) \quad , \quad \hat{\epsilon}_\pm = \frac{-1}{\sqrt{2}}(\pm 1, i, 0) , \tag{66d}$$

$$\tilde{\vec{p}} = \frac{\vec{p}}{(\omega_p + m)} . \tag{66e}$$

Factorization of the spin dependence is achieved through use of the Fierz identity

$$\left([\tfrac{1}{2}(1+\gamma^0)]^{(1)} [\tfrac{1}{2}(1-\gamma^0)]^{(2)T} \right)_{\alpha\beta,\gamma\delta} = [\tfrac{1}{2}(1+\gamma^0)]_{\alpha\gamma} [\tfrac{1}{2}(1-\gamma^0)]^T_{\beta\delta}$$
$$= \frac{1}{2} \sum_{s,m} (\Gamma_{sm})_{\alpha\beta} (\Gamma^\dagger_{sm})^T_{\gamma\delta} . \tag{67}$$

VI. CALCULATION OF ENERGY LEVELS TO ORDER $m\alpha^4$

As a first example of the use of this bound state formalism I will describe the calculation of the positronium ground state hyperfine splitting to $O(m\alpha^4)$. I will work in Coulomb gauge, so the orders of perturbation theory matrix elements are as given in (32). The perturbative expansion of (30) can be truncated after the first nontrivial term, leaving

$$E_n = E_n^0 + (\delta K) . \tag{68}$$

In (δK) only three terms are required, corresponding to a time-time Coulomb minus lowest order interaction (C-O), a space-space Coulomb (i.e., transverse) interaction (T), and an annihilation (A). These graphs are shown in Fig. 6(a). For example, the contribution of the C-O graph shown in Fig. 6(b) is

$$\Delta E_{\text{C-O}} = i \int (dp)'(dp')' \left\{ \text{tr}[(-ie\gamma^0)\bar{\Psi}^0(p)^T \right.$$

$$\left. \times (-ie\gamma^0)\Psi^0(p')] \frac{i}{(\vec{p}-\vec{p}\,')^2} - [K_0 \text{ part}] \right\}$$

$$\cong \frac{m}{W} \int (d^3p)'(d^3p')' \phi^*(\vec{p}) \frac{-4\pi\alpha}{(\vec{p}-\vec{p}\,')^2} \phi(\vec{p}\,')$$

$$\times \left\{ (1 - \frac{\vec{p}^{\,2}+\vec{p}^{\,\prime 2}}{8m^2} - \frac{\alpha^2}{16}) \frac{1}{2}\text{tr}[\chi^\dagger\chi + \frac{1}{4m^2}(\chi^\dagger\vec{\sigma}\cdot\vec{p}\,\vec{\sigma}\cdot\vec{p}\,'\chi + \chi\vec{\sigma}\cdot\vec{p}\,'\vec{\sigma}\cdot\vec{p}\,\chi^\dagger)] \right.$$

$$\left. - \frac{1}{2}\text{tr}[\chi^\dagger\chi] \right\}$$

$$= \frac{m\alpha^4}{16} \tag{69}$$

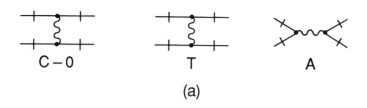

(a)

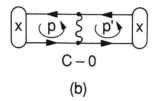

(b)

Fig. 6. Fig. 6(a) shows the three kernels which contribute to positronium energy levels at $O(m\alpha^4)$. Fig. 6(b) depicts the matrix element which (times i) gives $\Delta E_{\text{C-O}}$.

for both $s = 0$ and $s = 1$.

The results of all contributions are shown in Table I.

Table I Ground state energy level shifts (times $m\alpha^4$)

Kernel	1^1S_0	1^3S_1
C-O	$\frac{1}{16}$	$\frac{1}{16}$
T	$-\frac{3}{8}$	$-\frac{1}{24}$
A	0	$\frac{1}{4}$
$E^0 = 2m(1 - \frac{\alpha^2}{4})^{1/2}$	$-\frac{1}{64}$	$-\frac{1}{64}$
Total	$-\frac{21}{64}$	$\frac{49}{192}$

The ground state hyperfine splitting to this order is

$$\Delta E = E(1^3S_1) - E(1^1S_0)$$
$$= \frac{7}{12}m\alpha^4 \,. \tag{70}$$

VII. RENORMALIZATION OF THE BOUND STATE EQUATION

The BS equation derived in (17) is unsatisfactory when radiative corrections are taken into account. For one thing, electron self-energy graphs were left out of the noninteracting electron-positron propagator S. A complete BS equation, including radiative corrections, is

$$G' = S' + S'K'G' \tag{71}$$

where G' is the full two-to-two Green's function, S' is the full noninteracting electron-positron propagator including self-energy effects, and K' is the two-particle irreducible kernel [equal to the K of (17)]. The quantities G', S', and K' are pictured in Fig. 7. All vertices and propagators of K' are dressed except for the ones in the one-photon annihilation graph, because any dressed version of this graph would be two-particle reducible.

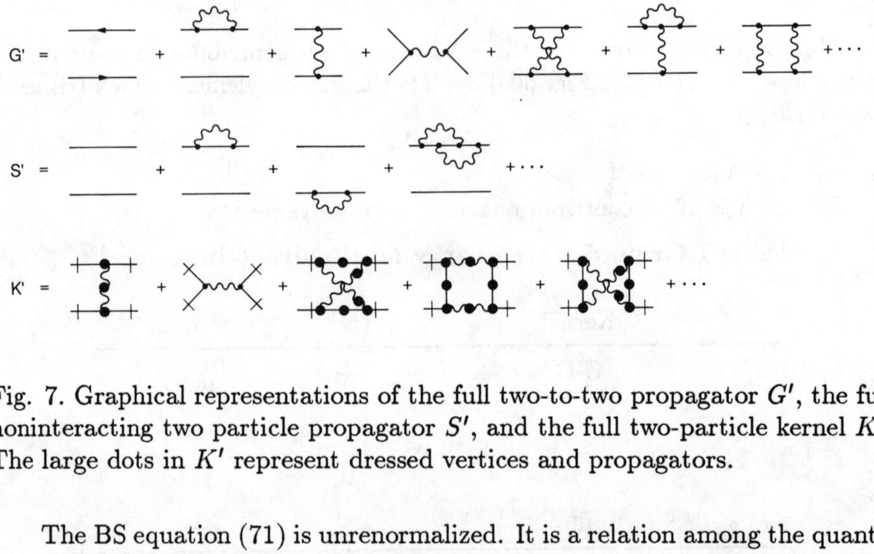

Fig. 7. Graphical representations of the full two-to-two propagator G', the full noninteracting two particle propagator S', and the full two-particle kernel K'. The large dots in K' represent dressed vertices and propagators.

The BS equation (71) is unrenormalized. It is a relation among the quantities G', S', and K', which are divergent functions of the unphysical parameters m_0 and e_0.[34] The BS equation is renormalized by multiplying by $(Z_2)^{-2}$. It becomes

$$G_R = S_R + S_R K_{``R"} G_R , \tag{72}$$

where

$$G_R = (Z_2)^{-2} G' , \tag{73a}$$

$$S_R = (Z_2)^{-2} S' , \tag{73b}$$

$$K_{``R"} = (Z_2)^{2} K' . \tag{73c}$$

We know from standard renormalization theory[35] that G_R, S_R, and the non-annihilation part of $K_{«R»}$ are finite functions of the physical parameters m and e, where

$$m = m_0 + \delta m \,, \tag{74a}$$

$$e = (Z_3)^{1/2} e_0 \,. \tag{74b}$$

The annihilation part of $K_{«R»}$ is not finite.[33] It can be written in terms of the physical charge e, but at the cost of a factor of $(Z_3)^{-1}$ as in Fig. 8(a). The explicit divergences in $K_{«R»}$ are cancelled in the product $S_R K_{«R»} G_R$ by new divergences such as those shown in Fig. 8(b), so that the product is in fact finite.

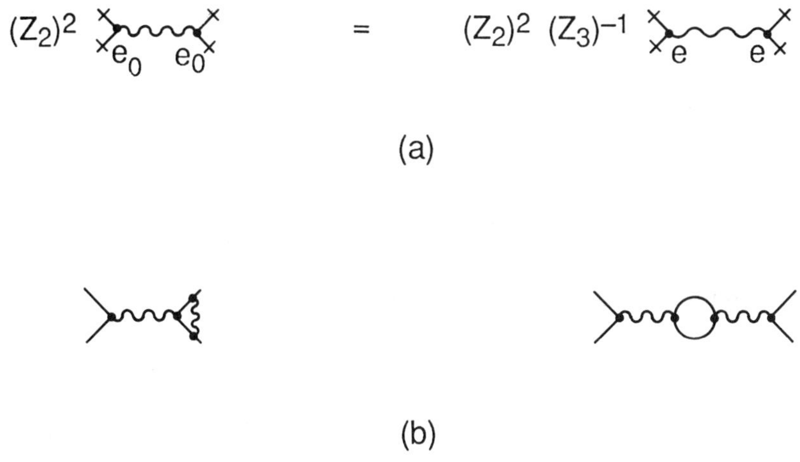

(a)

(b)

Fig. 8. Fig. 8(a) shows the one-photon annihilation part of $K_{«R»}$, written in terms of the bare electron charge e_0 and in terms of the physical electron charge e. Fig. 8(b) shows two graphs appearing in the product $S_R K_{«R»} G_R$ which contain new divergences.

The "renormalized" BS equation (72) cannot be directly related to the reference problem (21) because (72) has the renormalized full propagator S_R as the first term on the right hand side instead of the free propagator S. We can slip S into the renormalized BS equation at the cost of redefining the kernel, according to

$$G_R = S + S \tilde{K} G_R \tag{75}$$

where

$$\tilde{K} = K_{«R»} + (S^{-1} - S_R^{-1}) \,. \tag{76}$$

The extra contribution $(S^{-1} - S_R^{-1})$ to the kernel can be interpreted diagramatically. If we write

$$S_R(P;p) = \left[\frac{i}{\gamma(\frac{1}{2}P + p) - m - \Sigma_R(\frac{1}{2}P + p)}\right]^{(1)}$$
$$\times \left[\frac{i}{\gamma(-\frac{1}{2}P + p) - m - \Sigma_R(-\frac{1}{2}P + p)}\right]^{(2)T} \tag{77}$$

where Σ_R is the renormalized electron self-energy function, then

$$S^{-1}(P;p) - S_R^{-1}(P;p) = \frac{1}{i}\left[\gamma(\frac{1}{2}P + p) - m\right]^{(1)}\frac{1}{i}\left[\gamma(-\frac{1}{2}P + p) - m\right]^{(2)T}$$
$$- \frac{1}{i}\left[\gamma(\frac{1}{2}P + p) - m - \Sigma_R(\frac{1}{2}P + p)\right]^{(1)}$$
$$\times \frac{1}{i}\left[\gamma(-\frac{1}{2}P + p) - m - \Sigma_R(-\frac{1}{2}P + p)\right]^{(2)T}$$
$$= (-i)\left[\Sigma_R(\frac{1}{2}P + p)\right]^{(1)}\frac{1}{i}\left[\gamma(-\frac{1}{2}P + p) - m\right]^{(2)T}$$
$$+ \frac{1}{i}\left[\gamma(\frac{1}{2}P + p) - m\right]^{(1)}(-i)\left[\Sigma_R(-\frac{1}{2}P + p)\right]^{(2)T}$$
$$- (-i)\left[\Sigma_R(\frac{1}{2}P + p)\right]^{(1)}(-i)\left[\Sigma_R(-\frac{1}{2}P + p)\right]^{(2)T}. \tag{78}$$

A pictorial representation of $(S^{-1} - S_R^{-1})$ is given in Fig. 9.

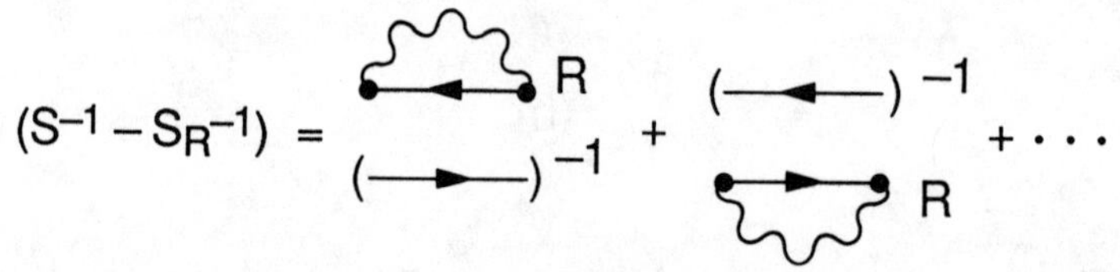

Fig. 9. The lowest order graphs of $S^{-1} - S_R^{-1}$. These graphs contribute to the kernel $\tilde{K}$.

VIII. CALCULATION OF ENERGY LEVELS TO ORDER $m\alpha^5$

In this section I will describe the calculation of the positronium ground state hyperfine splitting to $O(m\alpha^5)$. The perturbation expansion can be written as

$$E_n = E_n^0 + (\delta K) + (\delta K \hat{G}_0 \delta K) \tag{79}$$

since only these terms contribute at $O(m\alpha^5)$. The subtracted propagator $\hat{G}_0$ of (27) can be approximated by its $0P$ part S. After a bit of manipulation, the expression for the hyperfine splitting can be reduced to the difference between the 1^3S_1 and 1^1S_0 matrix elements of the kernels shown in Fig. 10. In this figure the factors of $(Z_2)^2(Z_3)^{-1}$ that accompany each one-photon annihilation graph are implicit. The homogeneous Barbieri-Remiddi equations shown in Fig. 11 have been used in the reduction of (79) to the terms in Fig. 10.

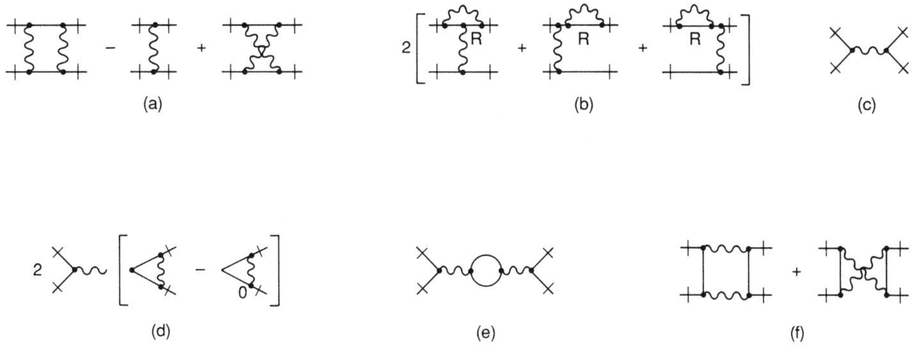

(a) (b) (c)

(d) (e) (f)

Fig. 10. Graphs contributing to the positronium hyperfine splitting at $O(m\alpha^5)$.

(a) (b)

Fig. 11. Homogeneous versions of the reference BS equation. Symbolically these equations are $SK_0\Psi_n^0 = \Psi_n^0, \bar{\Psi}_n^0 K_0 S = \bar{\Psi}_n^0$. They are obtained from the inhomogeneous reference BS equation (21) by taking the limit $E \to E_n^0$.

As an example of the evaluation of a graph at this order I will work out the contribution of Fig. 10(d), which is the ladder correction to the one-photon annihilation graph. Discussion of all of the graphs of Fig. 10 can be found in the original work of Karplus and Klein,[7] and also in Itzykson and Zuber.[36]

Graph 10(d) consists of two terms, a Coulomb photon vertex correction to the plain one-photon annihilation graph, and a "zeroth-order photon" vertex correction to the annihilation graph. The zeroth-order photon is just the Barbieri-Remiddi reference kernel of (56). Because of the homogeneous Barbieri-Remiddi reference equation shown in Fig. 11(a), the zeroth-order photon can be absorbed into the wave function, and the matrix element of the second graph in Fig. 10(d) is proportional to the matrix element of the one-photon annihilation graph, which has order $m\alpha^4$. The matrix element of the

86

first graph in Fig. 10(d), also has order $m\alpha^4$ because a Coulomb ladder photon contains the reference photon as its most important part. After the subtraction, graph 10(d) has order $m\alpha^5$.

The matrix element of each of the two terms in Fig. 10(d) factorizes into three pieces. The factors of, say, the matrix element of the second term in Fig. 10(d) are shown in Fig. 12(a), 12(b), and 12(c). It is not hard to evaluate these terms. One has

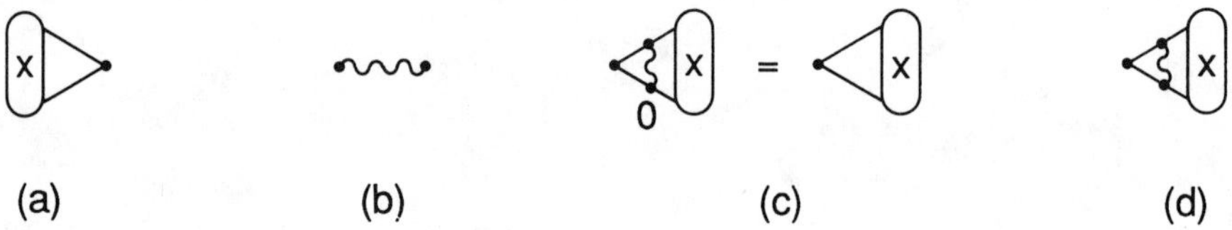

(a) (b) (c) (d)

Fig. 12. Factorized parts of the matrix element of the kernel shown in Fig. 10(d).

$$\int (dp)'\, \mathrm{tr}\left[(ie\gamma^j)\Psi^0(p)\right] = ie\sqrt{2}\,\phi_0\epsilon^j X_0 \tag{80}$$

where

$$X_0 = 1 + \frac{\alpha}{6} + O(\alpha^3) \tag{81}$$

for Fig. 12(c) and $-ie\sqrt{2}\,\phi_0\epsilon^{i*}X_0$ for Fig. 12(a), while terms involving $(-ie\gamma^0)$ vanish. The photon propagator Fig. 12(b) contributes $i\delta_{ij}/P^2$. Consequently, the matrix element of the second term in Fig. 10(d) is

$$(-1)i(2)\left[-ie\sqrt{2}\,\phi_0\epsilon^{i*}X_0\right]\left[i\delta_{ij}/(4W^2)\right]\left[ie\sqrt{2}\,\phi_0\epsilon^j X_0\right]$$

$$= 2\frac{m\alpha^4}{4}\left[\frac{m^2}{W^2}X_0^2\right]. \tag{82}$$

In (82) the (-1) is a fermionic minus sign and the i comes from definition (31a). Fig. 12(d) is a little harder to evaluate. One has

$$[\text{Fig. 12(d)}]^j = \int (dp)'\, \mathrm{tr}\left[(-ie\Lambda^j)\Psi^0(p)\right]$$
$$= ie\sqrt{2}\,\phi_0\epsilon^j X \tag{83}$$

where Λ^j is the one-loop Coulomb gauge vertex function.[24] One can show that the most important part of the vertex function is given by

$$\Lambda^j \to \gamma^j\left[\frac{2\gamma}{p}\tan^{-1}\left(\frac{p}{\gamma}\right) - \frac{2\alpha}{\pi} - B_{(1)}\right], \tag{84}$$

where $\gamma = m\alpha/2$ and where $B_{(1)}$ is the $O(\alpha^1)$ term of the vertex renormalization constant Z_1:

$$Z_1 = Z_2 = 1 + B_{(1)} + O(\alpha^2) \, . \tag{85}$$

In the dimensional regularization scheme with 2ω dimensions one has[23]

$$B_{(1)} = -\frac{\alpha}{4\pi}D \, , \tag{86a}$$

$$D = \frac{1}{2-\omega} - \gamma_E + \ln(4\pi) \, . \tag{86b}$$

Now in (83) the vertex function gets integrated with the wave function (48). In fact one has

$$\int (d^3p)' \frac{8\pi\gamma}{(\vec{p}^2 + \gamma^2)^2} = 1 \, , \tag{87a}$$

$$\int (d^3p)' \frac{2\gamma}{p} \tan^{-1}\left(\frac{p}{\gamma}\right) \frac{8\pi\gamma}{(\vec{p}^2 + \gamma^2)^2} = 1 \, , \tag{87b}$$

so that

$$X = 1 - \frac{2\alpha}{\pi} - B_{(1)} + O(\alpha^2) \, . \tag{88}$$

The arctangent term, which arises as an $O(\alpha)$ correction, leads to an $O(1)$ contribution to X.

The full result for Fig. 10(d) can now be obtained. It is

$$\begin{aligned} [\text{Fig. 10(d)}] &= 2\frac{m\alpha^4}{4}\left[\frac{m^2}{W^2}X_0(X - X_0)\right] \\ &= \frac{m\alpha^4}{4}\left[-\frac{4\alpha}{\pi} - \frac{\alpha}{3} - 2B_{(1)}\right] \, . \end{aligned} \tag{89}$$

The contributions to the hyperfine structure from all the graphs in Fig. 10 are given in Table II. Graphs 10(c), 10(d), and 10(e) only contribute to orthopositronium, while 10(f) only contributes to parapositronium. The quantity $C_{(1)}$ in rows (c) and (e) is related to the charge renormalization constant Z_3 by

$$Z_3 = 1 + C_{(1)} + O(\alpha^2) \, . \tag{90}$$

The imaginary part of the contribution on row (f) is related to the lowest order parapositronium decay rate (7) by $\Gamma = -2\,\mathrm{Im}\big(\Delta E(1\,^1S_0)\big)$.[37] The final result for the hyperfine splitting at this order is

$$\Delta E = m\alpha^4\left[\frac{7}{12} - \frac{\alpha}{\pi}\left(\frac{8}{9} + \frac{1}{2}\ln 2\right)\right] \, . \tag{91}$$

Table II. Contributions to the ground state hyperfine splitting

Graph	Contribution
Fig. 10(a)	$\dfrac{m\alpha^4}{3}\left[1 - \dfrac{3\alpha}{2\pi}\right]$
Fig. 10(b)	$\dfrac{m\alpha^4}{3}\left[\dfrac{\alpha}{\pi}\right]$
Fig. 10(c)	$\dfrac{m\alpha^4}{4}\left[1 + \dfrac{\alpha}{3} + 2B_{(1)} - C_{(1)}\right]$
Fig. 10(d)	$\dfrac{m\alpha^4}{4}\left[-\dfrac{4\alpha}{\pi} - \dfrac{\alpha}{3} - 2B_{(1)}\right]$
Fig. 10(e)	$\dfrac{m\alpha^4}{4}\left[-\dfrac{8\alpha}{9\pi} + C_{(1)}\right]$
Fig. 10(f)	$\dfrac{m\alpha^5}{2\pi}\left[1 - \ln 2 + \dfrac{i\pi}{2}\right]$

IX. CONCLUSION

In this report I have described one approach to the bound state problem
in QED, and I have discussed the use of that formalism for the calculation
of positronium energy levels. A central feature of this formalism is the role
played by the reference problem. The reference problem (49) has the form of a
BS equation with a modified kernel. The modified kernel is chosen so that the
reference problem can be exactly solved by comparison with the known solution
to the Schrödinger-Coulomb equation. A well-defined perturbation expansion
for the energies in terms of the reference quantities can then be used to find
the energy levels and decay rates of the positronium states.

The bound state formalism described in this report is not the only one
available. It is distinguished by having a four-dimensional reference problem
which uses free Dirac propagators to describe the electron propagation. Other
formalisms make use of modified electron propagation factors as well as modified
kernels, which are chosen so that the bound state equation can be written in a
three-dimensional form which is equivalent to an effective Schrödinger-Coulomb
or Dirac-Coulomb equation.[28,29,38,39] Formalisms with other gauges can also be
used. Fried-Yennie gauge has the good IR behavior of Coulomb gauge, but is
covariant and thus simpler to renormalize. Feynman gauge is also used. Often
a problem can be set up in Coulomb gauge and then rewritten so that Feynman
gauge can be used for the actual evaluations.[19,20,38] A completely new approach
to the bound state problem has been given recently by Caswell and Lepage,[40]
in which QED is rewritten in terms of a NR effective Lagrangian. Bound state

properties are computed in this approach using the ordinary techniques of NR quantum mechanics.

The new bound state formalisms are powerful enough to allow us to compute $O(\alpha^2)$ corrections to bound state properties. Experiment has already reached this level of accuracy for several $n = 1$ and $n = 2$ energy intervals and for the orthopositronium decay rate. It is important to complete these calculations to allow high precision comparisons between experimental and theoretical determinations of the energy intervals, and perhaps to resolve the existing "discrepancy" between experiment and theory for the orthopositronium decay rate.

ACKNOWLEDGEMENTS

I am grateful to W. Johnson, P. Mohr, and J. Sucher for their efforts in organizing this workshop, and to J. R. Schrieffer and I. Muzinich for the hospitality of the Institute for Theoretical Physics at Santa Barbara. This research was supported in part by the National Science Foundation under Grants No. PHY87-04324, PHY82-17853, PHY85-05682, and PHY86-04197, supplemented by funds from the National Aeronautics and Space Administration.

REFERENCES

1. Reviews have been given by S. Berko and H. N. Pendleton, Ann. Rev. Nucl. Part. Sci. **30**, 543 (1980), and by A. Rich, Rev. Mod. Phys. **53**, 127 (1981).

2. The conventions and natural units [$\hbar = c = 1$, $\alpha = e^2/4\pi \simeq (137)^{-1}$] of J. D. Bjorken and S. D. Drell, *Relativistic Quantum Mechanics*, (McGraw-Hill, New York, 1964) are used throughout. The symbol m represents the electron mass, $m \simeq 0.511$ MeV.

3. J. Pirenne, Arch. Sci. Phys. Nat. **28**, 233 (1946); **29**, 121, 207, and 265 (1947).

4. V. B. Berestetski, J. Exp. Theor. Phys. (USSR) **19**, 1130 (1949). See also V. B. Berestetski and L. D. Landau, J. Exp. Theor. Phys. (USSR) **19**, 673 (1949).

5. R. A. Ferrell, Phys. Rev. **84**, 858 (1951).

6. T. Fulton, Phys. Rev. A**26**, 1794 (1982).

7. R. Karplus and A. Klein, Phys. Rev. **87**, 848 (1852).

8. T. Fulton and P. C. Martin, Phys. Rev. **95**, 811 (1954).

9. H. A. Bethe, Phys. Rev. **72**, 339 (1947); H. A. Bethe, L. M. Brown, and J. R. Stehn, *ibid.* **77**, 370 (1950); J. M. Harriman, *ibid.*, **101**, 594 (1956); S. Klarsfeld and A. Maquet, Phys. Lett. **43B**, 201 (1973).

10. T. Fulton, D. A. Owen, and W. W. Repko, Phys. Rev. **A4**, 1802 (1971); R. Barbieri, P. Christillin, and E. Remiddi, Phys. Rev. **A8**, 2266 (1973); D. A. Owen, Phys. Rev. Lett. **30**, 887 (1973); Phys. Rev. **A16**, 452 (1977); V. K. Cung, T. Fulton, W. W. Repko, and D. Schnitzler, Ann. Phys. (N.Y.), **96**, 261 (1976); R. Barbieri and E. Remiddi, Phys. Lett. **65B**, 258 (1976); Nucl. Phys. **B141**, 413 (1978); G. P. Lepage, Phys. Rev. **A16**, 863 (1977); G. T. Bodwin and D. R. Yennie, Phys. Rep. **43**, 267 (1978); W. E. Caswell and G. P. Lepage, Phys. Rev. **A20**, 36 (1979); G. T. Bodwin, D. R. Yennie, and M. A. Gregorio, Rev. Mod. Phys. **57**, 723 (1985).

11. R. Barbieri, P. Christillin, and E. Remiddi, Phys. Rev. **A8**, 2266 (1973); M. A. Samuel, Phys. Rev. **A10**, 1450 (1974); M. Douglas, Phys. Rev. **A11**, 1527 (1975); V. K. Cung, T. Fulton, W. W. Repko, and D. Schnitzler, Ann. Phys. (N.Y.) **96**, 261 (1976); V. K. Cung, A. Devoto, T. Fulton, and W. W. Repko, Nuovo Cimento **43A**, 643 (1978); Phys. Rev. **A19**, 1886 (1979); W. E. Caswell and G. P. Lepage, Phys. Rev. **A18**, 810 (1978); Phys. Lett. **167B**, 437 (1986); W. Buchmüller and E. Remiddi, Cornell Report No. CLNS 80/450, (1980) (unpublished); J. R. Sapirstein, E. A. Terray, and D. R. Yennie, Phys. Rev. **D29**, 2290 (1984); G. S. Adkins, M. H. T. Bui, and D. Zhu, Phys. Rev. **A37**, 4071 (1988).

12. M. Deutsch and E. Dulit, Phys. Rev. **84**, 601 (1951).

13. M. Deutsch and S. C. Brown, Phys. Rev. **85**, 1047 (1952); R. Weinstein, M. Deutsch, and S. Brown, Phys. Rev. **94**, 758 (1954); *ibid.* **98**, 223 (1955); V. W. Hughes, S. Marder, and C. S. Wu, Phys. Rev. **106**, 934 (1957); E. D. Theriot, Jr., R. H. Beers, V. W. Hughes, and K. O. H. Ziock, Phys. Rev. **A2**, 707 (1970); E. R. Carlson, V. W. Hughes, M. L. Lewis, and I. Lindgren, Phys. Rev. Lett. **29**, 1059 (1972); A. P. Mills, Jr. and G. H. Bearman, Phys. Rev. Lett. **34**, 246 (1975); E. R. Carlson, V. W. Hughes, and I. Lindgren, Phys. Rev. **A15**, 241 (1977); P. O. Egan, V. W. Hughes, and M. H. Yam, Phys. Rev. **A15**, 251 (1977); A. P. Mills, Jr., Phys. Rev. **A27**, 262 (1983).

14. M. W. Ritter, P. O. Egan, V. W. Hughes, and K. A. Woodle, Phys. Rev. **A30**, 1331 (1984).

15. J. Pirenne, Arch. Sci. Phys. Nat. **29**, 265 (1947); J. A. Wheeler, Ann. N.Y. Acad. Sci. **48**, 219 (1946).

16. A. Ore and J. L. Powell, Phys. Rev. **75**, 1696 (1949).

17. C. I. Westbrook, D. W. Gidley, R. S. Conti, and A. Rich, Phys. Rev. Lett. **58**, 1328 (1987).

18. M. A. Stroscio and J. M. Holt, Phys. Rev. **A10**, 749 (1974); W. E. Caswell, G. P. Lepage, and J. Sapirstein, Phys. Rev. Lett. **38**, 488 (1977).

19. W. Caswell and G. P. Lepage, Phys. Rev. **A20**, 36 (1979).

20. G. S. Adkins, Ann. Phys. (N.Y.) **146**, 78 (1983).

21. J. D. Bjorken and S. D. Drell, *Relativistic Quantum Mechanics*, (McGraw-Hill, New York, 1964).

22. H. M. Fried and D. R. Yennie, Phys. Rev. **112**, 1391 (1958).

23. G. S. Adkins, Phys. Rev. **D27**, 1814 (1983).

24. G. S. Adkins, Phys. Rev. **D34**, 2489 (1986).

25. H. A. Bethe and E. E. Salpeter, Phys. Rev. **82**, 309 (1951); E. E. Salpeter and H. A. Bethe, Phys. Rev. **84**, 1232 (1951); J. Schwinger, Proc. Natl. Acad. Sci. U.S. **37**, 452, 455 (1951).

26. Radiative corrections to the individual fermion lines have been omitted for simplicity. They will be included in the discussion of Sec. VII.

27. M. Gell-Mann and F. Low, Phys. Rev. **84**, 350 (1951); S. Mandelstam, Proc. Roy. Soc. **A233**, 248 (1955); D. Lurié, A. J. Macfarlane, and Y. Takahashi, Phys. Rev. **140**, B1091 (1965).

28. G. P. Lepage, Phys. Rev. **A16**, 863 (1977).

29. W. E. Caswell and G. P. Lepage, Phys. Rev. **A18**, 810 (1978).

30. W. Buchmüller and E. Remiddi, Nucl. Phys. **B162**, 250 (1980).

31. E. Remiddi, in *Proceedings of the International School of Physics, "Enrico Fermi" Course LXXXI*, ed. by G. Costa and R. R. Gatto (North Holland, Amsterdam, 1982), p. 1.

32. The form of the Schrödinger-Coulomb Green's function used here was first written down by L. Hostler, J. Math. Phys. (N.Y.) **5**, 1235 (1964), and by J. Schwinger, J. Math. Phys. (N.Y.) **5**, 1606 (1964). It was rederived using the present method by G. S. Adkins, Nuovo Cimento **97B**, 99 (1987).

A more thorough discussion of the Schrödinger-Coulomb Green's function with references is given by M. Lieber, in The Coulomb Green's Function, this volume.

33. R. Barbieri and E. Remiddi, Nucl. Phys. **B141**, 413 (1978).

34. Strictly speaking, the Feynman rules in Section II should have been given in terms of the bare mass m_0 and bare charge e_0. That is, the fermion propagator is $\left(i(\gamma p - m_0)^{-1}\right)_{\alpha\beta}$ and the vertex is $-ie_0(\gamma^\mu)_{\alpha\beta}$.

35. J. D. Bjorken and S. D. Drell, *Relativistic Quantum Fields*, (McGraw-Hill, New York, 1965).

36. C. Itzykson and J.-B. Zuber, *Quantum Field Theory*, (McGraw-Hill, New York, 1980), Sec. 10-3.

37. Note that the hyperfine splitting involves minus the 1^1S_0 energy, so $-2\mathrm{Im}\left(\Delta E(1^1S_0)\right) = +m\alpha^5/2$.

38. G. T. Bodwin, D. R. Yennie, and M. A. Gregorio, Rev. Mod. Phys. **57**, 723 (1985).

39. D. R. Yennie, Introduction to Recoil Effects in Bound State Problems, this volume.

40. W. E. Caswell and G. P. Lepage, Phys. Lett. **167B**, 437 (1986).

INTRODUCTION TO RECOIL EFFECTS IN BOUND STATE PROBLEMS

D. R. Yennie
Laboratory of Nuclear Studies,
Cornell University, Ithaca, NY 14853

ABSTRACT

A description is given of some of the considerations which are necessary to treat two-body hydrogen-like systems when the electron's relativistic properties must be taken into account. Most modern approaches to the two-body problem are based on the physics of the Bethe-Salpeter approach, although the methodology is different. One uses some three-dimensional reference equation which incorporates as much physics as possible and the corrections appear as four-dimensional kernels which are treated by a procedure resembling usual perturbation theory. An example is given where both reduced mass effects and relativistic effects are treated in a unified approach.

The Reduced Mass Problem

The bound states I have in mind are hydrogen-like systems, for example, hydrogen, muonium (electron bound to μ^+), or positronium. For convenience, I'll use the symbol p to represent the nucleus. Every physicist knows how to rearrange the non-relativistic Schrödinger equation to express the kinetic energy in terms of the reduced mass, leading to a factor of the reduced mass in the Bohr expression for the energy levels. It is not very obvious how reduced mass dependence enters into more refined aspects of the energy level structure of hydrogen-like atoms. Getting this dependence straight can have very important practical consequences. For example, suppose that the hydrogen levels $2S_{1/2}$ and $2P_{1/2}$, which are degenerate in the infinite proton mass limit, were to have different powers of the reduced mass factor m_r/m_e ($m_r = m_e m_p/(m_e + m_p)$ is the reduced mass, m_e is the electron mass, m_p is the proton mass). This would lift the degeneracy by an amount of order $\alpha^4 m_r^2/M$, where M is the total mass ($M = m_e + m_p$). Such a term would be comparable to the Lamb shift, which is of order $\alpha^5 m_r^3/m_e^2$. In fact, there are no terms of this type, as was first shown by Breit and Brown[1] in 1948. Their absence is due to a subtle cancellation which appears to depend on properties of the Coulomb potential and is still not too well understood, even if it is easy to show.

One cannot introduce reduced mass effects in a relativistic way simply by changing the electron mass to the reduced mass in the Dirac equation. This is obvious from the fact that it would put the continuum threshhold at the energy $m_r c^2$ rather than at its correct position. Another conceivable approach would be to treat the kinetic energy of the proton ($\vec{p}^2/2m_p$) as a perturbation. That would be very bad since it would amount to expanding $m_r = m_e/(1 + m_e/m_p)$ as a power series in m_e/m_p, which would be wasted effort for the leading term which we already know. Of course, for positronium, such an expansion would be hopeless. As will be seen later, it is possible to get reduced mass effects into the basic relativistic problem. It is also important to do so since otherwise there will appear a plethora of spurious contributions which would have to be compensated by what appear to be dynamical recoil corrections. At some level of small corrections, of course, it may not be easy to distinguish true dynamical effects from those which should appear in the reduced mass dependence. Then it becomes a matter of individual taste how they should be defined.

Example: Hyperfine Splitting In The Ground State

To give an idea of the kind of corrections which arise in hydrogen-like systems and the present level of refinement, I'll describe the present situation in the hyperfine splitting (hfs). The hydrogen hfs is one of the most accurately measured quantities in physics[2,3] (having about 13 significant figures). It has many practical uses (perhaps the best known is that it corresponds to the 21 cm line). Unfortunately, not all of this accuracy is amenable to theoretical interpretation. The reason, aside from limitations of the fortitude of calculators, is that the internal structure of the proton has a part which is unknowable from measured proton (charge and magnetic moment) form factors. It is called the proton polarizability correction and it enters at the few parts per million level. Most other terms in the theory have been evaluated down to that level. Muonium hfs is known[4] to about one part in 10^7. The theoretical interpretation is limited mainly by the need for a more accurate value for the muon's mass, but there are also some small terms remaining to be evaluated. It does provide one of the serious tests of QED, roughly on the level of the solid state determinations of α, but not quite as good as the electron's anomalous moment. Positronium hfs is known experimentally at about the few ppm level.[5] While many of the theoretical terms are the same as for hydrogen and muonium, one cannot make an expansion using a small mass ratio, and also, annihilation kernels modify the expression considerably. Thus, terms similar to those presented in the following also appear in positronium, but the details differ.

Fermi[6] was the first one (in 1930) to estimate the energy splitting due to the interaction between the electron and proton spin. Making a non-relativistic reduction of the Dirac equation which incorporated a term from the current associated with the proton spin, he found an effective spin-spin operator

$$\frac{2}{3}\frac{\pi\alpha\vec{\sigma}_e \cdot \vec{\sigma}_p}{m_e m_p}(1+\kappa)|\psi_{nr}(0)|^2 \, ,$$

where $\psi_{nr}(0)$ is the non-relativistic wave function at the origin and κ represents the anomalous part of the proton magnetic moment. For the ground state splitting, this becomes (note: I use units with $\hbar = c = 1$)

$$E_F = \frac{8}{3}\alpha^4\frac{m_r^3}{m_e m_p}(1+\kappa) \, . \tag{1}$$

The reduced mass dependence here comes in through the wave function.

Breit[7] improved this result in 1930 by using Dirac wave functions. Note that the improvement is not completely trivial because those wave functions are weakly singular at the origin [like r^λ where $\lambda = (1-\alpha^2)^{1/2}-1$]. This means that one cannot first make a non-relativistic reduction which introduces a factor of $\delta(\vec{r})$. He obtained a correction factor of $(1+\frac{3}{2}\alpha^2+...)$, but could not incorporate reduced mass effects at that time. This illustrates a general feature of the analysis which prevailed for several decades. By making certain approximations, it is possible to obtain some contributions (such as reduced mass dependence in the leading term), but those approximations may be incompatible with others (such as the relativistic corrections where recoil is completely ignored). I believe it is important to treat *all* effects in a single consistent formalism. Then one can be certain that nothing has "fallen between the cracks." I'll describe one way of doing this later.

Next I'll describe various corrections to this simple picture which have been analyzed. We may express them as

$$\Delta E(\text{hfs}) = \frac{E_F}{1+\kappa} \left\{ (1+\kappa)[1 + \frac{3}{2}(Z\alpha)^2 + \text{ radiative corrections}] \right.$$

$$+ \text{ recoil corrections} + \text{radiative}-\text{recoil corrections}$$

$$\left. + \text{ miscellaneous small corrections} \right\} . \tag{2}$$

Here I have used the convention of introducing factors of Z ($=1$) to accompany powers of α associated with the binding.

Next let me show the various known corrections and describe them briefly. I shall not generally give references to this old established work unless it is particularly relevant for the present discussion. The radiative corrections are

$$a_e + \alpha(Z\alpha)(\ln 2 - \frac{5}{2}) - \frac{8\alpha(Z\alpha)^2}{3\pi} \ln Z\alpha(\ln Z\alpha - \ln 4 + \frac{281}{480})$$

$$+ \frac{\alpha(Z\alpha)^2}{\pi}(15.38 \pm 0.29) + \frac{\alpha^2(Z\alpha)}{\pi} D_1 . \tag{3}$$

Here a_e is the electron magnetic moment anomaly ($\alpha/2\pi+...$); it is known to better than $\pm 10^{-10}$. All terms to order 1 ppm have been calculated except the last term which represents two photon radiative corrections. The next uncalculated terms would be at the level of 10^{-8}, which is smaller than other uncertainties at present. Notice that the reduced mass dependence of these terms is already contained in the Fermi splitting. This can be justified by the fact that they all arise from momenta which are small compared to the proton mass and the dominant dependence arises from the value of the wave function at the origin, which scales as m_r^3. Another thing to observe is the presence of logarithms of $Z\alpha$. These show clearly the non-perturbative nature of bound state energy shifts. This dependence arises from the natural momentum scale in the wave function. There is no simple way of obtaining a power series in $Z\alpha$ by expanding kernels in powers of the external potential. However, one can still use the smallness of $Z\alpha$ to arrange contributions in order of size. The evaluation of these terms extends over three decades and represents many man years of effort. Most recently, the non-logarithmic term of order $\alpha(Z\alpha)^2/\pi$ was obtained numerically by Sapirstein;[8] earlier references may be found there. It is interesting to note that until recent years, most theoretical workers in this field were either radiative correction experts or recoil experts; the two types of calculations could, for the most part, be done independently.

The recoil corrections (shown to zeroth order in κ and ignoring proton structure effects) are

$$-\frac{3Z\alpha}{\pi} \frac{m_e m_p}{m_p^2 - m_e^2} \ln \frac{m_p}{m_e} + \frac{(Z\alpha)^2 m_r^2}{m_e m_p}[2\ln \frac{1}{2Z\alpha} - 6\ln 2 + 3\frac{11}{18}] . \tag{4}$$

The first term is associated with double photon exchange. The presence of the logarithm indicates that internal momenta up to the target mass are important. Clearly for this term it is essential to treat the proton motion relativistically. As in the case of the $Z\alpha$-logarithms, the presence of this term shows that there is no simple procedure to expand in inverse powers of m_p. On the other hand, the second term, which is associated mainly with three-photon exchange, arises from momenta of the order of the electron's mass and smaller. The logarithm indicates that very soft momenta are involved. The extraction of these terms is rather tricky since many kernels contribute to both of them. Notice that the terms presented here are still mass symmetric (this involves a little cheating on some small parts of the second term). The effect of κ and the proton's structure have recently been reviewed and extended by Bodwin and Yennie.[9] The result can be applied to positronium by letting $m_p \to m_e$; of course, positronium has still other terms of the same order to be computed (see Greg Adkins and Jonathan Sapirstein[10]).

The radiative-recoil corrections turn out to be

$$(\frac{\alpha}{\pi})^2 \frac{m_e}{m_p} \left[-2\ln^2 \frac{m_p}{m_e} + \frac{13}{12} \ln \frac{m_p}{m_e} + (18.18 \pm 0.63) \right] \ . \tag{5}$$

Here one has to give up the effort to retain complete mass symmetry. The leading term of this expression was first worked out by Caswell and Lepage.[11] The other terms by Sapirstein, Terray, and Yennie[12] required numerical calculation.

General Nature of the Problem of Treating Recoil in Hydrogen-Like Systems

Most modern approaches follow that of Bethe and Salpeter[13] in spirit, if not in detail. The starting point is to study the four point amplitude e,p→e,p as a function of the total energy (E) in the center of mass. One identifies the bound state energies as the positions of poles in this function. We shall not repeat their discussion except to note that it leads to a homogeneous four-dimensional integral equation for the Bethe-Salpeter wave function. It is a rather complicated and unfamiliar type of eigenvalue problem. In the case where the main kernel was the instantaneous Coulomb potential, Salpeter was able to reduce this further to a three-dimensional problem.[14] However, his equation was complicated by the presence of positive and negative energy projectors (based on plane waves). It also has the well-known defect (mentioned elsewhere in this volume by Joe Sucher) that it does not reduce to the Dirac equation when $m_p \to \infty$. This defect can be fixed up order by order (in α) by considering kernels with crossed photon lines, but the procedure is rather awkward and unnatural. It should be clear from this observation that the Bethe-Salpeter integral equation separates effects which should be treated together. By Herculean effort, researchers were able to obtain some of the terms recorded above; that was done in the 1950's. Let me illustrate the complications of this approach using an example from personal experience. Bodwin and I[15] pushed the recoil calculation of the hfs to terms of relative order $\alpha^2 \ln(1/2\alpha)m_r^2/(m_e m_p)$, confirming a contemporaneous calculation by Peter Lepage,[16] and helping to resolve a disagreement at that time. It was one of the most unpleasant calculations I've participated in. There were mysterious cancellations of apparently unrelated terms, and approximations made in the early stages to obtain leading terms later led to horribly complicated higher order terms which were very difficult to keep track of. There seemed to be no hope of evaluating the non-logarithmic terms which had become important for the comparison of theory and experiment.

There exist many formulations of two-body bound state theory in the literature. They all give equivalent results if treated to the same level of accuracy, but they differ in considerable detail in overall organization and ease of calculation. The one to be emphasized here seems especially adapted to the situation where there is a large mass ratio, but it can be used in the equal mass case to quite refined levels of contribution. The actual hydrogen system has additional complications because of nucleon structure, but these are generally quite small and are easily isolated. My discussion is based on that of Bodwin, Yennie, and Gregorio.[17] A more detailed account will appear in a review of bound state QED theory which is in preparation with Jonathan Sapirstein.

The four point amplitude G is described perturbatively by Feynman graphs with all possible interactions between the electron and the proton. It satisfies a four-dimensional integral equation which may be written formally

$$G = S + SKG , \qquad (6a)$$

where S is the product of two free propagators and K is the complete kernel, including photon exchanges and radiative corrections. In momentum space, the propagator product is given by

$$S = \frac{1}{\not{P}_e - m_e} \otimes \frac{1}{\not{P}_p - m_p} \qquad (6b)$$

where, in the center-of-mass frame,

$$(P_e + P_p)_\mu = E g_{\mu 0} .$$

Here E is the total energy of the system. It is convenient to split off a fixed energy piece of each four-momentum and write

$$P_{e\mu} = E' g_{\mu 0} + p_\mu , \quad P_{p\mu} = E'' g_{\mu 0} - p_\mu . \qquad (6c)$$

A particularly convenient separation of E into E' and E'' is given below; clearly, no result should depend on the energy routing selected.

Of course, bound states are never seen in such a perturbative picture, so we have to develop an integral equation for the amplitude and then study that inhomogeneous integral equation in the vicinity of the (presumed) poles in order to find a homogenous integral equation for the wave functions. While following this general idea, one starts with a sub-class of contributions which contain the dominant physics, and later adds back in the corrections which are (hopefully) small. In effect, one assumes that the true energies are close to the Bohr levels; if there were bound states or resonances in some entirely different place, they would be overlooked. At first, one considers only one-photon exchange kernels. In the Coulomb gauge, one starts with the instantaneous Coulomb interaction. In the Feynman gauge, one keeps only the g_{00} part of the interaction, which gives an effect close to that of the Coulomb potential.

We hope to arrange this so that we are led to a reference bound state equation which is easy to solve, includes as much physics as possible, keeps the mass symmetry to the extent possible, and has a (relatively) straightforward perturbative method for finding the energy shifts from physics not included in the reference equation. The perturbations consist of kernels not included in the original definition of the reference bound state equation as well as corrections to several other approximations which are required in simplifying the original integral equation to arrive at the reference equation. Naturally, there is no unique way of carrying through this program, and different approaches lead to considerably different amounts of labor in calculating some of the refined terms, such as those mentioned above.

There are many ways to accomplish this goal; all of them make use of the known solutions of the Schrödinger or Dirac equations with a given time-independent external potential. Different choices for the unperturbed problem simply produce different correction kernels. The trick is to make a choice that yields a fairly simple unperturbed problem *and* limits the number and complexity of the perturbation kernels required. Furthermore, one wants the perturbations to yield truly small effects so that one can get reasonable results in finite order. For example, even though one particle mass may be much larger than the other one, it would be poor strategy to start with the infinite mass limit. One should be able to get most of the reduced mass effects at the unperturbed level. This turns out to be a somewhat subtle problem, and in fact there is no precise distinction between reduced mass and genuine dynamical recoil effects. Nevertheless, we will be able to obtain much of the mass dependence exactly, and most of our recoil corrections *look* dynamical.

General Description of the Method of Analysis

I wish to present an overview which applies to most treatments of the two-body problem. Usually we are not interested in the wave function, except in some simplified approximation. We study instead the perturbation theory of the four-point function about some known reference. As will be evident, there is considerable flexibility in how to proceed through a calculation. The two known elements of (6a) are S and K (K is expressed as a sum, of course), and we must make approximations for both of them in order to find a solvable equation. In effect, we have the freedom to choose both the free Hamiltonian and the potential of the reference problem. Thus we write

$$S \equiv N\bar{S} + R \tag{7a}$$

where $\bar{S}$ is an effective propagator which must match exactly the behavior of S in the extreme non-relativistic limit. Namely, in that limit it should agree with the two-body Schrödinger propagator. In all practical cases, the relative energy p_0 of the two particles is fixed by some prescription. While the constant N is of course redundant, it usually occurs quite naturally; and it does give us some additional flexibility in putting formulas into a convenient form. Away from the region of non-relativistic momentum, $\bar{S}$ may be chosen in various ways depending on circumstances.

An understanding of why it is possible to make a decomposition like (7a) is provided by studying the pole structure of the two propagators in (6b) as a function of p_0. It turns out to be convenient to choose the separation of E into E' and E'' so that

$$E'^2 - m_e^2 = E''^2 - m_p^2 = -\gamma^2 . \tag{7b}$$

This is motivated by the fact that in a real scattering process the right hand side would be $\vec{p}^2$ in the center of mass. Any other choice would lead to the same final result, of course, but intermediate expressions would be a bit more complicated.

Final results cannot depend on the energy routing, but intermediate details may. With this choice, the p_0-dependence is made explicit in the following expression

$$\frac{1}{(2m_e p_0 + p_0^2 - \vec{p}^2 - \gamma^2 + i\delta)(-2m_p p_0 + p_0^2 - \vec{p}^2 - \gamma^2 + i\delta)}$$

$$\times \frac{1}{((p_0 - p_0')^2 - (\vec{p} - \vec{p}')^2 + i\delta)((p_0 - p_0'')^2 - (\vec{p} - \vec{p}'')^2 + i\delta)} . \tag{7c}$$

In order, the factors arise from: the electron propagator, the proton propagator, and the two photon propagators. This is less complicated than it appears. We use the fact that the characteristic momentum is $\gamma (\sim m_r \alpha)$. Then the electron has its positive energy pole below the axis at a distance $\approx (\gamma^2 + \vec{p}^2)/2m_e$ from the origin; its negative energy pole is above the axis near $-2m_e$. Similarly, the proton positive energy pole is above the axis at a distance $\approx -(\gamma^2 + \vec{p}^2)/2m_p$ and its negative energy pole is below the axis at $2m_p$. The photon poles are at a distance of order γ from the origin. The striking feature of this is that two poles nearly pinch the p_0-contour close to the origin. This dominates the behaviour of the integral. In comparison, Salpeter's approach for the Coulomb potential had no photon poles, and he could carry out the contour integral over p_0 exactly. Another approach by Gross[18] kept only the residue of the proton pole near the origin in the leading term. Both of these methods have the "infamous" square root operators in the leading approximation.

Our approach is to use the pinching to extract an effective $-2\pi i\delta(p_0)$ term in the integrand. For our purposes, this choice is independent of the spatial momentum; and it leads to great simplifications. Once this is done at each step in the ladder, all the photon interactions are converted to Coulomb potentials and we achieve a solvable equation. Of course, corrections to this leave a new class of kernels which must be dealt with along with the ones which occurred originally. Associated with this pinching region, one finds that the leading term in the numerator is the large component projector for each particle. Consequently, the behaviour of the dominant term from this region does agree with that of the Schrödinger equation for the large components. Any treatment must incorporate this dominant feature into the definition of $\bar{S}$, and various treatments differ primarily on how this is done and how the other poles are treated. To emphasize that p_0 is fixed at a small value, we write

$$\bar{S} = -2\pi i\delta(p_0)\bar{s} \tag{7d}$$

where $\bar{s}$ is now a three-dimensional propagator. It should be emphasized that we do not obtain $\bar{s}$ directly from S by setting $p_0 = 0$; instead, $\bar{s}$ corresponds to an operator residue associated with the dominant pole structure. The perturbation scheme for dealing with this separation is presented below.

Having settled on some choice of $\bar{S}$, we next turn our attention to K. Obviously, the most important term in K is the one photon exchange contribution (because of the small value of α), which is well-approximated by a Coulomb interaction. It turns out that it is sometimes possible to pack a little extra physics into our choice, so we simply call the approximate form $\mathcal{V}/N$. If we use this approximation and keep only the $N\bar{S}$ piece of S everywhere, we find that G is approximated by $N\bar{G}$ where $\bar{G}$ satisfies an integral equation

$$\bar{G} = \bar{S} + \bar{S}\mathcal{V}\bar{G} \tag{7e}$$

because of (7d), this is in reality a three-dimensional integral equation which has been "promoted" to a four-dimensional one. Since the solutions of the three-dimensional equation should be known in principle, this should provide a convenient starting point for the development of a perturbation procedure.

The next step is to develop an expansion of G in powers of $\bar{G}$. This differs from the usual type of expansion since both the free propagator and the interactions differ in (6a) and (7e). However, the result is very clear intuitively. First imagine iterating (6a) to arbitrarily large order. In this expansion, replace each factor of S using (7a). Between any two factors of $N\bar{S}$ replace K by its approximation $\mathcal{V}/N$ plus a correction $K - \mathcal{V}/N$. Sum up iterations of $\mathcal{V}$ using the integral equation for $\bar{G}$. Taking stock of the situation at this stage, we see that between any two factors of $N\bar{G}$ there now occurs a new kernel $\hat{K}$ defined by

$$\hat{K} = K - \mathcal{V}/N + KRK + KRKRK + \dots . \tag{8a}$$

Aside from some complicating technical details, one has a new Green's function $\hat{G}$ which satisfies the integral equation

$$\hat{G} = \bar{G} + \bar{G}N\hat{K}\hat{G} . \tag{8b}$$

As a preliminary to the perturbation theory, we consider the three dimensional Green's function $\bar{G}$. It is usually restricted to some subspace of the full 4×4 spinor space, for example by large component projectors, or by positive energy projectors (defined on a plane-wave basis). We do not wish to deal with these individual details at this point. Rather, we simply note that it is usually expressible in terms of eigenfunctions $|\bar{n}>$ and $<\bar{n}|$ through

$$\bar{G} = \sum_n \frac{|n><n|}{\mathcal{E} - \mathcal{E}_n} . \tag{9a}$$

The eigenfunctions are restricted to the appropriate subspace and have been "promoted" from three dimensions to four by appending a factor $-2\pi i\delta(p_0)$. $\mathcal{E}$ is a known function of the total energy E; similarly for $\mathcal{E}_n$. $<n|$ is not necessarily the adjoint of $|n>$, but the two functions satisfy an orthonormality relation

$$<n|L|n'> = \delta_{nn'} \tag{9b}$$

where L is an appropriate matrix.

Let us next describe how one finds a perturbation series for the energy levels. These are given by the poles of G, or equivalently, those of $\hat{G}$. To find them, we consider the series obtained by iterating (8b) and substituting the eigenfunction expansion (9a). Having done this, we take the expectation value of $\hat{G}$ in the state of interest to find

$$<0|L\hat{G}L|0> = \frac{1}{\mathcal{E} - \mathcal{E}_0} + \frac{\Sigma}{(\mathcal{E} - \mathcal{E}_0)^2} + \frac{\Sigma^2}{(\mathcal{E} - \mathcal{E}_0)^3} + \dots$$

$$= \frac{1}{\mathcal{E} - \mathcal{E}_0 - \Sigma(\mathcal{E})} , \tag{10a}$$

where

$$\Sigma(\mathcal{E}) = <0|N\hat{K}|0> + \sum_{n\neq 0} \frac{<0|N\hat{K}|n><n|N\hat{K}|0>}{\mathcal{E} - \mathcal{E}_n} + \dots \quad (10b)$$

Since a kernel typically produces a factor of α^4 or smaller, and may have a small mass ratio factor as well, terms which are not exhibited here are generally much too small ($O(\alpha^8 m_e)$ or smaller) to require further consideration at the present time. By expanding about $\mathcal{E}_0$, we find that the pole in $\mathcal{E}$ is located at

$$\hat{\mathcal{E}}_0 = \mathcal{E}_0 + \Sigma(\mathcal{E}_0) + \Sigma(\mathcal{E}_0)\frac{\partial\Sigma(\mathcal{E}_0)}{\partial\mathcal{E}} + \dots \quad (10c)$$

From this, the corrected energy can be found from the known relation between E and $\mathcal{E}$. Note the resemblance of this to Brillouin-Wigner perturbation theory.

Two Choices of Bound State Formalisms

We now turn to a particular implementation of the procedure just described. The first step is to analyze the structure of the denominators in (7c). Later, we'll specialize to two possibilites for the treatment of the numerators of the free propagators, giving us the Schrödinger equation and Dirac equation formulations.

With the choice (7b), (6b) takes the form

$$\frac{(m_e + E'\gamma_0 + \not{p})}{D_e(p)} \otimes \frac{(m_p + E''\gamma_0 - \not{p})}{D_p(-p)} \quad . \quad (11a)$$

where

$$D_e(\pm p) \equiv p^2 - \gamma^2 \pm 2E'p_0 + i\epsilon$$

$$\quad (11b)$$

$$D_p(\pm p) \equiv p^2 - \gamma^2 \pm 2E''p_0 + i\epsilon \quad .$$

One way to bring out the large contribution associated with the near pinching of the p_0 contour is to rearrange the product of denominators

$$\frac{1}{D_e(p)D_p(-p)} = \frac{1}{2E(-p_0 + i\epsilon)}\left\{\frac{1}{D_e(p)} - \frac{1}{D_p(-p)}\right\} \quad .$$

Note that while the complete expression on the rhs of the equation has no actual pole at $p_0 = 0$, it is convenient to introduce an imaginary part into the denominator to facilitate separating the two terms. Further rearrangement gives

$$\frac{1}{D_e(p)D_p(-p)} = \frac{-2\pi i\delta(p_0)}{-2E(\vec{p}^2 + \gamma^2)} - \frac{1}{2E}\left\{\frac{1}{(p_0 + i\epsilon)D_e(p)} + \frac{1}{(-p_0 + i\epsilon)D_p(-p)}\right\} \quad .$$

$$\quad (11c)$$

As will become apparent later, the presence of the factor $(2E)^{-1}$ in the δ-function term corresponds to the correct treatment of the reduced mass in the non-relativistic region. The δ-function term of (11c) is to be incorporated into the definition of $N\bar{S}$. The remaining terms are part of the definition of R; note that they do not produce a pinch of the contour. We should call attention to a possible problem which could occur if we do not use this result carefully. Term by term, the new form has less convergence at high momenta than the original form. Thus we should use this separation only when the resulting expressions remain well behaved.

Next consider the numerator structure of (11a). We can rewrite the two Dirac factors as

$$E''\gamma_0 + m_p - \not{p} = \tfrac{1}{2}(E'' + m_p)(1 + \gamma_0) + [-\tfrac{1}{2}(E'' - m_p)(1 - \gamma_0) - \not{p}] \qquad (12a)$$

and

$$E'\gamma_0 + m_e + \not{p} = \tfrac{1}{2}(E' + m_e)(1 + \gamma_0) + [-\tfrac{1}{2}(E' - m_e)(1 - \gamma_0) + \not{p}] . \qquad (12b)$$

As a matter of principle, it would be desirable to keep these complete numerators in $N\bar{S}$ if the resulting equation were solvable. The result is then actually very close to that obtained by rationalizing the Breit propagator.[19] In fact, the latter contains a little more physics. The trouble with either of these choices for the reference propagator is that in combination with the Coulomb potential they are too singular. In fact, for large $|\vec{p}\,|$, they approach a (direction-dependent) constant; when iterated with the Coulomb potential, one finds by simple power counting that the integrals become divergent in a non-renormalizable way. It turns out that we may keep one factor of $\not{p}$, but not two, without encountering this problem. We proceed to possible simplifications which seem practical. The first terms on the right side of (12a) and (12b) dominate the non-relativistic region, particularly for the heavier particle. We may choose to perturb about one or both of these terms.

At this stage, both the Coulomb gauge and the Feynman gauge lead to the same reference equation in the ladder approximation. Because there is at least one large component projector on both sides of the γ_μ-matrix in the interaction, only γ_0-terms survive. In the Coulomb gauge, the resulting interaction is directly the Coulomb potential (multiplied by $\gamma_0 \otimes \gamma_0$). For the Feynman gauge, the large component projectors together with the $-2\pi i\delta(p_0)$ factors produce the same result.

Schrödinger Equation Model

If we take the large component projectors in (12a) and (12b), we are lead immediately to the Schrödinger equation, which is symmetric in the two particles.

$$S \rightarrow N_s \bar{S}_s = N_s \bar{s}_s(\vec{p})(-2\pi i)\delta(p_0) , \qquad (13a)$$

where

$$N_s = \frac{(E'' + m_p)(E' + m_e)}{4Em_r} \approx 1 - \frac{\gamma^2}{2}\left(\frac{1}{m_e} - \frac{1}{m_p}\right)^2 + \dots \qquad (13b)$$

and

$$\bar{s}_s(\vec{p}) \equiv \frac{1}{\mathcal{E} - \vec{p}^{\,2}/2m_r}\frac{(1 + \gamma_0) \otimes (1 + \gamma_0)}{4} . \qquad (13c)$$

with

$$\mathcal{E} \equiv -\frac{\gamma^2}{2m_r} . \qquad (13d)$$

The interpretation is obvious. Note that the factor N_s is very close to unity, except for very small binding corrections. This small correction should be separated off and treated as a small perturbation (to avoid having the potential term in the reference Hamiltonian dependent on the eigenvalue). Clearly, this fits into the general scheme described earlier, since E can be expressed in terms $\mathcal{E}$ through (7b). Terms not incorporated into $\bar{S}_s$ become part of R. The obvious choice for $\mathcal{V}$ is the Coulomb potential V_c. With this choice, the three-dimensional Schrödinger equation is

$$\left(\frac{\vec{p}^2}{2m_r} + V_c \right) |n> = \mathcal{E}_n |n> \ . \tag{14}$$

The Dirac Equation Model

We take the large component projector term of (12a) with the complete (12b) to obtain

$$S \to N_d \bar{S}_d = N_d \bar{s}_d(\vec{p})(-2\pi i)\delta(p_0) \ , \tag{15a}$$

where

$$N_d = \frac{E'' + m_p}{2E} \approx \frac{m_p}{M} \equiv N \tag{15b}$$

and

$$\bar{s}_d(\vec{p}) = \frac{\frac{1}{2}(1 + \gamma_0^p)}{E'\gamma_0 + \not{p} - m_e} \ . \tag{15c}$$

The normalizing factor N_d, which is approximately m_r/m_e, turns out to give the main reduced mass effects. The two-body propagator $\bar{s}_d$ is simply the large component projector for the proton times the Dirac propagator for the electron.

The most obvious choice for $\mathcal{V}$ is given by

$$\mathcal{V} \equiv \gamma_0 V_c \frac{m_p}{M} = \gamma_0 V_c N \ . \tag{15d}$$

This is easy to implement and it leads directly to a Dirac equation in which the fine structure constant has been multiplied by m_p/M. There is a slightly different choice which includes a little more physics (correct to order $1/m_p$, see BYG); but since it slightly complicates the discussion, we stick to the simplest choice here. The difference between N_d and N is incorporated into a small perturbation.

The Dirac equation is now

$$(\vec{\gamma} \cdot \vec{p} + \mathcal{V} + m_e)|n> = E_n' \gamma_0 |n> \ . \tag{16a}$$

This equation can be solved in terms of the known solutions of the usual Dirac-Coulomb problem; one simply multiplies the fine structure constant by the factor N. It turns out that the Bohr energies have the correct reduced mass dependence. However, the fine structure, while preserving the Dirac degeneracy, does not have the correct reduced mass dependence. That dependence is restored by some of the perturbations. Alternatively, it could have been incorporated in a slightly modified Dirac equation.

In our analysis, E' plays the role of $\mathcal{E}$. A small shift $\Delta E'$ produces a shift in E given by

$$\Delta E = \Delta E' \frac{E}{E''} \, . \tag{16b}$$

There are two easily identifiable perturbations to be incorporated in R. One is the second term of (12a) with the original form of (11c) (left-hand side). The other is the first term of (12a) with the second term of (11c). The first of these perturbations is a recoil correction (vanishes as $m_p \to \infty$); the second is not. However, it can be shown that the non-recoil part of the perturbation is straightforwardly compensated by a contribution from the kernel in which two photon lines cross and so it need not be calculated.

How the One Photon Kernel Works in the Dirac Equation Model

At this point, our prescription appears somewhat unnatural in the way that the effects of the proton's small components are introduced through the higher kernels. As emphasized in the discussion of the Dirac equation model, either particle's Dirac properties can be incorporated fully into the reference wave function, but we cannot include those effects for both particles simultaneously. We now remark that it is possible to justify that in taking expectation values we can incorporate the small proton components in the wave function by appending the factors

$$\left[1 - \frac{\not{p}_i}{m_p + E''} \right] \tag{17a}$$

to the initial wave function and

$$\left[1 - \frac{\not{p}_f}{m_p + E''} \right] \tag{17b}$$

to the final wave function. These factors are appended to all first order kernels, and multi-photon kernels are themselves modified to avoid double counting. These appended factors come about from "borrowing" pieces from higher kernels.

Our reference model is independent of the choice of gauge, so the perturbations may be worked out in either the Feynman or the Coulomb gauge. In the study of recoil corrections to the hfs by BYG, it turned out that although the problem was initially set up in the Coulomb gauge, it was greatly simplified by transforming the kernels to the Feynman gauge. In that case, some awkward contributions simply cancelled without the necessity of being computed. Here we remain neutral about the choice of gauge as long as possible, but illustrate how the two gauges work for the simplest contributions.

The one-photon kernel, including the related $\mathcal{V}$ subtraction is given by

$$N\hat{K}(\text{one} - \text{photon}) = e^2 \frac{m_p + E''}{2E} \frac{\gamma_\mu \otimes \gamma_\nu G^{\mu\nu}}{q^2} - \mathcal{V} \tag{18}$$

where $G^{\mu\nu}$ defines the gauge of the virtual photon, and $q = p_i - p_f$ is the four-momentum transfer by the photon (in this case, with time component zero). To evaluate the expectation value of this kernel, we first include the appended proton small components. Then, using the fact that the wave function has large

component projectors on both sides of this expression, the proton factor may be
written

$$
\left[1 - \frac{\not{p}_f}{m_p + E''}\right] \gamma_\nu \left[1 - \frac{\not{p}_i}{m_p + E''}\right]
$$

$$
\doteq g_{\nu 0} - \frac{(p_f + p_i)_\nu}{m_p + E''} - \frac{\frac{1}{2}[\not{p}_f - \not{p}_i, \gamma_\nu]}{m_p + E''} - \frac{g_{\nu 0}\not{p}_f \not{p}_i}{(m_p + E'')^2} . \tag{19}
$$

Of course, ν must be spatial in the second and third terms on the right hand
side.

The first term yields the contribution $N\gamma_0 V_c$, which is cancelled by $-\mathcal{V}$.
The second term represents the convection current of the proton. Its ultimate
contribution depends on the choice of gauge. Starting our discussion with the
Feynman gauge, we find an effective electron interaction kernel

$$
\frac{-\not{p}_f - \not{p}_i}{2E} V_c = \frac{[m_e - E'\gamma_0 - \not{p}_f] + [m_e - E'\gamma_0 - \not{p}_i]}{2E} V_c - \frac{m_e - E'\gamma_0}{E} V_c . \tag{20a}
$$

The last term here could have been incorporated into an effective Dirac equation
(see BYG) or, alternatively, it may be treated perturbatively; see the remark
following (16a). The first term can be simplified using the Dirac equation; it
gives the effective contribution to $N\hat{K}$

$$
-2\frac{V_c \mathcal{V}}{2E} = -2\gamma_0 \frac{V_c^2}{2E} N \tag{20b} .
$$

This term by itself would give a sizable contribution to the Lamb shift, of order
$\alpha^4 m_e^2/M$. As is seen below, this leading order piece is precisely cancelled by
other terms of the same size.

Next consider the convection term in the Coulomb gauge. To the contributions discussed in the preceding paragraph must be added one from the transversality correction. This gives an effective electron interaction

$$
\frac{(\vec{p}_f{}^2 - \vec{p}_i{}^2)(\not{p}_i - \not{p}_f)\frac{1}{2}W_c}{2E} = \frac{\frac{1}{2}[\vec{p}^2, W_c]([\not{p}_i - m_e + E'\gamma_0] - [\not{p}_f - m_e + E'\gamma_0])}{2E} \tag{20c}
$$

where in momentum space $\frac{1}{2}W_c$ is $e^2/(\vec{q}^2)^2$ and its Fourier transform is $-\alpha r$.[20]
Again using the Dirac equation, we find that this reduces to

$$
-\frac{1}{4E}[\mathcal{V}, [\vec{p}^2, W_c]] \doteq \gamma_0 \frac{V_c^2}{2E} N . \tag{20d}
$$

This cancels one-half of (20b). The difference between the results in the two
gauges is compensated by a difference in the two-photon kernel calculation.
While not detailed here, the remainder is cancelled by another contribution
arising from the two-photon kernel so terms of this type do not actually occur.

There are some other interesting contributions in (19), one of which we shall
discuss here. The third term, which is not gauge sensitive, gives the leading
contribution to the hyperfine splitting. Now we are able to get the correct
reduced mass dependence of that effect. The proton's anomalous moment can

also be incorporated at this point. The net effect is to multiply the hfs term from (19) by $1 + \kappa$, where κ gives the anomalous part of the proton's magnetic moment. In momentum space, the hfs part of the one-photon kernel reduces to

$$N\hat{K}(\text{hfs}) \doteq \frac{4\pi\alpha(1+\kappa)}{2E} \frac{-i\vec{\sigma}_p \times \vec{q} \cdot \vec{\gamma}_e}{\vec{q}^2} \ , \tag{21a}$$

where $\vec{q} = \vec{p}_i - \vec{p}_f$. The expectation value of this kernel is to be evaluated using the Dirac equation wave functions.

For S-states, we find that the first order contribution of (21a) to the hfs is given by

$$\Delta E(T) = \frac{4\pi\alpha(1+\kappa)}{2m_p} < \frac{-i\vec{\sigma}_p \times \vec{q} \cdot \vec{\gamma}_e}{\vec{q}^2} > \ , \tag{21b}$$

Except for the modified fine structure constant, this is precisely the calculation done by Breit (1930). As an example, for the ground state it gives the result for the hfs

$$\Delta E(T) = \frac{8}{3}\frac{\alpha^4 N^3 m_e^2}{m_p}(1+\kappa)(1+\frac{3}{2}N^2\alpha^2 + O(N^4\alpha^4))$$

$$\approx E_F \left(1 + \frac{3}{2}\alpha^2 + O(\frac{\gamma^2}{m_e m_p})\right) \ , \tag{21c}$$

where E_F is given by (1.1), and terms of relative order $\alpha^2(m_e/m_p)^2$ have been dropped. Thus we have incorporated both reduced mass and relativistic corrections in a single formalism. Incidentally, most earlier treatments of the relativistic two-body problem do not do this. In fact, it appears as a second order (in $\hat{K}$) contribution in the other treatments. To our knowledge, the present treatment (from BYG) is the only unified discussion of $O(\alpha^2)$ recoil *and* non-recoil contributions.

Acknowledgments: In addition to support received from the program through ITP, this work was supported in part at Cornell by the National Science Foundation. The author expresses his thanks to the other members of the program for providing a stimulating atmosphere.

REFERENCES

1. Breit, G., and G. E. Brown, Phys. Rev. **74**, 1278 (1948).

2. H. Hellwig, R. F. C. Vessot, M. W. Levine, P. W. Zitzewitz, D. W. Allan, and D. J. Glaze, IEEE Trans. Instrum. **IM-19**, 200 (1970).

3. L. Essen, R. W. Donaldson, M. J. Bangham, and E. G. Hope, Nature **229**, 110 (1971).

4. Mariam, F. G., W. Beer, P. R. Bolton, P. O. Egan, C. J. Gardner, V. W. Hughes, D. C. Lu, P. A. Souder, H. Orth, J. Vetter, U. Moser, and G. zu Putlitz, Phys. Rev. Lett. **49**, 993 (1982).

5. Ritter, M., P. O. Egan, V. W. Hughes, and K. A. Woodle, Phys. Rev. A **30**, 1331 (1984).

6. E. Fermi, Z. Phys. **60**, 320 (1930).

7. G. Breit, Phys. Rev. **35**, 1447 (1930).

8. J. R. Sapirstein, Phys. Rev. Lett. **51**, 985 (1983).

9. G. T. Bodwin, and D. R. Yennie, Phys. Rev. D **37**, 498 (1988).

10. G. S. Adkins, and J. R. Sapirstein, work in progress.

11. W. E. Caswell and G. P. Lepage, Phys. Rev. Lett. **48**, 1092 (1978).

12. J. R. Sapirstein, E. A. Terray, and D. R. Yennie, Phys. Rev. Lett. **51**, 982 (1983); Phys. Rev. D **29**, 2990 (1984).

13. Salpeter, E. E., and H. A. Bethe, Phys. Rev. **84**, 1232 (1951).

14. Salpeter, E. E., Phys. Rev. **87**, 328 (1952).

15. G. T. Bodwin and D. R. Yennie, Phys. Rep. **43C**, 267 (1978).

16. G. P. Lepage, Phys. Rev. A **16**, 863 (1977).

17. G. T. Bodwin, D. R. Yennie, and M. Gregorio, Rev. Mod. Phys. **57**, 723 (1985). We refer to this as BYG.

18. Gross, F., Phys. Rev. **186**, 1448 (1969); Phys. Rev. C **26**, 2203 (1982).

19. Breit, G., 1929, Phys. Rev. **29**, 553.

20. There is no problem in the Fourier transform of such a singular function, provided one takes into account the effect of the factors of $\vec{q}$ in the numerator. The nervous reader may wish to study this by considering $\lim_{\epsilon \to 0} 1/(\vec{q}^2 + \epsilon^2)^2$.

STATUS REPORT ON THE THEORY

OF THE LAMB SHIFT IN HYDROGEN

H. Grotch

Department of Physics
Pennsylvania State University
University Park, Pennsylvania 16802

ABSTRACT

Recently calculated radiative recoil and pure recoil contributions to the hydrogen Lamb shift are reviewed. The former arise from lowest order radiative corrections on the electron line, treating the proton mass as finite, and result in contributions of order $\alpha(Z\alpha)^5 m^2/M(-2.5\text{kHz})$. The latter recoil corrections which come from double Coulomb, single transverse, and double transverse exchange processes lead to terms of order $(Z\alpha)^6 m^2/M(-3.3\text{kHz})$. An additional recoil correction, of order $(Z\alpha)^4 m^3/M^2(-2.2\text{kHz})$, is also discussed.

This recent work reduces the Lamb shift theory by 8kHz but the theoretical value is uncertain because of a discrepancy in the proton charge radius. For $(r_p^2)^{\frac{1}{2}} = 0.805(11)\text{fm}$ we obtain $1{,}057{,}849(11)$ kHz while for $(r_p^2)^{\frac{1}{2}} = 0.862(12)\text{fm}$ we find $1{,}057{,}867(11)$ kHz for the $n = 2$ splitting. This is to be compared to the measured values $1{,}057{,}849(9)$ kHz (Lundeen and Pipkin) and $1{,}057{,}851(2)$ kHz (Pal'chikov, *et al.*). Although the above numbers appear to support the smaller charge radius this conclusion is premature since there are still other theoretical contributions which have not been evaluated. These will be discussed.

1. INTRODUCTION

The most recent experimental determinations of the hydrogen $n = 2$ Lamb shift are

1,057,845 (9) kHz	Lundeen and Pipkin[1]
1,057,851 (2) kHz	Pal'chikov et al.[2]

In 1984, in a review of QED at the Ninth International Conference on Atomic Physics, held at the University of Washington, Kinoshita and Sapirstein provided an expression for the Lamb shift.[3] Since then, additional contributions have been calculated which reduce the Lamb shift by 8kHz.[4,5,6] The purpose of this paper is to discuss the origin of these additional terms, which in the expression below are denoted by g_i, to mention uncalculated expressions symbolized by h_i, and to review the current status of the theory of the Lamb shift.

The expression for the Lamb shift of the $n = 2$ state can be written as

$$\delta E_{\text{Lamb}} = \Delta E_{2S_{\frac{1}{2}}} - \Delta E_{2P_{\frac{1}{2}}}$$

$$
\begin{aligned}
= &\frac{\alpha(Z\alpha)^4 m}{6\pi} \left\{ \left(\frac{\mu}{m}\right)^3 \left[\frac{1}{8}\frac{m}{\mu} + ln(Z\alpha)^{-2} - 2.207909 \right. \right. \\
&+ \pi Z\alpha \left(\frac{427}{128} - \frac{3}{2}ln2\right) \\
&+ (Z\alpha)^2 \left\{ -\frac{3}{4}ln^2(Z\alpha)^{-2} + \left(4ln2 + \frac{55}{48}\right)ln(Z\alpha)^{-2} \right\} \\
&+ (Z\alpha)^2 \left\{-24.0 \pm 1.2\right\} + \alpha \left\{\frac{0.323}{\pi} + Z\alpha h_1\right\} \bigg] \\
&+ \frac{(Z\alpha)^5 m^2}{6\pi M} \left\{\frac{1}{4}ln(Z\alpha)^{-2} + 2.39977 + \frac{3\pi Z\alpha}{4}(g_1 + h_2)\right\} \\
&+ \frac{1}{12}(Z\alpha)^4 m^3 \langle r_p^2\rangle + \frac{(Z\alpha)^4 m^3}{M^2}g_2 \\
&+ \frac{\alpha(Z\alpha)^5 m^2}{8M}g_3
\end{aligned}
\tag{1}
$$

where

$$
\begin{aligned}
g_1 &= 3 - ln(2/Z\alpha) \\
g_2 &= -1/48 \\
g_3 &= \frac{35}{4}ln2 - \frac{39}{5} + \frac{31}{192} + (-0.415 \pm 0.004)
\end{aligned}
$$

g_1 is a pure recoil correction calculated by Erickson and Grotch[6] while g_2 is a quadratic recoil term obtained many years ago from the Breit-Fermi Hamiltonian.[7] g_3 is a radiative recoil correction calculated by Bhatt and Grotch.[4] These terms respectively give contributions of -3.3kHz, -2.2kHz and -2.5kHz for the $n = 2$ hydrogen Lamb shift. The unknown contribution h_1 is a pure recoil correction from Feynman graphs with three (maybe even more) photons exchanged between the electron and the proton while h_2 is the so-called two loop non-recoil binding correction.

Before discussing the various terms in greater detail we note that in addition to the uncertainties presented by uncalculated terms there is a major uncertainty resulting from the value of the proton radius $(r_p^2)^{\frac{1}{2}}$, which is extracted from electron proton scattering experiments.[8,9] From the known terms of Eq. (1) we obtain

$$1{,}057{,}849(11)\text{kHz} \qquad \text{if} \qquad (r_p^2)^{\frac{1}{2}} = 0.805(11)\text{fm}^{[8]}$$

or

$$1{,}057{,}867(11)\text{kHz} \qquad \text{if} \qquad (r_p^2)^{\frac{1}{2}} = 0.862(12)\text{fm}^{[9]}.$$

This discrepancy between measured values of the proton radius constitutes a large error (18kHz) in the theory of the Lamb shift. Although the theoretical result 1,057,849 seems to be in better agreement with experiment no conclusions should be drawn from this about the proton radius until the unknown contributions h_1 and h_2 are evaluated. Needless to say, it is important to directly resolve this discrepancy in the proton radius through additional more accurate experimental determinations. We now turn to a discussion of the contributions g_i which lead to the 8kHz correction.

2. QUADRATIC RECOIL CORRECTION

The term g_2 arises from the Breit Hamiltonian expanded to order v^2/c^2. When this expression is written in the form

$$H = H_\mu + \Delta H$$

where H_μ is the Hamiltonian with appropriate factors of reduced mass, such that

$$E^\mu_{n,j} = -\frac{\mu(Z\alpha)^2}{2n^2} - \frac{\mu(Z\alpha)^4}{2n^4}\left[\frac{n}{j+\frac{1}{2}} - \frac{3}{4}\right] + \cdots$$

there is a leftover piece

$$\Delta H = -\frac{Ze^2}{4\mu^2 r^3}\,\vec{\sigma}\cdot\vec{L}\left(\frac{m}{m+M}\right)^2 \tag{2}$$

which shifts the $2P_{\frac{1}{2}}$ state by $m^3(Z\alpha)^4/48M^2$. This gives rise to g_2.[7]

3. RADIATIVE RECOIL CORRECTION

We consider next the radiative recoil contribution g_3, but before we indicate how this is obtained we need to digress to discuss the matter of reduced mass corrections, a subject which is not at all transparent when the electron is a Dirac particle.

We have adopted an approach based on the equation given by Grotch and Yennie (GY)[10] many years ago. This equation to order M^{-1} (M is the nuclear mass) can be written in the form

$$(\not{\pi} - m)|n\rangle = 0 \tag{3}$$

where $\not{\pi} = \gamma_0 \pi_0 - \vec{\gamma}\cdot\vec{\pi}$ and

$$\pi_0 = E_n - V - \vec{p}^2/2M \ , \quad \cdot\vec{\pi} = \vec{p} - e\vec{A} \tag{4}$$

with V the Coulomb potential and $\vec{A}$ the vector potential produced by the nuclear motion. Here

$$e\vec{A} = -\frac{1}{2M}V\left(\vec{p} + \hat{r}\hat{r}\cdot\vec{p}\right). \tag{5}$$

Thus our starting point is a Dirac equation which includes the nuclear kinetic energy as well as the convection potential due to nuclear motion. All of this is done in the Coulomb gauge and it gives rise to a soluble Dirac equation which includes nuclear motion to order M^{-1}. In Fig. 1 we illustrate the graphs contributing to the potential which appears in the Dirac equation. The solutions of Eq. (3) contain the correct factors of reduced mass in the nonrelativistic energy as well as in the leading corrections.

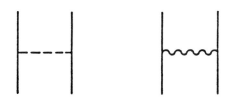

Figure 1: Coulomb and transverse photon interactions contained in Eq. (3).

To proceed to the Lamb shift calculation we now utilize the expansion techniques of Erickson and Yennie (EY).[11] The radiative energy shift can be expressed as

$$\Delta E_n = \frac{\alpha}{4\pi^3} \int \frac{d^4k}{ik^2} \langle n|\gamma_\mu \frac{1}{\not{p} - \not{k} - m + i\epsilon} \gamma^\mu |n\rangle - \langle n|\delta m|n\rangle \qquad (6)$$

where the state vectors and the operators now incorporate in an approximate way the finite nuclear mass. From Eq. (6), which we refer to as the external field approximation (EFA), we are able to carry out a systematic expansion which correctly leads to the reduced mass dependence of the leading terms of the Lamb shift ($\Delta E_n^{(4)}$ below) and also leads to some new corrections in $\Delta E_n^{(5)}$.[4] We find

$$\Delta E_n^{(4)} = \frac{4\alpha(Z\alpha)^4 m}{3\pi n^3} \left[\left(\frac{\mu}{m}\right)^3 \left\{ \frac{3}{8} + \frac{11}{24} + \ln(Z\alpha)^{-2} + \ln\left(1 + \frac{m}{M}\right) \right\} \delta_{l0} \right.$$
$$\left. + \frac{3}{8}\left(\frac{\mu}{m}\right)^2 C_{lj}\delta_{l1} + \left(\frac{\mu}{m}\right)^3 \ln\frac{\mu(Z\alpha)^2}{2\Delta\epsilon_n} \right] \qquad (7)$$

with μ the reduced mass, $\Delta\epsilon_n$ the average excitation energy, and $C_{1,\frac{1}{2}} = -1/3, C_{1,\frac{3}{2}} = 1/6$, and

$$\Delta E_n^{(5)} = \frac{4\alpha(Z\alpha)^5 m}{n^3}\left(\frac{\mu}{m}\right)^3 \left(1 + \frac{11}{28} - \frac{1}{2}\ln 2\right)\delta_{l0}$$
$$+ \frac{\alpha(Z\alpha)^5}{n^3}\frac{m^2}{M}\left(\frac{35}{4}\ln 2 - \frac{39}{5} + \frac{31}{192}\right)\delta_{l0}. \qquad (8)$$

We see that in Eq. (8) the first line gives a reduced mass correction to the relativistic correction to the Lamb shift calculated many years ago.[12] However, we also find that there are residual nuclear mass corrections which are given on the second line of Eq. (8). It should be stressed that the EFA we have adopted is a method which does not yield all the nuclear mass corrections we require. Consequently we needed to examine any remaining terms which might be present. This turned out to be a very formidable task which ultimately resulted in a very small correction. It is worthwhile to discuss briefly how these additional corrections arise.

Figure 2 contains typical terms included in the external field approximation approach. We have not indicated the associated mass counterterms but these are presumed to be included.

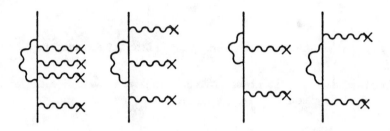

Figure 2: Representative radiative diagrams contained in the external field expansion of Eq. (6). Each interaction could be either Coulomb or convection.

Each diagram contains a single radiative correction on the electron line together with external field interactions. Each such interaction can be either a Coulomb or a transverse (same as convection) interaction with the nucleus. These interactions all give ladder graphs and an arbitrary number of interactions is allowed on external or internal electron lines. In practice two convection terms (order M^{-2}) or higher would be ignored. The actual calculation, based on Eq. (6), was not carried out by expanding in the external potential since the pitfalls of such a non-gauge-invariant procedure are known.[11] Nevertheless Fig. 2 is helpful in that it reveals which classes of diagrams are included. This procedure does incorporate all diagrams with a single exchange but it does not contain all two photon diagrams. It turns out that corrections which require more than two photon exchange are of too high an order to be of any consequence.

In Fig. 3 we illustrate all two photon exchange graphs with a single radiative correction on the electron line, including mass counterterm subtractions. Some of the internal lines have a "dot" on them. Shortly we will explain what this means. The diagrams of type (a) are internal and external self energies, those of type (b) are vertex functions, while those of type (c) are spanning diagrams.

We have already noted that the EFA contains ladder type diagrams corresponding to interactions with the proton and clearly Fig. 3 contains ladder graphs. The "dot" on certain propagators means that we must subtract away the corresponding ladder diagrams of the EFA in order to avoid double counting. In principle this is quite easy to carry out. Each diagram contains a loop integration due to the two exchanged photons. If this integration is carried out by picking up the residues from various poles, it is straightforward to show that the contribution from the positive energy proton pole in the ladder diagrams always gives exactly the term which must be subtracted away. Hence the "dot" reminds us to throw out this pole contribution.

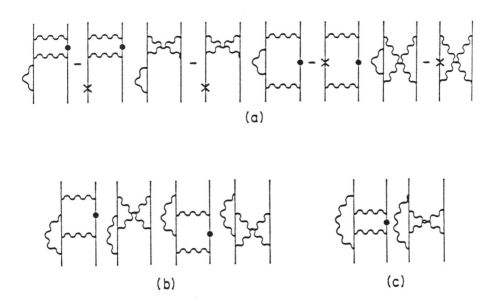

(a)

(b)   (c)

Figure 3: Corrections to the external field approximation. These are from two-photon exchange self-energy (a), vertex (b), and spanning (c) diagrams. The "dot" on the proton propagator implies subtraction of EFA, while the "cross" represents mass renormalization.

The calculation of these radiative recoil diagrams (they are all recoil after subtraction) was made somewhat easier by realizing that to the desired accuracy of $\alpha(Z\alpha)^5 m^2/M$ we could set the external three-momenta to zero. Using techniques of Sapirstein, Terray, and Yennie,[13] used for hyperfine calculations we then constructed an electron factor $L_{\nu\sigma}(p)$ which contained an incoming photon of momentum p and an outgoing photon of momentum $-p$ and a single radiative correction on the electron in all possible ways. We used the Fried-Yennie[14] gauge for the radiative photon since in this gauge, where the photon propagator is $[g^{\mu\tau} + 2k^\mu k^\tau/k^2]/k^2$, infrared singularities become much more manageable. Moreover, in this gauge it can readily be shown that exchanges of more than two photons are suppressed and that the external momenta can be set to zero. We have also directly checked $L_{\nu\sigma}$ to show that in the limit of vanishing momentum p the radiative corrections to the Compton amplitude vanish. This must be true since we know that the Thomson result must be obtained in this limit and consequently the radiative corrections on the electron line should drop out. Moreover this behavior of the electron factor is necessary to avoid infrared singular behavior in the p loop integral.

After combining the electron, photon, and proton factors we then carry out the remaining integrals using the program VEGAS.[15] The integrals are over parameters x, y associated with the electron line and $|\vec{p}|$ and p_4 associated with the loop. The net result obtained is

$$\frac{\alpha(Z\alpha)^5}{n^3}\frac{m^2}{M}(-0.415 \pm 0.004)\,\delta_{l0} \tag{9}$$

which for $n = 2$ contributes only about a half a kilohertz. Altogether the EFA and the above correction to it amount to a contribution of -2.5kHz. It would be worthwhile to repeat this calculation without the separation into EFA and a correction but thus far I have not attempted this.

The question of whether radiative corrections to the proton also contribute can be answered in the affirmative but such corrections are not due to photons. Instead the proton charge distribution is spread out due to its virtual cloud and these effects then appear as a proton size correction. If additional corrections beyond the form factor modification were important, it is doubtful they could be reliably computed since their origin would be from strong rather than electromagnetic effects. There is, however, a calculable correction which arises from vacuum polarization together with a form factor modification. This term was looked at by Lepage et al.[16] and they concluded that this effect is too small. In muonium one can examine directly the importance of radiative corrections on the muon line. The simple vertex correction gives too small an effect for muonium. I have not yet investigated whether two photon exchange terms would be sizable with radiative corrections on the muon but I expect they are not important at the present level of experimental accuracy.

4. PURE RECOIL CORRECTION

Another type of correction which we have investigated is the pure recoil correction to the Lamb shift. Equation (1) contains the pure recoil corrections of order $(Z\alpha)^5 m^2/M$ evaluated many years ago.[17] These recoil effects were obtained from:

i) double Coulomb exchange terms not already contained in the lowest order treatment of the bound state

ii) single transverse photon interactions in which the correction arises from retardation of the Breit potential

iii) double transverse or "seagull" contributions.

The correction g_1 which we have evaluated arises from a reconsideration of the above three terms. When these are treated more accurately it is found that an additional power of $Z\alpha$ is also present. It is possible, however, that further terms of order $(Z\alpha)^6 m^2/M$ are present and arise from diagrams not contained in the above three terms. We will come back to this point later. Now we will briefly describe how we obtain the additional -3.3kHz that the above terms contain.

From the paper of Grotch and Yennie[10] the recoil effect from double Coulomb exchange can be represented by Fig. 4, which depicts double Coulomb exchange as well as the subtraction of the second Born approximation of the Coulomb potential. It is only this difference which contains the double Coulomb pure recoil.

It is straightforward to obtain

$$\delta V_{dc} = -\int \frac{d^3 p'}{(2\pi)^3} \left[\frac{m\gamma_0 - E_{p'} + \vec{\alpha} \cdot \vec{p}'}{-2E_{p'}} \right] \frac{(-Ze^2)}{(\vec{p}' - \vec{p}_3)^2} \frac{(-Ze^2)}{(\vec{p}_1 - \vec{p}')^2}$$
$$\times \left[\frac{\vec{p}'^2 + \vec{p}_1 \cdot \vec{p}_3 + \vec{p}' \cdot (\vec{p}_1 + \vec{p}_3)}{M(m + E_{p'})^2} \right]. \tag{10}$$

To obtain the $(Z\alpha)^5 m^2/M$ correction it was sufficient to set the external momenta $\vec{p}_1$ and $\vec{p}_3$ to zero. The effect of δV_{dc} was then easily obtained by multiplying δV_{dc} by the square of the wave function at the origin. The corrections to this are not difficult to evaluate but they do require integration over the external momenta using appropriate wave functions. Including the lower order effect we find the pure recoil correction from double Coulomb exchange to be

$$\Delta E_{dc} = -\frac{4(Z\alpha)^2 |\psi(0)|^2}{3mM} \left(1 - \frac{9\pi Z\alpha}{8} \right). \tag{11}$$

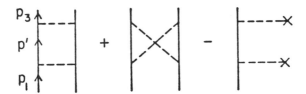

Figure 4: Double Coulomb exchange contributions to pure recoil.

Figure 5: Single transverse exchange contribution to pure recoil.

Next we consider the single transverse (st) contribution to obtain the higher order effect. To the required accuracy the expression used is[10]

$$\Delta E_{st} = \frac{Z\alpha}{2\pi^2 mM} \int \frac{d^3 k}{\vec{k}^2} \sum_m \frac{\langle n|\vec{p}_\perp \cdot e^{i\vec{k}\cdot\vec{r}}|m\rangle \langle m|\vec{p}|n\rangle (E_n - E_m)}{E_n - E_m - k}. \tag{12}$$

Note that instead of using a Feynman diagram approach it is necessary to use old-fashioned perturbation theory to incorporate all Coulomb exchanges which occur during the transit time between emission and absorption of the transverse photon. For $(Z\alpha)^5 m^2/M$ terms all the Coulomb exchanges are necessary. The calculated terms may be depicted in Fig. 5 in which any number of Coulomb exchanges occur. We must also subtract the Breit potential.

In the calculation of $(Z\alpha)^5 m^2/M$ it suffices to ignore $E_n - E_m$ as compared to k in the denominator of Eq. (12) except at low k where a Bethe log contribution occurs. To obtain the correction term this procedure is no longer valid and we need to maintain binding terms in the denominator. The energy difference in the numerator of Eq. (12) is handled by replacing $\vec{p}(E_n - E_m)$ by $[\vec{p}, V]$ and in the denominator E_m is replaced by the Hamiltonian. We can then use closure. By means of a careful separation of the resulting expression we are able to extract the lower order contribution and identify a higher order remainder. The details of the calculation are lengthy but straightforward. Altogether the pure recoil from a single transverse photon gives

$$\Delta E_{st} = \frac{8(Z\alpha)^2}{3mM}|\psi(0)|^2 \left[ln\frac{mZ\alpha}{\Delta\epsilon} + \frac{25}{12} + \frac{9\pi Z\alpha}{8} \right]. \tag{13}$$

The last term in this expression is the higher order correction.

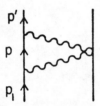

Figure 6: Double transverse or "seagull" contribution to pure recoil.

Finally let us consider the double transverse interaction. This arises from the Feynman diagram Fig. 6 with a "seagull" interaction at the proton. From Ref. 10

$$\Delta V_{dt} = \frac{(Z\alpha)^2}{mM} \int \frac{d^4p}{\pi^2 i} \frac{\vec{\delta}_{\perp 1} \cdot \vec{\delta}_{\perp 3}(-m)(p_0 - m)}{[(p-p_3)^2 + i\epsilon][(p-p_1)^2 + i\epsilon][p^2 - m^2 + i\epsilon]} \tag{14}$$

with $\vec{\delta}_{\perp 1} \cdot \vec{\delta}_{\perp 3} = 1 + \cos^2 \theta_{q_1 q_3}$.

The evaluation of the integral in Eq. (14) requires some delicacy. It is convenient to Euclideanize the integral by carrying out a Wick rotation. After this is done and the integrals over the bound state wave functions are explicitly included, the integrals have to be carried out. The results obtained are

$$\Delta E_{dt} = \frac{4(Z\alpha)^2}{3mM}|\psi(0)|^2 \left[\frac{3}{2}ln\frac{q}{m} - \frac{3}{2}\pi Z\alpha + 2(1 - ln2) - \frac{3}{4}\pi Z\alpha \left(ln\frac{2}{Z\alpha} - \frac{1}{2} \right) \right]$$

where

$$ln\frac{q}{m} = ln\left(\frac{2Z\alpha}{n}\right) + 1 + \frac{1}{2} + \ldots + \frac{1}{n} + \frac{n-1}{2n} \qquad (15)$$

When we combine Eqs. (11), (13), and (15) we obtain the pure recoil terms which appear in Eq. (1). The $(Z\alpha)^5 m^2/M$ terms agree with those previously given while the $(Z\alpha)^6 m^2/M$ terms give us the coefficient g_1 of Eq. (1).

5. CONCLUSION

Equation (1) contains an unknown quantity h_2, which if nonzero, will also contribute a pure recoil correction. If such a term is present it is a nontrivial task to evaluate it. It could conceivably arise from triple Coulomb exchange interactions (Fig. 7) beyond those inherent in the lower order treatment. Another possibility is to have "seagull" interactions as well as a single Coulomb exchange (Fig. 8). Again care is necessary to avoid double counting. We are presently studying such contributions with the hope of either convincing ourselves that such terms are negligibly small, or if this is not the case in actually calculating them. If there are infrared singularities in any of these graphs then the calculation may require multiple Coulomb exchange as occurs in the single transverse recoil correction. That would greatly complicate the analysis.

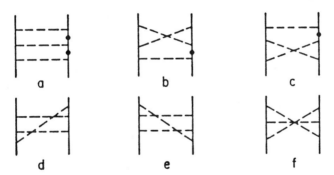

Figure 7: Three Coulomb exchange graphs which might contribute to order $(Z\alpha)^6 m^2/M$ pure recoil.

Equation (1) also contains a term h_1 which would give nonrecoil corrections of order $\alpha^2 (Z\alpha)^5 m$. Terms of this type are so-called two loop nonrecoil binding corrections. Some of the diagrams which contribute to this are shown in Fig. 9. The calculation of these terms appears to be a difficult task.

We have also studied recoil effects which also involve vacuum polarization and find these to be too small. It is, however, necessary to correct vacuum polarization terms by inclusion of reduced mass factors.

Even though research on the Lamb shift has continued for more than forty years, it appears that there is still much to do to further improve the theory.

118

The unknown contributions should all be calculated, and in addition new experiments should be done to better measure the proton charge radius. Until this issue is resolved a major uncertainty in the Lamb shift theory will persist.

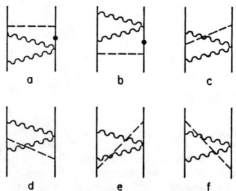

Figure 8: Double transverse with a Coulomb correction. These might contribute to order $(Z\alpha)^6 m^2/M$ pure recoil.

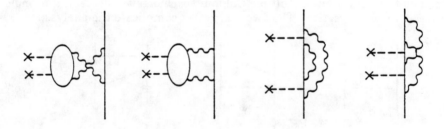

Figure 9: Typical two-loop nonrecoil binding correction graphs of order $\alpha^2(Z\alpha)^5 m$.

REFERENCES

1. S.R. Lundeen and F.M. Pipkin, *Phys. Rev. Lett.* **46**, 232 (1981); S.R. Lundeen and F.M. Pipkin, *Metrologia* **22**, 9 (1986).
2. V.G. Pal'chikov, Yu L. Sokolov, and V.P. Yakovlev, *Lett. J. Tech. Phys.* - **38** (#7), 347 (1983).
3. T. Kinoshita and J. Sapirstein, in "Proceedings of the Ninth International Conference on Atomic Physics, University of Washington, Seattle, WA, July 1984."
4. G. Bhatt and H. Grotch, *Phys. Rev.* **A31**, 2794 (1985); G. Bhatt and H. Grotch,*Phys. Rev. Lett.* **58**, 471 (1987); G. Bhatt and H. Grotch, *Annals of Phys.* **178**,1 (1987).
5. H. Grotch, *Physica Scripta* **T21**, 86 (1988).

6. G.W. Erickson and H. Grotch, *Phys. Rev. Lett.* **60**, 2611 (1988).
7. W.A. Barker and F.N. Glover, *Phys. Rev.* **99**, 317 (1955); see also D.A. Owen, *Ann. Phys. Fr.* **11**, 249 (1986).
8. D.J. Drickey and L.N. Hand, *Phys. Rev. Lett.* **9**, 521 (1962); L.N. Hand, D.J. Miller, and R. Wilson, *Rev. Mod. Phys.* **35**, 335 (1963).
9. G.G. Simon, Ch. Schmitt, F. Borkowski, and V.H. Walther, *Nulc. Phys.-* **A333**, 381 (1981).
10. H. Grotch and D.R. Yennie, *Rev. Mod. Phys.* **41**, 350 (1969).
11. G.W. Erickson and D.R. Yennie, *Ann. Phys. (N.Y.)* **35**, 271, 447 (1965).
12. M. Baranger, H.A. Bethe, and R. Feynman, *Phys. Rev.* **92**, 482 (1953); R. Karplus, A. Klein, and J. Schwinger, *Phys. Rev.* **86**, 288 (1952); see also Ref. 11.
13. J.R. Sapirstein, E.A. Terray, and D.R. Yennie, *Phys. Rev.* **D29**, 2290 (1984).
14. H.M. Fried and D.R. Yennie, *Phys. Rev.* **112**, 1391 (1958).
15. G.P. Lepage, *J. Comput. Phys.* **27**, 192 (1978).
16. G.P. Lepage, D.R. Yennie, and G.W. Erickson, *Phys. Rev. Lett.* **47**, 1640 (1980).
17. E.E. Salpeter, *Phys. Rev.* **87**, 328 (1952); T. Fulton and P.C. Martin, *Phys. Rev.* **95**, 811 (1954); see also Ref. 10 above.

TWO-ELECTRON ATOMS

HIGH PRECISION CALCULATION OF HELIUM ATOM ENERGY LEVELS

Jonathan Baker, Robert N. Hill, and John D. Morgan III
Department of Physics, University of Delaware, Newark, DE 19716

ABSTRACT

This work is concerned with the high-precision calculation of the energies of the ground and excited states of the helium atom (or other light helium-like ions) to match the recent advances in experimental laser spectroscopic studies of transitions between these states with a precision of better than 10^{-4} cm^{-1}. At this level of accuracy it is essential to include mass-polarization effects through 2nd order, relativistic effects properly scaled by appropriate powers of the reduced mass, and quantum electrodynamic (QED) effects. In recent work on excited states of helium we have attained this level of accuracy for all the non-QED effects. Our results are in good agreement with those obtained independently by G. W. F. Drake. We are proceeding with a high-precision evaluation of the Bethe logarithm, which is the principal source of uncertainty in the theoretically determined QED effects. The refinement of these calculations at the 10^{-5} cm^{-1} level and beyond is expected to stimulate further advances both theoretical and experimental, both in the calculations of $O(\alpha^4)$ Rydberg relativistic and QED effects and in the high-precision measurement of transition wavelengths.

OBJECTIVES

The objective of this research is the numerical calculation of the energies of the ground and excited states of the helium atom with an accuracy of about 10^{-4} cm$^{-1} \simeq 5 \times 10^{-10}$ atomic units. In recent work we have been able to achieve this level of accuracy for low-lying excited S-states of the helium atom, including finite nuclear mass effects exactly and relativistic effects in first-order perturbation theory. Extensions to other states with higher L are straightforward. Our next goal, which we are pursuing in collaboration with Robert Forrey and Abraham Feleke, is the calculation of QED (Lamb shift) corrections with similar accuracy to permit comparison of theory with the recent experimental measurement to better than 10^{-4} cm^{-1} of absolute wave numbers of transitions between excited states of the helium atom.[1,2] As has recently been emphasized by high precision experimental atomic spectroscopists, the attainment of this level of accuracy in theoretical calculations of helium atom energy levels is the current major obstacle to checking QED effects in two-electron atoms and obtaining a more precise value of the Rydberg constant.

124

MOTIVATION

Ever since the 1920's the helium atom has been a testing ground for the fundamentals of quantum theory. In the late 1950's agreement between theory and experiment was achieved with a level of accuracy of 0.15 cm^{-1}. Within the last few years wave numbers for transitions between low-lying triplet states have been measured[1,2] with an accuracy of better than 10^{-4} cm^{-1}. Values of Lamb shifts were deduced by subtracting theoretical nonrelativistic, mass-polarization, and relativistic contributions from the experimental wave numbers. In the words of W. M. Fairbank, Jr. and his coworkers, "Our experimental results for the Lamb shifts of the S states are presently an order of magnitude more precise than the QED calculations. With improvements in the nonrelativistic calculations for the 4S and 5S states, an additional order of magnitude can be obtained. Thus there is the potential for a very good check of QED theory for a two-electron system if better QED and nonrelativistic calculations can be done . . . If improvements can be made of an order of magnitude in the relativistic calculations for helium and two orders of magnitude in the QED contributions, our precision absolute wavelength measurements could be used to derive a value for the Rydberg constant accurate to 2 parts in 10^9. This would be not only an important check of the Rydberg value determined with hydrogen, but also a valuable check on the theory of two-electron atoms."[1] Moreover, in a recent survey W. C. Martin[3] notes the "relatively large discrepancies [in Lamb shifts for some states of helium] indicating the need for more accurate calculations."

Furthermore, the refinement of theoretical energy level calculations at the 10^{-4} cm^{-1} level and beyond will necessitate the development of improved procedures for systematically calculating higher-order relativistic and QED corrections. Hence high-precision theoretical calculations not only are essential for the full interpretation of recently obtained experimental results, but also they should help stimulate further advances in our understanding of how to perform higher-order calculations of relativistic and QED effects. This in turn can be expected to stimulate further advances in the high-precision laser spectroscopy of small atoms.

APPROACH

Five years ago David E. Freund, Barton D. Huxtable, and one of us (J.D.M.) published a study of the ground states of nonrelativistic two-electron atoms which included what was at that time the best-ever result on helium.[4] With 230 basis functions our variational energy was lower than even the <u>extrapolated</u> energy deduced by Frankowski and Pekeris[5] using up to 246 basis functions. Our estimated accuracy in the nonrelativistic energy was a few times 10^{-13} a.u. $\simeq$ several times 10^{-8} cm^{-1}. The key to this improvement was our careful choice of basis functions of the form

$$s^n t^{\ell} u^m (\ell ns)^j \, e^{-s/2}, \qquad (1)$$

where $s = r_1 + r_2$, $t = r_2 - r_1$, and $u = r_{12}$. These basis functions have the same analytic structure as that of the exact wavefunction at both two-particle[6] and three-particle coalescences,[7] and hence yield much more rapidly convergent variational energies.[8]

During the last year, on the University of Delaware's IBM 3081D and 3090 we have extended our calculations to the 2^3S, 3^3S, 4^3S, and 5^3S excited states of the helium atom using a mixed basis of functions of the type (1) together with others of the form

$$s^n t^{\ell} u^m (\ell ns)^j \begin{Bmatrix} \cosh(ct) \\ \sinh(ct) \end{Bmatrix} e^{-s/2} \qquad (2)$$

which were first employed by Frankowski.[9] Our nonrelativistic energies seem to be accurate to within about 10^{-10} a.u. $\simeq 2 \times 10^{-5}$ cm^{-1} or better. We have already generalized these basis functions to treat excited states of P symmetry, and the extension to states of higher total angular momentum is straightforward. Thus we can obtain nonrelativistic variational energies of comparable accuracy for these states also. We have carried out further programming of matrix elements to evaluate the mass-polarization correction $\frac{1}{M}\langle \vec{p}_1 \cdot \vec{p}_2 \rangle$ and the relativistic corrections, which involve $\langle p_1^4 + p_2^4 \rangle$, $\langle \delta(\vec{r}_1) + \delta(\vec{r}_2) \rangle$, $\langle \delta(\vec{r}_{12}) \rangle$, and $\langle H_2 \rangle$, where

$$H_2 = -\frac{1}{2} \frac{1}{r_{12}} \left[\vec{p}_1 \cdot \vec{p}_2 + r_{12}^{-2} \; \vec{r}_{12} \cdot (\vec{r}_{12} \cdot \vec{p}_1) \vec{p}_2 \right]. \qquad (3)$$

Indeed, absolute accuracies for theoretical non-radiative energies of excited states of helium have until recently been limited by the errors in Pekeris' calculation of the relativistic terms.[10] The principal reason is that the relativistic correction operators scale like (length)$^{-3}$ or (length)$^{-4}$ and so emphasize 'small-r' regions of configuration space, or equivalently, 'large-p' regions of momentum space. Hence it is especially important to use variational trial functions which have the correct behavior where $|\vec{p}|$ is large, i.e., at two- and three-particle coalescences, where the exact wavefunction has cusps.[6,7] Pekeris' basis functions had the correct two-particle cusp behavior but not the logarithmic cusp behavior at three-particle coalescences. In contrast, our basis functions, which also contain the correct logarithmic cusp behavior at three-particle coalescences, are particularly effective in evaluating these singular quantities with an accuracy of 10^{-4} cm^{-1} or better.

In fact, if one is trying to calculate binding energies with a relative error of 10^{-9} or less, it is generically the case that the error in calculating the relativistic corrections in first-order perturbation theory will exceed (in an absolute sense) the error in the variational estimate of the nonrelativistic energy, even though the former is multiplied by $\alpha^2 \simeq 5 \times 10^{-5}$. This fact, which might initially seem surprising, has a simple explanation. If one is evaluating an expectation value of an operator A using a variational trial function for a Hamiltonian H which does not commute with A, then the error $\delta\psi$ in the variational trial function appears linearly in the error in $\langle A \rangle$:

$$\delta\langle A \rangle = \langle \psi + \delta\psi | A | \psi + \delta\psi \rangle - \langle \psi | A | \psi \rangle$$

$$= \langle \delta\psi | A | \psi \rangle + \langle \psi | A | \delta\psi \rangle + \langle \delta\psi | A | \delta\psi \rangle,$$

while the error in $\langle H \rangle$ is quadratic in $\delta\psi$:

$$\delta\langle H \rangle = \langle \delta\psi | H | \delta\psi \rangle$$

since the Rayleigh-Ritz variational principle ensures that $\langle H \rangle$ is stationary with respect to small changes $\delta\psi$ in the trial function. Thus to leading order the error δE in the sum of the non-relativistic energy and the relativistic corrections is given by:

$$\delta E = \delta\langle H \rangle + \alpha^2 \, \delta\langle A \rangle$$

$$\simeq \langle \delta\psi | H | \delta\psi \rangle + 2 \, \alpha^2 \, \langle \delta\psi | A | \psi \rangle,$$

where A is the sum of the relativistic correction operators. The well-known hand-waving argument (which could be made rigorous if H and A were bounded operators) that if $\langle \delta\psi | H | \delta\psi \rangle$ is $O(\varepsilon^2)$, then $\langle \delta\psi | A | \psi \rangle$ is $O(\varepsilon)$, now suggests that

$$\delta E \simeq O(\varepsilon^2) + O(\alpha^2 \varepsilon).$$

If $\varepsilon < O(\alpha^2)$, then the second term, the error in the relativistic correction, actually dominates the first term, the error in the non-relativistic energy. Thus if one is seeking to reduce the relative error δE in the total energy to less than $O((\alpha^2)^2) \simeq O(10^{-9})$, the dominant contribution will be from the relativistic correction of $O(\alpha^2 \varepsilon)$. Furthermore, in this regime $\delta E = O(\alpha^2 \varepsilon)$, so $O(\varepsilon) = O(\delta E/\alpha^2)$ and $O(\varepsilon^2) = O((\delta E/\alpha^2)^2)$. For example, if one wants to compute the total

energy within a relative error δE of 10^{-11}, then one can expect
to need to know the non-relativistic energy to within a relative
error of

$$O(\varepsilon^2) = O((\delta E/\alpha^2)^2) \simeq O((10^{-11} \times 2 \times 10^4)^2) \simeq O(4 \times 10^{-14}).$$

This illustrates why calculating non-relativistic energies
accurate to more than 10 digits is not merely of academic
interest: the precise calculation of total energies including
relativistic corrections requires the use of highly accurate
wavefunctions obtained from extremely accurate Rayleigh-Ritz
variational calculations of the non-relativistic energy.

It has not been generally realized (with the notable
exception of G. W. F. Drake)[13,14] that at the 10^{-2} cm^{-1} level of
accuracy it is essential to take proper account of reduced mass
effects on the perturbative terms. Let H_μ be the non-relativistic
Hamiltonian

$$\frac{1}{2\mu} \left(p_1^2 + p_2^2 \right) + V(\vec{r}_1, \vec{r}_2) , \qquad (4)$$

where μ is the reduced mass of the electron-nucleus system and V
is the sum of Coulomb potentials, which are homogeneous of degree
-1. Let E_μ and Ψ_μ be an eigenvalue and a normalized eigenfunction
of H_μ. Then E_μ and Ψ_μ are related exactly to the eigenvalue E_1 and
the eigenfunction Ψ_1 of H_1, the Hamiltonian for which $\mu = 1$, by

$$E_\mu = \mu \, E_1$$

$$\Psi_\mu(\vec{r}_1, \vec{r}_2) = \mu^3 \, \Psi_1(\mu\vec{r}_1, \mu\vec{r}_2). \qquad (5)$$

(The factor of μ^3 arises from the normalization condition.) The
linear dependence of the energy on μ is commonly reflected in the
use of a reduced Rydberg constant $R_M = \mu \, R_\infty$.

Now if A is an operator which is homogeneous of degree $-n$,
then

$$\langle \Psi_\mu | A | \Psi_\mu \rangle = \mu^n \langle \Psi_1 | A | \Psi_1 \rangle, \qquad (6)$$

where $\Psi_\mu = \Psi_\mu(\vec{r}_1, \vec{r}_2)$ and $\Psi_1 = \Psi_1(\vec{r}_1, \vec{r}_2)$. For example, the
Coulomb potential expectation value scales linearly with μ:

$$\langle \Psi_\mu | V | \Psi_\mu \rangle = \mu \langle \Psi_1 | V | \Psi_1 \rangle , \qquad (7)$$

as does the kinetic energy expectation value because of the
factor of $1/\mu$ in the operator:

$$\langle \Psi_\mu | \frac{1}{2\mu} \left(p_1^2 + p_2^2 \right) | \Psi_\mu \rangle = \mu \langle \Psi_1 | \frac{1}{2} \left(p_1^2 + p_2^2 \right) | \Psi_1 \rangle , \qquad (8)$$

and hence the total unperturbed energy also scales linearly in μ.

However, the mass-polarization and the relativistic corrections
scale like higher powers of μ:

$$\langle \Psi_\mu | \vec{p}_1 \cdot \vec{p}_2 | \Psi_\mu \rangle = \mu^2 \langle \Psi_1 | \vec{p}_1 \cdot \vec{p}_2 | \Psi_1 \rangle \qquad (9)$$

$$\langle \Psi_\mu | p_1^4 + p_2^4 | \Psi_\mu \rangle = \mu^4 \langle \Psi_1 | p_1^4 + p_2^4 | \Psi_1 \rangle \qquad (10)$$

$$\langle \Psi_\mu | \delta(\vec{r}_1) + \delta(\vec{r}_2) | \Psi_\mu \rangle = \mu^3 \langle \Psi_1 | \delta(\vec{r}_1) + \delta(\vec{r}_2) | \Psi_1 \rangle \qquad (11)$$

$$\langle \Psi_\mu | H_2 | \Psi_\mu \rangle = \mu^3 \langle \Psi_1 | H_2 | \Psi_1 \rangle , \qquad (12)$$

where H_2 is given by eq. (3).

When Pekeris[11-12] evaluated these mass-polarization and
relativistic terms, he simply evaluated all expectation values
with his <u>unscaled</u> $\Psi_1(\vec{r}_1, \vec{r}_2)$ and then multiplied them by the
reduced Rydberg constant $R_M = \mu R_\infty$. The fact that this did not
properly reflect the true μ^2, μ^3, and μ^4 dependence of these terms
was not important at the level of experimental accuracy at that
time ($\sim 10^{-1}$ cm^{-1}). However, if one is aiming for a level of
accuracy of 10^{-4} cm^{-1}, as is the case in a recent article by Kono
and Hattori[15], it is essential to take account of these higher
powers of μ. E.g., for the 1^1S state of helium, Pekeris' estimate
of the mass-polarization term $1/M \langle \vec{p}_1 \cdot \vec{p}_2 \rangle$ was 4.7854 cm^{-1}
and Kono and Hattori's is 4.78551 cm^{-1}; the inclusion of the extra
power of $\mu = 1/(1 + 1/M) \simeq 1 - 1/M \simeq 1 - 1.371 \times 10^{-4}$ reduces this
correction by about 7×10^{-4} cm^{-1} to 4.78485 cm^{-1}. (Furthermore,
at this level of accuracy, it is essential to include the
mass-polarization correction in <u>second-order</u> perturbation theory.
The easiest way is just to include this operator in the
unperturbed Hamiltonian; this further reduces the total energy and
raises the ionization potential by about 2×10^{-3} cm^{-1}.) The
effect of scaling the relativistic corrections by the proper
powers of the reduced mass is even larger. The relativistic
correction to the ionization potential through $O(\alpha^2)$ a.u. is

$$E_j = \alpha^2 \left\{ \left[\frac{1}{8} \langle p_1^4 + p_2^4 \rangle - \frac{5}{8} Z^4 \right] \cdot \mu^4 \right.$$

$$\left. - \left[\frac{\pi}{2} Z \langle \delta(\vec{r}_1) + \delta(\vec{r}_2) \rangle + \pi \langle \delta(\vec{r}_{12}) \rangle + E_2 - \frac{1}{2} Z^4 \right] \cdot \mu^3 \right\} \qquad (13)$$

where $E_2 = \langle H_2 \rangle$, and the $-(5/8)Z^4 \alpha^2 \mu^4$ term is the $-(1/8)\alpha^2 \langle p^4 \rangle$
correction to the He$^+$ energy and the $+(1/2)Z^4 \alpha^2 \cdot \mu^3$ term is the
$(1/2)\pi Z \alpha^2 \langle \delta(\vec{r}) \rangle$ correction to the He$^+$ energy. Because this
relativistic correction is the difference of two terms which scale
with different powers of μ, it is remarkably sensitive to the use

of the proper scaling. For example, Pekeris' estimate of E_j was -0.5636 cm^{-1} and Kono and Hattori's is -0.5639 cm^{-1}; however, if one does the scaling properly using $(p_1{}^2 + p_2{}^2)/2 + V$ as the unperturbed Hamiltonian, then E_j is found to be -0.5693 cm^{-1}. Thus proper scaling alters this relativistic correction to the 1^1S binding energy by almost 6×10^{-3} cm^{-1}. This indicates that Pekeris' and Kono and Hattori's claims to have obtained a theoretical term value (excluding the Lamb shift) for the 1^1S state at the 10^{-4} cm^{-1} level of accuracy were premature.

A further point, which was independently noted by Drake and Makowski,[16] is that the calculation of the $\langle p^4 \rangle$ correction using the relation

$$
\begin{aligned}
\tfrac{1}{4} \langle p_1{}^4 + p_2{}^4 \rangle &= \langle ((p_1{}^2 + p_2{}^2)/2)^2 \rangle - \tfrac{1}{2} \langle p_1{}^2\, p_2{}^2 \rangle \\
&= \langle (H - V)^2 \rangle - \tfrac{1}{2} \langle p_1{}^2\, p_2{}^2 \rangle \\
&= E^2 - 2\,E\,\langle V \rangle + \langle V^2 \rangle - \tfrac{1}{2} \langle p_1{}^2\, p_2{}^2 \rangle \quad (14)
\end{aligned}
$$

needs to be modified to take account of the fact that the scaled unperturbed nonrelativistic Hamiltonian including the mass-polarization term is given by

$$
H = (p_1{}^2 + p_2{}^2)/2 + V + \tfrac{\mu}{M}\, \vec{p}_1 \cdot \vec{p}_2 \; . \tag{15}
$$

Hence

$$
((p_1{}^2 + p_2{}^2)/2)^2 = (H - V - \tfrac{\mu}{M}\, \vec{p}_1 \cdot \vec{p}_2)^2 \; , \tag{16}
$$

so that

$$
\begin{aligned}
\langle ((p_1{}^2 + p_2{}^2)/2)^2 \rangle ={}& E^2 - 2\,E\,\langle V \rangle + \langle V^2 \rangle \\
&+ \tfrac{\mu}{M} \{ -2\,E\, \langle \vec{p}_1 \cdot \vec{p}_2 \rangle + \langle V\, \vec{p}_1 \cdot \vec{p}_2 + \vec{p}_1 \cdot \vec{p}_2\, V \rangle \} \\
&+ \tfrac{\mu^2}{M^2} \langle (\vec{p}_1 \cdot \vec{p}_2)^2 \rangle .
\end{aligned} \tag{17}
$$

Since $\alpha^2 \simeq 5 \times 10^{-5}$ and for helium $\mu/M \simeq 1.4 \times 10^{-4}$, the $\alpha^2\, \mu^2/M^2$ term is on the order of 10^{-12} a.u. $\simeq 2 \times 10^{-7}$ cm^{-1} and can be neglected, but the $\alpha^2\, \mu/M$ term must be evaluated to make comparison with experiments at the 10^{-4} cm^{-1} level. For the 1^1S state of helium, we find (in agreement with Drake and Makowski) that the effect of this term is to decrease the ionization potential by 0.001455 cm^{-1}, on top of a decrease of the ionization potential of 0.000327 cm^{-1} due to the inclusion of the mass-polarization operator in the unperturbed Hamiltonian whose eigenfunction is used to calculate the relativistic correction using just equation (14).

Besides the $O(\alpha^2\, m/M)$ corrections which arise from proper reduced-mass scaling and the inclusion of the mass-polarization operator in the unperturbed Hamiltonian, there are other dynamical $O(\alpha^2\, m/M)$ corrections which were derived in 1963 by A. P. Stone[17] and had generally been neglected until Drake[13,14] drew attention to them. In Stone's notation, the additional operators are

$$\Delta_1 = \alpha^2 Z \frac{m}{M} \sum_{i \neq j} r_i^{-3} \left(\vec{r}_i \times \vec{p}_j \right) \cdot \vec{s}_i \qquad (18)$$

and

$$\Delta_2 = -\tfrac{1}{2} \alpha^2 Z \frac{m}{M} \sum_{i,j} r_i^{-1} \vec{p}_i \cdot \vec{p}_j + r_i^{-3} \vec{r}_i \cdot (\vec{r}_i \cdot \vec{p}_i) \vec{p}_j \ . \qquad (19)$$

(Stone's Δ_3 operator simply reflects the reduced mass scaling of $\langle H_2 \rangle$ and hence has already been included in our calculations.) For S-states, $\langle \Delta_1 \rangle$ vanishes but $\langle \Delta_2 \rangle$ is non-zero. For the 1^1S state of helium, we find that its effect is to increase the ionization potential by 0.01065 cm^{-1}.

Following Drake,[14] we correct for the fact that the nucleus is not pointlike by assuming a uniform spherical distribution of nuclear charge. Dropping higher-order terms in R/a_0, where R is the nuclear radius and a_0 is the Bohr radius, leads to a correction to the ionization potential of the form

$$\Delta E_{nucl} = \frac{2\pi Z}{3} \left(\frac{R}{a_0} \right)^2 \left\{ \langle \delta^{(3)}(\vec{r}_1) + \delta^{(3)}(\vec{r}_2) \rangle - \frac{(\mu Z)^3}{\pi} \right\} \qquad (20)$$

where the $(\mu Z)^3/\pi$ term takes account of the nuclear correction to the energy of a hydrogen-like ion of nuclear charge Z and reduced mass μ. With R for the α-particle of 1.673 fermi, this correction reduces the ionization potential of the 1^1S state of helium by 0.000987 cm^{-1}, in essential agreement with Drake's value 0.000988 cm^{-1}. It had not been included by Kono and Hattori[15] in their calculations.

We provide in Tables I and II the results of our calculation using 230 basis functions of the relativistic but non-radiative ionization potential of 1^1S helium. We have used the physical parameters

$$R_\infty = 109737.31573 \text{ cm}^{-1}$$

$$\alpha^{-1} = 137.035895$$

$$m_e/m_\alpha = 0.137093354 \times 10^{-3}.$$

The first column lists the unscaled expectation values, the second lists the expectation values scaled by μ^{n-1}, where $-n$ is the degree of homogeneity of the operator (we use μ^{n-1} instead of μ^n since the adjusted Rydberg $R_M = \mu R_\infty$ already contains one power of μ), and the third column lists the properly scaled expectation values in cm^{-1}. Our value for the ionization potential obtained by including the mass-polarization term in the unperturbed Hamiltonian is 198312.042299 cm^{-1}. It differs from Kono and Hattori's[15] value 198312.0364 cm^{-1} by ca. 6×10^{-3} cm^{-1}, but it agrees to within 5×10^{-6} cm^{-1} with Drake's[14] non-radiative value 198312.042223, obtained by subtracting out his Lamb shift contributions and including the linear term in our Eq. (17), once one has adjusted for our use of an infinite-mass Rydberg constant R_∞ which is 4×10^{-5} cm^{-1} larger than his. However, we emphasize that the uncalculated $O(\alpha^4)$ relativistic (but non-radiative) correction could well be as large as ca. 10^{-3} cm^{-1} for the ground state of helium, which highlights the need to develop a systematic procedure for calculating relativistic corrections which is more sophisticated than the straightforward expansions involving higher powers of p^2 and other singular operators, which would yield an infinite $O(\alpha^4)$ Rydberg correction.

For excited states the situation is much better. One can obtain a rough estimate of the $O(\alpha^4)$ a.u. = $O(\alpha^6)$ emu relativistic effect on the binding energy by examining the $O(\alpha^6)$ term in the power series expansion of the one-electron Dirac energy with $Z = 1$:

$$\left(1 + \frac{\alpha^2}{(n - \delta_j)^2} \right)^{-\frac{1}{2}} \simeq 1 + \alpha^2 \left(-\frac{1}{2n^2} \right) + \alpha^4 \left(\frac{3}{8n^4} - \frac{1}{n^3(2j+1)} \right)$$

$$+ \alpha^6 \left(-\frac{5}{16n^6} + \frac{3}{4}\frac{1}{n^5(j+\frac{1}{2})} - \frac{3}{8}\frac{1}{n^4(j+\frac{1}{2})^2} - \frac{1}{8}\frac{1}{n^3(j+\frac{1}{2})^3} \right) \quad (21)$$

For $n = 2$ and $j = \frac{1}{2}$, the $O(\alpha^6)$ term is about 10^{-10} Rydbergs $\simeq 10^{-5}$ cm^{-1}, which is entirely negligible at the 10^{-4} cm^{-1} level, and it is even smaller for higher values of n. It therefore makes sense to calculate binding energies of excited states of helium at the 10^{-5} cm^{-1} level, and to examine the effect of the QED corrections.

In Tables III and IV we have presented our results for the 2^3S state. The first entry in the first column of Table III, -2.17522937823678 a.u., is our value of the unscaled nonrelativistic infinite nuclear mass energy of the 2^3S state as computed with 265 basis functions; within rounding it agrees to all digits with Pekeris' value -2.175229378237 a.u. as computed with 1078 basis functions,[12] and to within 10^{-14} a.u. with Drake's[14] very recent and more accurate value $-2.1752293782367910(3)$ obtained by extrapolating results obtained with up to 616 basis functions. After finite nuclear mass, relativistic, and extended nuclear size effects have been included, our estimate in Table IV of its binding energy is 38454.829994 cm^{-1}. By comparison, upon

subtracting out Drake's values for the Lamb shift and including
the linear term in our eq. (17), Drake's estimate for this
ionization potential becomes 38454.829980, which agrees to all
digits with our value when the difference in our Rydberg constants
is taken into account.

In Tables V and VI we have given our results for the 2¹S
state of helium. Our nonrelativistic infinite nuclear mass energy
-2.14597404605419 agrees to within 10^{-13} a.u. with Drake's[14]
extrapolated estimate $-2.14597404605428(11)$ a.u. Inclusion of the
finite nuclear mass, relativistic, and extended nuclear size
effects results in an ionization potential of 32033.322468 cm-1.
Drake's value, upon removal of the Lamb shift correction and
including the linear term in our eq. (17), is 32033.322458 cm-1,
which agrees within 2×10^{-6} cm-1 with our value upon adjusting
for our use of a different R_{∞}.

To save space, we have listed in Table VII results for the
$n^{1,3}$S states with $3 \leq n \leq 6$ only for the non-relativistic infinite
nuclear mass energy (in a.u.) and the 'bottom-line' ionization
potential (in cm-1) including all finite-nuclear mass effects,
relativistic effects through $O(\alpha^2)$, relativistic recoil effects
through $O(\alpha^2\ m/M)$, and the nuclear size correction. For the
moment let us focus our attention on the 2^3S, 4^3S, and 5^3S states,
which are of especial interest because of the high-precision
experiment by Hlousek et al.[1] Our 'bottom-line' ionization
potentials in cm-1 for these state are:

2^3S	4^3S	5^3S
38454.82999	8012.56904	4963.68416

so we obtain for the non-QED contributions to the transition
energies between these states the estimates

2^3S $\rightarrow$ 4^3S		2^3S $\rightarrow$ 5^3S
38454.82999		38454.82999
8012.56904	and	4963.68416
30442.26095 cm-1		33491.14583 cm-1

which are accurate ca. 10^{-5} cm-1, where the uncertainty is
associated with the $O(\alpha^4)$ relativistic corrections and the
uncertainty in the infinite-nuclear-mass Rydberg constant, which
recently was determined by Zhao, Lichten, Layer, and Bergquist[18]
to be 109737.31573(3) cm-1 at the 1σ level. Comparison with the
experimental transition energies of Hlousek et al.

30442.26095(1) (theory)		33491.14583(1) (theory)
30442.13891(8) (expt.)	and	33491.01706(6) (expt.)
0.12204(8) cm⁻¹		0.12877(6) cm⁻¹

yields 'experimental' Lamb shift (QED) residuals for comparison with the best available independent theoretical calculations of this QED effect.

At the present time, the best estimate of the 2^3S Lamb shift is 0.1349 cm⁻¹, which Drake[14] has obtained using Goldman and Drake's 1/Z expansion for calculating the Bethe logarithm.[14] For the 4^3S and 5^3S Lamb shifts, Kono and Hattori[15] have obtained 0.015 cm⁻¹ and 0.0075 cm⁻¹, respectively, using Ermolaev and Swainson's hydrogenic approximation.[20] These result in theoretical Lamb shift residuals of 0.120 and 0.1274 cm⁻¹, respectively, for the $2^3S \to 4^3S$ and the $2^3S \to 5^3S$ transitions. Thus theory and experiment agree at the 2×10^3 cm⁻¹ level, but the errors in the theoretical Lamb shifts need to be reduced by about a factor of 30 to match the experimental results.

To improve the accuracy of helium Lamb shift calculations beyond the 10^{-3} cm⁻¹ level, one must accurately evaluate the Bethe logarithm, which for a state N is given by

$$\ln K_0 = \frac{\sum_{n \neq N} |\langle N | \vec{p}_1 + \vec{p}_2 | n \rangle|^2 (E_n - E_N) \ln |E_n - E_N|}{\sum_{n \neq N} |\langle N | \vec{p}_1 + \vec{p}_2 | n \rangle|^2 (E_n - E_N)}$$

$$= \frac{\langle N | (\vec{p}_1 + \vec{p}_2) \cdot (H - E_N) \ln |H - E_N| \ (\vec{p}_1 + \vec{p}_2) | N \rangle}{\langle N | (\vec{p}_1 + \vec{p}_2) \cdot (H - E_N)(\vec{p}_1 + \vec{p}_2) | N \rangle} ,$$

$$(22)$$

where the sums include both discrete and continuum states. The denominator can be expressed in terms of $\langle \delta(\vec{r}_1) + \delta(\vec{r}_2) \rangle$ and so can easily be evaluated accurately, but the numerator is much more cumbersome because of the logarithmic factor. The original method[21] of calculating the numerator involved a number of different approximations to the wavefunction of the state N and to those of the intermediate states labeled by n. Subsequently C. Schwartz[22] approximated the Hamiltonian in a finite basis of Kinoshita-type functions and expressed the logarithmic factor as an integral involving the resolvent of the Hamiltonian:

$$2\pi Z \ \psi_0^2(0) \ \ln K_0 = \lim_{K \to \infty} \left[\langle \nabla^2 \rangle \ K + 2\pi Z \ \psi_0^2(0) \ \ln K \right.$$

$$\left. - \int_0^K dk \ k \ \langle \vec{\nabla} \psi_0 | (E_0 - H - k)^{-1} | \vec{\nabla} \psi_0 \rangle \right] \quad (23)$$

By analytically cancelling the divergent terms and numerically performing the resulting integrals he calculated a Bethe logarithm for the ground state of helium of 4.370 ± 0.004, which yielded a Lamb shift for the ground state accurate to ± 0.005 cm^{-1}. One would like to reduce this theoretical uncertainty by a factor of 50 for the ground state, and also perform comparably accurate calculations for excited states.

Some time ago Charles Schwartz suggested to us that the best way of improving his results would be to take advantage of recent developments in numerical integration, and in the summer of 1987 we found a very potent technique for calculating Bethe logarithms, which leads to an integral related to Schwartz's but to which numerical integration methods are easier to apply, and which furthermore avoids having to use a variety of techniques to estimate the contribution to the integral from the high-k region. By the spectral theorem and the representation of projection operators as contour integrals of the resolvent, the awkward numerator can be written as

$$\langle \Psi_0 | (\vec{p}_1 + \vec{p}_2) \cdot (H - E_0) \, \ell n (H - E_0) \, (\vec{p}_1 + \vec{p}_2) | \Psi_0 \rangle$$

$$= \frac{1}{2\pi i} \int_C d\zeta \, \zeta \, \ell n \zeta \, \langle \Psi_0 | (\vec{p}_1 + \vec{p}_2) \cdot \left[\frac{1}{\zeta - (H - E_0)} - \frac{1}{\zeta} - \frac{H - E_0}{\zeta^2} \right] (\vec{p}_1 + \vec{p}_2) | \Psi_0 \rangle \tag{24}$$

where the extra terms in the integrand make the integral finite, and the contour C surrounds the spectrum of H-E₀:

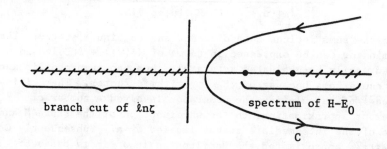

We then deform the contour so that it surrounds the branch cut, use the fact that

$$\int_{-1+i\varepsilon}^{-1-i\varepsilon} d\zeta \, \frac{\ell n \zeta}{\zeta} = \left[\frac{1}{2} \ell n^2 \zeta \right]_{-1+i\varepsilon}^{-1-i\varepsilon} \to 0 \text{ as } \varepsilon \to 0, \tag{25}$$

and evaluate the jump across the branch cut using

$$\ln(r\ e^{i\pi}) - \ln(r\ e^{-i\pi}) = 2\pi i \tag{26}$$

to obtain

$$\int_{-\infty}^{-1} d\zeta\ \zeta\ \langle\Psi_0|(\vec{p}_1+\vec{p}_2)\cdot\left[\frac{1}{\zeta-(H-E_0)} - \frac{1}{\zeta} - \frac{H-E_0}{\zeta^2}\right](\vec{p}_1+\vec{p}_2)|\Psi_0\rangle \tag{27}$$

$$+\int_{-1}^{0} d\zeta\ \zeta\ \langle\Psi_0|(\vec{p}_1+\vec{p}_2)\cdot\left[\frac{1}{\zeta-(H-E_0)} - \frac{1}{\zeta}\right](\vec{p}_1+\vec{p}_2)|\Psi_0\rangle$$

The substitution $\zeta = -k$ brings this expression into the form

$$\int_{0}^{1} dk\ \langle\Psi_0|(\vec{p}_1+\vec{p}_2)\cdot\left[\frac{k}{H-E_0+k} - 1\right](\vec{p}_1+\vec{p}_2)|\Psi_0\rangle \tag{28}$$

$$+\int_{1}^{\infty} dk\ \langle\Psi_0|(\vec{p}_1+\vec{p}_2)\cdot\left[\frac{k}{H-E_0+k} - 1 + \frac{H-E_0}{k}\right](\vec{p}_1+\vec{p}_2)|\Psi_0\rangle$$

which is very similar to Schwartz's expression, except that the divergent counter-terms are now themselves expressed as integrals over the same region, which has allowed the limit as Schwartz's $K \to \infty$ to be taken.

The first term, involving the integral of 0 to 1, is well-suited to numerical integration. To bring the second term into a like form, we make the substitution $k = z^{-2}$, which yields another integral from 0 to 1. Both integrals are now amenable to numerical integration based on the Whittaker cardinal function, as discussed at length in a recent article by F. Stenger.[23] As the number of mesh points N increases, the error in the numerical integration can be shown to be bounded from above by

$$(\text{constant})\ \exp(-\pi\sqrt{N}) \tag{29}$$

This is quite fast convergence. In the case of the hydrogen atom, in which the integrand can be evaluated in closed form in terms of the hypergeometric function and hence can easily be calculated to machine accuracy, with ca. 100 mesh points one can obtain a Bethe logarithm accurate to 10 digits, as shown in Table VIII. This calculation can be done on an IBM AT in less than one minute!

In proceeding to the more challenging case of the helium atom, we can continue to use the numerical integration technique, but the integrand, which involves matrix elements of the helium atom resolvent, can no longer be evaluated in closed form in terms of a known well-studied function. Instead, we must (as did Schwartz)[22] solve the linear equation

$$(H - E_0 + k)\Psi_1 = (\vec{p}_1 + \vec{p}_2)\Psi_0 \qquad (30)$$

and then calculate the inner product of the solution ψ_1 with $(\vec{p}_1 + \vec{p}_2)\Psi_0$. We are currently pursuing this approach in collaboration with Robert Forrey and Abraham Feleke, and hope to have results on which to report in the near future.

Looking further ahead, once we have implemented an algorithm for accurately calculating Bethe logarithms for the helium atom, one would then need to worry about the $O(\alpha^4)$ Rydberg corrections. There currently is no practicable method for evaluating these. What is required is a fresh approach, starting with QED and developing a systematic expansion which does not involve a series in powers of α whose coefficients are ever more pathological operators. We regard this problem as of interest not only for generating theoretical numbers for comparison with experiments, but also because it requires the addressing of a long-standing fundamental problem: how to formulate in a computationally tractable manner the dynamics of several relativistic interacting particles.

ACKNOWLEDGEMENTS

We are grateful to Gordon Drake for kindly directing us to A. P. Stone's article and for keeping us informed of his progress. It has been extremely helpful to be able to compare results of similar accuracy obtained with quite different basis functions. We are further indebted to a number of colleagues in the experimental world, in particular, William Fairbank, Jr., Gordon Berry, Ed Eyler, Bill Lichten, and Bill Martin, for telling us about those problems where more accurate theoretical results are needed to keep pace with experiment. This work has been supported by National Science Foundation Grants PHY-8707970, PHY-8608155, and PHY-8507907 and by a Precision Measurement Grant from the National Bureau of Standards.

REFERENCES

1. L. Hlousek, S. A. Lee and W. M. Fairbank, Jr., Phys. Rev. Lett. 50, 328 (1983); cf. also the comparably accurate experimental measurements of the $2^3S \to 3^3D$ transition by E. Giacobino and F. Biraben, J. Phys. B: At. Mol. Phys. 15, L385 (1982) and of the $2^3P \to 3^3D$ transition by P. Juncar, H. G. Berry, R. Damaschini, and H. T. Duong, J. Phys. B: At. Mol. Phys. 16, 381 (1983).

2. C. J. Sansonetti and W. C. Martin, Phys. Rev. A 29, 159 (1984).

3. W. C. Martin, Phys. Rev. A 35, 3575 (1987).

4. D. E. Freund, B. D. Huxtable and J. D. Morgan III, Phys. Rev. A 29, 980 (1984).

5. K. Frankowski and C. L. Pekeris, Phys. Rev. $\underline{146}$, 46 (1966); $\underline{150}$, 366E (1966).

6. T. Kato, Commun. Pure Appl. Math. $\underline{10}$, 151 (1957).

7. V. A. Fock, Izv. Akad. Nauk SSSR, Ser. Fiz. $\underline{18}$, 161 (1954); J. H. Macek, Phys. Rev. $\underline{160}$, 170 (1967); J. Leray, Actes du 6ème Congres du Groupement des Mathematiciens d'Expression Latine (Paris, Gauthier-Villars, 1982), pp. 179-182; Lecture Notes in Physics $\underline{195}$ (Heidelberg, Springer, 1985), pp. 235-247; J. D. Morgan III, Theoret. Chim. Acta, $\underline{69}$, 181-223 (1986).

8. For studies of how cusp behavior controls the rate of convergence, see C. Schwartz, Meth. Comp. Phys. $\underline{2}$, 241 (1963); B. Klahn and J. D. Morgan III, J. Chem. Phys. $\underline{81}$, 410 (1984); R. N. Hill, J. Chem. Phys. $\underline{83}$, 1173 (1985).

9. K. Frankowski, Phys. Rev. $\underline{160}$, 1 (1967).

10. See Ref. 3, Table VI: The theoretical uncertainty in the nonrelativistic energy of the 2^1S state is 4×10^{-5} cm^{-1}, while the uncertainty in the relativistic correction is 30×10^{-5} cm^{-1}, over seven times as large!

11. C. L. Pekeris, Phys. Rev. $\underline{112}$, 1649 (1958); $\underline{115}$, 1216 (1959).

12. C. L. Pekeris, Phys. Rev. $\underline{126}$, 1470 (1962).

13. G. W. F. Drake, Phys. Rev. Lett. $\underline{59}$, 1549 (1987).

14. G. W. F. Drake, Nucl. Inst. Meth. Phys. Research B $\underline{31}$, 7 (1988).

15. A. Kono and S. Hattori, Phys. Rev. A $\underline{34}$, 1727 (1986).

16. G. W. F. Drake and A. J. Makowski, J. Opt. Soc. Am. B $\underline{5}$, 2207 (1988).

17. A. P. Stone, Proc. Phys. Soc. $\underline{81}$, 868 (1963).

18. P. Zhao, W. Lichten, H. Layer, and J. Bergquist, Phys. Rev. Lett. $\underline{58}$, 1293 (1987).

19. S. P. Goldman and G. W. F. Drake, J. Phys. B: At. Mol. Phys. $\underline{17}$, L197 (1984).

20. A. M. Ermolaev and R. A. Swainson, J. Phys. B: At. Mol. Phys. $\underline{16}$, L35 (1982).

21. P. K. Kabir and E. E. Salpeter, Phys. Rev. $\underline{108}$, 1256 (1957).

22. C. Schwartz, Phys. Rev. $\underline{123}$, 1700 (1961).

23. F. Stenger, SIAM Review $\underline{23}$, 165 (1981).

TABLE I

HELIUM OUTPUT FOR S STATES CREATED MON. JUL 18 1988 21:13 : MASS POLARIZATION COMPUTED AS PERTURBATION

ICOORD= 0   ICOORD= 230   ORDER= 230   EXCHANGE SYMMETRY= 1   1=EVEN, -1=ODD
CHARGE = 0.20000000000Q+01   SCALE = 0.8000000000Q+00   EXPONSCALE = 0.2450000000Q+00

1^1S

EIGENVALUE NUMBER = 1

	UNSCALED VALUES	SCALED VALUES	SCALED VALUES 1/CM
ENERGY EIGENVALUE	-0.2903724377034Q+01	-0.2903724377034Q+01	-0.6372064807373Q+06
BINDING ENERGY	0.9037243770340Q+00	0.9037243770340Q+00	0.1983173865952Q+06
NUCLEAR SIZE CORRECTION	-0.4498170018111Q-08	-0.4496369369582Q-08	-0.9866282893377Q-03
-MASRAT*P1.P2	-0.2180736785851Q-04	-0.2180437862311Q-04	-0.4784851891114Q+01
-PI*ALPHA**2*Z*DELTA(R2)	-0.6057482301824Q-03	-0.6055821762174Q-03	-0.1328917063743Q+03
-PI*ALPHA**2*DELTA(R12)	-0.1779094543359Q-04	-0.1778606839577Q-04	-0.3903055723285Q+01
-E2 CORRECTION	0.7406981599613Q-05	0.7404951212694Q-05	0.1624976144890Q+01
-DEL1**2*DEL2**2*ALPHA**2/4	-0.9496994868602Q-04	-0.9493089832946Q-04	-0.2083206798656Q+02
(E-V(R))**2*ALPHA**2/2	0.8150357503474Q-03	0.8147006342803Q-03	0.1787816116881Q+03
-MU/M*ALPHA**2*Z*E*P1.P2	0.3372013924580Q-08	0.3370627467314Q-08	0.7396658179100Q-03
ALPHA**2/4*P1**4	0.7200658034777Q-03	0.7197697359509Q-03	0.1579495437011Q+03
-ALPHA**2*Z**4*5/8	-0.5325136196900Q-03	-0.5322946674920Q-03	-0.1168091622161Q+03
ALPHA**2*Z**4/2	0.4260109957501Q-03	0.4258941132426Q-03	0.9346014078074Q+02
-STONE DELTA2 TERM	0.1653418108759Q-06	0.1652964856699Q-06	0.3627341243027Q-01
-ALPHA**2*Z**4*MASRAT	-0.1168066525078Q-06	-0.1167745048665Q-06	-0.2562552832980Q-01
-STONE DELTA2 ION POTENTIAL	0.4853528579752Q-07	0.4852198080338Q-07	0.1064788410038Q-01
SCALE AS MU**3	-0.1900772611486Q-03	-0.1900251552054Q-03	-0.4169984116155Q+02
SCALE AS MU**4	0.1875521837877Q-03	0.1874750645889Q-03	0.4114038148507Q+02
RELATIVISTIC IONIZATION SHIFT	-0.2525077360905Q-05	-0.2550086746530Q-05	-0.5596026310840Q+00
CORRECTED IONIZATION POTENTIAL	0.9037000445888Q+00	0.9037000225687Q+00	0.1983120421406Q+06

138

TABLE II

$1\,^1S$

HELIUM OUTPUT FOR S STATES CREATED MON, JUL 18 1988 21:12 : MASS POLARIZATION INCLUDED IN HAMILTONIAN

ICORD= 0   ICOORD= 230   ORDER= 230   EXCHANGE SYMMETRY= 1   1=EVEN,-1=ODD
CHARGE = 0.2000000000Q+01   SCALE = 0.8000000000Q+00   EXPONSCALE = 0.2450000000Q+00

EIGENVALUE NUMBER = 1	UNSCALED VALUES	SCALED VALUES	SCALED VALUES 1/CM
ENERGY EIGENVALUE	-0.2903702581493Q+01	-0.2903702581493Q+01	-0.6372016978248Q+06
BINDING ENERGY	0.9037025814938Q+00	0.9037025814938Q+00	0.1983126036283Q+06
NUCLEAR SIZE CORRECTION	-0.4498065357978Q-08	-0.4496832301817Q-08	-0.9868053277266Q-03
-P1*ALPHA**2*2*DELTA(R2)	-0.6057440480376Q-03	-0.6055779952191Q-03	-0.1328907888770Q+03
-P1*ALPHA**2*DELTA(R12)	-0.1779066290069Q-04	-0.1778578594032Q-04	-0.3902993739976Q+01
-E2 CORRECTION	0.7402317761618Q-05	0.7400288561774Q-05	0.1623952971633Q+01
-DEL1**2*DEL2**2*ALPHA**2/4	-0.9496524690510Q-04	-0.9492620029024Q-04	-0.2083103703979Q+02
(E-V(R))**2*ALPHA**2/2	0.8150297669488Q-03	0.8146946533433Q-03	0.1787802992041Q+03
-MU/M*ALPHA**2*2*E*P1.P2	0.3369254995524Q-08	0.3367869668077Q-08	0.7390606339054Q-03
-2*ALPHA**2<(1/R1+1/R2)P1.P2>*MASRAT/2	-0.1027416995653Q-07	-0.1026994553610Q-07	-0.2253683550456Q-02
ALPHA**2*<1/R12(P1.P2)>*MASRAT/2	0.2741589825932Q-09	0.2740462537379Q-09	0.6013795682705Q-04
ALPHA**2/4*P1**4	0.7200578892874Q-03	0.7197618250147Q-03	0.1579478076889Q-03
-ALPHA**2*Z**4*5/8	-0.5325136196900Q-03	-0.5322946674922Q-03	-0.1168091622161Q-03
ALPHA**2*Z**4/2	0.4260108957520Q-03	0.4258941132642Q-03	0.9346014078074Q+02
-STONE DELTA2 TERM	0.1653351196969Q-06	0.1652897963252Q-06	0.3627194449005Q-01
-ALPHA**2*Z**4*MASRAT	-0.1168065250783Q-06	-0.1167745048665Q-06	-0.2562552832989Q-01
-STONE DELTA2 ION POTENTIAL	0.4852859461856Q-07	0.4851529145868Q-07	0.1064641616017Q-01
SCALE AS MU**3	-0.1900774668954Q-03	-0.1900253608958Q-03	-0.4170002925377Q+02
SCALE AS MU**4	0.1875442695974Q-03	0.1874671575226Q-03	0.4113864547325Q+02
RELATIVISTIC IONIZATION SHIFT	-0.2533197298003Q-05	-0.2558203373170Q-05	-0.5613637805408Q+00
CORRECTED IONIZATION POTENTIAL	0.9037000048296Q+00	0.9037000023290Q+00	0.1983120422990Q+06

140

TABLE III

HELIUM OUTPUT FOR S STATES CREATED MON, JUL 18 1988   22:33   : MASS POLARIZATION COMPUTED AS PERTURBATION

2^3S

ICORD= 265   ICOORD= 0   ORDER= 265   EXCHANGE SYMMETRY= -1   1=EVEN,-1=ODD
CHARGE = 0.2000000000Q+01   SCALE = 0.6705000000Q+00   EXPONSCALE = 0.2515000000Q+00

EIGENVALUE NUMBER = 1

	UNSCALED VALUES	SCALED VALUES	SCALED VALUES 1/CM
ENERGY EIGENVALUE	-0.2175229378236Q+01	-0.2175229378236Q+01	-0.4773422258276Q+06
BINDING ENERGY	0.1752293782367Q+00	0.1752293782367Q+00	0.3845313154070Q+05
NUCLEAR SIZE CORRECTION	-0.3945230709069Q-09	-0.3944149201655Q-09	-0.8655220351378Q-04
-MASRAT*P1.P2	-0.1020286859386Q-05	-0.1020126806781Q-05	-0.2238612650691Q+00
-PI*ALPHA**2*2*DELTA(R2)	-0.4417752009677Q-03	-0.4416540969841Q-03	-0.9691858327474Q+02
-PI*ALPHA**2*DELTA(R12)	0.0000000000000Q+00	0.0000000000000Q+00	0.0000000000000Q+00
-E2 CORRECTION	0.8671611916928Q-07	0.8669234765052Q-07	0.1902416296469Q-01
-DEL1**2*DEL2**2*ALPHA**2/4	-0.6499296368207Q-05	-0.6496624069936Q-05	-0.1425648726517Q+01
(E-V(R))**2*ALPHA**2/2	0.5634491785739Q-03	0.5632175068685Q-03	0.1235950106569Q+03
-MU/M*ALPHA**2*2*E*P1.P2	0.1181814937297Q-09	0.1181329013616Q-09	0.2592362103349Q-04
ALPHA**2/4*P1**4	0.5569498822057Q-03	0.5567208826151Q-03	0.1221693619304Q+03
-ALPHA**2/4*Z**4*5/8	-0.5325136196900Q-03	-0.5322946749202Q-03	-0.1168091622161Q+03
ALPHA**2*2*Z**4/2	0.4260108957520Q-03	0.4258941132426Q-03	0.9346014078074Q+02
-STONE DELTA2 TERM	0.1234158098429Q-06	0.1233819778256Q-06	0.2707550224068Q-01
-ALPHA**2*Z**4*MASRAT	-0.1168065250763Q-06	-0.1167745048865Q-06	-0.2562552832989Q-01
-STONE DELTA2 ION POTENTIAL	0.6609284764540Q-08	0.6607472959096Q-08	0.1449973910793Q-02
SCALE AS MU**3	-0.1567137434487Q-04	-0.1566707833785Q-04	-0.3438054909322Q+01
SCALE AS MU**4	0.2443626251574Q-04	0.2442621512316Q-04	0.5360199714362Q+01
RELATIVISTIC IONIZATION SHIFT	0.8764888180867Q-05	0.8759136787381Q-05	0.1922144805040Q+01
CORRECTED IONIZATION POTENTIAL	0.1752371228583Q+00	0.1752371172467Q+00	0.3845482982424Q+05

TABLE IV

HELIUM OUTPUT FOR S STATES CREATED MON, JUL 18 1988 22:30 : MASS POLARIZATION INCLUDED IN HAMILTONIAN

2^3S

ICORD= 265 ICOORD= 0	ORDER= 265	EXCHANGE SYMMETRY= -1	1=EVEN,-1=ODD
CHARGE = 0.2000000000Q+01	SCALE = 0.6705000000Q+00	EXPONSCALE = 0.2515000000Q+00	

EIGENVALUE NUMBER = 1	UNSCALED VALUES	SCALED VALUES	SCALED VALUES 1/CM
ENERGY EIGENVALUE	-0.2175228359190Q+01	-0.2175228359190Q+01	-0.4773420020586Q+06
BINDING ENERGY	0.1752283591902Q+00	0.1752283591902Q+00	0.3845290791651Q+05
NUCLEAR SIZE CORRECTION	-0.3945255893728Q-09	-0.3944174179489Q-09	-0.8655275163873Q-04
-PI*ALPHA**2*Z*DELTA(R2)	-0.4417753008012Q-03	-0.4416541967902Q-03	-0.9691860517665Q+02
-PI*ALPHA**2*DELTA(R12)	0.0000000000000Q+00	0.0000000000000Q+00	0.0000000000000Q+00
-E2 CORRECTION	0.8654437727568Q-07	0.8652065283590Q-07	0.1898648547401Q-01
-DEL1**2*DEL2**2*ALPHA**2/4	-0.6499231699773Q-05	-0.6496559428092Q-05	-0.1425634541217Q+01
(E-V(R))**2*ALPHA**2/2	0.5634491914527Q-03	0.5632175195586Q-03	0.1235950134820Q+03
-MU/M*ALPHA**2*2*E*P1.P2	0.1179311314451Q-09	0.1178826420178Q-09	0.2586870298514Q-04
-Z*ALPHA**2*<(1/R1+1/R2)P1.P2>*MASRAT/2	-0.2148763409598Q-09	-0.2147879908305Q-09	-0.4713405336409Q-04
ALPHA**2*<1/R12(P1.P2)>*MASRAT/2	-0.1682667855066Q-10	-0.1681975997033Q-10	-0.3691004608532Q-05
ALPHA**2/4*P1**4	0.5569498459811Q-03	0.5567208464054Q-03	0.1221683539844Q+03
-ALPHA**2*Z**4*5/8	-0.5325136196902Q-03	-0.5322946674920Q-03	-0.1168091622161Q+03
ALPHA**2*Z**4/2	0.4260108957520Q-03	0.4258941132426Q-03	0.9346014078074Q+02
-STONE DELTA2 TERM	0.1234154142030Q-06	0.1233815822942Q-06	0.2707541544346Q-01
-ALPHA**2*Z**4*MASRAT	-0.1168065250783Q-06	-0.1167745048665Q-06	-0.2562552832989Q-01
-STONE DELTA2 ION POTENTIAL	0.6608889124655Q-08	0.6607077427668Q-08	0.1449887113577Q-02
SCALE AS MU**3	-0.1567164630840Q-04	-0.1566735023475Q-04	-0.3438114576069Q-01
SCALE AS MU**4	0.2443622629110Q-04	0.2442617891342Q-04	0.5360191768331Q-01
RELATIVISTIC IONIZATION SHIFT	0.8765479982699Q-05	0.8758828678664Q-05	0.1922077192262Q+01
CORRECTED IONIZATION POTENTIAL	0.1752371237702Q+00	0.1752371180189Q+00	0.3845482999370Q+05

TABLE V

HELIUM OUTPUT FOR S STATES CREATED MON, JUL 18 1988 21:25 ; MASS POLARIZATION COMPUTED AS PERTURBATION

2^1S

ICORD= 265   ICORD= 0   ORDER= 265   EXCHANGE SYMMETRY= 1   1=EVEN,-1=ODD
CHARGE = 0.2000000000Q+01   SCALE = 0.6400000000Q+00   EXPONSCALE = 0.2450000000Q+00

EIGENVALUE NUMBER = 2

	UNSCALED VALUES	SCALED VALUES	SCALED VALUES 1/CM
ENERGY EIGENVALUE	-0.2145974046054Q+01	-0.2145974046054Q+01	-0.4709223025625Q+06
BINDING ENERGY	0.1459740460541Q+00	0.1459740460541Q+00	0.3203320642048Q+05
NUCLEAR SIZE CORRECTION	-0.3032994235269Q-09	-0.3032162799544Q-09	-0.6653915921917Q-04
-MASRAT*P1.P2	-0.1302916649471Q-05	-0.1302738052742Q-05	-0.2858787619362Q+00
-PI*ALPHA**2*Z*DELTA(R2)	-0.4381300974839Q-03	-0.4380099927337Q-03	-0.9611890446804Q+02
-PI*ALPHA**2*DELTA(R12)	-0.1446841128768Q-05	-0.1446445057258Q-05	-0.3174143594223Q+00
-E2 CORRECTION	0.4927376160182Q-06	0.4926025416907Q-06	0.1080989416473Q+00
-DEL1**2*DEL2**2*ALPHA**2/4	-0.1901355870309Q-04	-0.1900574094911Q-04	-0.4170706214328Q+01
(E-V(R))**2*ALPHA**2/2	0.5664199366693Q-03	0.5661870433263Q-03	0.1242466592802Q+03
-MU/M*ALPHA**2*E*P1.P2	0.1488921560669Q-09	0.1488309364781Q-09	0.3266013744560Q-04
ALPHA**2/4*P1**4	0.5474063779907Q-03	0.5471813023772Q-03	0.1200759530659Q+03
-ALPHA**2*Z**4*5/8	-0.5325136196900Q-03	-0.5322946674920Q-03	-0.1168091622161Q+03
ALPHA**2*Z**4/2	0.4260108957520Q-03	0.4258941132426Q-03	0.9346014078074Q+02
-STONE DELTA2 TERM	0.1216019007484Q-06	0.1215685650787Q-06	0.2667755889927Q-01
-ALPHA**2*Z**4*MASRAT	-0.1168065250738Q-06	-0.1167745048665Q-06	-0.2562552832990Q-01
-STONE DELTA2 ION POTENTIAL	0.4795375670066Q-08	0.4794061112131Q-08	0.1052030569382Q-02
SCALE AS MU**3	-0.1306881316846Q-04	-0.1306523061034Q-04	-0.2867093613665Q+01
SCALE AS MU**4	0.1489275830074Q-04	0.1488663488522Q-04	0.3266790849800Q+01
RELATIVISTIC IONIZATION SHIFT	0.1823945132279Q-05	0.1821404274884Q-05	0.3996972361351Q+00
CORRECTED IONIZATION POTENTIAL	0.1459745670826Q+00	0.1459745672041Q+00	0.3203332223896Q+05

TABLE VI

2^1S

HELIUM OUTPUT FOR S STATES CREATED MON, JUL 18 1988 21:22 : MASS POLARIZATION INCLUDED IN HAMILTONIAN

ICORD= 265   ICORD= 0   ORDER= 265   EXCHANGE SYMMETRY= 1   1=EVEN,-1=ODD
CHARGE = 0.2000000000Q+01   SCALE = 0.6400000000Q+00   EXPONSCALE = 0.2500000000Q+00

EIGENVALUE NUMBER = 2

	UNSCALED VALUES	SCALED VALUES	SCALED VALUES 1/CM
ENERGY EIGENVALUE	-0.2145972274585790+01	-0.2145972274585790+01	-0.4709220172415SQ+06
BINDING ENERGY	0.1459727245857910+00	0.1459727245857910+00	0.3203292309950040+05
NUCLEAR SIZE CORRECTION	-0.3032989623863320-09	-0.3032158189402090-09	-0.6653905805211230-04
-PI*ALPHA**2*Z*DELTA(R2)	-0.4381300790577780-03	-0.4380099743126330-03	-0.9611890042562700+02
-PI*ALPHA**2*DELTA(R12)	-0.1446881672331220-05	-0.1446485038174300-05	-0.3174232540471800+00
-E2 CORRECTION	0.4916992218068640-06	0.4915644321346800-06	0.1078711341660200+00
-DEL1**2*DEL2**2*ALPHA**2/4	-0.1901307152877370-04	-0.1900525397506700-04	-0.4170599350528400+01
(E-V(R))**2*ALPHA**2/2	0.5664195960467200-03	0.5661867028192900-03	0.1224468455782000+03
-MU/M*ALPHA**2*E*P1.P2	0.1463110637031800-09	0.1482500830405500-09	0.3253267233607600-04
-Z*ALPHA**2*<(1/R1+1/R2)P1.P2>*MASRAT/2	-0.7361729139815600-09	-0.7358702237372900-09	-0.1614827079510800-03
ALPHA**2*<1/R12(P1.P2)>*MASRAT/2	0.5295775231849100-11	0.5293597782138500-11	0.1161651167676700-05
ALPHA**2/4*P1**4	0.5474059419519100-03	0.5471808665176800-03	0.1200758574189000+03
-ALPHA**2*Z**4*5/8	-0.5325136196960020-03	-0.5322946674920200-03	-0.1168091622161100+03
ALPHA**2*Z**4/2	0.4260108957520100-03	0.4258941113242640-03	0.9346014078074800+02
-STONE DELTA2 TERM	0.1216008401393370-06	0.1215673057152100-06	0.2667728234131800-01
-ALPHA**2*Z**4*MASRAT	-0.1168065250783800-06	-0.1167745048665800-06	-0.2562552832989000-01
-STONE DELTA2 ION POTENTIAL	0.4794115060991140-08	0.4792800848625900-08	0.1051754011428300-02
SCALE AS MU**3	-0.1306987494019000-04	-0.1306629209100000-04	-0.2867326549807300+01
SCALE AS MU**4	0.1489232226189910-04	0.1488619902565500-04	0.3266695202793990+01
RELATIVISTIC IONIZATION SHIFT	0.1822447321701200-05	0.1819906934654S0-05	0.3993686529866070+00
CORRECTED IONIZATION POTENTIAL	0.1459745683052400+00	0.1459745657648S0+00	0.3203332246815700+05

TABLE VII

State	Non-relativistic Infinite-Nuclear Mass Energy (in a.u.)	Ionization Potential (in cm^{-1}) Including Finite-Nuclear Mass and Relativistic Effects through $O(\alpha^2)$ and $O(\alpha^2\, m/M)$
3^1S	-2.061271989974061	13445.865230
4^1S	-2.033586717002784	7370.453040
5^1S	-2.021176851568409	4647.162084
6^1S	-2.014563098420080	3195.803262
3^3S	-2.068689067472336	15073.911369
4^3S	-2.036512083099804	8012.569037
5^3S	-2.022618872301605	4963.684156
6^3S	-2.015377452993134	3374.552940

TABLE VIII

Evaluation of $\ln K_0$ for hydrogen. $N_f = 2N+1$ is the number of function evaluations, β_1 is the integral from 0 to 1, and β_2 is the integral from 1 to ∞.

$\beta_1 = -0.646081193149311803...$

$\beta_2 = +5.22804394356041641...$

$\ln K_0 = \frac{1}{2} (\beta_1 + \beta_2) = 2.290981375205555230...$

N	$N_f = 2N+1$	Error in β_1	Error in β_2
49	99	3.3×10^{-10}	-3.2×10^{-9}
64	129	1.5×10^{-11}	-1.4×10^{-10}
81	163	6.4×10^{-13}	-6.3×10^{-12}
100	201	2.8×10^{-14}	-2.8×10^{-13}
121	243	1.2×10^{-15}	-1.2×10^{-14}
144	289	5.4×10^{-17}	-5.3×10^{-16}
169	339	2.4×10^{-18}	-2.3×10^{-17}

146

NEW VARIAITONAL TECHNIQUES AND HIGH PRECISION EIGENVALUES FOR HELIUM

Gordon W. F. Drake

Department of Physics, University of Windsor
Windsor, Ontario, Canada N9B 3P4

ABSTRACT

We discuss in these lecture notes the general problem of obtaining high precision eigenvalues for helium. Recently developed variational techniques are described which improve the accuracy by several orders of magnitude over what was previously available for the nonrelativistic energies. The calculation of mass polarization, relativistic, relativistic reduced mass and quantum electrodynamic (QED) corrections are also discussed. A comparison of total transition frequencies with experiment yields well–defined discrepancies which can reasonably be attributed to as–yet uncalculated QED effects.

INTRODUCTION

The study of the helium atom goes back to the early days of quantum mechanics, where it provided one of the first comfirmations that the Schrödinger equation provides the correct (nonrelativistic) description of atomic structure. It is the simplest system which incorporates the many–body nature of more complex atoms. The Hamiltonian

$$H = -\tfrac{1}{2}\nabla_1^2 - \tfrac{1}{2}\nabla_2^2 - \frac{Z}{r_1} - \frac{Z}{r_2} + \frac{1}{r_{12}} \qquad (1.1)$$

would be separable if it weren't for the last term arising from the electron––electron Coulomb interaction, and one could then write the wave function in the product form

$$\Psi(\vec{r}_1, \vec{r}_2) = \varphi_1(\vec{r}_1)\varphi_2(\vec{r}_2) \pm \text{exchange} \qquad (1.2)$$

The Hartree–Fock method in fact corresponds to finding the best possible solution that can be written in the form of (1.2) (with possibly some additional central field assumptions). For the $1s^2\ {}^1S$ ground state of helium, the Hartree–Fock energy is about -2.863 a.u., whereas the exact energy is $-2.903724 \cdots$ a.u. The difference of about 0.04 a.u. or $1\ \text{eV}$ is called the correlation energy.

Hylleraas[1,2] in 1928 first suggested writing the ground state trial wave function in the explicitly correlated form

$$\Psi(\vec{r}_1, \vec{r}_2) = \sum_{i,j,k} a_{ijk}\, r_1^i r_2^j r_{12}^k\, e^{-\alpha r_1 - \beta r_2} \pm \text{exchange} \qquad (1.3)$$

where the a_{ijk} are linear variational parameters for each combination of powers in (1.3). The same trial function is often expressed in terms of the equivalent variables

$$s = r_1 + r_2, \quad t = r_2 - r_1, \quad u = r_{12}$$

with $r_{12} = |\vec{r}_1 - \vec{r}_2|$. The usual procedure is to include all combinations of i, j, k such that $i + j + k \leq N$, where N is an integer, and then study the convergence of the calculation as N is increased. One can show that this basis set eventually becomes complete as $N \to \infty$ so that, unlike (1.2), the exact solution can be expanded in terms of the functions in (1.3). For any finite N, the a_{ijk} are determined by Schrödinger's variational principle

$$H(\Psi) = \frac{\langle \Psi \mid H \mid \Psi \rangle}{\langle \Psi \mid \Psi \rangle} = \text{Min.} \tag{1.4}$$

which gives the system of homogeneous linear equations

$$\frac{\partial E}{\partial a_{ijk}} = 0, \quad \text{all } a_{ijk} \ . \tag{1.5}$$

The solutions to (1.5) can be regarded from a more general point of view. If we think of the functions in (1.3) as the members of a basis set

$$\chi_l = r_1^i r_2^j r_{12}^k \, e^{-\alpha r_1 - \beta r_2} \tag{1.6}$$
$$l = 1,2,\cdots,P$$

where l denotes the l'th distinct combination of values for i, j, k, then the solutions to (1.5) correspond to finding the linear combinations

$$\Phi_m = \sum_{l=1}^{P} r_l^{(m)} \chi_l \tag{1.7}$$

which satisfy

$$\langle \Phi_m \mid \Phi_n \rangle = \delta_{m,n} \tag{1.8}$$

$$\langle \Phi_m \mid H \mid \Phi_n \rangle = \varepsilon_m \delta_{m,n} \ . \tag{1.9}$$

Thus the solutions to (1.5) are the same as what one would obtain by diagonalizing the Hamiltonian in the orthonormal basis set constructed from the same set of functions. If there are P linearly independent functions, then one obtains P variational eigenvalues ε_m ($m=1,2,\cdots,P$).

An important property of the eigenvalues obtained above follows from the MATRIX INTERLEAVING THEOREM, which says that when an extra row and column is added to a matrix, the new eigenvalues fall between the old, with the new highest higher than the old highest and the new lowest lower that the old lowest. Since by (1.4) the lowest eigenvalue is bounded from below by the true ground state, the higher eigenvalues must similarly lie above the corresponding excited states, and move progressively downward as P is increased. The result is summarized by the HYLLERAAS–UNDHEIM THEOREM[3]:

When a Hamiltonian operator whose spectrum is bounded from below is diagonalized in a P–dimensional finite basis set, then the P eigenvalues are upper bounds to the first P energies of the actual spectrum.

Table 1. Results of the calculation of Accad *et al.*[4] for the nonrelativistic ionization energy (J_{nr}), relativistic correction (ΔJ_{rel}) and $<\delta(\vec{r}_{12})>$ for the n ^{1}P states of helium.

State	J_{nr} (cm^{-1})	ΔJ_{rel} (cm^{-1})	$\langle \delta(r_{12}) \rangle$ (a.u.)
2 ^{1}P	27176.688	0.46772	0.000736
3 ^{1}P	12101.57	0.1734	0.000253
4 ^{1}P	6818.1	0.080	0.00011
5 ^{1}P	4368.2	0.04	0.0001

The optimization of the nonlinear parameters in (1.6) is not so easily carried out because the equations

$$\partial E/\partial\alpha = 0, \quad \partial E/\partial\beta = 0$$

are transcendental. One must resort to a process of recalculating the variational eigenvalues for different values of α and β in order to locate the variational minimum for a given state. This problem is further discussed below.

The Hylleraas method has been applied with great success by many authors to the low–lying states of helium and helium–like ions, culminating in the 1960's and early 1970's with the extensive calculations of C. L. Pekeris and co–workers[4]. Their work covers the singlet and triplet n S and n P states up to $n = 5$ for nuclear charge Z in the range $2 \leq Z \leq 10$.

Despite this large body of work, there remain important problems to be solved as follows.

(1) The calculations of Pekeris *et al.*[4] give energies for the lowest–lying states accurate to about 10^{-9} a.u., using basis sets containing up to 560 terms. This leads to uncertainties in energy differences of ±0.0002 cm^{-1}, which is larger than the current levels of experimental precision[5]. The comparison between theory and experiment is limited as much by a lack of knowledge of the nonrelativistic energies as it is by higher order relativistic and QED effects. One might try to obtain higher accuracy simply by increasing N. The problem is that the number of terms is given by

$$P = (N+1)(N+2)(N+3)/6 \tag{1.10}$$

and so grows rapidly with N. In addition, numerical problems of near linear dependence in the basis set become progressively more severe as N increases.

(2) An even more serious problem is that the accuracy of Pekeris' calculations (as measured by the rate of convergence with N) seriously deteriorates as one goes up the Rydberg series to more highly excited states. The point is illustrated by the data in Table 1. Approximately one significant figure after the decimal is lost each time n (the principal quantum number) is increased by one. There is clearly no point going beyond the 5 ^{1}P and 5 ^{3}P states, and even here, the accuracy is far short of what is required for spectroscopic precision. The large body of high precision experimental data for transitions among high n states[6,7] remains unanalysed in terms of *a priori* calculations. The asymptotic

calculations of Drachman[8] become useful when the angular momentum L is also large, but there is as yet no overlap region where the asymptotic expansions can be compared with the bounds provided by direct variational calculations.

The initial aim of these notes is to describe new variational techniques which now make it possible to improve the precision of calculated eigenvalues by several orders of magnitude. An important advantage of these techniques is that they do not suffer from a loss of accuracy as one goes up the Rydberg series to moderately high values of n (~10). We will then turn to the calculation of a number of relativistic, QED and finite nuclear mass corrections to the nonrelativistic eigenvalues. Of particular interest here is a newly derived[9] connection between two–electron QED corrections for low–lying states, and long range Casimir–Polder type retardation corrections for high–lying states.

NEW VARIATIONAL TECHNIQUES

As pointed out in the introduction, traditional variational calculations are difficult to improve in precision beyond $\pm 10^{-10}$ a.u. Various strategies have been tried in the past to hasten the rate of convergence. For example, Fock[10] showed that an analytic expansion of the wave function about the point $r_1 = 0$, $r_{12} = 0$ contains logarithmic terms and half–integral powers. Recently, Freund et al.[11] obtained ground state energies accurate to a few parts in 10^{13} by including terms of this type in relatively small 230–term wave functions. For the 2 ^{3}P state, Schwartz[12] found a significant improvement in convergence when he included terms of the form $(r_1 + r_2)^{1/2}$, but his final accuracy was limited to about one part in 10^{10} with 439 terms. In addition, he found it necessary to carry 52 decimal digits in the calculations, indicating possible problems of near linear dependence in the basis set. Recent work on the extension of these methods to more highly excited states is described in the article by J. Baker et al. in this volume.

A key ingredient of the new techniques which allow a dramatic improvement in accuracy is to double the basis set so that (1.3) becomes

$$\Psi(\vec{r}_1, \vec{r}_2) = \sum_{i,j,k} [a_{ijk}\, \chi_{ijk}(\alpha_1, \beta_1) + b_{ijk}\, \chi_{ijk}(\alpha_2, \beta_2)] \times (\text{angular function})$$

$$\pm \text{exchange} . \tag{2.1}$$

Here, (angular function) denotes a vector coupled product of solid spherical harmonics for the two electrons to form a state of total angular momentum L, and, as before,

$$\chi_{ijk}(\alpha,\beta) = r_1^i r_2^j r_{12}^k\, e^{-\alpha r_1 - \beta r_2} .$$

Each combination of powers i, j, k is now included twice in (2.1) with different nonlinear parameters α_1, β_1 and α_2, β_2. At first sight, one might think that this would lead to problems of linear dependence, but in fact a complete optimization of the energy with respect to all four nonlinear parameters leads to well–defined and numerically stable values for the parameters, with the two sets being well separated from each other. For the first set of terms in (2.1), the optimum values of α_1 and β_1 are close to their screened hydrogenic values $\alpha_1 \simeq Z$ and $\beta_1 \simeq (Z-1)/n$. These terms describe the asymptotic behavior of the wave function. For the second set of terms the optimum values of α_2 and β_2

are much larger. These terms describe the complex inner correlation effects. The complete optimization therefore has the effect of producing a natural division of the basis set into two sectors with quite different distance scales – an asymptotic sector (Sector A) and an inner correlation sector (Sector B). Recent work by Kono and Hattori[13] also makes use of doubled basis sets, but the lack of complete optimization and other constraints they place on the basis set limits their accuracy to a few parts in 10^{10}.

Nonrelativistic Wave Functions and Energies

We now describe in somewhat greater detail how the basis sets are constructed and the nonlinear parameters optimized. One can take advantage of the near screened hydrogenic nature of the excited states of helium by writing the nonrelativistic Hamiltonian for infinite nuclear mass in the form $H = H_0 + V$ where (in atomic units)

$$H_0 = -\tfrac{1}{2}\nabla_1^2 - \tfrac{1}{2}\nabla_2^2 - \frac{Z}{r_1} - \frac{(Z{-}1)}{r_2} \qquad (2.2)$$

$$V = \frac{1}{r_{12}} - \frac{1}{r_2} \qquad (2.3)$$

and $Z{-}1$ is the screened nuclear charge. Although the above decomposition is unsymmetric in r_1 and r_2, the total Hamiltonian H remains symmetric. One simply interchanges r_1 and r_2 in H_0 and V when operating on the exchange part of the wave function. The advantage gained is that the solutions to the zero order problem

$$H_0 \psi_0(1snl) = E_0 \psi_0(1snl) \qquad (2.4)$$

are known exactly, and the eigenvalues

$$E_0 = -\frac{Z^2}{2} - \frac{(Z{-}1)^2}{2n^2} \qquad (2.5)$$

give correctly the first few figures of the true energy E. It is numerically advantageous to include $\psi_0(1snl)$ in the basis set and to cancel algebraically the E_0 contribution to the matrix elements so that the variational principle applied to $H - E_0$ yields directly the correction to E_0. For example, matrix elements involving ψ_0 are simply

$$\langle \chi_{ijk} | H - E_0 | \psi_0 \rangle = \langle \chi_{ijk} | V | \psi_0 \rangle . \qquad (2.6)$$

The final results would be the same without the subtraction, but the above procedure saves several significant figures in the evaluation of matrix elements, particularly for the more highly excited states.

With the ψ_0 term included, the complete variational trial function becomes

$$\Psi_{tr} = \psi(\vec{r}_1, \vec{r}_2) \pm \psi(\vec{r}_2, \vec{r}_1) \qquad (2.7)$$

with

$$\psi(\vec{r}_1, \vec{r}_2) = a_0 \psi_0(1s, nl)$$

$$+ \sum_{i,j,k} [a_{ijk} \chi_{ijk}(\alpha_1, \beta_1) + b_{ijk} \chi_{ijk}(\alpha_2, \beta_2)] \times (\text{angular function}) \ .$$

$$(2.8)$$

The a_{ijk} terms of (2.8) represent the A Sector and the b_{ijk} terms the B Sector as described earlier. Not all terms need be included in both Sectors such that $i + j + k \leq N$. In fact, one can show with some experimentation that truncations of the form $i \leq N_1$, $k \leq N_2$ in Sector A have little effect on the energies. It is physically reasonable that high powers of r_1 and r_{12} are not important in the asymptotic region, and one can progressively reduce the maximum values of N_1 and N_2 as one goes to more highly excited states. Also the single term χ_{000} should be omitted from the first summation since it is nearly the same as $\psi_0(1s, nl)$ when α_1 and β_1 are close to the screened hydrogenic values. No significant truncation appears to be possible for Sector B without loss of precision, and so all terms are included.

We next describe an efficient scheme for the complete optimization of all four α's and β's. (There may be more than four for states of higher angular momentum where additional types of angular functions are required). Since differentiation of (2.8) with respect to an α or β just brings down a factor of $-r_1$ or $-r_2$, one begins by calculating analytically the derivatives

$$\frac{\partial E}{\partial \alpha_t} = -2 \langle \Psi_{tr} | H - E | r_1 \psi(\vec{r}_1, \vec{r}_2; \alpha_t) \pm r_2 \psi(\vec{r}_2, \vec{r}_1; \alpha_t) \rangle \qquad (2.9)$$

$$\frac{\partial E}{\partial \beta_t} = -2 \langle \Psi_{tr} | H - E | r_2 \psi(\vec{r}_1, \vec{r}_2; \beta_t) \pm r_1 \psi(\vec{r}_2, \vec{r}_1; \beta_t) \rangle \qquad (2.10)$$

$$t = 1, 2$$

with the assumed normalization $\langle \Psi_{tr} | \Psi_{tr} \rangle = 1$. Here $\psi(\vec{r}_1, \vec{r}_2; \alpha_t)$ denotes the terms in (2.8) which depend explicitly on α_t. An important simplification occurs because there is no contribution to the derivatives from the implicit dependence of E on α_t or β_t through the linear coefficients a_{ijk} and b_{ijk}. This follows because the energy is already stationary with respect to first order variations of the linear coefficients. Once the first derivatives are known, the second derivatives can be estimated by changing the α_t's and β_t's in the direction of lower energy and taking differences. Newton's method can then be applied to locate the zero's of the first derivatives. Provided that the initial α's and β's are chosen close to a minimum, the procedure converges in a few iterations. As an indication of the numerical stability, ordinary double precision arithmetic (about 16 decimal digits) is just adequate for the evaluation of derivatives and optimization of the basis sets up to the largest ones used. However as a check, the final calculations of wave functions and energies are done in quadruple precision (about 32 decimal digits). This does not typically change the variational eigenvalues by more than a few parts in 10^{14}.

Mass Polarization Corrections.

To account for mass polarization effects, all of the calculations (including the nonlinear optimization) are repeated with the Hamiltonian (2.2) and (2.3) extended to include the mass polarization operator $(\mu/M)\vec{p}_1 \cdot \vec{p}_2$, where μ is the electron reduced mass and M is the nuclear mass. First order perturbation

Table 2. Variational Eigenvalues for the P States of Helium (a.u.)

N	Number of terms	2 ^{1}P	2 ^{3}P
4	70	-2.1238430609942	-2.1331641578114
5	112	-2.1238430837930	-2.1331641842173
6	168	-2.1238430861444	-2.1331641903258
7	240	-2.1238430864417	-2.1331641906805
8	328	-2.1238308648797	-2.13316419076479
9	432	-2.1238308649648	-2.13316419077680
10	552	-2.1238308649750	-2.13316419077784
11	688	-2.1238308649800	-2.13316419077883
12	840	-2.1238308649807	-2.13316419077905
extrapolation		-2.12384308649808(1)	-2.13316419077910(5)
Lewis and Serafino[15] (455 terms)			-2.1331641814
Schiff et al.[14] (560 terms)		-2.1238430858	-2.1331641905
Kono and Hattori[13] (284 terms)		-2.1238430862	-2.1331641906
Schwartz[12] (439 terms)			-2.133164190626
Morgan and Baker[16] (213 terms)		-2.1238430864025	-2.1331641906735

theory gives the linear dependence of the energy on μ/M, but higher order terms are too large to be neglected at this level of accuracy. The higher order terms could be calculated directly, but it is simpler just to include the mass polarization term in the Hamiltonian from the outset, and extract the higher order dependence on μ/M by differencing.

Numerical Results

As a typical example, the variational eigenvalues for the 1s2p ^{1}P and ^{3}P states are listed in Table 2. It can be seen that the accuracy of the Schiff *et al.*[14] 560–term calculation is achieved with only 240 terms. The results are also substantially more accurate than those of Kono and Hattori[13] for basis sets of comparable size. The extrapolated values are obtained by assuming that ratios of successive differences continue decreasing at a constant rate. The uncertainty of 1 part in 10^{14} for the 2 ^{1}P state and 5 parts in 10^{14} for the 2 ^{3}P state is conservatively estimated to be the entire amount of the extrapolation.

The optimum values of the nonlinear parameters are listed in Table 3. For the smaller basis sets, the optimum values are sharply defined and independent of the values used to start the iterative optimization process. However, for the largest basis sets, the final one or two figures are not significant because the derivatives become very small in the neighborhood of the minimum and cannot be reliably calculated in double precision. This indeterminacy does not affect the accuracy of the final results because the variational eigenvalues become correspondingly insensitive to the precise values of the nonlinear parameters. It is nevertheless important to carry the optimization up to the largest basis sets for two reasons. First, the change is large enough in going from one basis set to the next that it does affect the energy, and may produce a false impression of convergence if the optimization is not done. Second, it is clear from the table that, while α_1 and β_1 remain nearly constant, α_2 and β_2 increase nearly linearly

Table 3. Values of the optimized nonlinear parameters (in units of Z/a_0).

N	α_1	2 ^{1}P β_1	α_2	β_2	α_1	2 ^{3}P β_1	α_2	β_2
				infinite nuclear mass				
4	0.8132	0.4348	1.2691	0.9855	0.8737	0.3957	1.0731	0.9483
5	0.8500	0.4573	1.2170	1.1702	0.9601	0.4843	1.1307	1.0051
6	0.8170	0.4363	1.3242	1.2932	0.8370	0.4095	1.2160	1.3002
7	0.8040	0.4527	1.3872	1.4554	0.7627	0.3597	1.3228	1.2168
8	0.7988	0.4335	1.3679	1.5165	0.8174	0.4081	1.3786	1.4448
9	0.7636	0.4668	1.5936	1.7152	0.8070	0.4282	1.4777	1.6807
10	0.7863	0.4731	1.7324	1.8809	0.8232	0.4327	1.7656	2.0173
11	0.7979	0.4653	1.9190	2.0705	0.8461	0.4369	2.0199	2.2724
12	0.7921	0.4786	2.1403	2.2933	0.8362	0.4290	2.2859	2.5432
				finite nuclear mass				
4	0.8087	0.4336	1.2720	0.9749	0.8707	0.3949	1.0702	0.9498
5	0.8424	0.4573	1.2063	1.1567	0.9577	0.4852	1.1264	0.9955
6	0.7989	0.4278	1.3447	1.2804	0.8273	0.4073	1.2100	1.2959
7	0.7891	0.4581	1.4194	1.3861	0.7499	0.3538	1.3361	1.1840
8	0.7500	0.4111	1.3970	1.4636	0.7664	0.3938	1.3905	1.3922
9	0.7560	0.4715	1.6095	1.6980	0.7626	0.3935	1.4554	1.6332
10	0.7523	0.4685	1.7148	1.8994	0.7649	0.4214	1.7416	1.9917
11	0.7601	0.4707	1.9384	2.0500	0.8178	0.4450	2.0199	2.3185
12	0.7845	0.4980	2.1520	2.3398	0.8210	0.4495	2.4511	2.6489

with N. This behavior ensures that the two sectors of the doubled basis set remain linearly independent, and is in fact necessary to preserve the numerical stability of the wave function.

The optimum nonlinear parameters with mass polarization included differ only slightly from the infinite nuclear mass case (see Table 3) and the eigenvalues show a parallel pattern of convergence. For the P–state example discussed above, the energy shifts due to mass polarization are conveniently expressed in the form

$$\Delta E_{\text{M}}(\text{He } 2\ ^1\text{P}) = 0.0460445247(3)(\mu/M) - 0.168271(1)(\mu/M)^2 \quad (2.11)$$

$$\Delta E_{\text{M}}(\text{He } 2\ ^3\text{P}) = -0.0645724250(1)(\mu/M) - 0.204958(1)(\mu/M)^2 \quad (2.12)$$

in units of $2R_{\text{M}}$, where $R_{\text{M}} = [1 - (\mu/M)]R_\infty$ and μ is the reduced electron mass $mM/(m+M)$. For ^{4}He, $(\mu/M) = 1.370745663 \times 10^{-4}$. The leading coefficient above is the first order perturbation coefficient $\langle \vec{p}_1 \cdot \vec{p}_2 \rangle$ calculated for infinite nuclear mass, and the next coefficient is obtained by subtracting the leading term from the directly calculated total energy shift due to mass polarization. This second order term may therefore contain a small contamination from the next term of order $(\mu/M)^3$. However, the above formulas are certainly adequate to calculate accurate mass polarization shifts for values of μ/M in the neighborhood of the one used for ^{4}He.

As shown by Drachman[17] the order of magnitude of the average of the second order terms in (2.11) and (2.12) can be understood as a kinematical effect as follows. The outer electron is pictured as moving in an approximately hydrogenic orbit about an effective nucleus with charge +e and mass $M+m$ consisting of the actual nucleus of mass M together with the inner electron. This results in a slightly larger effective reduced mass for the outer electron and hence stronger binding.

RELATIVISTIC CORRECTIONS

Corrections of $O(\alpha^2)$.

The relativistic corrections of $O(\alpha^2)$ are calculated by evaluating the matrix elements of the nonrelativistic Pauli form of the Breit interaction H_1, H_2, $\cdots$, H_5 as given by Bethe and Salpeter[18]. Since these terms are well known and have been extensively discussed previously, the detailed formulas will not be repeated here. However, it is important to realize that they contain one–electron contributions from the nonrelativistic reduction of the Dirac equation, as well as two–electron terms from $e^2/r_{12} + B$, where B is the full 16–component Dirac form of the Breit interaction. For S–states, the only terms which contribute are

$$\Delta E_{\text{rel}} = -\frac{\alpha^2}{4}\langle p_1^4 \rangle + \langle H_2 \rangle + \pi\alpha^2 \langle Z\delta(\vec{r}_1) + \delta(\vec{r}_{12})\rangle \qquad (3.1)$$

where H_2 is the orbit–orbit interaction. It arises in part from retardation effects in B.

For states of higher angular momentum, one requires the spin–orbit, spin–other–orbit and spin–spin matrix elements from the Breit interaction terms H_3 and H_5. These are conveniently expressed in terms of reduced matrix elements according to (Edmonds 1960)

$$\langle \gamma'L'S'JM \,|\, H_{\text{so}} \,|\, \gamma LSJM \rangle = (-1)^{L+S'+J} \begin{Bmatrix} J & S' & L' \\ 1 & L & S \end{Bmatrix} \langle \gamma'L'S'\|H_{\text{so}}\|\gamma LS \rangle \qquad (3.2)$$

and similarly for H_{soo}. For P–states, the values of the multiplying factor are $-1/3$, $-1/6$ and $1/6$ for the diagonal elements with $S = S' = 1$ and $J = 0,1,2$ respectively. For the off–diagonal element with $S' = 1$, $S = 0$ and $J = 1$, the factor is $-1/3$. For the spin–spin interaction, which can be written in terms of irreducible tensors of rank 2, the corresponding formula is

$$\langle \gamma'L'S'JM \,|\, H_{\text{ss}} \,|\, \gamma LSJM \rangle = (-1)^{L+S'+J} \begin{Bmatrix} J & S' & L' \\ 2 & L & S \end{Bmatrix} \langle \gamma'L'S'\|H_{\text{ss}}\|\gamma LS \rangle \qquad (3.3)$$

with the P–state multiplying factor being $1/3$, $-1/6$ and $1/30$ for $J = 0,1,2$. The reduced matrix elements are therefore related to the quantities C_Z, C_e and D defined by Schwartz (1964) according to

$$\langle 2\ ^3\text{P}\|H_{\text{so}}\|2\ ^3\text{P} \rangle = 6C_Z$$

$$\langle 2\ ^3\text{P}\|H_{\text{soo}}\|2\ ^3\text{P} \rangle = 6C_e$$

$$\langle 2\ {}^3\mathrm{P}\|H_{\mathrm{ss}}\|2\ {}^3\mathrm{P}\rangle = -10D \ .$$

The diagonal relativistic correction for infinite nuclear mass is then

$$\Delta E_{\mathrm{rel}} = -\tfrac{\alpha^2}{4}\langle p_1^4\rangle + \langle H_2\rangle + \pi\alpha^2\langle Z\delta(\vec{r}_1) + \delta(\vec{r}_{12})\rangle$$
$$+ \langle H_{\mathrm{so}}\rangle + \langle H_{\mathrm{soo}}\rangle + \langle H_{\mathrm{ss}}\rangle \ . \tag{3.4}$$

Relativistic Reduced Mass Corrections.
    The relativistic reduced mass corrections are corrections of $O(\alpha^2 m/M)$ arising from two sources. The first is from a transformation to center–of mass and relative co–ordinates in the Breit interaction itself, together with the mass scaling of these terms as discussed by Stone[20], and Douglas and Kroll[21]. The second can be thought of as a second–order cross term between the Breit interaction and mass polarization corrections to the wave functions. This contribution is straight–forward in principle to evaluate since one need only repeat the calculation of matrix elements with mass polarization included and subtract the results to obtain the higher order correction. However, one important point concerns the calculation of

$$H_1 = -(p_1^4 + p_2^4)/(8m^3c^2) \tag{3.5}$$

when mass polarization effects are included. The usual procedure is to use the fact that the nonrelativistic wave function satisfies

$$\frac{1}{2m}(p_1^2 + p_2^2)\Psi = f\Psi \tag{3.6}$$

with
$$f = E + \frac{Ze^2}{r_1} + \frac{Ze^2}{r_2} - \frac{e^2}{r_{12}} \tag{3.7}$$

to transform $\langle H_1\rangle$ into $\langle H_1'(\infty)\rangle$, where

$$H_1'(\infty) = \frac{1}{4m^3c^2}\Big[p_1^2 p_2^2 - 2m^2 f^2\Big] \tag{3.8}$$

is the form appropriate for infinite nuclear mass. However, if finite mass effects are included, there is an additional contribution. It is convenient first to transform to reduced mass atomic units where distance is measured in units of $a_\mu = (m/\mu)a_0$ and energy in units of $e^2/a_\mu = 2(\mu/m)R_\infty$. Then (3.6) becomes

$$(-\tfrac{1}{2}\nabla_1^2 - \tfrac{1}{2}\nabla_2^2)\Psi = (f + \tfrac{\mu}{M}\nabla_1\cdot\nabla_2)\Psi \tag{3.9}$$

and

$$H_1'(M) = \tfrac{\alpha^2}{4}\Big[\tfrac{\mu}{m}\Big]^3\Big[\nabla_1^2\nabla_2^2 - 2\Big[f + \tfrac{\mu}{M}\nabla_1\cdot\nabla_2\Big]^2\Big]\frac{e^2}{a_\mu} \ . \tag{3.10}$$

If terms quadratic in (μ/M) are neglected, then the additional contribution is

Table 4. Contributions to the matrix elements of $<p_1^4>/4 = <F_0> + (\mu/M)[<F_1> + <G>]$ as defined by equations (3.13) and (3.14) the text (in atomic units).

State	$<F_0>$	$<F_1>$	$<G>$
H$^-$ 1s^2 ^{1}S	0.61563964(3)	−0.00063343(3)	−0.0238770(2)
He 1s^2 ^{1}S	13.5220168(1)	−0.1758280(1)	−0.9082723(1)
1s2s ^{1}S	10.2796689(1)	0.0200726(2)	−0.0797993(1)
1s2s ^{3}S	10.45888519(1)	0.01062391(4)	−0.01558432(1)
1s2p ^{1}P	10.0292513215(1)	0.3393(1)	−0.0767530999(1)
1s2p ^{3}P	9.912093697(4)	−0.5923(1)	0.11975737(1)

$$(\Delta H_1')_M = -\frac{\alpha^2}{2}\left[\frac{\mu}{m}\right]^3 \frac{\mu}{M}(f\nabla_1\cdot\nabla_2 + \nabla_1\cdot\nabla_2 f)\ \frac{e^2}{a_\mu}\ . \tag{3.11}$$

In calculating $\langle(\Delta H_1')_M\rangle$, delta function singularities involving $\delta(\vec{r}_{12})$ can be avoided by letting $\nabla_1\cdot\nabla_2$ operate to the right in the first term of (3.11) and to the left in the second term so that f is just a multiplicative factor. One can show by direct calculation that this yields the same expectation values for the complete matrix element $\langle H_1'(M)\rangle$ as the equivalent form[18]

$$\langle H_1''\rangle = -\frac{\alpha^2}{8}\left[\frac{\mu}{m}\right]^3 [(\nabla_1^2\Psi,\nabla_1^2\Psi) + (\nabla_2^2\Psi,\nabla_2^2\Psi)]\ \frac{e^2}{a_\mu}\ . \tag{3.12}$$

However, (3.12) appears to be much more slowly convergent with basis set size than (3.10), and so (3.10) is the more useful form. As an example, Table 4 gives the expectation values of the operators

$$F = (2f^2 - \nabla_1^2\nabla_2^2)/4 \tag{3.13}$$

and

$$G = (f\nabla_1\cdot\nabla_2 + \nabla_1\cdot\nabla_2 f)/2 \tag{3.14}$$

with $\langle F\rangle$ expressed in the form

$$\langle F\rangle = \langle F_0\rangle + \langle F_1\rangle(\mu/M)\ . \tag{3.15}$$

The first term of (3.15) is the matrix element calculated for infinite nuclear mass, and the second is the change in $\langle F\rangle$ when the mass polarization term is included in the Hamiltonian. Then, in atomic units

$$\langle p_1^4\rangle/4 = \langle F_0\rangle + (\mu/M)[\langle F_1\rangle + \langle G\rangle] \tag{3.16}$$

up to terms linear in (μ/M). The $\langle G\rangle$ term raises the ground state energies of H$^-$ and He by 0.000152 cm^{-1} and 0.001455 cm^{-1} respectively. For the 1s2s ^{1}S

Table 5. Values for various matrix elements required to calculate relativistic and QED corrections to the energy. Each quantity is expressed in the form $\langle T \rangle = \langle T_0 \rangle + \langle T_1 \rangle (\mu/M)$ a.u., where $\langle T_1 \rangle (\mu/M)$ is the change in the matrix element when the mass polarization term $\vec{p}_1 \cdot \vec{p}_2 (\mu/M)$ is included explicitly in the Hamiltonian $H = -(\nabla_1{}^2 + \nabla_2{}^2)/2 - Z/r_1 - Z/r_2 + 1/r_{12}$.

matrix element	$\langle T_0 \rangle$	$\langle T_1 \rangle$
He 1s2p ^{1}P		
$\langle p_1^4 \rangle/4$	10.0292513215(1)	0.2625(1)
$\langle H_2 \rangle/\alpha^2$	$-0.02033047408(2)$	0.1045056(3)
$\pi\langle \delta(\vec{r}_1) \rangle$	4.00362331908(2)	0.12376(3)
$\pi\langle \delta(\vec{r}_{12}) \rangle$	0.002309601(1)	$-0.010853(2)$
$\Delta_2/(\alpha^2 m/M)$	$-16.286503967(1)$	
Q	0.003374498(1)	
He 1s2p ^{3}P		
$\langle p_1^4 \rangle/4$	9.912093697(4)	0.4728(1)
$\langle H_2 \rangle/\alpha^2$	0.03508088684(1)	0.1523726(2)
$\pi\langle \delta(\vec{r}_1) \rangle$	3.954827224(1)	$-0.22525(3)$
$\langle ^3\mathrm{P} \| H_{\mathrm{so}} \| ^3\mathrm{P} \rangle/\alpha^2$	0.207955244(1)	0.69869(1)
$\langle ^3\mathrm{P} \| H_{\mathrm{soo}} \| ^3\mathrm{P} \rangle/\alpha^2$	$-0.3088684536(5)$	$-0.959469(3)$
$\langle ^3\mathrm{P} \| H_{\mathrm{ss}} \| ^3\mathrm{P} \rangle/\alpha^2$	$-0.1351209953(1)$	$-0.3244130(1)$
$\langle ^3\mathrm{P} \| \Delta_1 \| ^3\mathrm{P} \rangle/(\alpha^2 m/M)$	$-0.59732701(1)$	
$\Delta_2/(\alpha^2 m/M)$	$-15.75868217(1)$	
Q	0.00381391791(1)	
He 1s2p ^{3}P $-$ 1s2p ^{1}P		
$\langle ^3\mathrm{P} \| H_{\mathrm{so}} \| ^1\mathrm{P} \rangle/\alpha^2$	0.1073194807(3)	$-0.054970(1)$
$\langle ^3\mathrm{P} \| H_{\mathrm{soo}} \| ^1\mathrm{P} \rangle/\alpha^2$	$-0.03876204224(2)$	$-0.0320412(3)$
$\langle ^3\mathrm{P} \| \Delta_1 \| ^1\mathrm{P} \rangle/(\alpha^2 m/M)$	$-0.07646528(2)$	

and ^{3}S states, the shifts are 0.000128 cm^{-1} and 0.000025 cm^{-1}.

To summarize, the finite nuclear mass corrections can be written in the form

$$\Delta E_{\mathrm{RR}} = (\Delta E_{\mathrm{RR}})_{\mathrm{M}} + (\Delta E_{\mathrm{RR}})_{\mathrm{X}} \qquad (3.17)$$

where

Table 6. Comparison of matrix elements of $\delta(\vec{r}_1)$ for infinite nuclear mass obtained by use of the Hiller *et al.*[22] global operator (3.23), and by direct calculation. The quantity tabulated is $(\pi<\delta(\vec{r}_1)> - 4) \times 10^3$.

	No. of	1s3p ³P		1s3p ¹P	
N	terms	global	direct	global	direct
4	70	−1.293291	−1.290964	0.122234	0.120471
5	110	−1.293440	−1.292690	0.121486	0.120348
6	160	−1.293611	−1.293491	0.121852	0.121968
7	220	−1.293755	−1.294345	0.121982	0.122600
8	290	−1.293663	−1.293936	0.121898	0.122212
9	345	−1.293676	−1.293892	0.121790	0.121645
10	411	−1.293639	−1.293731	0.121820	0.121832
11	489	−1.293634	−1.293672	0.121805	0.121743
12	580	−1.293632	−1.293668	0.121812	0.121796
13	685	−1.293629	−1.293637	0.121811	0.121788
14	805	−1.293630	−1.293646	0.121813	0.121808
Accad *et al.* (1971).		−1.2942		0.1220	

$$(\Delta E_{RR})_M = \Delta_1 + \Delta_2 - (m/M)\{-\frac{3\alpha^2}{4}\langle p_1^4 \rangle + 2[\langle H_2 \rangle + \pi\alpha^2 \langle Z\delta(\vec{r}_1) + \delta(\vec{r}_{12}) \rangle$$
$$+ \langle H_{soo} \rangle + \langle H_{ss} \rangle]\} \tag{3.18}$$

with

$$\Delta_1 = \sum_{k \neq l} \left[\frac{Ze^2}{mMc^2 r_k^3} \right] \vec{r}_k \times \vec{p}_l \cdot \vec{s}_k \tag{3.19}$$

$$\Delta_2 = -\sum_{k,l} \left[\frac{Ze^2}{2mMc^2 r_k^3} \right] [r_k^2 \vec{p}_k \cdot \vec{p}_l + \vec{r}_k \cdot (\vec{r}_k \cdot \vec{p}_k) \vec{p}_l] \tag{3.20}$$

and $(\Delta E_{RR})_X$ is the second–order cross term induced in ΔE_{rel} by mass polarization corrections to the wave function. For the 2 ³P state, the various contributions to $(\Delta E_{RR})_X$ listed in Table 5 agree with the $E_2^{(i,7)}$ $(i=1,\cdots,6)$ calculated by Lewis and Serafino[15], but the results here converge to several more significant figures. The terms Δ_1 and Δ_2 are finite mass corrections to the Breit interaction discussed above.

It is instructive to investigate what happens to the above terms in the one–electron case. Here, Δ_1 does not contribute and Δ_2 reduces to

$$\Delta_2 = -(Z/2mMc^2)r^{-1}(p^2 + p_r^2) \tag{3.21}$$

where $p_r^2 = r^{-2}\vec{r} \cdot (\vec{r} \cdot \vec{p})\vec{p}$. This term, when combined with the other terms in

(3.18), gives the one–electron operator H_b in eq. (42.2) of Bethe and Salpeter[19]. Its expectation value is the well–known relativistic reduced mass shift

$$E_b = - \left[\frac{Z\alpha}{2n}\right]^2 \frac{m}{M}|E_0| \ . \tag{3.22}$$

The values of the various quantities appearing in (3.4) and (3.18) are summarized in Table 5. The uncertainties are estimated from the degree of convergence with basis set size. All of these quantities are substantially more accurate than any previous calculation. For $\delta(\vec{r}_1)$, the expectation values are calculated by means of the more rapidly convergent global operator derived by Hiller et al.[22], modified to take account of the $(\mu/M)\vec{p}_1\cdot\vec{p}_2$ term in the Hamiltonian.[23] The modified operator is

$$\langle \delta(\vec{r}_1)\rangle = \frac{1}{2\pi}\left\langle \frac{Z}{r_1^2} - \frac{1}{r_{12}^2}\frac{\partial r_{12}}{\partial r_1} - \frac{1}{r_1^3}l_1^2 \right.$$
$$+ \frac{\mu}{Mr_1}\left[\nabla_1\cdot\nabla_2 - \frac{1}{r_1^2}\vec{r}_1\cdot(\vec{r}_1\cdot\nabla_1)\nabla_2 - \frac{1}{r_1^2}\vec{r}_1\cdot\nabla_2\right] \right\rangle \tag{3.23}$$

The results in Table 6 for the 1s3p ^{3}P and ^{1}P states illustrate that the global operator (3.23) gives about one more significant figure than a direct evaluation of $\langle\delta(\vec{r}_1)\rangle$.

Anomalous magnetic moment corrections to H_{so}, H_{soo} and H_{ss} are included in the following section, along with other QED terms.

QUANTUM ELECTRODYNAMIC CORRECTIONS

Results for Low–lying States.

As discussed by Araki[24], Kabir and Salpeter[25] and Sucher[26], the two–electron QED energy shift is to lowest order in αZ the one–electron energy shift corrected for the electron density at the nucleus, together with explicit two–electron terms dependent on $\langle\delta(\vec{r}_{12})\rangle$ and Q. In this approximation, the energy shift is

$$\Delta E_L = \Delta E_{L,1} + \Delta E_{L,2} \tag{4.1}$$

where
$$\Delta E_{L,1} = \tfrac{4}{3}Z\alpha^3\{\ln(Z\alpha)^{-2} + \ln[Z^2\mathrm{Ryd}/\epsilon(nLS)] + \tfrac{19}{30}$$
$$+ 3\pi Z\alpha(\tfrac{427}{384} - \tfrac{1}{2}\ln2) + (Z\alpha)^2[-\tfrac{3}{4}\ln^2(Z\alpha)^2 + C_{61}\ln(Z\alpha)^2 + C_{60}]$$
$$+ \tfrac{\alpha}{\pi}0.4042\}\langle\delta(\vec{r}_1) + \delta(\vec{r}_2)\rangle \tag{4.2}$$

$$\Delta E_{L,2} = \alpha^3(\tfrac{14}{3}\ln\alpha + \tfrac{164}{15})\langle\delta(\vec{r}_{12})\rangle - \tfrac{14}{3}\alpha^3 Q \tag{4.3}$$

and
$$Q = \frac{1}{4\pi}\lim_{a\to 0}\langle r_{12}^{-3}(a) + 4\pi(\gamma + \ln a)\delta(\vec{r}_{12})\rangle \ . \tag{4.4}$$
Here, γ is Euler's constant and a is the radius of a sphere centered at $r_{12} = 0$

Table 7. Contributions to the P–state energies of helium (cm^{-1}), using $R_\infty =$ 109737.31569 cm^{-1}, $\alpha^{-1} = 137.03596$, $\mu/M = 1.370745633\times10^{-4}$ and $R_M =$ 109722.273495 cm^{-1}. $\Delta E_M^{(1)}$ and $\Delta E_M^{(2)}$ are the first and second order mass polarization corrections given by equations (2.11) and (2.12), and ΔE_{st} is the singlet–triplet mixing correction.

	$2\,^1P_1$	$2\,^3P_0$	$2\,^3P_1$	$2\,^3P_2$
E_{nr}	−27176.690015	−29222.155521	−29222.155521	−29222.155521
$\Delta E_M^{(1)}$	1.385032	−1.942356	−1.942356	−1.942356
$\Delta E_M^{(2)}$	−0.000694	−0.000845	−0.000845	−0.000845
ΔE_{rel}	−0.467728	1.300852	0.314817	0.237533
ΔE_{anom}	0.0	0.001203	−0.000620	0.000131
ΔE_{st}	0.000158	0.0	−0.000158	0.0
$(\Delta E_{RR})_M$	−0.000284	−0.000014	0.000208	0.000132
$(\Delta E_{RR})_X$	0.000126	0.000593	0.000263	0.000228
ΔE_{nuc}	0.000002	−0.000026	−0.000026	−0.000026
$\Delta E_{L,1}$	0.002009	−0.041087	−0.041095	−0.041111
$\Delta E_{L,2}$	−0.002096	−0.001518	−0.001518	−0.001518
Total	−27175.773489	−29222.838720	−29223.826851	−29223.903353

which is excluded from the integration over r_{12}. The Bethe logarithms can be calculated correct to terms of relative order Z^{-1} from the screened hydrogenic form

$$\ln[\epsilon(nLS)/Z^2\mathrm{Ryd}] = \ln[\epsilon_0(nLS)(Z-\sigma)^2/Z^2\mathrm{Ryd}] \qquad (4.5)$$

where $\ln\epsilon_0(nLS)$ is determined from the hydrogenic Bethe logarithms according to

$$\ln\epsilon_0(nLS) = [\ln\epsilon(1s) + n^{-3}\ln\epsilon(nl)]/(1 + n^{-3}\delta_{l,0}) \ . \qquad (4.6)$$

The hydrogenic Bethe logarithms are tabulated by Klarsfeld and Maquet[27]. The screening constants σ in eq. (4.5) have been calculated by Goldman and Drake[28,29] for the low–lying S– and P–states from the leading terms in the $1/Z$ expansion of the two–electron Bethe logarithm. All of the screening constants turn out to be quite small, indicating that (4.5) may be quite accurate even for neutral helium. As an example, the final results for the P–states are

$$\ln[\epsilon(2\,^1P)/\mathrm{Ryd}] = \ln[19.6952298(Z + 0.00600)^2] \qquad (4.7)$$

$$\ln[\epsilon(2\,^3P)/\mathrm{Ryd}] = \ln[19.6952298(Z + 0.00475)^2] \ . \qquad (4.8)$$

The coefficints C_{61} and C_{60} in eq. (4.2) are state dependent. The values of C_{61}

Table 8.   Comparison of calculated energy levels (in cm^{-1}) with Martin's[5] tabulation of experimental values.   The calculated energies are adjusted by 198310.773489 cm^{-1} to bring the 2 ^{1}P reference state into correspondence.

Term	calculated value	Martin	difference
1 $^1S_0$	0.110688	0.00±0.15	0.110688
2 $^1S_0$	166277.541501	166277.542(3)	−0.0005(30)
2 $^3S_1$	159856.078387	159856.07760(50)	0.00077(50)
2 $^1P_1$	171135.000000	171135.00000(11)	0.00000(11)
2 $^3P_2$	169086.870151	169086.869782	0.00037
2 $^3P_1$	169086.946636	169086.946208	0.00043
2 $^3P_0$	169087.934769	169087.934120	0.00064
3 $^1D_2$	186105.071296	186105.06984(9)	0.00146(9)
3 $^3D_3$	186101.650790	186101.64950(3)	0.00129(3)
3 $^3D_2$	186101.653305	186101.65204(3)	0.00126(3)
3 $^3D_1$	186101.697503	186101.69622(3)	0.00128(3)
4 $^3D_1$	191444.605211	191444.60399(6)	0.00122(6)
5 $^3D_1$	193917.265921	193917.26469(8)	0.00123(8)

for the 1s, 2p$_{1/2}$ and 2p$_{3/2}$ states are 3.964530, 0.429167 and 0.241667 respectively.   In parallel with eq. (4.6) for the Bethe logarithm, the two–electron value for C_{61} is taken to be

$$C_{61}(nLS) = [C_{61}(1s) + n^{-3} C_{61}(nl)]/(1 + n^{-3}\delta_{0,l}) \ . \qquad (4.9)$$

After transforming from jj to LS coupling, this gives the values

$$C_{61}(2\ ^3P_0) = 3.964530, \quad C_{61}(2\ ^3P_1) = 4.010363, \quad C_{61}(2\ ^3P_2) = 3.994738$$

and $C_{61}(2\ ^1P_1) = 4.002551.$

For C_{60}, the values obtained from an equation analagous to (4.9) do not differ significantly from the value −24 for the 1s state[30,31].

In addition to the above, there are anomalous magnetic moment corrections to the spin–dependent operators H_{so}, H_{soo} and H_{ss} discussed in the previous section.   It is convenient to calculate and tabulate these separately because the statistically weighted sum of the corrections vanishes for each manifold of fine structure states.   The shift is[20]

$$\Delta E_{anom} = (\alpha/\pi)(\langle H_{so}\rangle + \tfrac{2}{3}\langle H_{soo}\rangle + \langle H_{ss}\rangle) \ . \qquad (4.10)$$

This replaces the anomalous magnetic moment correction which normally appears in the one–electron Lamb shift for $l \neq 0$.

Finally, the correction due to finite nuclear size is given in lowest order by

$$\Delta E_{\text{nuc}} = \frac{2\pi Z \left(R/a_0\right)^2}{3} \left\langle \delta(\vec{r}_1) + \delta(\vec{r}_2) \right\rangle \qquad (4.11)$$

where R is the root–mean–square radius of the nuclear charge distribution and a_0 is the Bohr radius. The accurate value $R = 1.673$ fm[32] is known from fine structure measurements in muonic helium.. Numerical values for the above corrections are summarized in Table 7 for the P–states. All quantities are given relative to the corresponding term for the He$^+$(1s) state so that its negative is a contribution to the ionization potential.

The final results of the new high precision calculations for the low–lying S– and P–states, and the D–states up to $n = 5$ are summarized in Table 8. For ease of comparison with the tabulations of Martin[5], the calculated energies are re–expressed with the 1s2p ^{1}P state fixed at 171135 cm^{-1} in conformity with his convention. The required constant shift for all the states is 198310.773489 cm^{-1}. Since all quantities have converged to more than the number of figures quoted, the differences shown in the last column of the table must be attributed either to terms not included in the calculation, or to errors in the measurements. For example, it is known from the calculations of Lewis and Serafino[15] that there are spin–dependent corrections of about 0.0004 cm^{-1} to the 1s2p ^{3}P$_J$ fine structure from the second–order mixing effects of the Breit interaction, and other higher order terms. These terms account for the differences between theory and experiment for the 2 ^{3}P$_J$ fine structure. A full derivation of all the $O(\alpha^4)$ spin–independent operators in the two–component nonrelativistic approximation has not yet been carried out, but these terms could reasonably account for the remaining discrepancies among the $n = 2$ states. What is harder to understand is why the discrepancies seem to increase for the higher–lying states, tending to a constant value of about 0.0013 cm^{-1} for the n ^{3}D sequence. The fine structure splittings for these states are well reproduced but there is a further displacement of the 3 ^{1}D$_2$ state relative to 3 ^{3}D$_2$ of 0.0002 cm^{-1}. Since uncertainties arising from uncalculated relativistic and QED effects are much reduced for these states, their positions can be taken as reliable relative to the lower–lying states. It may be that a downward shift as large as 0.0013 cm^{-1} will emerge from future calculations of higher order relativistic and QED corrections for the 2 ^{1}P state.

Relationship to Asymptotic Expansions for Rydberg States.

If one regards a Rydberg atom as a distinguishable outer electron moving in the field of a polarizable core consisting of the inner electron and the nucleus, then the instantaneous Coulomb interactions lead to an effective potential of the form[33]

$$V_{\text{eff}} = -\frac{1}{2} \left[\frac{a_0}{R}\right]^4 \left[\frac{\alpha_d}{a_0^3}\right] + \frac{1}{2} \left[\frac{a_0}{R}\right]^6 \left[\frac{-\alpha_q + 6\beta}{a_0^5}\right] + O(R^{-7})$$

$$+ \text{ finite nuclear mass corrections} \qquad (4.12)$$

in units of e^2/a_0. Here, α_d is the dipole polarizability, α_q is the quadrupole polarizability, β is a nonadiabatic correction etc. All of these instantaneous effects are automatically included in a complete Hylleraas–type variational calculation for a Rydberg state if done to sufficient accuracy.

In studies of retardation effects in the long range interactions of Rydberg electrons with the inner core[34], Au, Feinberg and Sucher[35], and more recently Au[9] have shown that there are important corrections to V_{eff}. The form of the corrections depend on the dominant range of $R \simeq r_2$ values for the outer electron. For $R < a_0/\alpha$, the correction is

$$\Delta V = \frac{\alpha^2}{4}\left[\frac{a_0}{R}\right]^4 - \frac{7\alpha^3}{6\pi}\left[\frac{a_0}{R}\right]^3 + O[\alpha^4(a_0/R)^2] \qquad (4.13)$$

The first term is automatically included in a Hylleraas–type calculation as a correlation correction to matrix elements of the Breit interaction. The second term is identical to the "Q" term in (4.3) and (4.4) if one makes the replacement $r_{12} \to R$ and neglects the $\delta(\vec{r}_{12})$ contact terms. The latter give rise to exponentially decreasing potentials in the asymptotic region, and so do not contribute to a $1/R^n$ type expansion. However, the contact terms may still make a significant contribution in this inner asymptotic range. Also the replacement $r_{12} \to R$ introduces errors of several percent for only moderately excited states. For example, the percentage difference between $\langle 1/r_{12}^3 \rangle$ and $\langle 1/R^3 \rangle$ increases slowly from 4.36% for the 1s3d state to 5.70% for the 1s8d state.

For $R >> a_0/\alpha$, the terms in (4.13) display what Au[9] calls "Casimir behavior" due to the long range retardation effects. Each term changes its R–dependence and the two combine to give the single Kelsey–Spruch correction

$$\Delta V = \frac{11\alpha}{4\pi}\left[\frac{a_0}{R}\right]^5\left[\frac{\alpha_d}{a_0^3}\right] . \qquad (4.14)$$

The difference between this and eq. (4.13) is what one might define as a genuine long range retardation correction. The difference represents effects which would not be included in a straight–forward extension of Hylleraas–type calculations to very high n states, no matter how accurately they are done.

OUTLOOK FOR THE FUTURE

High precision calculations of the kind described here have now been completed for the low–lying S– and P–states[23,36], and the D–states up to $n = 8$.[37] Preliminary results have been obtained for the higher–lying P–states up to $n = 10$ with no significant loss of accuracy. This indicates that there should be no serious problems in extending the D–state calculations up to $n = 10$, and then proceeding on to study the G– and H–states where a direct comparison with the asymptotic calculations of Drachman[8,17] will become meaningful, in addition to very interesting comparisons with Lundeen's high precision measurements.

There is clearly a need for a more thorough study of the $O(\alpha^4 Z^4)$ relativistic corrections. The spin–dependent terms of this order are known in a non-relativistic Pauli form from the work of Douglas and Kroll[21], but no similar analysis exists for the spin–independent parts.

ACKNOWLEDGEMENTS

The author is grateful to the Santa Barbara Institute of Theoretical Physics for its hospitality during the preparation of these notes, and providing the opportunity for extensive discussions with other participants, especially C.

K. Au and J. Sucher concerning the asymptotic expansions for Rydberg states. Research support by the Natural Sciences and Engineering Research Council of Canada is gratefully acknowledged.

1. E. A. Hylleraas, Z. Phys. **48**, 469 (1928).
2. E. A. Hylleraas, Z. Phys. **54**, 347 (1929).
3. E. A. Hylleraas and B Undheim, Z. Phys. **65**, 759 (1930).
4. Y. Accad, C. L. Pekeris and B. Schiff B., Phys. Rev. A **4**, 516 (1971).
5. W. C. Martin, Phys. Rev. A **36**, 3575 (1987).
6. J. W. Farley, K. B. MacAdam and W. H. Wing, Phys. Rev. A **20**, 1754 (1979).
7. E. A. Hessels, W. G. Sturrus, S. R. Lundeen and D. R. Cok, Phys. Rev. A **35**, 4489 (1987).
8. R. J. Drachman, Phys. Rev. A. **31**, 1253 (1985) and earlier references therein
9. C. K. Au, Phys. Rev. A, submitted (1988) and private communication.
10. V. A. Fock, Izv. Akad. Nauk SSSR Ser. Fiz. **18**, 161 (1954).
11. D. E. Freund, B. D. Huxtable and J. D. Morgan III, Phys. Rev. A **29**, 980 (1984).
12. C. Schwartz, Phys. Rev. **134**, A1181 (1964).
13. A. Kono and S. Hattori, Phys. Rev. A **34**, 1727 (1986).
14. B. Schiff, H. Lifson, C. L. Pekeris and P. Rabinowitz, Phys. Rev. **160**, A1104 (1965).
15. M. L. Lewis and P. H. Serafino, Phys. Rev. A **18**, 867 (1978).
16. J. D. Morgan III and J. Baker, University of Deleware, private communication.
17. R. J. Drachman, Phys. Rev. A **37**, 979 (1987).
18. A. R. Edmonds, *Angular Momentum in Quantum Mechanics* (Princeton University Press, Princeton 1960).
19. H. A. Bethe and E. E. Salpeter, *Quantum Mechanics of One– and Two–Electron Atoms* (Springer–Verlag, Berlin, 1957).
20. A. P. Stone, Proc. Phys. Soc. (Lond.) **81**, 868 (1963).
21. M. Douglas and N. M. Kroll, Ann. Phys. (N.Y.) **82**, 89 (1974).
22. J. Hiller, J. Sucher and G. Feinberg, Phys. Rev. A **18**, 2399 (1978).
23. G. W. F. Drake, Nucl. Instrum. and Meth. B **31**, 7 (1988).
24. H. Araki, Prog. Theo. Phys. **17**, 619 (1957).
25. P. K. Kabir and E. E. Salpeter, Phys. Rev. **108**, 1256 (1957).
26. J. Sucher, Phys. Rev. **109**, 1010 (1958).
27. S. Klarsfeld and A. Maquet, Phys. Lett **43B**, 201 (1973).
28. S. P. Goldman and G. W. F. Drake, J. Phys. B **16**, L183 (1983).
29. S. P. Goldman and G. W. F. Drake, J. Phys. B **17**, L197 (1984).
30. P. J. Mohr, Ann. Phys. (N.Y.) **88**, 26 and 521 (1974).
31. J. Sapirstein, Phys. Rev. Lett. **47**, 1723 (1981).
32. E. Borie and G. A. Rinker, Phys. Rev. A **18**, 324 (1978).
33. R. J. Drachman, Phys. Rev. A. **26**, 1228 (1982).
34. E. J. Kelsey and L. Spruch, Phys. Rev. A **18**, 15, 845 and 1055 (1978).
35. C. K. Au, G. Feinberg and J. Sucher, Phys. Rev. Lett. **53**, 1145 (1984).
36. G. W. F. Drake, Phys. Rev. Lett. **59**, 1549 (1987).
37. G. W. F. Drake and A. J. Makowski, J. Opt. Soc. Am., in press (1988).

TIME-ORDERED DIAGRAMMATIC APPROACH TO THE TWO-PHOTON EXCHANGE POTENTIAL: AN EXAMINATION OF THE CASIMIR EFFECTS IN LONG RANGE EM INTERACTIONS

C.K. Au

Physics Department, University of South Carolina, Columbia, SC 29208

ABSTRACT

Electromagnetic interactions between charged or neutral Coulombic systems are mediated by the exchange of virtual photons. In the Coulomb-gauge, the virtual photon can be broken down into a Coulomb photon which is transmitted instantaneously and a transverse photon which propagates at the finite speed of light. The latter is the retarded photon, inclusion of which leads to the Casimir effect in long range electromagnetic interactions. We review the dispersion theory of these dispersion forces and its relation to time ordered diagramatic calculations. Specific examples include the charged-neutral and neutral-neutral systems. Its relevance to other calculations on the energy levels in the electronic fine structures of Rydberg states of helium is also discussed.

1. INTRODUCTION

From the gravitational interaction that governs the motion of heavenly bodies to the electromagnetic interaction that determines the properties of atoms and molecules and to a certain extent affects the properties of nuclei, long range potentials play a very important role in the study of physics. Besides the familiar Newtonian gravitational potential and the Coulombic electrical potential, long range potentials include all that behave asymptotically as some powers of $1/R$, such as the van der Waals $1/R^6$ interaction. While the quantum theory of gravity is still in its infancy stage, the quantum theory of electromagnetism is well formulated. Quantum Electrodynamics (QED), as it is called, gives an almost complete theory of the interaction of charged particles with photons. Electromagnetic interactions between charged particles are mediated by the exchange of photons. Radiative corrections or electromagnetic self interactions occur when a charged particle emits a photon and then reabsorbs the same photon. Single photon exchange between charged particles leads to the Coulomb potential and its relativistic correction the Breit potential. When used in a Schrödinger or a Dirac equation, the effect of this single photon exchange potential will be iterated, to the degree that the equation is successfully solved. Multiphoton exchange effect is usually small because they are down by the additional powers of the coupling constant. When one of the interacting particles is neutral, multi-photon exchange must be considered, because even though the overall neutral particle is made up of charged constituents, the effect of single photon exchange is almost completely suppressed by the cancellation among positive and negative charges, leaving only over all magnetic and short range interactions.

These multiphoton exchange forces between neutral systems are also known as dispersion forces. In his work *Optiks*, Newton[1] raised the interesting question that the very property of matter to reflect, refract and inflect light is also responsible for the force between small entities of matter which in turn is responsible for the great part of the phenomena of nature. Over a century ago J.D. van der Waals,[2] based on experimental observed deviation from the Ideal Gas Law correctly concluded that there exists a weak attractive force between the gas molecules. The first successful attempt to explain the origin of these forces was given by London[3] who showed that these forces could be related to the polarizabilities of the molecules. London's analysis, however, was incomplete because it only included the exchange of instantaneous Coulomb photons. The Coulombic interaction between a pair of netural atoms A and B in the dipole approximation is:

$$V_d = e^2 \left(\vec{r}_A \cdot \vec{r}_B - 3\vec{r}_A \cdot \hat{R}\vec{r}_B \cdot \hat{R} \right) /R^3 \equiv \frac{d_{AB}}{R^3} \tag{1.1}$$

where $\vec{r}_A$ and $\vec{r}_B$ are the atomic electron coordinates from their respective atomic nuclei and $\vec{R}$ is the displacement vector from the mass center of atom A to that of B. For systems without permanent electric dipole moments, the first order energy correction due to V_d vanishes because of either parity or rotational invariance. The second order energy correction does not vanish and gives the London potential:

$$V_L \equiv -C_{AB}/R^6, \tag{1.2}$$

where

$$C_{AB} = \sum_{n_A m_B} \frac{|\langle 0_A 0_B | d_{AB} | n_A m_B \rangle|^2}{E_A^n + E_B^m - E_A^o - E_B^o} \tag{1.3}$$

is the London constant for the pair of atoms A and B.

In the late 1940's, experiments[4] involving colloidal systems appeared to indicate that the intermolecular potential might be weaker than that expressed in the London form. In a classical paper, published in 1948, Casimir and Polder[5] derived a generalization of the London potential by including also the contribution of transverse photon exchange in addition to the pure Coulomb exchange in London's calculation. For most systems, transverse photon exchange leads only to corrections of order v/c, which vanish in the nonrelativistic limit when c becomes infinite. The Casimir-Polder two photon exchange potential assumes the London form for small separation distances of the order of a few atomic radii. However for separation distances larger than α^{-1} times the atomic radii, the Casimir-Polder potential assumes a different power form and asymptotically behaves as:

$$\lim_{R \gg a/\alpha} V_{CP} \sim \alpha_A \alpha_B / R^7. \tag{1.4}$$

Here α is the fine structure constant and α_A and α_B are respectively the dipole polarizabilities of the atoms A and B. In the nonrelativistic limit, α goes to zero and so the London form persists for all distances.

Casimir and Polder obtained their result for the interaction energy between the two atoms A and B by considering the interaction of the electrons in atom B with the retarded vector potential at atom B due to the electronic current in atom A and then adding this retarded interaction energy to that due to the instantaneous electrostatic interaction. In the same paper, Casimir and Polder also considered the interaction between an atom and a conducting wall. They obtained this result by placing the atom at a finite distance R from one wall of a box of length L. They then compared the energy of the quantum mode sums between this situation and that when the atom is placed in the center of the box. They found that as L goes to infinity, while the mode sum in each case diverges, their difference remains finite. Furthermore, the system also exhibits retardation behavior. At short distances, the interaction energy behaves like R^{-3} while at distances large compared to the typical wavelength of the radiation spectrum of the atom, the interaction behaves like R^{-4}. In a NATO Advanced Study Institute Conference on the Physics of Strong Fields and the Casimir Effect held in Maratea, Italy in the summer of 1986, Professor Casimir referred to his method of calculation as the "Poor Men's Renormalization." He also recalled that when he informed Heisenberg of his results, Heisenberg murmured, "It all sounds like quantum fluctuation to me."

Today, we understand quantum fluctuation to be a consequence of the emission and absorption of virtual quanta, whether it is by the same or different particles within the system. The van der Waals interaction is a two virtual photon exchange process. Since the classic work of Casimir and Polder, a number of other calculations of retarded dispersion forces, using a variety of approaches have been carried out. Among these is a phenomenological approach, pioneered by Lifshitz[6] and collaborators, in which the neutral systems are described by an effective frequency dependent dielectric constant. There have also been calculations using covariant Feynman techniques in which the Casimir and Polder result is rederived for all separation distances. In the late 1960's, a dispersion relation approach to the two photon exchange process was developed by Feinberg and Sucher.[7] Besides being model independent, this method treats electric and magnetic effects on equal footing and expresses the retarded dispersion potential in terms of quantities that are in principle experimentally measurable. Furthermore, the covariance of the method allows relativistic corrections to be easily incorporated.

Yet almost 40 years after the fascinating prediction of Casimir and Polder, no direct experimental evidence for the retarded van der Waals potential between a pair of atoms has been found. However, recent advances in precision measurements using lasers have stimulated a lot of interest, both theoretical[8] and experimental,[9] in the study of the Casimir effect in high Rydberg states of helium. Using a time ordered perturbational approach, Kelsey and Spruch obtained the effective potential between the Rydberg electron and the core He^+ ion valid only in the asymptotic region.

Later, Feinberg and Sucher[10] applied the dispersion relation approach they have developed for the retarded dispersion potential between a pair of atoms to that between a charged and a neutral system, valid for all distances larger than a few times the size of the neutral complex. They found that while their result agreed with the Kelsey-Spruch result at large distances, their result is significantly smaller at shorter distances. By identifying the contributions from two Coulomb photon exchange to the effective potential, Au, Feinberg and Sucher[11] obtained the retardation correction to the two photon exchange potential (the Casimir correction potential). This potential was then used to calculate the corrections to the energy levels of high Rydberg states of helium and it was found that it gave a significantly smaller correction than that predicted by the Kelsey-Spruch result.

Since the retardation correction to the energy levels was to be added on to existing nonrelativistic and nonretarded calculations then available, it was then extremely desirable to bridge the gap between the nonrelativistic and nonretarded calculation with the covariant calculation. To this end, an explicit time-ordered diagramatic analysis was carried out in the Coulomb gauge.[12] It was then verified that the two Coulomb exchange graph yields all the terms in the existent nonrelativistc and nonretarded calculations[13,14] attributable to two photon exchange, and indeed the graphs involving transverse photon exchange lead to a correction potential that we first identified in a covariant analysis.[11]

Below, a brief review of the dispersion relation approach and its relation to the time ordered diagrammatic calculation is given without going into too much mathematical detail. The relevance of the present approach to other calculations on the energy levels in the electronic fine structures of Rydberg states of helium is also discussed. The latest experimental results by Lundeen et al. as reported in the April 1988 APS Meeting in Baltimore and also at a seminar here at ITP on May 4, 1988, appear to be in agreement with the retardation correction calculations.

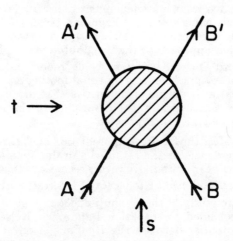

Figure 1: Feynman diagram for the scattering of two particles A and B with indication for the s and t channels.

2. REVIEW OF DISPERSION RELATION APPROACH TO RETARDED DISPERSION FORCES

In this approach, pioneered by Feinberg and Sucher,[7] the effective potential in a two-body interaction is obtained as the Fourier transform of the two-body scattering amplitude with respect to the momentum transfer $\vec{Q}$ at threshold energy. The scattering amplitude is written as a function of the Mandelstam variables s and t and a dispersion relation is written in the t-channel for the atom A – atom B scattering amplitude [Fig. 1]. Here, s is the square of the center of mass energy and t is the square of the 4-momentum transfer. For elastic scattering, t is simply $-|\vec{Q}|^2$. Thus we have:

$$F_{2\gamma}(s,t) = \frac{1}{\pi} \int_{t_0}^{\infty} \frac{\rho_{2\gamma}(s,t')dt'}{t'-t} + \text{left hand cut contribution,} \qquad (2.1)$$

and

$$V_{2\gamma}(R) = \frac{1}{4m_A m_B} \int \frac{d\vec{Q}}{(2\pi)^3} e^{i\vec{Q}\cdot\vec{R}} \int_0^{\infty} \frac{dt'}{\pi} \frac{\rho_{2\gamma}(s_0,t')}{t'-t}, \qquad (2.2)$$

$$= \frac{1}{16\pi^2 m_A m_B R} \int_0^{\infty} dt \, \rho_{2\gamma}(s_0,t) \exp(-\sqrt{t}R). \qquad (2.3)$$

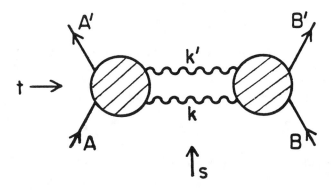

Figure 2: Feynman diagram for the scattering of particles A and B via 2 photon exchange.

In Eq. (2.2), a normalization factor has been included so that the first Born amplitude would coincide with $-F_{2\gamma}/(8\pi\sqrt{s_0})$ where $\sqrt{s_0} = (m_A + m_B)$. We have only retained the contribution from the right hand cut in Eq. (2.1) because in the two photon exchange channel, the threshold starts at $t_0 = 0$ which leads to a long range potential according to Eq. (2.3) whereas the left hand cut has a threshold starting at around $(-4m_A m_B)$ thereby leading to a very short ranged potential. From the structure of Eq. (2.3), it is clear that $V_{2\gamma}$ will be known if the spectral function $\rho_{2\gamma}$ is determined. According to dispersion relation theories, the spectral function is simply given by the discontinuity of the scattering amplitude across the t-cut. One can apply standard Feynman rules to write down the two photon exchange scattering amplitude [Fig. 2]:

$$(-i)F_{2\gamma}(s,t) = \frac{1}{2}(-i)^2 \int \frac{d^4k d^4k'}{(2\pi)^4} \delta(Q - k - k') \frac{(-i)}{k^2} \frac{(-i)}{k'^2} (\Gamma^A : \Gamma^B) \quad (2.4)$$

where Q, k and k' are the 4-momentum transfer and the 4-momenta of the virtual photons and $\Gamma^A_{\mu\nu}$ is the two photon emission amplitude for the atom A and $\Gamma^A : \Gamma^B$ indicates a contraction between the tensorial amplitude. Then according to Cutkosky's theorem on generalized unitarity[15] the discontinuity across the t-cut is simply given by the scattering amplitude if one puts the intermediate particles in the t-channel on the mass shell. Mathematically, this is equivalent to making the following replacements:

$$\frac{1}{k^2} \rightarrow -2\pi i \delta(k^2)\theta(k_0), \quad (2.5a)$$

and

$$\frac{1}{k'^2} \rightarrow -2\pi i \delta(k'^2)\theta(k'_0). \quad (2.5b)$$

An extremely interesting consequence of this maneuver is that Eq. (2.3) can now be expressed in terms of on-shell photon emission amplitudes, even though it is necessary to know these amplitudes in the unphysical region. Upon employing the replacements (2.5) in Eq. (2.4), we have:

$$\rho_{2\gamma}(t) = N \int d\Phi \, \Gamma^A : \Gamma^B \quad (2.6)$$

where N is a normalization constant, $d\Phi$ is the two virtual photon phase space factor. A further simplification can now be made since $\Gamma^A_{\mu\nu}$ and $\Gamma^B_{\mu\nu}$ are now the on-shell two photon emission amplitudes by atoms that are themselves on the mass shell. This enables us to write

$$\Gamma^A_{\mu\nu} = \sum_X F^A_X T^A_{X;\mu\nu} \quad (2.7)$$

where $X = E$ or M for electric and magnetic, F^A_X is the electric or magnetic form factor and $T^A_{X;\mu\nu}$ is the $\mu\nu$ component of the tensor T^A_X constructed from the characteristic vectors in the two photon emission process by the atom A.

A relation similar to Eq. (2.7) can be written down for atom B. On the assumption that the two photon emission amplitude is itself an analytic function in terms of its own Mandelstam variables within its region of analyticity, the two photon emission amplitude can then be obtained from the elastic photon scattering amplitude by taking advantage of the crossing relations.

Here, I would like to digress a little to point out that for elastic photon scattering off a heavy atom, the amplitude is

$$M = \Gamma_{\mu\nu}\epsilon_\mu\epsilon'_\nu$$

$$= \frac{-\omega^2}{8\pi^2}\epsilon_i\epsilon'_j \left[F_E\delta_{ij} - (\hat{k}'_i\hat{k}_j - \delta_{ij})F_M \right] \qquad (2.8)$$

in a gauge where the time component of the photon polarization vector is zero and ω is simply the photon energy. For a heavy atom in a rotationally invariant state, when spin is neglected, the form factors F_E and F_M are not even independent. This follows trivially from an examination of the Kramer-Heisenberg-Waller matrix element one can write down for the Rayleigh scattering amplitude:[16]

$$M = \epsilon_i\epsilon'_j \sum_n \langle 0|p_i e^{-i\vec{k}\cdot\vec{r}}|n\rangle \frac{1}{E_0 - E_n \pm \omega}\langle n|p_j e^{i\vec{k}'\cdot\vec{r}}|0\rangle. \qquad (2.9)$$

$$= \epsilon_i\epsilon'_j \frac{\partial}{\partial k_i}\frac{\partial}{\partial k'_j} \int\int d\vec{p}_1 d\vec{p}_2 F\left[(\vec{p}_1 - \vec{k})^2\right] G\, F\left[(\vec{p}_2 - \vec{k}')^2\right] \qquad (2.10)$$

$$= \frac{1}{2}J_t\vec{\epsilon}\cdot\vec{\epsilon}' + J_{tt}(\vec{\epsilon}'\cdot\vec{k})(\vec{\epsilon}\cdot\vec{k}') \qquad (2.11)$$

where $J = J(t, \vec{k}\cdot\vec{k}, \vec{k}'\cdot\vec{k}', \omega)$ is the integral in Eq. (2.9) and a subscript t indicates a partial derivative with respect to $t \equiv -2\omega^2(1 - \cos\theta)$ and $\cos\theta \equiv \hat{k}\cdot\hat{k}'$. In Eq. (2.10), F is the atomic wave function in momentum space and rotational invariance is asserted by saying that the wave function depends only on the momentum magnitude and G is the atomic Green's function. The factors $e^{-i\vec{k}\cdot\vec{r}}$ and $e^{i\vec{k}'\cdot\vec{r}}$ displace the atomic wave function in momentum space by amounts $\vec{k}$ and $\vec{k}'$ respectively. On comparing Eqs. (2.8) and (2.11) we see how the two form factors are related. The magnetic form factor thus obtained accounts for the diamagnetic part.[16] Of course, when spin is included, there is a paramagnetic part and F_E and F_M are indeed independent.

On using Eq. (2.7) in Eq. (2.6), one gets:

$$\rho_{2\gamma}(t) = N \int \sum_{X,Y} d\Phi_{XY}\, F_X^A F_Y^B \qquad (2.12)$$

where

$$d\Phi_{XY} \equiv d\Phi\, T_X^A : T_Y^B \qquad (2.13)$$

172

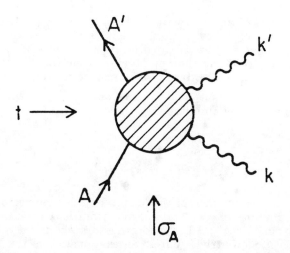

Figure 3: When the graph in Figure 2 is cut open, each side appears as a Compton scattering process. The left hand side is depicted here with the Compton channel designated as the σ-channel.

and the colon again indicates a contraction. We shall refer the photon-atom scattering channel as the σ-channel [Fig. 3] and a dispersion relation is then written in the variable σ:

$$F_X^A(\sigma_A, t) = \int_{m_A^2}^{\infty} \frac{d\sigma'}{\pi} \rho_X^A(\sigma', t) \left[\frac{1}{\sigma' - \sigma_A} + \frac{1}{\sigma' - \bar{\sigma}_A} \right] \qquad (2.14)$$

where the two terms inside the square bracket refer to the right hand and left hand cut contributions. A similar relation is written for F_Y^B. Then upon using Eq. (2.14) in Eq. (2.12) and then in Eq. (2.3), we finally have:

$$V_{2\gamma}(R) = \sum_{X,Y} \int_0^{\infty} \frac{dk_A}{\pi} \int_0^{\infty} \frac{dk_B}{\pi} \rho_X^A(k_A, t) \rho_Y^B(k_B, t) U_{XY}(k_A, k_B; R) \qquad (2.15)$$

where

$$k_A \equiv \frac{\sigma_A' - m_A^2}{2m_A} \qquad (2.16)$$

and a similar definition holds for k_B. We have grouped the rest of the complex integral in the factor U_{XY}:

$$U_{XY} \sim \frac{1}{R} \int dt \, e^{-\sqrt{t}R} \int d\Phi_{XY} \phi(k_A, k_B, t) \qquad (2.17)$$

where ϕ is a function of the kinematic factors. Within its region of analyticity, the spectral function $\rho_X(k,t)$ can be expanded as a series in t:[17]

$$\rho_X(k,t) = \sum_{N=0}^{\infty} \rho_{X,N}(k)t^N. \qquad (2.18)$$

Using this in Eq. (2.15) we finally get:

$$V_{2\gamma}(R) = \sum_{X,Y} \sum_{N,M} \frac{d^{2(N+M)}}{dR^{2(N+M)}} \int dk_A \int dk_B \rho_{X,N}^A(k_A)\rho_{Y,M}^B(k_B) \qquad (2.19)$$

$$U_{XY}(k_A,k_B;R).$$

It turns out that the expansion index N in Eq. (2.18) signifies the order of the multipole contribution to the dispersion potential starting with $N = 0$ being the dipole contribution.[17] In a few instances, the spectral function $\rho_X(k,t)$ can be directly calculated. These include the case of the nonrelativistic Rayleigh scattering by a hydrogen atom ignoring spin, the Compton scattering by a point charge of spin 0 or 1/2. In cases where it cannot be calculated directly, on the assumption that the electric effect dominates over the magnetic effect, $\rho_{E,0}(k)$ can be obtained from the Optical Theorem since it is essentially the imaginary part of the forward scattering amplitude which is in turn related to the total photon scattering cross section. The evaluation of the factor U_{XY} depends on the masses of the particles A and B in question[7,10] since in a practical

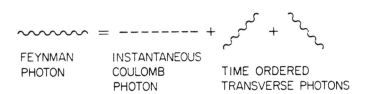

FEYNMAN   INSTANTANEOUS
PHOTON   COULOMB
             PHOTON   TIME ORDERED
                                TRANSVERSE PHOTONS

Figure 4: One Feynman photon manifests as one instantaneous Coulomb photon plus two transverse photons propagating in opposite direction with the finite speed of light in a time-ordered Coulomb gauge analysis. The wavy line indicates a transverse photon and the dashed line indicates a Coulomb photon.

calculation of Eq. (2.17), ϕ is expanded in terms of $\sqrt{t}/m, k/\sqrt{t}, t/(km)$, etc. and the order to which such expansion be kept critically depends on the mass

m. I refer the readers to the original papers of Feinberg and Sucher[7,10)] in the evaluation of the factor U_{XY} when both A and B are massive atoms and when one of them is a light charged particle such as an electron and the other is a heavy atom. Once U_{XY} is evaluated, the calculation of $V_{2\gamma}$ from Eq. (2.19) is rather straight forward.

3. THE TIME ORDERED DIAGRAMMATIC APPROACH TO THE THEORY OF RETARDED DISPERSION FORCES

While the dispersion relation approach is a powerful one and can be modified to accommodate many different systems, it is still worthwhile to reexamine the theory of dispersion forces from a time-ordered diagrammatic analysis in a specific noncovariant gauge like the Coulomb gauge.[12)] The advantage is that this provides a bridge between the powerful covariant calculation and other calculations that have included the Coulomb and maybe even the Breit interaction to any desired order. However, there are certain contributions in the multiphoton exchange potential that are not reproducible by iterating the Breit and/or the Coulomb potential. A time-ordered diagrammatic analysis allows each process to be examined independently, and if desired, a corresponding effective potential can be obtained. The disadvantage lies in the complexity of the calculation and that an atomic model is to be used in the outset. Also, the case of the dispersion force between a pair of atoms has to be treated quite differently from the case of the dispersion force between a light charged particle and an atom. The complexity arises because one covariant Feynman photon is equivalent to the sum of one instantaneous (Coulomb) photon plus two transverse photons propagating in opposite directions with the finite speed of light [Fig. 4]. We shall only discuss the dispersion force in electron atom scattering. It is interesting to point out that while the retardation (Casimir) effect in dispersion forces is a relativistic QED effect in the sense that such effect would vanish if the velocity of light is infinite, the essence of the Casimir effect will be obtained by treating the electrons nonrelativistically and the photon relativistically in a second quantised scheme. Even more surprisingly, some of the elements of a relativistic electron theory, such as the creation and annihilation of virtual electron and positron pairs, are contained in such a treatment where the nonrelativistic electron is coupled to the photon via the minimal coupling scheme. Natural units with $\hbar = c = 1$ will be used. $\alpha = e^2$ is the fine structure constant and a is the Bohr radius.

In this approach, the T-matrix element corresponding to each time ordered diagram is written and the effective potential is obtained as the Fourier transform of the on-shell T-matrix element in the momentum transfer $\vec{Q}$. For simplicity, we consider the atom to be infinitely massive and let $\vec{p}$ and $\vec{p}'$ denote the incident and emergent momenta of the scattering electron, and let $\vec{Q} \equiv \vec{p}' - \vec{p}$ be the momentum transfer [Fig. 5]. Rotational invaraince demands that the T-matrix element be a function of scalars formed out of the vectors $\vec{p}$ and $\vec{Q}$, namely, p^2, Q^2 and $\vec{p} \cdot \vec{Q}$. Off the energy shell, all these three scalars are independent. On the energy shell, because $p^2 = p'^2$ and so we have

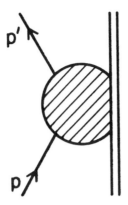

Figure 5: The scattering of an electron from momentum $\vec{p}$ to $\vec{p}'$ by an infinitely massive atom. The double line indicates the atom.

$\vec{p} \cdot \vec{Q} = -Q^2/2$. If the effective potential is obtained as the Fourier transform of the off-energy-shell T-matrix element, then momentum dependent terms occur as mathematical artefacts in the effective potential. When the on-shell condition is duly imposed, these pathological terms disappear. The nice feature is that it is not necessary to carry the inverse Fourier transform of the momentum dependent effective potential in its entirety before the on-shell constraint is imposed. Below, in Sec. 3.1, I give the essence of the mathematical prescription[18] to obtain the effective potential, including the nonadiabatic corrections, from the T-matrix. Then in Secs. 3.2–3.4, I give the effective potentials in the two-Coulomb, one-Coulomb-one-transverse, and two-transverse, photon exchange channels respectively. The mathematical prescription described in Sec. 3.1 has been used whenever appropriate.

3.1 THE EFFECTIVE POTENTIAL: GENERAL PROPERTIES AND NONADIABATIC CORRECTIONS

The effective potential V is related to the scattering T-matrix amplitude M by the Fourier transform in the momentum transfer $\mathbf{Q}$ with respect to the position vector $\mathbf{R}$:

$$V(R, p^2) \equiv \int \frac{d^3 Q}{(2\pi)^3} e^{i\mathbf{Q} \cdot \mathbf{R}} M(Q^2, p^2). \qquad (3.1)$$

The amplitude M depends on the structure of the target. Equation (3.1) clearly indicates that the effective potential should not have momentum dependent terms. In a practical computation of the scattering amplitude, various approximation procedures, for example, a multipole expansion, are taken. To ensure that the subsequent scattering amplitude remains finite afer taking these procedures, various cutoffs need to be introduced. However, if one is interested only in the effective potential as defined by Eq. (3.1), such cutoffs may be avoided by exchanging the orders of integration. The price one pays for this is that the Fourier transform in Q in Eq. (3.1) is taken before the completion of the evaluation of the scattering amplitude. In this event, one misses replacing $\mathbf{p} \cdot \mathbf{Q}$ by $-Q^2/2$ and the scattering amplitude is left off the energy shell and manifests itself as a function $\bar{M}$ of the scalar quantities p^2, Q^2 and $\mathbf{p} \cdot \mathbf{Q}$. Consequently, the effective potential $\bar{V}$, obtained as the Fourier transform of $\bar{M}$, becomes a function of p^2, R and $i\mathbf{p} \cdot \mathbf{R}$. Besides being ambiguous in the ordering of $\mathbf{p}$ and $\mathbf{R}$ since these variables do not commute, these $(i\mathbf{p} \cdot \mathbf{R})$-dependent terms lead to non-Hermitian terms in the effective potential. It is possible to eliminate these non-Hermitian terms by a suitable transformation,[12,19] leading to effective local potentials. To see this from a slightly different angle, let us examine the term that goes like $i\mathbf{p} \cdot \mathbf{R}f(R)$. This resembles the coupling to a vector potential having a radial component only and whose dependence is entirely radial. Hence this vector potential seems not to lead to any physical magnetic field and such terms in the effective potential can be "gauged" away by a suitable gauge transformation. However, because of the factor "i", the gauge function one needs to use is purely imaginary and the subsequent gauge transformation thus has a physical effect on the scattering wave function. Indeed, such a gauge transformation with a purely imaginary gauge function is equivalent to the scattering wave function transformation that Norcross[19] employs to eliminate these non-Hermitian terms in the effective potential. However, such "gauge" transformations cannot be used to convert into local forms terms that depend on higher powers of $(i\mathbf{p} \cdot \mathbf{R})$.

These $(i\mathbf{p} \cdot \mathbf{R})$-dependent terms occur when one considers the nonadiabatic contributions to the effective potential, whose origin lies in the finite mass of the projectile. This makes $\mathbf{R}$, the position vector of the projectile relative to the target, a dynamical variable, not just a parameter. So is the momentum conjugate $\mathbf{p}$. Hence any projectile recoil effect and any nonadiabatic term have the same origin, and one must not emphasize one aspect while partially or completely neglecting the other.

The momentum dependent terms and the energy (p^2)-dependent terms are very different in origin since $\mathbf{p} \cdot \mathbf{Q}$ can be identified with $-Q^2/2$ and Q^2 and p^2 are independent scalar variables of the scattering amplitude. We have given a brief argument above why the $(i\mathbf{p} \cdot \mathbf{R})$-dependent terms necessarily have a physical effect on the wave function irrespective of the energy. At this point, we wish to emphasize that while all the p^2 and higher powers of (p^2) dependent terms can be neglected at threshold energy, none of the $(i\mathbf{p} \cdot \mathbf{R})$-dependent terms which appears in the off-shell potential $\bar{V}$ can be neglected since threshold energy does not imply $\vec{p} = 0$ when off the energy shell; they must be suitably converted into local forms. The mathematical prescription for such a procedure is given shortly. Slightly less general procedures, developed explicitly for the first and second nonadiabatic corrections, have been given previously.[12]

In pursuing an effective interaction potential in the study of the scattering of a projectile by a target atom, one has, from the outset, classified the relevant variables into two groups: the atomic electron coordinates as the fast variables and the relative coordinates from the target to the projectile as the slow variables. For such an effective potential to be meaningful, one is necessarily confined to the low projectile energy region. We begin by assuming that the off-shell scattering amplitude $\bar{M}$ can be expressed as a power series in $\mathbf{p} \cdot \mathbf{Q}$ and p^2:

$$\bar{M}(Q^2, \mathbf{p} \cdot \mathbf{Q}, p^2) = \sum_{k,j} M_{kj}(Q^2)(\mathbf{p} \cdot \mathbf{Q})^k p^{2j} \tag{3.2}$$

which on the energy shell reduces to:

$$M(Q^2, p^2) = \sum_{k,j} M_{kj}(Q^2)(-Q^2/2)^k p^{2j}. \tag{3.3}$$

The truly local effective interaction potential is the Fourier transform of the on-shell scattering amplitude M in the variable $\mathbf{Q}$ with respect to $\mathbf{R}$:

$$
\begin{aligned}
V(R, p^2) &\equiv \int \frac{d^3Q}{(2\pi)^3} e^{i\mathbf{Q}\cdot\mathbf{R}} M(Q^2, p^2) \\
&= \sum_j p^{2j} \sum_k \left(\frac{\nabla^2}{2}\right)^k F_{kj}(R) \\
&= \sum_j p^{2j} \sum_k \left(\frac{1}{2R^2}\frac{\partial}{\partial R}R^2\frac{\partial}{\partial R}\right)^k F_{kj}(R)
\end{aligned}
\tag{3.4}
$$

where

$$F_{kj}(R) \equiv \int \frac{d^3Q}{(2\pi)^3} e^{i\mathbf{Q}\cdot\mathbf{R}} M_{kj}(Q^2) \tag{3.5}$$

is a function of $R = |\mathbf{R}|$ only. The effective potential as it appears in Eqs. (3.4) and (3.5) has energy dependence but no momentum dependence. At threshold energy $p^2 = 0$ and only the $j = 0$ terms survive in the double sum in Eq. (3.4):

$$V(R, p^2 = 0) = \sum_k \left(\frac{1}{2R^2}\frac{\partial}{\partial R}R^2\frac{\partial}{\partial R}\right)^k F_{ko}(R). \tag{3.6}$$

The $k = 0$ term in Eq. (3.6) reproduces the adiabatic potential and the $k \neq 0$, $j = 0$ terms give the nonadiabatic corrections to the effective potential at threshold energy. Thus, a very convenient interpretation is that k is the nonadiabatic correction index and j is the energy dependence index.

If one had taken an effective potential as the Fourier transform of the off-shell scattering amplitude $\bar{M}$ as given in Eq. (3.1), the off-shell effective potential $\bar{V}$ would have an $(i\mathbf{p} \cdot \mathbf{R})$ dependence:

$$\bar{V}(R, \mathbf{p} \cdot \mathbf{R}, p^2) \equiv \int \frac{d^3Q}{(2\pi)^3} e^{i\mathbf{Q}\cdot\mathbf{R}} \bar{M}(Q^2, \mathbf{p} \cdot \mathbf{Q}, p^2). \tag{3.7}$$

$$= \sum_j p^{2j} \sum_k \int \frac{d^3Q}{(2\pi)^3} e^{i\mathbf{Q}\cdot\mathbf{R}} M_{kj}(Q^2)(\mathbf{p} \cdot \mathbf{Q})^k. \tag{3.8}$$

A tedious but staightforward computation leads to:

$$\int \frac{d^3Q}{(2\pi)^3} e^{i\mathbf{Q}\cdot\mathbf{R}} M_{kj}(Q^2)(\mathbf{p} \cdot \mathbf{Q})^k$$

$$= \sum_{m=0}^{[k/2]} (-i\mathbf{p} \cdot \mathbf{R})^{k-2m} p^{2m} C_{2m}^k \frac{(2m+1)!!}{(2m+1)} (-i)^{2m} \left(\frac{1}{R} \frac{\partial}{\partial R} \right)^{k-m} F_{kj} \tag{3.9}$$

where

$$[k/2] = \begin{cases} k/2 & \text{if } k \text{ is even} \\ (k-1)/2 & \text{if } k \text{ is odd} \end{cases} \tag{3.10}$$

and

$$C_{2m}^k = \frac{k!}{(2m)!(k-2m)!} \tag{3.11}$$

is the usual combinatorics factor. Because this combinatoric factor is automatically zero when $m > [k/2]$, the upper limit in the summation in Eq. (3.9) can be removed. Then Eq. (3.8), in conjuction with Eq. (3.9), can be written as:

$$\bar{V}(R, \mathbf{p} \cdot \mathbf{R}, p^2) = \bar{V}(\mathbf{R}, \mathbf{p}) \tag{3.12}$$

$$= \sum_{j,k,m} p^{2(j+m)} (-i\mathbf{p} \cdot \mathbf{R})^{k-2m} C_{2m}^k \frac{(2m+1)!!}{(2m+1)} (-i)^{2m} \left(\frac{1}{R} \frac{\partial}{\partial R} \right)^{k-m} F_{kj}(R).$$

We now proceed to show how all the F_{kj}'s can be obtained from a knowledge of $\bar{V}(R, \mathbf{p} \cdot \mathbf{R}, p^2)$ as a power series expansion in $(-i\mathbf{p} \cdot \mathbf{R})$ and p^2. We have:

$$\bar{V}(R, \mathbf{p} \cdot \mathbf{R}, p^2) = \sum_{k,j} \bar{V}_{kj}(R)(-i\mathbf{p} \cdot \mathbf{R})^k p^{2j}. \tag{3.13}$$

For convenience, we define the differential operator

$$D \equiv \frac{1}{R} \frac{d}{dR}, \tag{3.14}$$

and its inverse $I \equiv D^{-1}$ such that for any function $f(R)$

$$I f(R) \equiv \int_\infty^R x \, f(x) \, dx. \tag{3.15}$$

Then on comparing Eqs. (3.12) and (3.13), we obtain:

$$\bar{V}_{kj}(R) = \sum_{m=0}^{j} C_{2m}^{k+2m} \frac{(2m+1)!!}{(2m+1)} (-i)^{2m} D^{k+m} F_{k+2m,j-m}(R). \tag{3.16}$$

In general, Eq. (3.16) can be inverted in sequence and a general solution can be written in recurrent form:

$$F_{kj} = I^k \bar{V}_{kj} - \sum_{m=1}^{j} C_{2m}^{k+2m} \frac{(2m+1)!!}{(2m+1)} (-i)^{2m} D^m F_{k+2m,j-m} \tag{3.17}$$

where we have suppressed the functional dependence on R of the F_{kj}'s and $\bar{V}_{kj}$'s without fear of confusion. It is clear that by substituting Eq. (3.17) into itself, F_{kj} can be expressed in terms of $\bar{V}_{k+2m,j-m}$ for $m = 0, 1, \ldots j$. We write down the solutions explicitly for $j = 0, 1$ and 2:

$$F_{ko}(R) = I^k \bar{V}_{ko}(R) \tag{3.18}$$

$$F_{kl}(R) = I^k \bar{V}_{kl}(R) + C_2^{k+2} I^{k+1} \bar{V}_{k+2,0}(R), \tag{3.19}$$

and

$$F_{k2} = I^k \bar{V}_{k2} + C_2^{k+2} I^{k+1} \bar{V}_{k+2,1} + \left[C_2^{k+2} C_2^{k+2} - 3 C_2^{k+4} \right] I^{k+2} \bar{V}_{k+4,0}. \tag{3.20}$$

Then the energy dependent effective potential in a truly local form, when expanded as a power series in p^2 (the collision energy) is:

$$V(R, p^2) = \sum_j p^{2j} V_j(R) \tag{3.21}$$

with $V_j(R)$ given by

$$V_j(R) = \sum_k \left(\frac{\nabla^2}{2} \right)^k F_{kj}(R) \equiv \sum_k V_{kj}(R) \tag{3.22}$$

according to Eq. (3.4). The function F_{kj} is determined by Eq. (3.17) where the function $\bar{V}_{kj}$ is in turn determined by the power series in Eq. (3.13). Here V_{kj} is the k-th order nonadiabatic correction to the effective potential jth order in the collision energy (in p^{2j}). In particular, at threshold energy, the effective potential is given by $V_0(R) = \sum_k V_{ko}(R)$, that is, all the V_{ko} terms survive, not just V_{oo} (which is identical to $\bar{V}_{oo}$). Thus we have, from either Eq. (3.22) or Eq. (3.6):

$$V(R, p^2 = 0) = V_0(R) = \sum_k \left(\frac{\nabla^2}{2} \right)^k I^k \bar{V}_{ko}(R). \tag{3.23}$$

The operator I, when acted on by the Laplacian operator ∇^2, obeys the equation:

$$\nabla^2 R^n I^k = n(n+1)R^{n-2}I^k + (3+2n)R^n I^{k-1} + R^{n+2}I^{k-2}. \qquad (3.24)$$

This equation can be used iteratively to generate $(\nabla^2)^k I^k$ needed in Eq. (3.23). As an illustration, we write down the explicit results for the first few cases:

$$\nabla^2 I^k = 3I^{k-1} + R^2 I^{k-2}, \qquad (3.25)$$
$$\nabla^2\nabla^2 I^k = 15I^{k-2} + 10R^2 I^{k-3} + R^4 I^{k-4}, \qquad (3.26)$$

and

$$(\nabla^2)^3 I^k = 105I^{k-3} + 105R^2 I^{k-4} + 21R^4 I^{k-5} + R^6 I^{k-6}. \qquad (3.27)$$

This will enable us to obtain nonadiabatic corrections up to the third order. In general $(\nabla^2)^k I^k$ has the form $\sum_{j=0}^{k} C_j I^{-j}$ where the coefficients C_j's can be determined from the repeated applications of Eq. (3.24). We remind our readers that

$$I^{-N} = (I^{-1})^N = D^N = \left(\frac{1}{R}\frac{d}{dR}\right)^N \qquad (3.28)$$

and the operator I is defined in Eq. (3.15).

We conclude this subsection by pointing out that in the usual cases where an expansion in nonadiabatic corrections and collisional energy dependence is valid, the effective potential $V_{kj}(R)$ usually decreases as the order of k and j increase. But there exist exceptions where these corrections become important compared to $V_{oo}(R)$, even at threshold energy, thus making the effective potential $V_o(R) = \sum_k V_{ko}(R)$ drastically different from $V_{oo}(R)$. This corresponds to a breakdown of the decomposition of the effective potential into its adiabatic approximation plus nonadiabatic corrections.

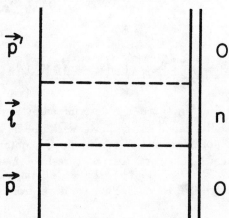

Figure 6: The two Coulomb photon exchange process in electron atom scattering.

3.2 TWO COULOMB EXCHANGE

There is only one graph corresponding to the two Coulomb photon exchange process [Fig. 6]. In the dipole approximation, the scattering T-matrix element is:

$$M_{C-C} = \frac{2e^4}{\pi} \int d^3k d^3q d^3l \frac{k_\sigma q_\rho}{k^2 q^2} \sum_n \frac{\langle 0|r_\sigma|n\rangle\langle n|r_\rho|0\rangle}{\Delta + X} \delta(\vec{p}' - \vec{q} - \vec{l})\delta(\vec{l} - \vec{k} - \vec{p}).$$

(3.29)

Here σ and ρ are the space component indices and repeated indices are summed over. In contrast to Sec. 2, k^2 and q^2 here are 3-vector squares. In Eq. (3.29),

$$\Delta \equiv E_n - E_0 \qquad (3.30a)$$

is the excitation energy of the atom to the nth state, and

$$X \equiv \frac{l^2 - p^2}{2m} \qquad (3.30b)$$

is the recoil energy of the scatteirng electron. The effective potential obtained as the Fourier transform of the on-shell amplitude $M_{C-C}|_{\vec{p}\cdot\vec{Q}=-Q^2/2}$ [see last subsection] can be expressed as:

$$V_{C-C}(p^2, R) \equiv \sum_{k,j} V_{kj}^{C-C}(R)p^{2j}. \qquad (3.31)$$

At threshold energy, only the $j = 0$ terms survive. This corresponds to evaluating the effective potential at $s = s_0$ in Sec. 2. In this case, we have

$$V_{k,0}^{C-C}(R) = \frac{e^2(-1)^{k+1}\beta_{k+1}^1(2k+2)!(2k)!}{R^{2k+4}2^{2k+2}(k!)^2} \qquad (3.32)$$

where

$$\beta_{k+1}^L \equiv 4me^2 \sum_n \frac{|\langle 0|r^L P_L|n\rangle|^2}{[2m(E_n - E_0)]^{k+1}} \qquad (3.33)$$

is the kth order nonadiabatic 2^Lth multipole atomic structure constant. The $k \neq 0$ terms in Eq. (3.32) are the nonadiabatic corrections. For the lowest energy dependent correction $(j = 1)$, we exhibit the $k = 0$ term:

$$V_{01}^{C-C} = (-1)e^2\beta_3^1 \frac{3}{R^6}. \qquad (3.34)$$

The result in Eq. (3.32) which reflects only the dipole contribution has been successfully extended to include all multipoles. The general result is:[20]

$$V_{k0}^{C-C,L}(R) = \frac{e^2(-1)^{k+1}\beta_{k+1}^L}{R^{2(k+L+1)}} \frac{(2k+2L)!}{(2L)!} \frac{(2k!)}{(k!)^2 2^{2k+1}}. \qquad (3.35)$$

Here, L denotes the order of the multipole and for $L = 1$, Eq. (3.35) reduces to Eq. (3.32).

182

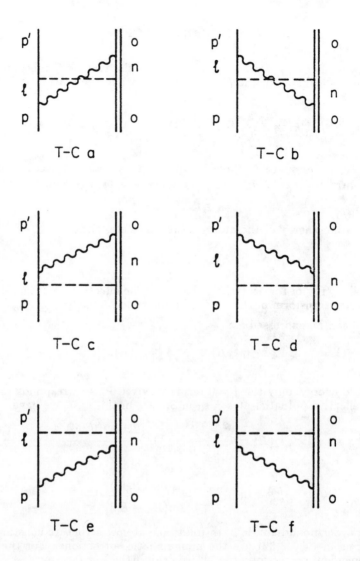

Figure 7: Time-ordered Feynman diagrams for the one-Coulomb-and-one-transverse-photon exchange in electron-atom scattering.

3.3 ONE COULOMB AND ONE TRANSVERSE PHOTONS EXCHANGE

There are a total of six topologically different time-ordered diagrams contributing to the one Coulomb and one transverse photon exchange process [Fig. 7]. The scattering transition matrix element in the dipole approximation corresponding to the sum of all these six diagrams is:

$$M_{T-C} = \frac{4\alpha^2}{\pi m} \int d^3k d^3q d^3l \sum_n \frac{\Delta |\langle 0|z|n\rangle|^2}{(\Delta + X)(\Delta + \omega)(X + \omega)} \frac{q_\sigma p_\tau}{q^2} (\delta_{\sigma\tau} - \hat{k}_\sigma \hat{k}_\tau)$$

$$\times \delta(\vec{p}' - \vec{q} - \vec{l})\delta(\vec{l} - \vec{k} - \vec{p}). \tag{3.36}$$

where

$$\omega = |\vec{k}|, \tag{3.37}$$

and Δ and X are defined in Eqs. (3.30a) and (3.30b). We can expand the energy denominator in Eq. (3.36) as:

$$[(\Delta + \omega)(\Delta + X)(\omega + X)]^{-1} = \frac{1}{(\Delta + \omega)\Delta\omega} - \frac{X}{(\Delta\omega)^2} + \cdots, \tag{3.38}$$

and then write the corresponding effective potential according to this expansion as:

$$V_{T-C} = V_{T-C}^{(1)} + V_{T-C}^{(2)} + \cdots \tag{3.39}$$

where $V_{T-C}^{(1)}$ is the effective potential obtained from M_{T-C} if the energy denominator is replaced by the first term on the right hand side of Eq. (3.38) *etc.* After a detail calculation, we have:

$$V_{T-C}^{(1)} = \frac{4\alpha^2}{\pi m R^3} \int dk \, k \, j_2(kR) \sum_n \frac{|\langle 0|z|n\rangle|^2}{\Delta + k}, \tag{3.40}$$

and

$$V_{T-C}^{(2)} = \frac{-\alpha\beta_1^1}{8m^2 R^4} \left[\frac{21}{R^2} + p^2\right]. \tag{3.41}$$

In Eq. (3.41), we encounter the first nonadiabatic and energy dependent corrections to the retardation potential. They come from diagrams T-C,c to T-C,f.

It is also possible to evaluate the effective potential contribution to $V_{T-C}^{(1)}$ corresponding to each graph independently. We give the results here. Below, the subscripts T-C,a to T-C,f refer to the diagrams in Fig. 7. The neglect of the nonadiabatic corrections in the transverse photon exchange effective potential is equivalent to setting the recoil energy factor X [defined by Eq. (3.30a)] equal to zero. This is consistent with the approximation used in obtaining the Breit interaction from one transverse photon exchange. I define the following functions:

$$P_{T-C,X}(z) \equiv \frac{\pi}{4}\left(1 + \frac{6}{z^2}\right) - \frac{3}{z} + f(z)\left[1 - \frac{3}{z^2}\right] - \frac{3g(z)}{z}; \tag{3.42}$$

$$P_{T-C}(z) \equiv \frac{\pi}{4} - P_{T-C,X}(z); \tag{3.43}$$

$$J_E(z) \equiv \frac{8}{z}\left[\frac{13}{4}z - \frac{z^3}{2} + \left(3 - 5z^2 + z^4\right)f(2z) + \left(6z - 2z^3\right)g(2z)\right] \tag{3.44}$$

and

$$J_E^{corr}(z) \equiv J_E(z) - \frac{12\pi}{z} + 22. \tag{3.45}$$

In the above, the functions $f(z)$ and $g(z)$ are the auxilliary functions for the sine and cosine integrals, given in Gautschi and Cahill.[21] The functions $J_E(z)$ and $J_E^{corr}(z)$ are the same functions defined in AFS. The subscript 'X' in Eq. (3.42) indicates the crossing of the Coulomb and transverse photons in diagrams T-C,a and b in Fig. 7. In terms of these functions and at the level of approximation employed in Ref. 11, after carrying out the procedures described in Subsection 3.1, I find the following (on neglecting the nonadiabatic corrections):

1)

$$V_{T-C,X}(R) \equiv V_{T-C,a} + V_{T-C,b}\ , \tag{3.46}$$

$$= \frac{-2\alpha^2}{\pi m R^4}\sum_n |\langle 0|z|n\rangle|^2 P_{T-C,X}(\Delta R), \tag{3.47}$$

where the sum is over the electronic states of the atom and $\Delta \equiv E_n - E_0$ is the atomic electron excitation energy. Equation (3.47) exhibits the Casimir behavior, the appearance of an additional power of R^{-1} at large distances:

$$V_{T-C,X}(R) = \begin{cases} \dfrac{-\alpha^4 a^2}{3\pi R^3} & \text{for small } R \sim \text{ few } a, \\[2ex] -\dfrac{1}{2}\dfrac{\alpha^2\langle z^2\rangle}{m R^4} & \text{for large } R > a/\alpha. \end{cases} \tag{3.48}$$

In Eq. (3.48), $\langle z^2\rangle \equiv \langle 0|z^2|0\rangle = \frac{1}{3}\langle 0|r^2|0\rangle$.

2)

$$V_{T-C,B}(R) \equiv V_{T-C,d} + V_{T-C,e} \tag{3.49}$$

$$= \frac{\alpha^2}{2m R^4}\sum_n |\langle 0|z|n\rangle|^2$$

$$= \frac{1}{2}\frac{\alpha^2\langle z^2\rangle}{m R^4}. \tag{3.50}$$

Notice that Eq. (3.50) does not exhibit any Casimir behavior.

3)

$$V_{T-C,NB}(R) \equiv V_{T-C,c} + V_{T-C,f} \tag{3.51}$$

$$= \frac{2\alpha^2}{\pi m R^4}\sum_n |\langle 0|z|n\rangle|^2\left[\frac{\pi}{4} - P_{T-C,X}(\Delta R)\right] \tag{3.52}$$

$$= V_{T-C,B} + V_{T-C,X}. \tag{3.53}$$

Equation (3.53) is a consequence of neglecting the nonadiabatic corrections in the transverse photon exchange effective potential. It can also be verified by studying the energy denominators in Eqs. (3.4)–(3.6) in Au1 (note the typo correction in Ref. 12) upon setting the recoil energy factor X equal to zero. The further subscripts B and NB in Eqs. (3.49) and (3.51) are used to denote "Breit" and "No Breit." $V_{T-C,NB}$ also exhibits the Casimir behavior:

$$V_{T-C,NB} = \begin{cases} \dfrac{1}{2}\dfrac{\alpha^2 \langle z^2 \rangle}{mR^4} & \text{for small } R \sim \text{few } a, \\[3mm] \dfrac{4\alpha}{\pi m R^5}\,\alpha_{\text{dipole}} & \text{for large } R > a/\alpha, \end{cases} \tag{3.54}$$

where α_{dipole} in Eq. (3.54) is the static electric dipole polarizability of the atom. Then on putting these together, we have:

$$\begin{aligned} V_{T-C}^{(1)} &= V_{T-C,X} + V_{T-C,B} + V_{T-C,NB} \\ &= 2\left[V_{T-C,B} + V_{T-C,X}\right]. \end{aligned} \tag{3.55}$$

This can also be verified by directly performing the integral on the right hand side of Eq. (3.40).

3.4 TWO-TRANSVERSE PHOTON EXCHANGE

There are twenty topologically different time ordered diagrams contributing to the two transverse photon exchange process [Fig. 8]. In a theory in which the electron (treated nonrelativistically) is coupled to the photon (always relativistic) via the minimal coupling scheme, we must include processes involving the simultaneous emission or absorption of two photons (the seagull vertices). In a relativistic picture, these simultaneous two photon interaction (seagull interaction) corresponds to the appearance of electron-positron pairs. To allow for crossing, which is more evident in the diagrams displaying the pairs [Fig. 9], we must assign a weighing factor of 2 to the graphs involving one or two seagull vertices. It is convenient to group the two-transverse photon exchange diagrams into two groups: those that involve a seagull vertex on the scattering electron line [diagrams (a)–(e)] and the rest [diagrams (f)–(t)]. The first set of diagrams [(a)–(e)] do not involve an intermediate electron scattering state and so will not give rise to any nonadiabatic corrections. After some rather laborious calculations, the effective potential corresponding to diagrams [(a)–(e)] is found to be:

186

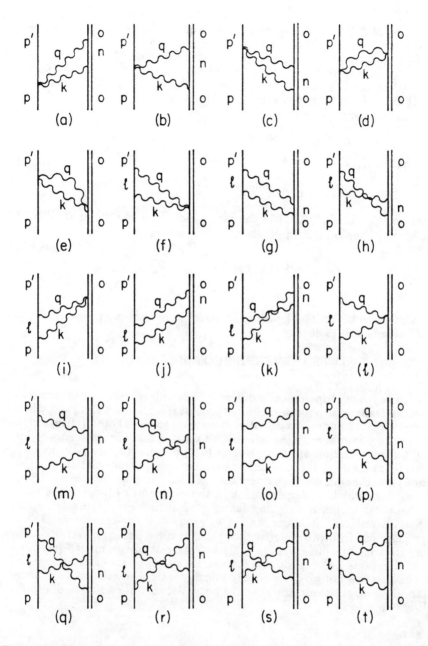

Figure 8: Time-ordered Feynman diagrams for the two-transverse-photon exchange in electron-atom scattering.

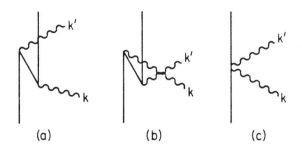

Figure 9: The graph involving an electron-positron pair (a) and its crossed counterpart (b) in a relativistic spinor theory both appear as a simultaneous two photon interaction term in a nonrelativistic theory (c).

$$V_{T-T}^{(a)-(e)}(R) \equiv 2\left(V_{T-T,a} + V_{T-T,b} + V_{T-T,c} + V_{T-T,d} + V_{T-T,e}\right) \quad (3.55)$$

$$= \frac{2\alpha^2}{\pi m R^4} \sum_n |\langle 0|z|n\rangle|^2 \left[\frac{J_E^{corr}(\Delta R)}{8\Delta R} - \frac{\pi}{2} + 2P_{T-C,X}(\Delta R)\right]. \quad (3.56)$$

The factor 2 in the right hand side of Eq. (3.56) takes into account of crossing mentioned earlier. $V_{T-T}(R)$ exhibits double Casimir behavior, i.e., the large distance behavior differs from the small distance behavior by two extra powers of R^{-1}. This traces its origin to the appearance of the factor $(\Delta + k)(\Delta + q)$ in the denominator in Eq. (5.3) in Au1 [again, note the typo correction in Ref. 12]. I find:

$$V_{T-T}(R) = \begin{cases} -\dfrac{\alpha^4 a^2}{2\pi R^3} & \text{for small } R \sim \text{ few } a, \\[3mm] -\dfrac{5}{4}\dfrac{\alpha}{\pi m R^5}\,\alpha_{\text{dipole}} & \text{for large } R > a/\alpha. \end{cases} \quad (3.57)$$

On consolidating the results of Eq. (3.55) and (3.57), we find:[22]

$$V_{T-C}^{(1)} + V_{T-T}^{(a)-(e)} = \frac{2\alpha^2}{\pi m R^4} \sum_n |\langle 0|z|n\rangle|^2 \frac{J_E^{corr}(\Delta R)}{\delta \Delta R}, \quad (3.58)$$

$$= V_{2\gamma,AFS}^{corr}.$$

The result in Eq. (3.58) is the leading correction to the two Coulomb exchange potential due to retardation (the Casimir correction) and is identical to the correction potential $V_{2\gamma}^{corr}$ derived from a dispersion relation analysis described in Sec. 2 that Au, Feinberg and Sucher[11] (AFS) used to calculate the retardation (Casimir) corrections to the energy levels in high Rydberg states of helium. A power counting procedure can be used to show that the rest of the two-transverse-photon exchange diagrams leads to a correction of order $(\lambdabar/R)^2$, where $\lambdabar$ is the Compton wavelength of the electron, compared to the leading retardation potential expressed in Eq. (3.58).

4. LATEST EXPERIMENTAL STATUS ON THE SEARCH FOR THE CASIMIR CORRECTIONS AND RELEVANCE OF THE PRESENT ANALYSIS TO OTHER CALCULATIONS

It has long been recognized that the Casimir correction to the van der Waals potential between a pair of atoms is hopelessly small to be directly observed. According to Feinberg,[23] the best candidate for detection of such corrections may be in a pair of spherical superconductors. Even here, the effect is discouragingly small. Advances in laser spectroscopy in the late 1970's have stimulated a lot of interest in the study of high Rydberg states in atoms. Among the few experimental high precision spectroscopic studies on Rydberg states of helium is one headed by Lundeen[24] in the University of Notre Dame. He and his group were studying the fine structures of helium in situations where one electron is in the $1s$ ground state and the other electron is in a high n (principal quantum number, $n = 10$) and high l (angular momentum quantum number, $l = 4, 5, \ldots 9$) state. The fine structure in these states has been studied theoretically by Drachman who first only considered the Coulombic interactions and the finite mass effects in these systems.[14] Later on, relativistic corrections and nuclear recoil corrections were included.[25] One can add the Casimir corrections to the energy levels onto Drachman's calculation by evaluating the expectation value of the Casimir correction potential given in Eq. (3.58) for the Rydberg states. Our calculations showed that such corrections are of the order of 10's of kilohertz for the Rydberg states studied and are much less than the expectation values of the asymptotic form of the retardation correction potential.[11]

This QED prediction seems to have been verified in some recent precision microwave spectroscopy experiments by Hessels, Sturrus and Lundeen.[26] The Casimir correction potential used in AFS $[V_{2\gamma,AFS}^{corr}]$ was also later rederived by Babb and Spruch[22] (BS1) explicitly for the helium Rydberg state system and the calculated retardation corrections in AFS $[(\Delta E_{ret})_{AFS}]$ were affirmed in a calculation by these authors[27] using the "pseudostate" method (BS2). Kelsey and Spruch[8] (KS) were the first to point out the feasiliblity of detecting the Casimir correction in helium Rydberg state system and they also derived the asymptotic form of the Casimir correction potential. The theoretical fine structure splittings in helium Rydberg states with which Hessels et al.[26] compare their experimental measurements are obtained by adding the AFS retardation corrections to the theoretical fine structure splittings obtained by Drachman.[14,25]

Recently, Drake[28] has developed a new variational technique to obtain very accurate nonrelativstic wave functions for helium. Mass polarization corrections are included in the nonrelativistic Hamiltonian. Relativistic corrections of order α^2 Rydbergs are calculated by evaluating the matrix elements of the nonrelativistic Pauli form of the Breit interaction as given by Bethe and Salpeter[29] with the nonrelativistic variational wave functions thus obtained [Drake sets α/a, the Hartree, equal to one in his paper whereas natural units are used in the present paper]. Relativistic reduced mass corrections, one-electron energy shifts corrected for the electron density at the nucleus,[30] and two-electron corrections to the Bethe logarithm are included. For the further QED corrections due to transverse photon exchange between the two electrons, Drake uses the Araki[31]-Sucher[32] potential. Apart from contact type terms which fall off exponentially and hence are ignored here, as we are primarily concerned with moderate to large distance behavior, this potential is given by $Q' = -\frac{7}{6}\frac{\alpha^4 a^2}{\pi R^3}$ which approximates the Q-term derived by Araki. I shall henceforth refer to this as the Q-correction. Using the above scheme, Drake[28] obtained energies of the low-lying states of helium in good agreement with experiments.[33,34]

In view of the current interest in the comparison between theory and experiment,[26] it is important to examine the situation where one extends Drake's method to high-ℓ helium Rydberg states and asks what remaining transverse photon exchange effects, if any, are to be added to such an analysis. As we shall see, the correction to the Coulomb-Breit interaction obeys different power laws at small and large distances. I refer to such behavior as the Casimir behavior. At short distances, this correction does reduce to the Q-correction term in Drake's calculation. However, this Q-term will not be the correct form to use in Rydberg states where the moderate and large distance behavior becomes important. Hence the proper correction to the Coulomb-Breit interaction will be given by the correction to the Coulomb-Breit interaction identified shortly, less the Q-correction QED term Drake has included. For the states that Drake has studied, this change leads to negligible corrections.

A suitable point to begin is to realize that the Breit interaction is the effective one transverse photon exchange potential derived in the approximation that the electronic excitation energy is small compared to the energy of the exchanged photon.[29] I shall refer to this as the Breit approximation. Hence, one must be careful not to iterate the Breit interaction to second order. In any case, in Drake's calculation, the Breit interaction is taken only to first order and so the difficulty associated with iterating the Breit interaction is not present. For convenience, I shall denote the transverse-photon exchange correction to the Coulomb-Breit interaction as V'. Then the proper correction to be added on to Drake's calculation is given by $V'' \equiv V' - Q'$.

The Coulomb variation calculation, to the degree it is satisfactorily solved, is equivalent to summing up the Coulomb ladders up to the order in perturbation theory that yields a correction comparable to the margin of accuracy in the variational calculation. Given the expected accuracy quoted in Drake's calculation, it is certainly a few orders beyond two Coulomb exchange in the dipole approximation. Hence, within the framework of the effective potential due to two photon exchange in the dipole approximation, Drake's calculation, excluding the Q-correction, in the language of a time-ordered diagrammatic approach such as that employed in KS, BS1, Au1 and Au2, only includes those diagrams with one-Coulomb and one-transverse photon exchange in ladder form, *i.e.*, where the Coulomb photon and the transverse photon lines do not cross. The effective potential arising from the irreducible crossed Coulomb-transverse diagram is distinct from those obtained by iterating the Coulomb-Breit interaction.[35] The topologically different time-ordered one-Coulomb and one-transverse photon exchange processes are depicted in diagrams T-C,a through T-C,f in Fig. 7 and the effective potential corresponding to each diagram is given in Sec. 3.3. According to the discussions above, Drake's account of the Breit interaction would have incorporated the processes depicted in diagrams T-C,c through T-C,f of Fig. 7, but not the processes depicted in the crossed diagrams T-C,a and T-C,b. These two correspond to part of the Q-correction with the rest coming from two-transverse-photon exchanges. The representation of the non-crossed transverse photon line by the Breit interaction is equivalent to neglecting the electronic excitation energy relative to the energy of the exchanged transverse photon. Under this approximation, the effective potential corresponding to each of the four diagrams T-C,c through T-C,f becomes equal, to the extent that we neglect the nonadiabatic corrections to the Casimir correction potential, which is equivalent to neglecting corrections of order $\lambdabar/R$ where $\lambdabar$ is the electron Compton wavelength. The leading nonadiabatic corrections to the Casimir correction potential have been identified in Au2 but these were not included in the AFS calculations. Also, the neglect of these nonadiabatic corrections is equivalent to the neglect of the excitation energy of the Rydberg electron relative to either the excitation energy of the core electron or the energy of the exchanged transverse photon, a fact consistent with the approximations under which the Breit interaction potential is obtained. More important, however, is the neglect of the core electron excitation energy relative to the exchanged photon energy. This approximation prevents the effective potential corresponding to diagrams T-C,c and T-C,f from exhibiting the Casimir behavior — namely the behavior to change from a R^{-n}-type potential at short distances to a $R^{-(n+1)}$-type potential at large distances. However, for diagrams T-C,d and T-C,e, the same approximation leads only to neglecting the nonadiabatic corrections. In addition to the crossed one-Coulomb and one-transverse photon exchange diagrams, all the processes corresponding to two-transverse-photon exchanges need to be examined. For this, we refer ourselves back to the discussions in Secs. 3.3 and 3.4.

Based on the results in Eqs. (3.42–3.57), I make the following very interesting observations:

i) At short distances, the potential due to two-transverse photon exchange V_{T-T} and that due to the irreducible one-Coulomb and one-transverse photon exchange $V_{T-C,X}$ are both of order $\frac{\alpha^4 a^2}{\pi R^3}$.

ii) At short distances, the potential $V_{T-C,NB} \equiv V_{T-C,c} + V_{TC,f}$ and the potential $V_{T-C,B} \equiv V_{TC,d} + V_{T-C,e}$ both behave like $\frac{1}{2}\frac{\alpha^2 \langle z^2 \rangle}{mR^4}$ and their sum accounts for the majority of the Casimir retardation correction potential at short distances.

iii) If the approximation used in obtaining the Breit interaction is employed in the derivation of $V_{T-C,NB}$ by neglecting the core ionic electron excitation energy relative to the exchanged photon energy, then $V_{T-C,NB}$ will not exhibit the Casimir behavior and in fact $V_{T-C,NB}$ becomes equal to $V_{T-C,B}$.

iv) If the Breit interaction is included as a first order correction after taking into account of the Coulombic correlation of the electrons, then the corresponding processes in the time-ordered diagrammatic approach are the ones depicted in diagrams T-C,c through T-C,f and the corresponding effective potential is evaluated under the approximation mentioned in (iii) above. In this case, the sum of all these effective potentials is $\frac{1}{4}\frac{\alpha^3 a^3}{R^4}$ and does not exhibit the Casimir behavior. Here we have explicitly used the result $\langle z^2 \rangle = a^2/4$ for the He$^+$ ion core. This sum is a fair representation of the Casimir correction potential at short distances but fails badly at large distances. This sum corresponds to what Drake would have included in his treatment of the Coulomb-Breit interaction at this level of approximation in his calculation to the extent that such potential is meaningful.

v) At large distances, the effective potential from the irreducible one-Coulomb one-transverse crossed graphs exactly cancels $V_{T-C,B}$ in the present approximation where the nonadiabatic corrections are neglected, leaving behind only $V_{T-C,NB}$. This and V_{TT} both behave like $\sim \left(\frac{\alpha}{\pi m R^5}\right)\alpha_{\text{dipole}}$ at large distances and together correctly yield the $\frac{11\alpha}{4\pi m R^5}\alpha_{\text{dipole}}$ potential.[8] This potential was first obtained by Bernabeau and Tarrach for electron-atom scattering.[36]

vi) Based on (iv) above, the remaining transverse photon exchange effects in potential form that need to be added on to the Coulomb-Breit interaction are given by $2V_{T-C,X}$ and V_{T-T}. At short distances, both these potentials behave like $\frac{\alpha^4 a^2}{\pi R^3}$ and their sum is $-\frac{7}{6}\frac{\alpha^4 a^2}{\pi R^3}$.

vii) If one assumes that in the helium states studied by Drake, the transverse photon exchange effective potential is fairly represented by its short distance form, then the leading correction to the Coulomb-Breit interaction is given by $-\frac{7}{6}\frac{\alpha^4 a^2}{\pi R^3}$. The present derivation is based on the assumption that the Rydberg electron is far away from the core. Hence R can be identified with r_{12}, the interelectronic distance. In this case, this leading correction is identical to Q-correction term that Drake includes in his calculation. This leaves correction of order $\alpha^5 a/R^2$ leading to energy corrections only of order α^4 Rydberg which are smaller than the level of accuracy asserted by Drake. Indeed, the small distance form of the effective potential should hold very well for the states studied by Drake.

viii) In general, because all the noncrossed one-Coulomb one-transverse photon exchange effective potentials evaluated in the Breit approximation together yield a potential that differs from the entire Casimir correction potential by the amount $V' \equiv [2V_{T-C,X} + V_{T-T}]$ which are given in Eqs. (3.47) and (3.56), the remaining part of the retardation effect that needs to be added on to Drake's method of calculation is simply given by subtracting from V' the Q-correction that Drake has already included.

This is given by:

$$V'' \equiv V' - \left(-\frac{7\alpha^4 a^2}{6\pi R^3}\right)$$

$$= V_{2\gamma,AFS}^{corr} - \frac{\alpha^3 a^3}{4R^4} + \frac{7\alpha^4 a^2}{6\pi R^3}. \qquad (4.1)$$

ix) The leading nonadiabatic corrections to the transverse photon exchange potential identified in Au1 and Au2 come from the noncrossed one-Coulomb and one-transverse photon exchange graphs. Since the Coulombic variational calculation of Drake takes into account the electrons' recoil, the effects of these nonadiabatic terms are already incorporated in Drake's calculation.

Finally, I would like to explain why the effective potentials, derived here for electron-atom (hydrogenic atom without spin) scattering, can be applied to the calculation of electronic fine structures of helium Rydberg state energy levels. First, we observe that in the approach where the zeroth order approximation is given by an inner electron in the hydrogenic $1s$ state in a He$^+$ ion and an outer electron in the hydrogenic nl state in a neutral H atom, the leading Coulombic interaction perturbation is $e^2 \vec{r} \cdot \vec{R}/R^3$ which is identical to that in electron-atom scattering. Second, the coupling of the transverse photon to the electron is the same whether the electron is in a bound or continuum state. The neglect of the Rydberg electron transition energy relative to the photon energy results in corrections of order $(\alpha a/R)$ to the total retardation potential. In a relativistic treatment for the electron, the simultaneous two-transverse photon interaction is absent but the leading effects in the processes involving electron-positron pairs (see Fig. 9) just reduce to these simultaneous two photon interaction (seagull-type) terms, leaving corrections of order p/m. Such corrections, in terms of the two-transverse photon exchange potential, are of order $(\alpha a/R)$ if the simultaneous two-photon interaction is used to replace the interaction involving pairs on the Rydberg electron line and are of order α if the same replacement is applied to the core electron. The only place I see a difference in the effective potential due to transverse photon exchange in the case of electron-atom scattering and that of Rydberg states is in the calculation of the second nonadiabatic type corrections in the transverse photon exchange potential. But here, the correction is of order $(a/R)^4$ relative to the Casimir correction potential.

Thus the correction V'' pointed out in the present paper can be added on to a calculation such as the one by Drake in which the Coulombic interaction is incorporated very accurately and the transverse photon exchange is included as a first order Breit correction together with the remaining QED correction given by the Q-correction [the Araki-Sucher potential]. The correction V'' is quite independent of the way the Coulombic problem is solved with the understanding that corrections to the energy levels due to V'' are simply given by its expectation values for the Rydberg states under study. It is also derived under the assumption that exchange effect is ignorable in the transverse photon exchange diagrams depicted in Figs. 7 and 8. The present method of identifying V'' as the correction potential to be added on to Drake's type of calculation is parallel to the identification of $V^{corr}_{2\gamma,AFS}$ in AFS as the correction potential to be added on to Drachman's type of calculation on helium Rydberg state electronic fine structures. Then according to Eq. (4.1), the correction that needs to be added on to Drake's calculated values is given by (in Rydbergs):

$$\langle V'' \rangle = (\Delta E_{ret})_{AFS} - \frac{\alpha^2}{4} \frac{[3n^2 - l(l+1)]}{n^5(l+3/2)(l+1)(l+1/2)l(l-1/2)}Ryd$$

$$+ \frac{7}{3}\frac{\alpha^3}{\pi} \frac{1}{n^3 l(l+1/2)(l+1)}Ryd \tag{4.2}$$

where $(\Delta E_{ret})_{AFS}$ is the Casimir corrections for the Rydberg state $(1s\ nl)$ expressed in Rydbergs as calculated in AFS. For the Rydberg states where the small distance behavior of the Casimir correction potential is a fair representation of the total retardation correction, the second and third term in Eq. (4.2) (with the opposite signs) give almost all of the Casimir correction. In this case $\langle V'' \rangle$ will be negligibly small and hence would explain why Drake's method yields values in excellent agreement with experiments. One could have obtained V' by subtracting from $V^{corr}_{2\gamma,AFS}$ the effective potentials from the noncrossed one-Coulomb-one-transverse-photon exchange diagrams evaluated in the Breit approximation. The present analysis shows how various cancellations occur in different limits and why the Casimir behavior is not exhibited in the Breit approximation. V'' is obtained by subtracting from V' the two-electron QED correction already included by Drake. The challenge to the experimentalist would be whether the correction indicated by Eq. (4.2) could be seen if Drake would extend his calculation to Ryberg states.[37]

Since V'' goes as R^{-2} for small R, $\langle V'' \rangle$ will be finite even if the Rydberg electron is in an s state. Thus it would be very interesting to test this in the muonic-electron helium $(\mu^- e^- \alpha^{++})$ system[38] because the electron, even in the 1s and 2s states, will be far away enough from the $\mu^- \alpha^{++}$ core for retardation effect to be important. In this case, the small argument expansion of $V^{corr}_{2\gamma,AFS}$ will not be valid and hence the last two terms in Eq. (4.1) will not offset much of the first term, thereby leaving a rather significant V''. If Drake's method can be successfully applied to this system, then this very fine prediction can be tested in high precision two photon spectroscopy experiments.

ACKNOWLEDGEMENT

I would like to thank G. Drake, M. Lieber, J. Sucher and other participants in the ITP Workshop for many helpful discussions. Much greater details of this work will be presented in a review paper currently under preparation for *Physics Reports* by Feinberg, Sucher and the author. This work is supported by the National Science Foundation under Grant No. PHY87-10118 at the University of South Carolina.

REFERENCES

1. I. Newton, *Optiks*, 4th ed. (London, 1730; Dover Publ. reprint, 1952).

2. J.D. van der Waals, Ph.D. Thesis (Leiden, 1873).

3. F. London, *Z. Physik* **63**, 245 (1930).

4. E.J.W. Verwey, *J. Phys. and Colloid Chem.* **51**, 631 (1947).

5. H.B.G. Casimir and D. Polder, *Phys. Rev.* **73**, 360 (1948).

6. E.M. Lifshitz, *Soviet Phys. JETP* **2**, 73 (1956).

7. G. Feinberg and J. Sucher, *Phys. Rev.* **139**, B1619 (1965); *Phys. Rev.* **A2**, 2395 (1970).

8. E.J Kelsey and L. Spruch, *Phys. Rev.* **A18**, 1055 (1978). This paper will be referred to as KS.

9. K.B. MacAdam and W.H. Wing, *Phys. Rev.* **A13**, 2163 (1976); D.R. Cok and S.R. Lundeen, *Phys. Rev.* **A19**, 1830 (1979).

10. G. Feinberg and J. Sucher, *Phys. Rev.* **A27**, 1958 (1983). This paper will be referred to as FS.

11. C.K. Au, G. Feinberg and J. Sucher, *Phys. Rev. Lett.* **53**, 1145 (1984). This paper will be referred to as AFS. G. Feinberg, J. Sucher and C.K. Au, *Ann. Phys. (N.Y.)* **173**, 355 (1987).

12. C.K. Au, *Phys. Rev.* **A34**, 3568 (1986); and *Phys. Rev.* **A38**, 7 (1988). These two papers will be referred to as Au1 and Au2. Equations (3.5) and (3.6) in Au1 are incorrect by an overall sign; the factor '3' in the numerator in Eq. (5.3) in Au1 should be '2'.

13. R.J. Drachman, *J. Phys.* **B12**, L699 (1979).

14. R.J. Drachman, *Phys. Rev.* **A26**, 1228 (1982).

15. R.E. Cutkosky, *J. Math. Phys.* **1**, 429 (1960).

16. C.K. Au, *Phys. Rev.* **A31**, 1310 (1985).

17. C.K. Au and G. Feinberg, *Phys. Rev.* **A6**, 2433 (1972).

18. C.K. Au, University of South Carolina Preprint, submitted to *Phys. Rev. A* (1987).

19. C.J. Kleinman, Y. Hahn and L. Spruch, *Phys. Rev.* **165**, 53 (1968); D.W. Norcross, *Phys. Rev.* **A28**, 3095 (1983).

20. C.K. Au, *Phys. Rev.* **A37**, 292 (1988).

21. W. Gautschi and W.F. Cahill, in *Handbook of Mathematical Functions*, edited by M. Abramowitz and I.A. Stegun (Dover, New York, 1964) Chap. 5.

22. These results are also obtained directly for helium Rydberg states by J.F. Babb and L. Spruch, *Phys. Rev.* **A36**, 456 (1987). This paper will be referred to as BS1.

23. G. Feinberg, *Phys. Rev.* **B9**, 2490 (1974).

24. S.L. Palfrey and S.R. Lundeen, *Phys. Rev. Lett.* **53**, 1141 (1984).

25. R.J. Drachman, *Phys. Rev.* **A31**, 1253 (1985); *Phys. Rev.* **A33**, 2780 (1986).

26. E.A. Hessels, W.G. Sturrus, and S.R. Lundeen, University of Notre Dame preprint (1988); S.R. Lundeen, private communication (1988).

27. J.F. Babb and L. Spruch, *Phys. Rev.* **A38**, 13 (1988). This paper will be referred to as BS2.

28. G.W.F. Drake, *Phys. Rev. Lett.* **59**, 1549 (1987).

29. H.A. Bethe and E.E. Salpeter, *Quantum Mechanics of One- and Two-Electron Atoms* (Springer-Verlag, Berlin, 1957).

30. P.K. Kabir and E.E. Salpeter, *Phys. Rev.* **108**, 1256 (1957).

31. H. Araki, *Prog. Th. Phys. (Japan)* **17**, 619 (1957).

32. J. Sucher, Ph.D. Thesis, Columbia University, unpublished (1958), available from University Microfilms, Ann Arbor, Michigan; *Phys. Rev.* **109**, 1010 (1958).

33. C.J. Sansonetti and W.C. Martin, *Phys. Rev.* **A29**, 159 (1984); W.C. Martin, *Phys. Rev.* **A29**, 1883 (1984); *Phys. Rev.* **A36**, 3575 (1987).

34. L. Hlousek, S.A. Lee and W.M. Fairbank, Jr., *Phys. Rev. Lett.* **50**, 328 (1983).

35. G. Feinberg and J. Sucher, preprint (1988).

36. J. Bernabeau and R. Tarrach, *Ann. Phys. (N.Y.)* **102**, 323 (1976).

37. As pointed out in item (ix), Drake's method would include the recoil effects of the Rydberg electron in the noncrossed one-Coulomb- and one-transverse-photon exchange graphs. The potentials corresponding to these corrections are given in Eq. (3.8) in Au2. These effects have not been taken into account in the comparison between theory and experiment in Ref. 2. However, for the states studied in Ref. 2, these nonadiabatic corrections to the Casimir potential only lead to negligibly small corrections.

38. The test of retardation effect in the $\mu^- e^- \alpha^{++}$ system was first suggested by G. Feinberg.

EVALUATION OF TWO–PHOTON EXCHANGE GRAPHS FOR THE HELIUM ISOELECTRONIC SEQUENCE

J. Sapirstein
University of Notre Dame
Notre Dame, IN 46556

ABSTRACT

Two–photon exchange graph contributions to the ground state energy of two–electron ions are evaluated in Furry picture QED. The natural separation of correlation effects from negative energy state effects is demonstrated. Some of these latter effects are evaluated both analytically and numerically, and shown to contribute at the level $(Z\alpha)^3$ a.u..

INTRODUCTION

A problem that has been recognized as a central one in atomic physics is that of finding a way to join the theory that one assumes underlies all of atomic and molecular physics, Quantum Electrodynamics(QED), with the calculational techniques employed by many–electron atomic theorists. It is useful to recall the situation in the simplest atom, hydrogen. The analytic solutions available for the Schrödinger or Dirac equation give of course an extremely accurate description of the atom by the standards characteristic of many–electron systems. This is because the Schrödinger equation accounts for all effects of order 1 a.u., and the Dirac equation also includes all effects of order α^2 a.u.. In addition, the effect of the finite mass of the nucleus enters in a simple way, allowing one to replace the electron mass with the reduced mass. The interest in hydrogen then shifts to finding the corrections to the picture of an electron with reduced mass in a constant nuclear Coulomb field afforded by QED: we know that the nucleus is not simply a source of that Coulomb field, and in addition know that a variety of effects specific to field theory, such as the fact that electron–positron pairs can be formed virtually from the vacuum, and the electron can emit and reabsorb virtual photons, will produce small but measurable corrections to the above picture. Because these effects are very small in hydrogen, entering with higher powers of α and m_e/m_p, the use of the Dirac Hamiltonian has a well defined meaning. However, in order to evaluate the small corrections to the Dirac spectrum predicted by QED, one must use some form of the powerful calculation techniques developed for relativistic quantum field theory, which in the case of the Bethe–Salpeter equation involve the study of four point Green's functions, which seem quite removed from Hamiltonian formalisms. However, the latter formalisms provide a highly accurate description of the atom, so the basic equations that follow from them must in some sense 'live inside' the more fundamental QED formalisms. It is this concept that I would like to discuss here in the framework of the multi–electron problem.

THE ONE–ELECTRON PROBLEM

Before leaving the one–electron problem, let me illustrate the idea for the Bethe–Salpeter approach. The basic idea of this method is to analyze the scattering of a proton and an electron in the center of mass frame: a one–loop contribution is illustrated below.

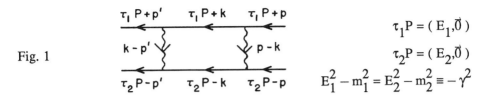

Fig. 1

$$\tau_1 P = (E_1, \vec{0})$$

$$\tau_2 P = (E_2, \vec{0})$$

$$E_1^2 - m_1^2 = E_2^2 - m_2^2 \equiv -\gamma^2$$

While this can be analyzed with the elegant techniques described by Yennie[1] in his lectures, it will be useful to introduce a breakup of the two electron propagators in terms of projection operators,

$$\Lambda_+(\vec{p}) \equiv (E_p + m \gamma_o + \vec{\alpha} \cdot \vec{p})/ 2E_p \qquad (1.a)$$

$$\Lambda_-(\vec{p}) \equiv (E_p - m \gamma_o - \vec{\alpha} \cdot \vec{p})/ 2E_p \quad , E_p \equiv \sqrt{m^2 + \vec{p}^2}, \qquad (1.b)$$

$$\frac{1}{\not{p} - m + i\varepsilon} = \frac{\Lambda_+(\vec{p}) \, \gamma_o}{p_o - E_p + i\varepsilon} + \frac{\Lambda_-(\vec{p}) \, \gamma_o}{p_o + E_p - i\varepsilon} \qquad (2)$$

Now, before going on to a detailed analysis of Fig. 1, consider the physics it should represent. If particle 2 is very massive, we expect it to behave primarily as a source of a Coulomb field, in the presence of which the electron scatters. But if the electron is nonrelativistic, we know already from nonrelativistic quantum mechanics that its propagation is described by the Coulomb Green's function, which obeys

$$[\frac{\vec{p}^2}{2m} - \frac{Z\alpha}{r} + \varepsilon] \, G(\vec{r}, \vec{r}\,'; \varepsilon) = - \frac{1}{2m} \delta^3(\vec{r} - \vec{r}\,') \qquad (3)$$

where the energy of the electron is $m - \varepsilon$. Very elegant solutions to this equation have been found by Schwinger[2] and Hostler[3] (and applied to high accuracy evaluation of Bethe logarithms by Lieber[4] and Huff[5]): however, at this point I only want to observe that G admits an expansion in the parameter $Z\alpha$, which in momentum space reads ($\gamma^2 \equiv 2m\varepsilon$)

$$G(p,p\,';\varepsilon) = - \frac{\delta^3(\vec{p} - \vec{p}\,')}{\vec{p}^2 + \gamma^2} - \frac{mZ\alpha}{\pi^2} \frac{1}{\vec{p}^2 + \gamma^2} \frac{1}{|\vec{p} - \vec{p}\,'|^2} \frac{1}{\vec{p}\,'^2 + \gamma^2}$$

$$- \frac{(mZ\alpha)^2}{\pi^4} \frac{1}{\vec{p}^2 + \gamma^2} [\int d^3k \frac{1}{|\vec{p} - \vec{k}|^2} \frac{1}{\vec{k}^2 + \gamma^2} \frac{1}{|\vec{k} - \vec{p}\,'|^2}] \frac{1}{\vec{p}\,'^2 + \gamma^2} + ...] \qquad (4)$$

I want to show that the third term 'lives' inside the Feynman diagram of Fig. 1. To do so, work in Coulomb gauge with both Coulomb photons so that there is no k_0 dependence in the photon propagators. Then applying the breakup of the two fermion propagators indicated in Eqn. 2 we can perform the k_0 integration with Cauchy's theorem. As usual, possibly dangerous terms with one positive energy and one negative energy projector cannot contribute, and we are left with two positive or two negative energy terms. Now, the denominator of the positive–positive term becomes

$$D(\vec{k}) = E_1 + E_2 - \sqrt{m_1^2 + \vec{k}^2} - \sqrt{m_2^2 + \vec{k}^2} \approx E_1 - m_1 + E_2 - m_2 - \vec{k}^2/2m_r$$
$$\approx -(\vec{k}^2 + \gamma^2) / 2m_r \tag{5}$$

But now, after adding the Coulomb photon propagators together with the remaining three–dimensional momentum integration, a structure has appeared from our Feynman diagram,

$$\int d^3k \; \frac{1}{|\vec{p} - \vec{k}|^2} \; \frac{1}{\vec{k}^2 + \gamma^2} \; \frac{1}{|\vec{k} - \vec{p}'|^2} , \tag{6}$$

that we can recognize as part of nonrelativistic Coulomb Green's function. It is not hard to carry out a similar analysis for higher order 'ladder' diagrams with Coulomb photons as the rungs, finding terms that build up the full propagator. Thus the full relativistic Green's function has in it a nonrelativistic description of electron propagation in a Coulomb field. It has however, a great deal more: we had to make several approximations, such as Taylor expanding $E_1(k)$ and dropping the negative–negative term to get to the NR level. However, the nature of these approximations is such as to introduce corrections that are higher order in α and m_1/m_2 than the basic NR result. The correction terms behave very differently from what one normally encounters in atomic calculations: the presence of the nuclear Coulomb field becomes relatively unimportant when the electrons have energies of the order of or greater than the electron rest mass: the effect of negative energy states becomes important, the most striking example of which is their role in softening the linear self energy divergences of classical electromagnetism to the logarithmic ones of QED. It is this feature of QED that makes its application to atomic systems of particular difficulty: part of a typical diagram 'wants' to be treated as a fully relativistic Feynman graph, while another, usually dominant part, 'wants' to be treated with the techniques of nonrelativistic atomic physics.

THE MANY–ELECTRON PROBLEM

Despite these difficulties, it has been possible in the two–body problem to make a great deal of progress, as described in Yennie's lectures[1]. What are the prospects of applying QED to many–electron systems? Unfortunately, the fundamental problem immediately faced is that it is no longer possible, with the exception of helium, to control all contributions of order 1 a.u.: the computer power simply does not exist to directly solve a many–particle Schrödinger or Dirac equation. However, there exists an approach that offers at least the hope of evaluating the dominant part of these contributions, that also fits neatly into a QED framework. The approach, relativistic many–body perturbation theory, provides a systematic expansion of physical observables in terms of a quantity V_c, a difference between the effect of electron Coulomb repulsion and some kind of central potential that should be chosen to make perturbations in V_c small. How well does it work? As the example to be concentrated on in the following, consider the ground state of two electrons in the presence of a nucleus charge Z, and let V_c be chosen to account for only the Coulomb repulsion of the electrons. This is a very well known expansion: the first few orders behave in the NR limit as follows:

Table 1: Perturbation Expansion of the Ground State
Energy of Two–electron Systems

Order	Contribution in a.u.	Z=2 result	Cumulative Z=2 result
0	$-Z^2$	–4.00000	–4.00000
1	5/8 Z	1.25000	–2.75000
2	–.1576664	–0.15767	–2.90767
3	.00869903/Z	0.004350	–2.90332
4	–.0008887/Z^2	–0.00022	–2.90354
			[exact = –2.90372]

Note that the coefficients of the expansion get smaller all by themselves: however, it is clear that the convergence of the expansion can be greatly hastened at high Z. This then offers one line of attack on atomic systems that may get around the principal difficulty of many–body physics, the need to consider very large numbers of graphs in order to control even 1 a.u. effects: a small parameter, 1/Z, is added to our list of quantities to perturb in, so one can expand in α, 1/Z, and m_e/m_{nuc}. However, as soon as one goes to high Z, relativistic and QED effects become enhanced, so now we have to examine QED graphs to see if the above kind of rapidly convergent result is contained inside them, and also what pure QED effects survive. Before going on to the main discussion, it is important to realize that all orders methods such as described by Lindgren[6] and Mårtensson–Pendrill[7], offer a great deal of hope for accounting for a large enough fraction of 1 a.u. contributions in order to get to the QED effect level. I believe that these techniques can also be subsumed into a QED approach based on coupled Schwinger–Dyson equations[8]. However, for the present I want to concentrate on low order perturbation theory.

QED PERTURBATION THEORY

The standard many–body perturbation theory expression that gives rise to the relativistic generalization of the second order result for helium listed above is

$$E_2 = \frac{1}{2} \sum_{abmn} \frac{g_{mnab}\, \tilde{g}_{abmn}}{\varepsilon_a + \varepsilon_b - \varepsilon_m - \varepsilon_n} + \sum_{abcm} \frac{\tilde{g}_{bmab}\, \tilde{g}_{c\,amc}}{\varepsilon_a - \varepsilon_m} \tag{7}$$

where $\tilde{g}_{abcd} \equiv g_{abcd} - g_{abdc}$ and the basic Coulomb matrix element is defined by

$$g_{abcd} \equiv \alpha \int \frac{d^3r\, d^3r'}{|\vec{r} - \vec{r}'|} \psi_a^\dagger(\vec{r})\, \psi_c(\vec{r})\, \psi_b^\dagger(\vec{r}')\, \psi_d(\vec{r}'). \tag{8}$$

This is valid for any closed shell atom: for our case the sums over the core a and b range only over the magnetic quantum numbers of the 1s state. The summations over m and n are restricted to be positive energy states <u>above</u> the core; however, the second part of E_2 can be absorbed into the first by relaxing this restriction to allow the sums to range into the core, excluding only the case n=m=1s. The convention used for unrestricted sums is to use i and j in place of m and n, so

$$E_2 = \frac{1}{2} \sum_{abij}' \frac{g_{ijab}\, \tilde{g}_{abij}}{\varepsilon_a + \varepsilon_b - \varepsilon_i - \varepsilon_j} \tag{9}$$

One method of evaluating E_2 with high accuracy is the use of finite basis sets: one can then find that E_2 changes appreciably along the isoelectronic sequence: a shortened table follows[9]:

Table 2: Second–order Correction to the Ground State Energy of Two–electron Ions along the Isoelectronic Sequence. (Units of a.u.)

Z	E_2
2	−.15768
10	−.15808
20	−.15952
40	−.16661
60	−.18210
80	−.21269
100	−.27687

The reason for the variation with Z is that although nonrelativistically $E_2 = Z^0 \cdot C$, with C a constant, relativistically $C \to C(Z\alpha)$. One can fit this function at low Z by $−.15767 − .070 \, (Z\alpha)^2$, which is in agreement with the nonrelativistic approach in which $(Z\alpha)^2$ is picked up as the expectation of a relativistic operator such as the relativistic mass increase $p^4/8m^3$. However, a great advantage of purely numerical evaluations is that such terms are picked up along with all higher order $(Z\alpha)$ terms. From experience with a similar kind of function, $F(Z\alpha)$ in the one–electron Lamb shift, we know it is possible for such functions to become nonperturbative in the sense that the power series expansion carried to a given order applied at high Z can disagree totally with the expansion carried to the next order.

The expression for E_2 falls out in a very clean way from QED. As described by Mohr[10] in his lectures, the use of the Gell–Mann Low formula[11] in the context of the Furry representation[12] allows a systematic graphical attack on energy levels of high–Z few electron atoms. We wish to use

$$\Delta E = \lim_{\varepsilon \to 0} \frac{i\varepsilon}{2} \frac{\partial}{\partial \lambda} \ln <0_C|S|0_C> \Big|_{\lambda=1} \quad , \quad |0_C> \equiv b^{\dagger}_{1s\uparrow} \, b^{\dagger}_{1s\downarrow} \, | \, 0 >, \qquad (10)$$

where we are interested here in a fourth order contribution

$$S^{(4)} = \frac{(-ie\lambda)^4}{4!} \int d^4x \, d^4y \, d^4z \, d^4w \, e^{-\varepsilon(|x_0| + |y_0| + |z_0| + |w_0|)}$$

$$<0_C| \, T \, [\, \overline{\psi}(x) \, \gamma_{\mu} \, A^{\mu}(x) \, \psi(x) \quad \overline{\psi}(y) \, \gamma_{\nu} \, A^{\nu}(y) \, \psi(y) \qquad (11)$$

$$\overline{\psi}(z) \, \gamma_{\alpha} \, A^{\alpha}(z) \, \psi(z) \, \overline{\psi}(w) \, \gamma_{\beta} \, A^{\beta}(w) \, \psi(w) \,] \, |0_C>$$

Applying Wick's theorem to the product of four interaction Hamiltonians gives rise to a large number of terms: we concentrate here on the term in which there are two fermion contractions connected to the photon propagators in the topology of a box or crossed box. There are 12 ways to do this, leading to

$$S^{(4)} = \frac{(e\lambda)^4}{2} \int d^4x \, d^4y \, d^4z \, d^4w \; e^{-\epsilon(|x_o| + |y_o| + |z_o| + |w_o|)}$$
$$(D_{\mu\alpha}(x-z) \, D_{\nu\beta}(y-w) + D_{\mu\beta}(x-w) \, D_{\nu\alpha}(y-z)) \tag{12}$$
$$< 0_C \mid : \overline{\psi}(x) \, \gamma^\mu \, S_F(x,y) \, \gamma^\nu \, \psi(y) \, \overline{\psi}(z) \, \gamma^\alpha \, S_F(z,w) \, \gamma^\beta \, \psi(w) : \mid 0_C>$$

To make contact with MBPT I now choose to work in Coulomb gauge, and consider only Coulomb photons: the very interesting contributions involving transverse photons are under active investigation by Walter Johnson, Peter Mohr, Steven Blundell, and myself, but will not be discussed further here. The propagators are then

$$D_{\rho\sigma}(x-y) = \delta_{\rho 0} \, \delta_{\sigma 0} \, \frac{\delta(x_o - y_o)}{|\vec{x} - \vec{y}|}$$

$$S_F(x,y) = \int \frac{dE}{2\pi} e^{-iE(x_o - y_o)} \left[\sum_{n_+} \frac{\psi_{n_+}(\vec{x}) \, \overline{\psi}_{n_+}(\vec{y})}{E - \varepsilon_{n_+} + i\delta} + \sum_{n_-} \frac{\psi_{n_-}(\vec{x}) \, \overline{\psi}_{n_-}(\vec{y})}{E - \varepsilon_{n_-} - i\delta} \right]$$

$$\equiv \int \frac{dE}{2\pi} e^{-iE(x_o - y_o)} \sum_i \frac{\psi_i(\vec{x}) \, \overline{\psi}_i(\vec{y})}{E - \varepsilon_i \pm i\delta} \tag{13}$$

Upon inserting these forms into $S^{(4)}$, notice we can simplify the expression considerably by recognizing that two Coulomb matrix elements are being formed. This comes about because we can write ($\varepsilon_0 = \varepsilon_{1s}$)

$$< 0_C \mid : \overline{\psi}(x) \, \psi(y) \, \overline{\psi}(z) \, \psi(w) : \mid 0_C> =$$
$$- \sum_{mnrs} < 0_C \mid b_m^\dagger \, b_r^\dagger \, b_n \, b_s \mid 0_C> \; \overline{\psi}_m(x) \, \psi_n(y) \, \overline{\psi}_r(z) \psi_s(w) \tag{14}$$

$$= \sum_{ab} e^{-i\varepsilon_0(-x_o + y_o - z_o + w_o)} [\, \overline{\psi}_a(\vec{x}) \, \psi_a(\vec{y}) \, \overline{\psi}_b(\vec{w}) \, \psi_b(\vec{z}) -$$
$$\overline{\psi}_a(\vec{x}) \, \psi_b(\vec{y}) \, \overline{\psi}_a(\vec{w}) \, \psi_b(\vec{z}) \,]$$

Then the QED expression for the box becomes

$$S_{Box}^{(4)} = \frac{\lambda^4}{2} \int dx_o \, dy_o \; e^{-2\epsilon(|x_o| + |y_o|)} \int \frac{dE_1}{2\pi} \int \frac{dE_2}{2\pi} \; e^{i(x_o - y_o)(2\varepsilon_0 - E_1 - E_2)} \cdot$$

$$\sum_{abij} \cdot \frac{\tilde{g}_{baij} \, g_{ijba}}{(E_1 - \varepsilon_i \pm i\delta) \, (E_2 - \varepsilon_j \pm i\delta)} \tag{15}$$

$$= -\lambda^4 \varepsilon \frac{\partial}{\partial \varepsilon} \sum_{abij} \frac{1}{[4\varepsilon^2 + (2\varepsilon_0 - E_1 - E_2)^2]}$$

$$\frac{\tilde{g}_{baij}\, g_{ijba}}{(E_1 - \varepsilon_i \pm i\delta)\,(E_2 - \varepsilon_j \pm i\delta)}$$

The term in the sum where $\varepsilon_i = \varepsilon_j = \varepsilon_0$ gives a double pole in ε that can be shown to equal $-\lambda^4 E_1{}^2/\varepsilon^2$, where E_1 is the first order energy shift, corresponding to the 5/8 Z term above. This precisely cancels with a denominator term in the Gell–Mann Low formula, so this term is to be dropped, as it is in the MBPT case. Evaluating the E_2 integration with Cauchy's theorem then gives a contribution to S of order $1/\varepsilon$, which then leads to the energy shift

$$\Delta E_{Box} = -\frac{i}{2} \int \frac{dE}{2\pi} \sum_{abij} \frac{\tilde{g}_{baij}\, g_{ijba}}{(E - \varepsilon_i \pm i\delta)\,(2\varepsilon_0 - E - \varepsilon_j \pm i\delta)} \qquad (16)$$

Now, at this point the E integration can be performed exactly as it was in the earlier discussion of electron–nucleus scattering. However, at this point note that it can also be treated as an integral to be performed numerically. In Figure 2 below, a contour rotation is indicated that allows such an integration, with the slight complication that the semicircles around the ground state poles indicated give non–vanishing contributions that are actually related to the second term in the MBPT expression of Eqn. (7). Gaussian integration then recovers the more standard result arising from application of Cauchy's theorem to high accuracy. The reason for mentioning this other integration method is that when transverse photons are considered, their pole structure considerably complicates the Cauchy approach, while the contour rotation method is equally applicable: this is the technique the above mentioned collaboration is using for the full problem.

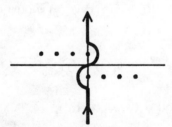

Fig. 2. The complex k_0 plane, showing poles from the electron propagators and a contour rotation that allows for exact evaluation of the box graph even when transverse photons are present.

The result of the integration is

$$\Delta E_{box} = \frac{1}{2} \sum_{abij} \frac{\tilde{g}_{abij}\, g_{ijab}}{2\varepsilon_0 - \varepsilon_i - \varepsilon_j} F_{ij} \quad ; \qquad (17)$$

$$F_{++} = -F_{--} = 1,\ F_{+-} = F_{-+} = 0$$

But the + + term is precisely Eqn. 9, a familiar expression from many–body theory. This is an even simpler result than the example from Bethe–Salpeter considered above because we need make no further NR reductions, given that relativity is correctly built into this form of MBPT. However, the $F_{-\,-}$ term must be evaluated once we get to a certain level of precision. The $F_{+\,+}$ term can either be evaluated directly, or grouped together with similar terms in an all–orders approach: it is conceivable that the $F_{-\,-}$ term could also be put in an all orders framework, but because it is such a small quantity perturbation theory should suffice.

Before I conclude with the analytic and numerical evaluation of Eqn. 17 and its crossed box counterpart, I want to illustrate the reason we are interested in such terms. We will see that their order is $(Z\alpha)^3$ a.u.. Now, the Lamb shift in one–electron atoms has the order $Z^4\alpha^3$. Thus these terms will be of interest in situations where experiment and theory are sensitive to small fractions of the one–electron Lamb shift. Johnson, Blundell, and I have found that the lithium isoelectronic sequence provides an example of this situation[13]: for example, at Z=28, the measured 2s–2p* splitting differs from our atomic theory <u>without</u> QED (although including the analog of $F_{+\,+}$) by .0174 a.u.. The one–electron Lamb shift would contribute .0199 a.u.: some effect is acting to screen it at about the 12% level. We believe that the $F_{-\,-}$ term along with a well–defined set of QED effects of the same order will account for the discrepancy.

The crossed box contribution can be evaluated analogously to the box, and contributes

$$\Delta E_{Xbox} = \frac{1}{2} \sum_{abij} \frac{g_{bjia}\ g_{iabj} - g_{ajia}\ g_{ibbj}}{\varepsilon_i - \varepsilon_j} F_{ij} \; ;$$

$$F_{+\,-} = -F_{-\,+} = 1, \; F_{+\,+} = F_{-\,-} = 0 \tag{18}$$

The denominator of this expression is now of order 2m instead of being of the order of an atomic unit. This makes the behavior of the crossed box graph very different from that of the box graph. Firstly, the larger denominator tends to make the contribution smaller by a factor of $(Z\alpha)^2$ than Eqn. 17. However, this naive power counting argument does not take into account the fact that different integration regions are important for this graph. In particular, the effect of the Coulomb interaction is now relatively unimportant because the integral is dominated by high momentum intermediate states. This fact allows us to make approximations that allow for an analytic evaluation of Eqn. 18, and the correct order will turn out to be $(Z\alpha)^3$ a.u.. Therefore this is an example of a term that gives the screening of the Lamb shift discussed above. To carry out the i and j summations we approximate the bound state propagators with free propagators, specifically the Fourier transform of the first and second parts of Eqn. 2 for positive and negative energy state summations respectively. This will take the summations into integrals over momentum p and q. In addition, dropping the $\alpha \cdot p$ and $\alpha \cdot q$ term and replacing γ_0 with unity in Eqn. 1 leads to corrections of higher order in $Z\alpha$ owing to those approximations coupling in lower components. Evaluating our expression in momentum space gives a complicated expression that simplifies greatly when we make the final approximation of dropping wave function momenta compared to p and q. The result is

$$\Delta E_{XBox} \approx \alpha^2 \int_0^\infty \frac{dp}{p^2} \frac{(E_p^2 - m^2)}{E_p^3} \sum_{ab} \int d^3r \, [\, \psi_a^\dagger(\vec{r})\psi_a(\vec{r}) \, \psi_b^\dagger(\vec{r})\psi_b(\vec{r}) -$$

$$\psi_a^\dagger(\vec{r})\psi_b(\vec{r}) \, \psi_b^\dagger(\vec{r})\psi_a(\vec{r}) \,] \tag{19}$$

$$= \frac{(Z\alpha)^3}{4\pi} \text{ a.u.}$$

At $Z = 20$ this is .000248 a.u.. At 92 the result increases to .0241 a.u.. However, because the relativistic basis set we use includes negative energy states, it is a simple matter to do the exact numerical calculation. The results are .000193 a.u. and .0124 a.u.. We attribute the deviation at $Z=20$ to $(Z\alpha)^4$ terms; in fact, the results for $Z=10,20$, and 30 are fit well by

$$\Delta E_{XBox} = [\, \frac{(Z\alpha)^3}{4\pi} - .12 \, (Z\alpha)^4 \,] \text{ a.u.} \tag{20}$$

The large discrepancy at $Z=92$ is explained by the fact that at this large Z the power series expansion is basically meaningless: a great advantage of numerical methods that becomes evident in this context is their ability to evaluate such nonperturbative quantities.

CONCLUSION

In conclusion, the box and crossed box play a dual role in the physics of the helium isoelectronic sequence. They contain the results of MBPT, but also contain corrections of order $(Z\alpha)^3$ a.u. and higher that are pure QED effects. There are other Feynman graphs that contribute at this same order, such as the vertex correction to one–photon exchange. These are discussed in more detail in reference 13. The complete evaluation of all terms in this order has the potential of complementing the beautiful tests of QED in helium made by Sucher[14] and later Douglas and Kroll[15] by extending the application of the theory to the spectroscopy of high–Z ions, where one also tests the behavior of QED in intense Coulomb fields.

Acknowledgements: I would like to thank Steve Blundell, Walter Johnson, Peter Mohr, and Don Yennie for valuable discussions. This work was supported in part by NSF grants PHY–85–03417 and PHY–86–08101.

REFERENCES

1. D. R. Yennie, in this volume.
2. J. Schwinger, J. Math Phys. 5, 1235 (1964)
3. L. Hostler, J. Math Phys. 5, 1235 (1964)
4. M. Lieber, in this volume.
5. R. Huff, Phys. Rev. 186, 1367 (1969)
6. I. Lindgren, in this volume.
7. A. M. Mårtensson–Pendrill, in this volume.
8. V.M. Tolmachev, Litovskii Fizicheskii Sbornik 3, 47 (1963), A. Klein and R. Prange, Phys. Rev. 112, 994 (1958), and T. Fulton, in this volume.

9. W. R. Johnson and J. Sapirstein, Phys. Rev. Lett. $\underline{57}$, 1126 (1986).
10. P. Mohr, in this volume.
11. M. Gell–Mann and F. Low, Phys. Rev. $\underline{84}$, 350 (1951), J. Sucher, *ibid.*, $\underline{107}$ 1448 (1957).
12. W. H. Furry, Phys. Rev. 81, $\underline{115}$ (1951).
13. W. R. Johnson, S. A. Blundell, and J. Sapirstein, Phys. Rev. $\underline{A37}$, 2764 (1988).
14. J. Sucher, Phys. Rev. $\underline{109}$, 1010 (1958).
15. M. Douglas and N. M. Kroll, Ann. Phys. N.Y. $\underline{82}$, 89 (1974).

MANY-ELECTRON ATOMS: STRUCTURE CALCULATIONS

APPLICATIONS OF RELATIVISTIC MBPT TO LITHIUM–LIKE ATOMS

W.R. Johnson
University of Notre Dame, Notre Dame, In. 46556

ABSTRACT

The relativistic "no–pair" hamiltonian is used as a starting point for applications of many–body perturbation theory to the structure of alkali atoms. Detailed calculations of energies of 2s and 2p states of lithium–like ions are presented and it is shown that these calculations are of sufficiently high quality to permit one to extract screening corrections to the one–electron Lamb–shift from high–precision measurements of energy levels.

INTRODUCTION

We present some relativistic many–body perturbation theory (MBPT) calculations of energy levels of ions with high nuclear charge. For few–electron ions such as those of the helium or lithium isoelectronic sequences comparison of calculated energy levels with experiment provides tests of QED in the strong fields of high Z nuclei. Experimental studies of low–lying levels of helium–like and lithium–like ions using accelerator–based beam–foil spectroscopy have been carried on for more than two decades and are a major source of data on levels of few electron ions. These beam–foil data are complemented by data from Tokamak plasmas, from laser–produced plasmas, and from spectra of solar flares. Precise experimental data are available for nuclear charges Z up to about 36 for the helium and lithium isoelectronic sequences.[1] We concentrate here on the lithium sequence since it is in some ways simpler to treat using perturbation theory.

Our aim is to describe how the electron–electron interaction can be accounted for lithium–like ions at a level of precision adequate to examine the residual QED effects along the Li sequence. For the very high Z members of this sequence a description of the ions using Dirac–Coulomb wave functions and evaluating the electron–electron interaction in lowest–order perturbation theory is sufficiently accurate to permit one to study QED effects. Near the neutral end of the sequence electron–electron correlation corrections are relatively large and the tools of MBPT are required to treat the interaction at the desired level of accuracy.

It should be mentioned that our ultimate goal is to carry out a calculation of the parity–violating 6s → 7s electric dipole transition amplitude in cesium at the one–percent level of accuracy. A precise theoretical value for this amplitude will make it possible to extract coupling constants for the neutral weak interaction from atomic physics experiments a with precision comparable to that obtainable in high energy accelerator–based measurements of W and Z_0 masses.[2] Examining Li–like ions should help us develop, in a relatively simple three–electron system, the tools necessary to attack our more ambitious goal in the fifty–five electron cesium atom. We start our calculations using a "no–pair" configuration–space hamiltonian of the type described by Sucher.[3] We calculate the atomic structure as precisely as we can using this Hamiltonian; later we evaluate terms omitted from the no–pair Hamiltonian as perturbations.

210

BASIC FORMULAS

As mentioned above, our point of departure is the Hamiltonian written down by Sucher in his second lecture,[3] viz.:

$$H = H_0 + V , \tag{1a}$$

$$H_0 = \sum_i h_i , \tag{1b}$$

$$h = c\boldsymbol{\alpha} \cdot \mathbf{p} + (\beta - 1)mc^2 + V_{nuc}(\mathbf{r}) + U(\mathbf{r}) , \tag{1c}$$

$$V = \frac{1}{2} \sum_{i,j} \Lambda_+ \left[\frac{\alpha}{r_{ij}} + b_{ij} \right] \Lambda_+ - \sum_i \Lambda_+ U(\mathbf{r}_i)\Lambda_+ . \tag{1d}$$

In Eqn.(1c), $V_{nuc}(\mathbf{r})$ is the nuclear potential and $U(\mathbf{r})$ is a central potential introduced to account approximately for the electron–electron interaction. The quantity b_{ij} in Eqn.(1d) is the Breit interaction and Λ_+ is a positive energy projection operator, constructed from solutions to Eqn.(1c). In our studies of the lithium–like ions we choose U to be the Hartree–Fock potential for the two 1s electrons of the helium–like core.

We follow Lindgren and Morrison[4] and express H in second quantized form. If we let a_k and $a_k^\dagger$ denote electron annihilation and creation operators, respectively, then we may write

$$H_0 = \sum_k \epsilon_k \, a_k^\dagger a_k , \tag{2a}$$

$$V = \frac{1}{2} \sum_{ijkl} (g_{ijkl} + b_{ijkl}) \, a_i^\dagger a_j^\dagger a_l a_k - \sum_{ik} U_{ik} \, a_i^\dagger a_k . \tag{2b}$$

In Eqn.(2a) the quantity ϵ_k is the eigenvalue of the one–electron Dirac equation

$$h \, u_k(\mathbf{r}) = \epsilon_k \, u_k(\mathbf{r}) . \tag{3}$$

For the important case that U is the Hartree–Fock potential, V_{HF}, one solves the set of coupled Eqns.(3) self–consistently to obtain core orbitals $u_a(\mathbf{r})$ and uses these self–consistent core orbitals to determine the potential and the orbitals for excited states. The quantities g_{ijkl} in Eqn.(2b) denote two–electron Coulomb matrix elements

$$g_{ijkl} = \langle ij| \frac{1}{r_{12}} |kl\rangle , \tag{4}$$

while the quantities b_{ijkl} denote two–electron matrix elements of the Breit interaction,

$$b_{ijkl} = <ij| -\frac{1}{2r_{12}} (\alpha_1 \cdot \alpha_2 + \alpha_1 \cdot \hat{r}_{12} \, \alpha_2 \cdot \hat{r}_{12}) \, |kl> . \tag{5}$$

The terms U_{ij} in Eqn.(2b) are matrix elements of the central potential $U(r)$,

$$U_{ij} = <i| \, U \, |j> . \tag{6}$$

For the special case that $U(r)$ is the HF potential one has the relation

$$U_{ij} = (V_{HF})_{ij} = \sum_a (g_{iaja} - g_{iaaj}) , \tag{7}$$

where the sum extends over occupied core states.

In zeroth approximation the Fock–space many–electron wave function for lithium is

$$\Psi_0 = a_v^\dagger \, a_{1s\uparrow}^\dagger \, a_{1s\downarrow}^\dagger \, | \, 0 > ; \tag{8}$$

this is an eigenstate of H_0 with energy, $E_v^{(0)}$,

$$E_v^{(0)} = \epsilon_v + 2\epsilon_{1s} . \tag{9}$$

The HF eigenvalue ϵ_v has the physical significance of being the removal energy of an electron in the state v; namely, the energy of the state v relative to the He–like core. This removal energy is well known experimentally for low–Z ions. The zeroth approximation to the removal energies for neutral lithium agree with measured values at the one–percent level as shown in Table I. It is the one–percent difference between the values in the rows labeled $E^{(0)}$ and "Exp." that we calculate using many–body perturbation theory.

The corresponding situation for the heavier alkali atoms is not nearly as good as for lithium. For example, in neutral cesium the zeroth–order theoretical removal energy is $-0.127\cdots$ au, while the corresponding experimental energy is $-0.143\cdots$ au; a difference of about 14%.[5]

Since we base our calculations on single–particle HF orbitals, there is no first–order correction to the **removal** energy caused by the Coulomb field part of the interaction. Indeed, the lowest non–vanishing Coulomb correction is the second–order correction given by:

$$E_v^{(2)} = -\sum_{anm} \frac{g_{vanm} (g_{nmva} - g_{nmav})}{(\epsilon_n + \epsilon_m - \epsilon_v - \epsilon_a)}$$

$$+ \sum_{abm} \frac{g_{vmab} \left(g_{abvm} - g_{abmv} \right)}{\left(\epsilon_v + \epsilon_m - \epsilon_a - \epsilon_b \right)} . \tag{10}$$

In Eqn.(10) the indices a and b designate core orbitals, and n and m electron orbitals outside the core, while v designates a valence orbital. These multiple sums are difficult to evaluate directly since the excited state sums are infinite sums over the bound state spectrum of the Dirac equation plus integrals over the positive energy continuum states.

Table I. Contributions to the valence electron removal energy for n = 2 levels of lithium.

Term	$2s_{1/2}$	$2p_{1/2}$	$2p_{3/2}$
$E^{(0)}$	−0.196320	−0.128638	−0.128636
$E^{(2)}$	−0.001649	−0.001375	−0.001374
$E^{(3)}$	−0.000125	−0.000145	−0.000145
$B^{(1)}$	0.000005	0.000003	0.000001
$B^{(2)}$	−0.000002	−0.000001	−0.000001
R.M.	0.000015	0.000008	0.000008
Total	−0.198076	−0.130148	−0.130147
Exp.	−0.198142	−0.130236	−0.130235
CPA	−0.198143	−0.130225	−0.130224

In the nonrelativistic case, a single sum over excited states, such as that occurring in the second term in Eqn.(10) can be converted into an integral over the solution to an inhomogeneous radial differential equation, the "single–particle equation", using the Sternheimer[6] or Dalgarno–Lewis[7] method. In a similar way the double sum occurring in the first term in Eqn.(10) can be converted into a double integral over the solution to a partial differential equation in two radial variables, the "pair equation". Numerical methods to treat the pair equation were developed by McKoy and Winter[8] and by Musher and Schulman[9] in the 1960's. Single–particle and pair equations have been used extensively by Lindgren and his co–workers[4] to treat the sums occurring in MBPT. In the relativistic case, the sums over excited states in Eqn.(10) include positive energy contributions only so projection operators occur in the inhomogeneous terms of the associated differential equations. The difficulty of constructing these projection operators numerically make the problem of solving the relativistic single–particle and pair equations considerably more complicated than the corresponding nonrelativistic problem. Nevertheless, techniques have been developed to solve the pair equations in the relativistic case and approximate solutions have been recently published by Lindroth.[10]

An alternative to the differential equation approach is to introduce a finite set of basis functions to approximate the actual spectrum of the Dirac equation, and to replace sums over the actual spectrum of the Dirac equation by sums over the finite pseudo spectrum. Finite basis sets for the Dirac equation constructed from Slater–type orbitals have been discussed extensively by Drake and Goldman,[11] and also by Grant.[12] We have made use of basis sets constructed from B–splines which are relative easy to generate, and have very nice numerical properties.[13] Our B–spline pseudo spectrum typically consists of 80 to 100 functions. Half of these functions belong to the positive energy spectrum and half to the negative energy spectrum. The low–lying terms in the positive energy spectrum agree very precisely with the exact bound states. The sums in Eqn.(10) can be carried out to any desired accuracy by simply increasing the number of terms in the basis. We found that 40 positive energy basis functions was sufficient to evaluate the sums in Eqn.(10) to the accuracy required for the present study of the Li–isoelectronic sequence. Results of our evaluation of $E^{(2)}$ from Eqn.(10) for the n=2 states of lithium are given in the second row of Table I. One finds that this second–order correction accounts for approximately 90% of the correlation energy. The situation is similar for the remaining alkalis; the second–order Coulomb correction accounts for all but a few percent of the correlation energy.

For neutral lithium the next most important correction is the third–order Coulomb correction, $E^{(3)}$, which contains 56 terms. A typical term in the third–order energy is

$$-\sum \frac{g_{abnm}\, g_{nmrs}\, g_{rsab}}{(\epsilon_n + \epsilon_m - \epsilon_a - \epsilon_b)(\epsilon_r + \epsilon_s - \epsilon_a - \epsilon_b)} . \tag{11}$$

The indices n,m,r,s denote excited electron states, as before, and the sum extends over all such states. If we make use of the spherical symmetry of the one–electron orbitals, then each index m, $\cdots$ corresponds to a double sum over an angular momentum quantum number, κ_m, and a principal quantum number, n_m. If we restrict the number of angular momentum values to $|\kappa_m| \leq 10$ and the number of spline functions corresponding to the principal quantum number to $n_m \leq 40$, then there are 4×10^{11} terms to accumulate, each of which is the product of Slater integrals. Accumulating quadruple sums of this type is the slowest single step in our calculation. Our third–order calculations were carried out on the CRAY XMP/48 computer at NCSA. In the third row of Table I the corrections obtained by evaluating all 56 terms in the third–order energy are presented. Because of the rapidly increasing complexity of perturbation theory with order, it is really impractical for us to contemplate a complete evaluation of the fourth–order energy at the present time.

To avoid mistakes in signs or in factors of two in the many terms contributing to the perturbation theory expression for the energy, we have developed a computer code in REDUCE to implement Wick's theorem and to determine the formulas for $E^{(n)}$. In addition, we have developed a code in REDUCE to carry through the logic of the JLV theorems[4] and to produce formulas for the various terms in perturbation theory after angular momentum analysis. These codes are very valuable aids in avoiding simple mistakes. After this analysis there remain complicated numerical steps to be carried out before one arrives at the numerical values in Table I. The values given in the table were determined independently by at least two of us in order to further reduce the chance of error.

In the fourth and fifth rows of Table I we list the contributions to the energy of the n=2 states from the Breit interaction. The term $B^{(1)}$ is just the first—order exchange correction

$$B^{(1)} = -\sum_a b_{avva} \, , \tag{12}$$

and $B^{(2)}$ is the second—order correction corresponding to one Breit interaction and one Coulomb interaction. The sixth row contains the corrections associated with reduced mass and mass polarization. The mass—polarization correction for lithium—like systems was worked out using screened hydrogenic wave functions by Hughes and Eckhart in 1930.[14]

In the seventh row of Table I we show the total of all of the corrections considered and in the eighth. row the experimental values from the NBS tables of atomic spectra.[15] The differences of less than a part in a thousand between the theory and experiment is primarily due to fourth—order Coulomb corrections. The most important of these fourth—order terms and terms of higher—order can be picked up using coupled pair approximation (CPA) techniques similar to those described by Lindgren in his third lecture.[16] In the last column of Table I we list the values of the energy obtained for the three n = 2 states of lithium by Z.W. Liu who solved the relativistic CPA equations.[17] Liu's calculations include the same Breit interaction corrections and reduced mass corrections as are included in perturbation theory; they also include triple—excitation corrections from third—order perturbation theory. With the CPA approach the difference between theory and experiment is reduced to better than a part in 10^4.

For ions of higher charge along the isoelectronic sequence one expects (and indeed finds) the perturbation theory to converge more rapidly than for neutral lithium because of the increasing dominance of the nuclear charge. The QED effects not included in the no—pair hamiltonian such as the electron self—energy and vacuum—polarization corrections grow much more rapidly with increasing Z than the Coulomb correlation corrections. A comparison between theory and experiment along the isoelectronic sequence therefore allows one to isolate and study these QED corrections. In Table II we show theoretical, experimental, and "theory—experiment" for the $2p_{1/2} - 2s_{1/2}$ transition energy along the lithium isoelectronic sequence. It is seen that the difference is approximately equal to the one—electron Lamb—shift shown in the last column of the table.

Table II. Differences between theory and experiment for the $2p_{1/2} - 2s_{1/2}$ energy interval (a.u.) in Li–like ions are compared with one–electron Lamb shift calculations.

Z	Theory	Experiment[a]	Th.– Exp.	–Lamb Shift
3	0.067928(4)	0.067906	0.000022(4)	0.000010
4	0.14554 (1)	0.14548	0.00006 (1)	0.00003
5	0.22041 (1)	0.22034	0.00007 (1)	0.00006
6	0.29391 (1)	0.29381	0.00010 (1)	0.00012
7	0.36676 (1)	0.36662	0.00014 (1)	0.00021
8	0.43934 (1)	0.43912	0.00022 (1)	0.00033
9	0.51186 (1)	0.51150 (1)	0.00036 (2)	0.00051
10	0.58442 (1)	0.58390 (1)	0.00052 (1)	0.00074
11	0.65716 (1)	0.65640 (1)	0.00076 (2)	0.00103
13	0.80339 (1)	0.80197 (2)	0.00142 (2)	0.00186
15	0.95101 (1)	0.94849 (5)	0.00252 (6)	0.00306
17	1.10041 (1)	1.09652 (5)	0.00389 (6)	0.00473
20	1.32862 (1)	1.32156 (4)	0.00706 (4)	0.00831
22	1.48403 (1)	1.47420 (4)	0.00983 (4)	0.01154
24	1.64239 (1)	1.6290 (1)	0.0134 (1)	0.01556
26	1.80406 (1)	1.7861 (1)	0.0180 (1)	0.02046
28	1.96940 (1)	1.9460 (4)	0.0234 (4)	0.02636
30	2.13877 (1)			0.03336
33	2.40122 (1)			0.04608
36	2.67485 (1)	2.6181 (4)	0.0568 (4)	0.06200
41	3.15960 (1)			0.09640
54	4.63506 (2)			0.2458
74	7.80652 (2)			0.7278
92	11.87525 (5)			1.5724

a.) Edlén, Ref.1

This later value is just the value expected for a hydrogen–like ion of the corresponding nuclear charge. One sees that the "theory–experiment" value is systematically lower than the one–electron Lamb–shift, an effect referred to as screening of the Lamb shift. One can identify those terms in QED responsible for screening of the Lamb–shift, and the challenge that now remains is to evaluate those terms and check the extent to which one really understands a simple relativistic many–body system such as a highly charged lithium–like ion.

References

1. B. Edlén, Physica Scripta **28**, 51 (1983).
2. S.A. Blundell, article in these proceedings.
3. J. Sucher, article in these proceedings.
4. I. Lindgren and J. Morrison, *Atomic Many–Body Theory*, 2nd Ed. (Springer,Berlin,1986).
5. W.R. Johnson, M. Idrees, and J. Sapirstein, Phys. Rev. A **35**, 3218 (1987).

216

6. R.M. Sternheimer, Phys. Rev. **80**, 102 (1950).
7. A. Dalgarno and J.T. Lewis, Proc. R. Soc. London, Ser. A **233**, 70 (1955).
8. V. McKoy and N.W. Winter, J. Chem. Phys. **48**, 5514 (1968).
9. J. I. Musher and J.M. Schulman, Phys. Rev. **173**, 93 (1968).
10. Eva Lindroth, Phys. Rev. **37**, 316 (1988).
11. G.W.F. Drake and S.P. Goldman, Phys. Rev. A**23**, 2093 (1981).
12. I.P. Grant, Phys. Rev. A**25**, 1220 (1982).
13. W.R. Johnson, S.A. Blundell, and J. Sapirstein, Phys. Rev. A**37**, 307 (1988).
14. D.S. Hughes and Carl Eckart, Phys. Rev. **36**, 694 (1930).
15. C.E. Moore, *Atomic Energy Levels*, Natl. Bur. Stand. (U.S.) Circ. No. 35 (U.S. GPO, Washington,D.C.,1971), Vol.I.
16. I. Lindgren, article in these proceedings.
17. Zuwei Liu, Thesis, University of Notre Dame, unpublished.

NON-RELATIVISTIC APPLICATION OF THE FINITE BASIS SETS CONSTRUCTED FROM B SPLINES

T. N. CHANG

Department of Physics, University of Southern California
Los Angeles, CA 90089-0484

ABSTRACT

We present a simple extension of the use of finite basis sets constructed from B Splines to the non-relativistic configuration interaction (CI) and perturbation calculations following the recent relativistic many-body perturbation application by Johnson et al [Phys. Rev. A37, 307 (1988)]. With the *continuum* contribution from the multi-electron interactions explicitly included in the B-Spline basis, the numerical accuracy of our non-relativistic calculations is improved substantially. A combined application of perturbation and CI calculation to the fine structure level-splittings of He 1s4f multiplets will be presented first. The effectiveness of the finite B-Spline basis set is further demonstrated in its application to the determination of the energy levels and the oscillator strengths of Be atom in a simple CI calculation.

INTRODUCTION

The ability of the finite basis sets constructed from B Splines to account for the effects of the many-electron interaction has been demonstrated successfully in a recent relativistic many-body perturbation application by Johnson et al.[1] Unlike the Slater-type orbits employed extensively in the construction of finite basis set in many other theoretical calculations,[2] the B-Spline basis set does not depend on the somewhat arbitrary procedure in selecting the nonlinear parameters in the exponential functions. The bell shape B Splines with similar amplitude also tend to treat the entire physical region included in the chosen finite volume more uniformly than the Slater-type orbits which generally favor the small r region. Many of the basic properties of B Splines can be found elsewhere[1,3,4] and we will limit our discussion to those related to the construction of the *non-relativistic* B-Spline basis set only.

In this paper, we will first outline briefly the basic procedure in constructing the *finite* set of non-relativistic one-particle radial function in terms of the B Splines. Second, we will present the result of a recent application of finite B-Spline basis set to a combined perturbation and CI calculation for the fine structure level-splittings of the He 1s4f $^{1,3}F$ multiplets.[5] Third, we will report the results of another recent application of finite B-Spline basis in a simple non-relativistic CI calculation for the energy levels and the oscillator strengths of Beryllium atom.[6]

THE NON-RELATIVISTIC B-SPLINE BASIS SET

The non-relativistic radial functions χ usually satisfy an eigenequation in the form of

$$-\frac{1}{2}\frac{d^2\chi}{dr^2} + V(r)\chi = \epsilon\chi \tag{1}$$

where the one-particle interaction potential $V(r)$ is non-local in general. The potential $V(r)$ is chosen to represent the effective interaction for the atomic system in each individual application as appropriate. The solution χ is expanded in terms of a set of B Splines[1,3,4] of order k and total number n defined by an exponential knot sequence confined between $r = 0$ and $r = R$, i.e.,

$$\chi(r) = \sum_{i=1}^{n} c_i B_i(r). \tag{2}$$

We have omitted the index of n and k from the functions B_i for simplicity. The value of R is selected to be sufficiently large to cover the estimated physical size of the atomic states of our interest. At the endpoints $r = 0$ and $r = R$, all B Splines equal to zero except B_1 and B_n, i.e.,

$$\begin{cases} B_1(r = 0) = 1 & \& \quad B_i(r = 0) = 0, \ i = 2, 3, \ldots, n; \\ B_n(r = R) = 1 & \& \quad B_i(r = R) = 0, \ i = 1, 2, \ldots, n-1. \end{cases} \tag{3}$$

The non-relativistic boundary conditions $\chi(0) = \chi(R) = 0$, therefore, require that

$$c_1 = c_n = 0. \tag{4}$$

Substitution of Eqs. (2) and (4) into Eq. (1) leads to a $(n-2) \times (n-2)$ symmetric generalized eigenvalue equation

$$HC = \epsilon AC. \tag{5}$$

where H and A are $(n-2) \times (n-2)$ symmetric matrices given by

$$H_{ij} = -\frac{1}{2} < B_i \mid \frac{d^2}{dr^2} \mid B_j > + < B_i \mid V \mid B_j >; i\&j = 2, \ldots, (n-1) \tag{6}$$

$$A_{ij} = < B_i \mid B_j >; \qquad i\&j = 2, \ldots, (n-1). \tag{7}$$

The eigenfunction χ_ν corresponding to the energy eigenvalue ϵ_ν is given by

$$\chi_\nu = \sum_{i=2}^{n-1} c_i B_i(r). \tag{8}$$

where the set of n-2 coefficients c_i forms the eigenvector

$$C_\nu = (c_2, c_3, \ldots, c_{n-1}). \tag{9}$$

corresponding to ϵ_ν. The calculated ϵ_ν of the first few lowest negative energy solutions with their eigenfunctions completely confined in R should agree with the numerical results of Eq. (1) from direct integration. The complete set of n-2 radial eigenfunctions χ_ν of Eq. (1) constructed from B Splines can then be used to form the finite basis set for either the CI or perturbation calculations.

FINE-STRUCTURE LEVEL-SPLITTINGS OF THE He 1s4f MULTIPLETS

In this section, we present a combined application of perturbation and configuration interaction calculation (CI) for the non-relativistic electrostatic energy for the He 1s4f ^{1,3}F multiplets. The quantitative results from our study will show that the numerical efficiency and the quantitative accuracy of our *ab initio* calculation are greatly improved with the use of the *finite* basis set constructed from the B Splines.

The high-precision microwave-optical resonance measurement in the early 70's by Wing and his co-workers[7] on the fine strucutre level-splittings (FSLS) for the high angular momentum Rydberg states of helium with one highly excited electron in a high-ℓ orbit screened by a tightly bound inner 1s electron has inaugurated a series of similar experiments[8-9] with ever increasing precision which is capable of testing smaller physical effects beyond the usual spin-dependent interactions.[10] For state with its total orbital angular momentum L$\leq$2, the spin-dependent contribution to FSLS is small compared to the contribution from the electrostatic exchange interaction. As L increases, the exchange interaction decreases much more rapidly than the spin-dependent interactions. In particular, for F-state with L=3, the spin-dependent contribution become comparable to that from the electrostatic exchange interaction. For states with L$\geq$4, the FSLS is overwhelmingly dominated by the contribution from the spin-dependent interactions.[5]

Within the quasi-hydrogenic approximation, the theoretical interpretation of the FSLS for a two-electron atomic system, takings into account both the electrostatic and spin-dependent interactions (i.e., the spin-orbit, the spin-other-orbit, and the spin-spin interactions in the Breit-Pauli approximation[11]), was first developed by Araki[12] and later extended by Parish and Mires.[13] Following the derivation of Parish and Mires, the relative energies of the four fine structure levels (i.e., 1s$n\ell$ ^{1,3}L$_J$ multiplets) may be expressed by a set of simple analytical formulas, i.e.,[5]

$$\epsilon_1 \simeq -K - (\ell+1)(1 - \frac{(6\ell-1)}{Z(2\ell-1)})\lambda; J = L - 1 \tag{10}$$

$$\epsilon_2 \simeq -\frac{1}{2}[(2K + \lambda\frac{Z-1}{Z})^2 + 4\ell(\ell+1)(\frac{Z+1}{Z})^2\lambda^2]^{\frac{1}{2}} - \frac{1}{2}\lambda\frac{Z-1}{Z}; J = L \tag{11}$$

$$\epsilon_3 \simeq -K + \ell(1 - \frac{(6\ell+7)}{Z(2\ell+3)})\lambda; J = L + 1 \tag{12}$$

$$\epsilon_4 \simeq \frac{1}{2}[(2K + \lambda\frac{Z-1}{Z})^2 + 4\ell(\ell+1)(\frac{Z+1}{Z})^2\lambda^2]^{\frac{1}{2}} - \frac{1}{2}\lambda\frac{Z-1}{Z}; J = L \tag{13}$$

where

$$\lambda = \frac{1}{2}\alpha^2 Z(Z-1)^3/[n^3\ell(\ell+1)(2\ell+1)] \tag{14}$$

is in atomic unit, α is the fine structure constant, and K is the electrostatic exchange energy. Eqs. (10-12) represent the relative energies for the three triplet levels and Eq. (13) represents the energy for the singlet level in the usual LS coupling. Eqs. (10-13) are derived from the theoretical results of

Parish and Mires[13] for the L$\geq$2 states [i.e., Eqs. (48-51) in Ref. 13] by the approximation that

$$\xi \simeq \lambda/Z \tag{15}$$
$$\eta \simeq \lambda/[Z(2\ell-1)(2\ell+3)] \tag{16}$$

and neglecting the small contributions from ξ' and η'.

The FSLS of He 1snf ^{1,3}F multiplets is characterized by the strong singlet-triplet mixing as a result of comparable contributions from the exchange interaction and the spin-dependent interactions. All earlier works have shown that the first order exchange energy is about a factor of 2 larger than its expected value. The large cancellation to the first order exchange energy due to the second order contribution found in our earlier perturbation calculation[14] suggests that a more definitive calculation should at least include an estimation of the higher order contribution. Experimentally, in a recent rf electric resonance experiment, Schilling et al [15] have derived the exchange energy 2K for the He 1s4f multiplets from the measured fine structure intervals between the ^{1}F$_3$ and the ^{3}F$_J$ levels. Their experimental 2K value is in agreement with the most recent variational CI-Hy (configuration interaction-Hylleraas) result by Sims and Martin[16] but about 13% larger than the value from our earlier perturbation calculation.[14] In the following we will report the result of our renewed theoretical effort by using the finite basis set constructed from B Splines which has led to a substantial improvement in the quantitative accuracy of the electrostatic exchange energy 2K over our previous perturbation calculation.

Similar to our early study,[14] we have selected the one-electron hydrogenic wave functions subject to a nuclear charge Z for the inner electron and a charge of (Z-1) for the outer electron in a quasi-hydrogenic approximation. The hydrogenic wave function $\mathcal{U}_{n\ell mm_s}$ is expressed in terms of the product of the radial part $\chi_{n\ell}(r)$, the angular part $Y_{\ell m}$, i.e., the spherical harmonics, and the spin function $\sigma(m_s)$

$$\mathcal{U}_{n\ell mm_s} = \frac{\chi_{n\ell}(r)}{r} Y_{n\ell}(\Omega)\sigma(m_s). \tag{17}$$

The first order electrostatic exchange energy K_1 is given by

$$K_1 = \frac{1}{2\ell+1} < 1s \mid Y_\ell(n\ell,1s;r) \mid n\ell > \tag{18}$$

where $< n_1\ell_1 \mid Y_k(n_2\ell_2,n_3\ell_3;r) \mid n_4\ell_4 >$ is the coulomb energy integral given elsewhere.[5] The second order electrostatic exchange energy K_2 is given by

$$K_2 = 2 \sum_{n_1\ell_1} \sum_{n_2\ell_2} \frac{V_d(n_1\ell_1,n_2\ell_2)V_{ex}(n_1\ell_1,n_2\ell_2)}{\Delta D(n_1\ell_1,n_2\ell_2)} \tag{19}$$

where the complete sets of intermediate states $n_1\ell_1$ and $n_2\ell_2$, each representing both the $bound$ and the $continuum$ components, are included in the summation. The double summation is taken over all allowed two-electron configurations except 1snℓ. The direct and exchange integrals V_d and V_{ex} are given by

$$V_d(n_1\ell_1,n_2\ell_2) = \begin{cases} \eta_{\ell_1}(0\ell\ell_1\ell_2;L) < 1s \mid Y_{\ell_1}(n\ell,n_2\ell_2;r) \mid n_1\ell_1 >; n_1\ell_1 \neq 1s \\ \delta_{\ell\ell_2} < n\ell \mid Y_0(1s,1s;r) - \frac{1}{r}) \mid n_2\ell_2 >; n_1\ell_1 = 1s \end{cases} \tag{20}$$

and

$$V_{ex}(n_1\ell_1, n_2\ell_2) = \eta_{\ell_2}(0\ell\ell_2\ell_1; L) < 1s \mid Y_{\ell_2}(n\ell, n_1\ell_1; r) \mid n_2\ell_2 > \qquad (21)$$

where the angular factor η_k is given elsewhere.[17] The energy denominator ΔD is given by

$$\Delta D(n_1\ell_1, n_2\ell_2) = \epsilon_{1s}(Z) + \epsilon_{n\ell}(Z-1) - \epsilon_{n_1\ell_1}(Z) - \epsilon_{n_2\ell_2}(Z-1) \qquad (22)$$

with the choice that $\ell_2 > \ell_1$ and $n_1\ell_1$ and $n_2\ell_2$ represent the orbitals for the inner and outer electron respectively.

Like all many-body perturbation calculations, the most time-consuming part of the numerical works is the summing over the complete intermediate basis included in the double summation in Eq. (19). In the present calculation, we replace the complete intermediate basis with the *finite* basis set constructed from B Splines. Within the quasi-hydrogenic approximation, the finite basis set is generated in a hydrogenic potential given by

$$V(r) = -\varsigma/r + \frac{1}{2}\ell(\ell+1)/r^2 \qquad (23)$$

where $\varsigma = Z$ for the inner electron and $\varsigma = Z-1$ for the outer electron. In the present calculation, the numerical eigenvalues of the first few lowest negative energy solutions agree to the hydrogenic values to 10^{-10} Ryd. The *quasi-complete* set of eigenfunctions constructed from B Splines are then used in the calculation of the second order exchange energy $2K_2$.

The calculated exchange interactions up to second order are listed in Table I. The basis functions χ are constructed from sets of B Splines of n ranging from 27 to 62 and k ranging from 7 to 11. The numerical results listed in Table I are the converged individual contribution from each $\ell_1\ell_2$ pair, i.e., $\ell_1 = \ell_i$ for the *inner* electron and $\ell_2 = \ell_o$ for the *outer* electron, with various combinations of n and k as we vary the radius R from $90a_0$ to $200a_0$. We note that the combined contribution for the first four second order terms up to $(\ell_i\ell_o) = (2,3)$ combination agree well with our early perturbation calculation[14] which did not include the contributions from other higher $(\ell_i\ell_o)$ combinations. The present calculation also confirms the substantial contribution from the *continuum* part of the intermediate basis set as indicated by the large difference between the *more complete* second order calculation and the one which includes only the *bound* component shown in Table I. With an uncertainty of about 0.18 MHz or less, the sum of the first and second order exchange energies is about 95.11 MHz. This is substantially smaller than the measured value of 158±3 MHz by Schilling *et al.*[15]

This larger discrepancy in the exchange energy between the experimental value and the more complete first and second order perturbation results clearly indicates the need to carry out theoretical calculation beyond second order. The obvious choice would be an extension of the perturbation calculation to include higher order contributions. Such an approach would require an enormous numerical effort with little advanced assurance that the perturbation series would indeed converge when third or fourth order terms are included. Alternatively, we could include contributions up to infinite orders in a configuration interaction (CI) calculation if a high numerical accuracy can be achieved to about 10^{-10} Ryd. or better in evaluating the energy difference between separate calculations for the ^{1}F and ^{3}F states. A straightforward *non-variational* CI calculation would also require enormous numerical effort with no assurance that the contributions from two-electron configurations corresponding to higher $(\ell_i \ell_o)$ combination could be calculated at the same level of numerical accuracy.

Table I. Calculated first and second order perturbation contribution to the electrostatic exchange energy 2K (in 10^{-8} Ryd.) for the He 1s4f multiplets. (From Ref. 5)

Energy	Orbital Angular Momenta		Electrostatic Exchange Energy 2K (in 10^{-8} Ryd.)	
	$\ell_i(Z=2)$	$\ell_o(Z=1)$	B-Spline Basis	H-like bound states (up to n=18)
1st order($2K_1$)			10.2555	10.2555
2nd order	0	3	0.2934	0.0001
	1	2	-4.8483	-3.6293
	1	4	-1.1877	-0.0075
	2	3	-0.6570	-0.0072
	2	5	-0.3299	-0.0001
	3	4	-0.1963	0.0000
	3	6	-0.1307	0.0000
	4	5 & 7	-0.1412	0.0000
	5	6 & 8	-0.0705	0.0000
	6	7 & 9	-0.0388	0.0000
	7	8 & 10	-0.0226	0.0000
	8	9 & 11	-0.0139	0.0000
	9	10 & 12	-0.0092	0.0000
	10	11 & 13	-0.0061	0.0000
	11	12 & 14	-0.0041	0.0000
	12	13 & 15	-0.0027	0.0000
Total ($2K_2$)			-7.3641	-3.6440
$2K_1 + 2K_2$			2.8914 (95.11 MHz)	6.6115 (217.48 MHz)

To minimize the numerical effort in the estimation of higher order contributions, we started exploring the possibility of a combined perturbation and CI calculation after recognizing that the larger individual second order contribution comes predominantly from configurations with $\ell_i=1$. Typically, the direct integrals V_d involving such configurations are few orders of magnitude larger than those with ℓ_i other than 1. In particular, we have examined the second order contribution from each configuration series $n_i \ell_i \ell_o$ [i.e., a set of configurations with a fixed inner electron orbit $n_i \ell_i$ combined with outer electron in orbits consisting of the complete set of solutions χ of Eq. (1) corresponding to the fixed ℓ_o] and concluded that the contributions from the first few configuration series with a $(\ell_i \ell_o) = (1,2)$ or pd combination and the first few configuration series with a $(\ell_i \ell_o) = (1,4)$ or pg combination are about two orders of magnitude larger than the individual contribution from other configuration series. In our combined perturbation and CI calculation, we first carried out separate CI calculation for both 1s4f ^1F and 1s4f ^3F states by including a selected number of dominating npd and npg configuration series in addition to the 2sf series which also contributes significantly. The singlet-triplet energy separation from the CI calculation is then substracted from the calculated first order value $2K_1$ to determine the second and higher order contribution to $2K$ due to the configuration series included in the CI calculation. The second order perturbation contribution from all other configurations *not* included in the CI calculation (listed in Table II) are then added to the contribution from CI calculation to get our final calculated $2K$ value.

Table II. The combined CI and perturbation contribution to the exchange energy $2K$ for He 1s4f multiplets. (From Ref. 5)

Contributing term	Electrostatic exchange energy $2K$	
	in 10^{-8} Ryd.	in MHz
First order $(2K_1)$	10.2555	337.34
CI contribution	-3.8714	-127.35
2nd order excluding configurations in CI calculation	-1.5868	-52.19
Total	4.7973	157.80
Sims and Martin[16]		158.02
Experiment[15]		158±3

As the number of configuration series included in the CI calculation increases, the calculated $2K$ value slowly converges to the experimental value. The calculated singlet-triplet separation Δ from CI calculation including contribution from 2sf series and all npd and npg series is determined by extrapolating the calculated Δ_N with increasing number of configuration series N according to the limiting expression[5]

$$\Delta_N = \Delta \left(1 - e^{-\mu(N-\delta)}\right) \tag{24}$$

where μ and δ are fitting parameters which are adjusted to get the best fit as N approaches the limiting value. The numerical fit to Eq. (24) from the calculated

Δ_N for N ranging from 27 to 38 leads to a value of $\Delta = 6.38406\times10^{-8}$ Ryd. which in turn yields a CI contribution of -3.8714×10^{-8} Ryd. The estimated uncertainty of our CI calculation is approximately 0.10 MHz.[5]

Table II lists the first order value $2K_1$, the CI contribution and the second order contribution from all other configurations *not* included in the CI calculation. The sum of these three contributions leads to a value of $2K = 157.80\pm0.28$ MHz which compares very well with the theoretical value from Sims and Martin[16] and the experimental value from Schilling *et al.*[15] Finally, Table III summarizes the FSLS values from the present calculation which are in excellent agreement with the observed values and the CI-Hy result.[15,16]

Table III. Comparison between the theoretical and experimental electrostatic exchange energy 2K and the FSLS values (in MHz) for He 1s4f multiplets.

Term	Present[5]	Sims & Martin[16]	Experiment[15]
2K	157.80	158.02	158 ± 3
ν_{12}	231.88	232.40	232.8 ± 2.0
ν_{14}	490.37	490.39	490.8 ± 2.0
ν_{13}	703.46	704.11	704.3 ± 1.0

The present calculation has shown that the combined application of the perturbation and CI calculation can indeed offer an effective alternative to a separate perturbation or CI calculation if the dominating contributions can be identified in detail. Our calculation has also shown that the *finite* basis set constructed from B Splines can be applied very effectively to the non-relativistic perturbation and CI calculations.

THE ENERGY LEVELS AND OSCILLATOR STRENGTHS of Be ATOMS

In this section, we present the results of a recent simple CI calculation on the energy levels and the oscillator strengths of Be atom using the finite B-Spline basis set.[6] The numerical procedures in a simple CI calculation are already presented in detail in our earlier works.[6,18] Following this simple CI procedure, the only component in the basis set that is not entirely fixed is the radial part χ of the one-particle orbital wave function. As a result, the success of the simple CI calculation depends primarily on the choice of the radial function χ and the size of the basis set included in the calculation of the Hamiltonian matrix. Most of the existing applications of simple CI procedure to the alkaline earth atoms,[19,20] including our recent works on Mg,[18,21] have employed basis set constructed from one-particle *bound* orbits subject to a potential V(r) corresponding to the ^1S rare gas-like frozen core. In spite of the success of these calculations, the use of *incomplete bound* basis set has inevitably left out the possible contribution from the *continuum* component which in principle should also be included as a part of *complete* basis set constructed from the one-particle orbits generated in the potential $V(r)$. In addition to this theoretical consideration, the accuracy of the simple CI calculation should also improve if the *incomplete bound* basis set is replaced by a *quasi-complete*[6] basis set such as the one constructed from the B Splines. In fact, the results presented in this section will show that the accuracy of the simple CI calculation with finite B-Spline basis set is at least comparable to and in most cases better than many of the most elaborate theoretical procedures.

In constrast to most of the more elaborate theoretical approaches, the simple CI calculation does not require the *separate* application of the optimization procedure to the initial and final states of an atomic transition. This particular feature together with the high numerical accuracy offer a substantial advantage in computational efficiency, thus, allowing for a more comprehensive calculation involving a large number of transitions in a *single* calculation by using a *single* basis set.

Similar to most of our recent works,[18,21] the radial function χ in the present calculation satisfies the eigenequation, i.e.,

$$h_\ell^{HF}(r)\chi_{n\ell}(r) = \epsilon_{n\ell}\chi_{n\ell}(r) \tag{25}$$

where h_ℓ^{HF} is the one-particle frozen core Hartree-Fock (FCHF) hamiltonian defined explicitly by Eq. (7) in Ref. 18. More specifically, the basis set in our simple CI calculation consists of a number of two-electron *configuration series* $n\ell\ell'$. Each *configuration series* consists of a series of configuration wavefunctions representing configurations with one electron in a fixed orbit $n\ell$ and the other ℓ'-electron in orbitals of varying energy corresponding to the entire set of solutions χ of Eq. (25) constructed from B Splines.

The B Splines in the Be calculation is confined in a fixed radius $R=150a_0$ to ensure the numerical *reliability* for the first four to five excited states in each SL series.[6] We have tested the accuracy of our numerical calculation by first varying the values of R, n and k which define the B Splines set. The first seven figures (i.e., up to 10^{-6} Ryd.) of the energy eigenvalues of the first five excited states in each SL series remain unchanged in our calculation when the value of R is varied from $130a_0$ to $160a_0$. We have also varied the value of n from 27 to 42 and the value of k from 7 to 9 and, again, the first seven figures of the calculated energy eigenvalues remain unchanged. After we selected $R=150a_0$, $n=27$, and $k=7$, we start examining the variation of energy eigenvalues as we increase the size (i.e., number of configuration series) of the basis set. The energy eigenvalue converges rapidly to 10^{-3} Ryd. when the number of contributing configuration series increases from one to 5. For a convergence up to 10^{-4} Ryd. or better, it requires typically 15 to even larger numbers of configuration series.

In Figure 1, we compare the observed quantum defects[22] of the Be 2snp ^{1}P series leading to the Be II 2s ionization limit with the theoretical values from few selected calculations.[6,23,24] Theoretical values from few other calculations[20,25-27] are also listed in Table IV for comparison. The results from the present simple CI calculation with finite B-Spline basis set (open square in Figure 1 and B-Spline FCHF in Table IV) compares very well with the experimental data (solid line).[22] The MCHF quantum defect by Saha and Froese Fischer[25] (Table IV) agrees to the B-Spline value to less than 1%. The results from the earlier close-coupling calculation (solid circle) by Norcross and Seaton,[23] though not as accurate as the B-Spline FCHF values, also agree well with the experimental

Table IV. Quantum defects for the 2snp ^{1}P series leading
to Be II 2s limits.

Theory	2s2p	2s3p	2s4p	2s5p	2s6p
B-Spline FCHF[6]	0.1569	0.2888	0.3274	0.3451	0.3554
Bound FCHF[6]	0.1262	0.2533	0.2949	0.3145	0.3260
Saha & Fischer[25]	0.1582				
Norcross & Seaton[23]	0.1506	0.2834	0.3226	0.3405	0.3505
Weiss[26]	0.1504	0.2715	0.3104		
Hibbert[27]	0.1481				
Laughlin & Victor[20]	0.1182	0.2413	0.2804	0.2877	0.2874
O'Mahony & Greene[24]	0.185	0.235	0.319		
Observed[22]	0.1661	0.2958	0.3324	0.3487	0.3574

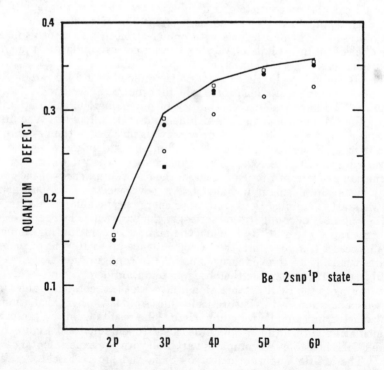

Figure 1. Comparison of quantum defects for the 2snp ^{1}P series
of Be I atom leading to the Be II 2s limit. (From Ref. 6)

data. The B-Spline FCHF results are in general more accurate than the results from more elaborate CI calculations by Weiss[26] and Hibbert[27] shown in Table IV. The model potential result by Laughlin and Victor[20] should improve to the bound FCHF value if the size of their basis set is enlarged. The R-matrix result (solid square) by O'Mahony and Greene[24] does not agree with the experiment as well as other calculations, but it should improve if its basis set is expanded. The *continuum* contribution excluded in the simple CI calculation with *incomplete bound only* basis is represented in Figure 1 by the difference between the open square and the open circle (or between B-Spline FCHF and Bound FCHF results in Table IV).

Figure 2 and Table V compare the quantum defects of the $2p^2$ and 2snd ^{1}D series of Be. Again, the B-Spline FCHF results[6] (open square in Figure 2) are in close agreement with the experimental data[22] (solid line) and the MCHF results.[25] The results of the close-coupling calculation[23] (solid circle) are slightly less accurate than the B-Spline FCHF results whereas the R-matrix method by O'Mahoney and Watanabe[28] (solid square) would again require an enlarged basis set if its numerical accuracy is to be improved.

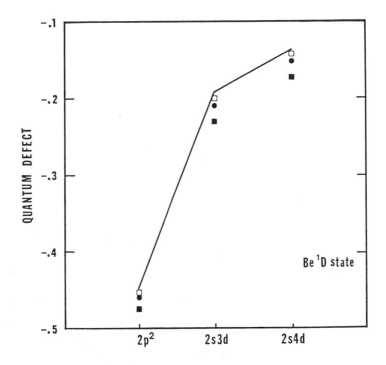

Figure 2. Comparison of quantum defects for the $2p^2$ and 2snd ^{1}D series of Be I atom leading to the Be II 2s limit. (From Ref. 6)

Table V. Quantum defects for the $2p^2$ and 2snd ^{1}D series leading to Be II 2s limits.

Theory	$2p^2$	2s3d	2s4d
B-Spline FCHF[6]	-0.4544	-0.1998	-0.1431
Bound FCHF[6]	-0.4846	-0.2322	-0.1711
Saha & Fischer[25]	-0.4516	-0.2013	-0.1466
Norcross & Seaton[23]	-0.4594	-0.2095	-0.1526
Hibbert[27]	-0.4708		
O'Mahony & Watanabe[28]	-0.475	-0.231	-0.173
Observed[22]	-0.4480	-0.1928	-0.1369

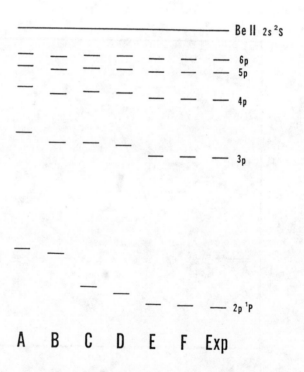

Figure 3. Comparison of the Be 2snp ^{1}P spectra leading to the Be II 2s ^{2}S ionization threshold. The experimental data are taken from Ref. 22. The configuration series included in the calculated spectra (A-F) are listed in Table VI. (From Ref. 6)

Figures 3 and 4 illustrate the energy corrections to the 2snp ^{1}P and the 2p^2 and 2snd ^{1}D series due to the configuration interaction. The configuration series included in each calculated ^{1}P and ^{1}D energy spectrum in Figures 3 and 4 are tabulated in Table VI. Spectrum A in both Figures 3 and 4 represent the calculated energy spectrum when the Hamiltonian matrix includes only the dominant configuration series 2sℓ. For the 2snp ^{1}P series, the spectrum B in Figure 3 indicates the relatively small energy correction to the 2s2p ^{1}P state and the fairly substantial corrections to the higher 2snp ^{1}P states due to the 2ps series. In contrast, the lowest 2s2p ^{1}P state is strongly affected by the 2pd series as shown by the spectra C and D in Figure 3. The energy correction to the 2s3p ^{1}P state due to the 2pd series remains large and for even higher 2snp ^{1}P states the energy corrections become small comparing to the effect due to the 2ps series. The *continuum* contribution from the 2pd configuration series is shown by the difference between the spectra C and D. Our calculation has also shown that the continuum contribution to the energy correction of the 2snp ^{1}P states from the 2ps configuration series is much smaller than that from the 2pd series. The convergence of the energy eigenvalue to 10^{-4} Ryd. is illustrated by the calculated spectra E and F in Figure 3 as the number of configuration series in our calculation increases from 16 to 20. For the 2snd ^{1}D states, the energies are strongly affected by the presence of the 2p^2 configuration as shown by the spectrum B in Figure 4. The *continuum* contribution from the 2pp series, represented by the difference between the spectra B and C, is relatively small. The contributions from the 2pf and 3dd series are noticeable as shown by the spectrum D. Again, the energy converges to about 10^{-4} Ryd. when the number of the configuration series increases from 16 to 20 as shown by the spectra E and F in Figure 4.

Similar to our earlier oscillator strengths (OS) study on the Mg atom,[29] the oscillator strengths for transitions involving states strongly mixed with the doubly excited perturber are substantially affected by the configuration interaction. For example, Figures 5 and 6 illustrate the variation of the calculated OS in both length and velocity approximations for the 2s2p ^{1}P to 2s3d ^{1}D and 2s4d ^{1}D transitions as the number of configuration series included in the initial and final states increases at different stage of approximation indicated in Table VII. As expected, the calculated OS is strongly affected by the presence of the 2ps and 2pd configuration series in the 2s2p ^{1}P state and the 2pp series in the 2snd ^{1}D states as shown by the large variation between the configuration combinations I and II in Figures 5 and 6. The calculated OS also increase significantly when the 2pf and 3dd series are added to the calculation of the 2snd ^{1}D states as shown by the difference between the configuration combinations II and III. The redistribution[29] of the oscillator strength is also evident as the OS value increases for the 2s2p ^{1}P - 2s4d ^{1}D transition whereas it decreases for the 2s2p ^{1}P - 2s3d ^{1}D transition. Further inclusion of the configuration series will only change the calculated OS slightly as shown by the calculations IV - VII in Figures 5 and 6. In general, the length results with a fluctuation of about 0.1% converge more smoothly than the velocity results with a fluctuation of about 1%. The agreement between the length and velocity results improves to about 3% or less except for transitions with OS values much less than 0.01 as the calculated excitation energy for each transition converges to its experimental value.

The accurate experimental determination of the oscillator strengths is ex-

Table VI. The configuration series n$\ell\ell'$ included in each of the calculated ^{1}P and ^{1}D spectrum of Be I shown in Figures 3 and 4. (From Ref. 6)

Calculated Spectrum	Configuration series	
	2snp ^{1}P	2p^2 and 2snd ^{1}D
A	2sp	2sd
B	2sp + 2ps	2snd + 2pnp (Bound orbits only)
C	2snp + 2pnd (Bound orbits only)	2sd + 2pp
D	2sp + 2pd	2sd + 2pp + 2pf + 3dd
E	16 series	16 series
F	20 series	20 series

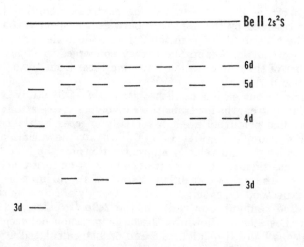

Figure 4. Comparison of the Be 2p^2 and 2snd ^{1}D spectra leading to the Be II 2s ^{2}S ionization threshold. The experimental data are taken from Ref. 22. The configuration series included in the calculated spectra (A–F) are listed in Table VI. (From Ref. 6)

Table VII. The configuration series $n\ell\ell'$ included in the initial and final states of the 2s2p ^{1}P to 2snd ^{1}D transitions in Be at different stage of approximztion shown in Figures 5 and 6. (From Ref. 6)

Calculated Spectrum	Configuration series	
	2s2p ^{1}P	2snd ^{1}D
I	2sp	2sd
II	2sp+2ps+2pd	2sd+2pp
III	2sp+2ps+2pd	2sd+2pp+3dd+2pf
IV	16 series	2sd+2pp+3dd+2pf
V	16 series	16 series
VI	16 series	20 series
VII	20 series	20 series

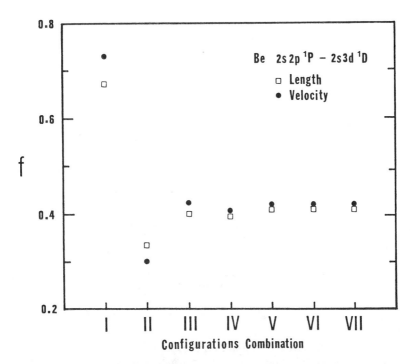

Figure 5. The variation of the calculated oscillator strengths f in length (open square) and velocity (solid circle) approximation for the 2s2p ^{1}P to 2s3d ^{1}D transition in Be. The configuration combinations representing the initial and final states of the transitions are listed in Table VII. (From Ref. 6)

232

tremely difficult and most of the available observed OS values are derived from early lifetime measurements using the beam-foil technique.[30] As pointed out by Saha and Froese Fischer,[25] the uncertainty of the measured OS could increase when the effect due to other less dominant decay channels are not included in converting the lifetime data into the oscillator strengths. Although the OS values from the present calculation are generally in good agreement with all available observed data, the accuracy of our calculated OS is measured primarily by i) the small fluctuation of the calculated OS as the size of the basis set in the simple CI calculation increases, ii) the close agreement between the measured and the calculated excitation energy of the transition, and iii) the less than 3% agreement between the *converged* length and velocity OS values except for transitions with OS much less than 0.01. More detailed discussions on the calculated energy levels and the oscillator strengths involving many other excited states are presented in Ref. 6.

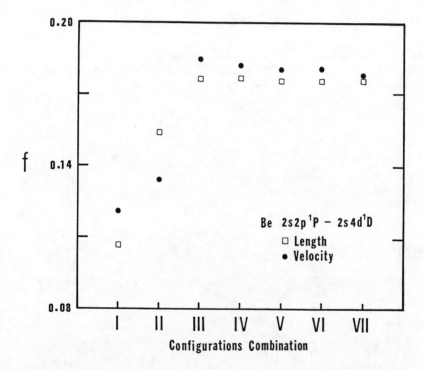

Figure 6. The variation of the calculated oscillator strengths f in length (open square) and velocity (solid circle) approximation for the 2s2p ^{1}P to 2s4d ^{1}D transition in Be. The configuration combinations representing the initial and final states of the transitions are listed in Table VII. (From Ref. 6)

The simplicity of the simple CI procedure together with the ability to take into account the *continuum* contribution by finite B-Spline basis set have made the simple CI approach one of the most efficient computational procedures with numerical accuracy at least comparable to and often better than more elaborate theoretical approaches. Applications to other atomic processes dominated by the strong multi-electron interactions are currently in progress.

ACKNOWLEDGMENTS

This work is supported in part by the National Science Foundation under Grant No. PHY84-08333. I would like to thank Walter R. Johnson for an introduction to the B-Spline technique and many stimulating discussions. I would also like to acknowledge the hospitality of the Institute of Theoretical Physics at the University of California at Santa Barbara during my part-time participation at the workshop on "Relativistic, QCD, and Weak Interaction Effects in Atoms" where the present work was first conceived.

REFERENCES

1. W. R. Johnson, S. A. Blundell, and J. Sapirstein, Phys. Rev. A37, 307 (1988).
2. For example, B. Gao and A. F. Starace, Phys. Rev. Lett. 61, 404 (1988); T. Mukoyama and C. D. Lin, Phys. Rev. A35, 4942 (1987).
3. C. deBoor, *A Practical Guide to Splines* (Springer, New York, 1978).
4. C. Bottcher and M. R. Strayer, Ann. Phys. (N. Y.) 175, 64 (1987).
5. T. N. Chang, Phys. Rev. A39, xxxx (1989).
6. T. N. Chang, Phys. Rev. A39, xxxx (1989); preprint, to be published in J. Chinese Phys. (1989).
7. W. H. Wing and W. E. Lamb, Jr., Phys. Rev. Lett. 28, 265 (1972); W. H. Wing, K. R. Lea, and W. E. Lamb, Jr., *Atomic Physics 3*, edited by S. J. Smith and G. K. Walters (Plenum, New York, 1973), p. 119.
8. E. A. Hessels, W. G. Sturrus, and S. R. Lundeen, Phys. Rev. A35, 4489 (1987); D. R. Cok and S. R. Lundeen, Phys. Rev. A23, 2488 (1981).
9. J. W. Farley, D. A. Weinberger, S. S. Tarng, N. D. Piltch, and W. H. Wing, Phys. Rev. A25, 1559 (1982); J. W. Farley, K. B. MacAdam, and W. H. Wing, Phys. Rev. A20, 1754 (1979).
10. S. L. Palfrey and S. R. Lundeen, Phys. Rev. Lett. 53, 1141 (1984); C. K. Au, G. Feinberg, and J. Sucher, Phys. Rev. Lett. 53, 1145 (1984).
11. H. A. Bethe and E. E. Salpeter, *Quantum Mechanics of One- and Two-electron Atoms* (Springer-Verlag, Berlin, 1957).
12. G. Araki, Proc. Phys. Math. Soc. Jap. 19, 128 (1937).
13. R. M. Parish and R. M. Mires, Phys. Rev. A4, 2145 (1971).
14. T. N. Chang and R. T. Poe, Phys. Rev. A10, 1981 (1974); A14, 11 (1976); R. T. Poe and T. N. Chang, *Atomic Physics 3*, edited by S. J. Smith and G. K. Walters (Plenum, New York, 1973), p. 143.
15. W. Schilling, Y. Kriescher, A. S. Aynacioglu, and G. v. Oppen, Phys. Rev. Lett. 59, 876 (1987).
16. J. S. Sims and W. C. Martin, Phys. Rev. A37, 2259 (1988).
17. T. N. Chang, Phys. Rev. A37, 4090 (1988).
18. T. N. Chang and Y. S. Kim, Phys. Rev. A34, 2609 (1986).

234

19. R. N. Zare, J. Chem Phys. $\underline{45}$, 1966 (1966); $\underline{47}$, 3561 (1967).
20. C. Laughlin and G. A. Victor, *Atomic Physics 3*, ed. S. J. Smith and G. K. Walters (Plenum, NY, 1973), p. 247.
21. T. N. Chang, Phys. Rev. A$\underline{34}$, 4550 (1986).
22. C. E. Moore, *Atomic Energy Levels*, Natl. Bur. Stand. Circ. No. 467 (US GPO, Washington, D. C., 1949), Vol. I; L. Johansson, Phys. Scripta $\underline{10}$, 236 (1974); Ark. Fys. $\underline{23}$, 119 (1962).
23. D. W. Norcross and M. J. Seaton, J. Phys. B$\underline{9}$, 2983 (1976).
24. P. F. O'Mahony and C. H. Greene, Phys. Rev. A$\underline{31}$, 250 (1985).
25. H. P. Saha and C. Froese Fischer, Phys. Rev. A$\underline{35}$, 5240 (1987); C. Froese Fischer, Phys. Rev. A$\underline{30}$, 2741 (1984).
26. A. W. Weiss, Phys. Rev. A$\underline{6}$, 1261 (1972).
27. A. Hibbert, J. Phys. B$\underline{9}$, 2805 (1976); J. Phys. B$\underline{7}$, 1417 (1974).
28. P. F. O'Mahony and S. Watanabe, J. Phys. B$\underline{18}$, L239 (1985).
29. T. N. Chang, Phys. Rev. A$\underline{36}$, 447 (1987).
30. I. Martinson, A. Gaupp, and L. J. Curtis, J. Phys. B$\underline{7}$, L463 (1974); S. Hontzeas, I. Martinson, P. Erman, and R. Buchta, Phys. Scripta $\underline{6}$, 55 (1972); J. Bromander, Phys. Scripta $\underline{4}$, 61 (1971); T. Andersen, K. A. Jessen, and G. Sorensen, Phys. Rev. $\underline{118}$, 76 (1969).

Notes on Basis Sets
for Relativistic Atomic Structure and QED

I.P.Grant

Department of Theoretical Chemistry, University of Oxford,
Oxford OX1 3UB, UK

ABSTRACT

This note attempts to assess the suitability of a number of types of basis set used to approximate solutions of Dirac operator equations in relativistic atomic structure and QED. Numerical evidence of basis set performance needs to be supported by mathematically rigorous theorems, most of which do not yet exist. The note outlines some attempts to supply such theorems.

INTRODUCTION

The idea of expanding wavefunctions as linear combinations of the members of a suitable set of square integrable basis functions is an old one which has been much exploited in quantum chemistry. Recently several groups represented at this Workshop have begun to use such methods for relativistic atomic structure calculations which include the effects of electron correlation and radiative corrections; they have often achieved considerable success. An investigation into the mathematical foundations of these methods of approximation is therefore timely, as it is essential to have some assurance that basis set approximations converge in some well understood mathematical sense for each observable quantity being computed. Numerical evidence may encourage a belief that the approximation is convergent; this may easily be an illusion if there is no properly formulated mathematical justification.

Some background to my interest in these questions may help motivate my approach. I and my colleagues published a relativistic atomic structure package in 1980[1]; this consisted of a number of independent modules (MCP, MCT, MCBP, MCDF, BENA, QED) communicating by disc files, provided with an interface to make the system easy to use by non-specialists. In order to emphasize that this is not confined to multiconfiguration Dirac-Fock calculations, we now call the system GRASP — General-purpose Relativistic Atomic Structure Program. We are about to publish a new version (Version 1), due primarily to Ken Dyall (now at IBM, San Jose, CA), which eliminates a number of problems in the older codes and introduces some new features. The MCDF module relies on numerical integration and iterative techniques to solve coupled Dirac equations. The iteration can fail to converge to any solution in unfavourable cases, and the main intention of Version 1 is to use improved algorithms to overcome some of the problems. Farid Parpia is working on Version 2, which uses strategies more in line with those of Charlotte Fischer[2] which are expected to help with problems of complex spectra in which elaborate correlation calculations are required. Version 2 is also needed to provide high quality wavefunctions for states of heavy atoms and ions (such as Pb and Hg) to serve as targets in relativistic close-coupling electron-scattering and photoionization calculations using the R-matrix method. Here, the intention is to study the resonances encountered at near-threshold energies in measured elastic and inelastic electron scattering cross-sections, analyse polarization properties, and predict cross-sections, angular distribution and polarization parameters in photoionization cross-sections.

Basis set methods fit into this program in a number of ways:

i) As an alternative method of computing atomic structures to finite difference methods. (A new closed shell Dirac-Fock code, SWIRLES, has been constructed to assess performance.)

ii) As a method of computing the electronic structure of molecules with heavy atom constituents for which, in general, basis set methods are the favoured approach. (A relativistic version, RATMOL, of the nonrelativistic ATMOL electronic structure code is being tested.)

iii) The fact that diagonalization of the Hamiltonian in a finite basis set yields a complete set of eigenvectors of the matrix representation naturally suggests their use to represent the electron/positron operators of the Furry picture of QED. This makes it easy to carry out perturbation theory calculations in the no-pair approximation, equivalent to non-relativistic MBPT[3], and code is already available in SWIRLES to evaluate diagrams to second order. One would like to include diagrams which contain contributions from negative energy states within the computational scheme, including those which require renormalization, and preliminary work on these topics is under way.

It is particularly in the context of calculating radiative corrections that the need for mathematical foundations becomes most acute. We know, for example, that finite dimensional representations fail to reproduce properties of unbounded operators such as the familiar canonical commutator, $[p, q] = -ih/2\pi$. In a situation in which we have to be careful in taking limits of numerical approximations, this example serves as an awful warning. It would be a pity to put a lot of effort into searching for limits which can be shown not to exist!

BASIS SETS FOR RADIAL DIRAC EQUATIONS

After eliminating angular variables, the radial Dirac operator for symmetry κ reduces to a 2×2 matrix differential operator of the form

$$h_{DR} := \begin{bmatrix} V(r) & c\left(-\dfrac{d}{dr} + \dfrac{\kappa}{r}\right) \\ c\left(\dfrac{d}{dr} + \dfrac{\kappa}{r}\right) & -2mc^2 + V(r) \end{bmatrix} \qquad (1)$$

where we use atomic units, and have shifted the zero of energy to coincide with that of a nearly free electron. $V(r)$ is now a central potential. The operator (1) acts on the 2-component vector

$$u(r) = \begin{bmatrix} P(r) \\ Q(r) \end{bmatrix}, \tag{2}$$

defined such that the spherically averaged probability density is

$$\rho(r) = \left\{ |P(r)|^2 + |Q(r)|^2 \right\} / 4\pi r^2, \qquad 0 < r < \infty. \tag{3}$$

We introduce basis sets of square integrable functions $\{\pi_\nu(r)\}_{\nu=1}^\infty, \{\rho_\nu(r)\}_{\nu=1}^\infty$, and form finite approximations of the form

$$u_N(r) = \begin{bmatrix} P_N(r) \\ Q_N(r) \end{bmatrix}, \tag{4}$$

where

$$P_N(r) = \sum_{\nu=1}^N p_\nu \pi_\nu(r), \qquad Q_N(r) = \sum_{\nu=1}^N q_\nu \rho_\nu(r). \tag{5}$$

We shall see that it is important that the two expansions contain the same number of terms. Substituting (5) into the Rayleigh quotient

$$W[u_N] := (u_N|h_{DR}|u_N)/(u_N|u_N) \tag{6}$$

we can make this stationary with respect to variations of the expansion coefficients $\{p_\nu, q_\nu\}_{\nu=1}^N$, leading to a generalized matrix eigenvalue problem

$$\begin{bmatrix} \mathbf{V}_N & c\boldsymbol{\Pi}_N \\ c\boldsymbol{\Pi}'_N & \mathbf{V}'_N - 2mc^2\mathbf{S}'_N \end{bmatrix} \begin{bmatrix} \mathbf{p}_N \\ \mathbf{q}_N \end{bmatrix} = \epsilon \begin{bmatrix} \mathbf{S}_N\mathbf{p}_N \\ \mathbf{S}'_N\mathbf{q}_N \end{bmatrix} \tag{7}$$

in which vectors $\mathbf{p}_N$, $\mathbf{q}_N$ are $N \times 1$, and matrix blocks are all $N \times N$. The elements of the $\mathbf{V}$, $\mathbf{V}'$, $\mathbf{S}$ and $\mathbf{S}'$ matrices are given by

$$V_{\mu\nu} = \int_0^\infty \pi_\mu(r) V(r) \pi_\nu(r)\, dr, \qquad V'_{\mu\nu} = \int_0^\infty \rho_\mu(r) V(r) \rho_\nu(r)\, dr,$$

$$S_{\mu\nu} = \int_0^\infty \pi_\mu(r) \pi_\nu(r)\, dr, \qquad S'_{\mu\nu} = \int_0^\infty \rho_\mu(r) \rho_\nu(r)\, dr, \tag{8}$$

and $\boldsymbol{\Pi}_N$ and $\boldsymbol{\Pi}'_N$ by

$$\Pi_{\mu\nu} = \int_0^\infty \pi_\mu(r) \left(-\frac{d}{dr} + \frac{\kappa}{r} \right) \rho_\nu(r)\, dr,$$

$$\Pi'_{\mu\nu} = \int_0^\infty \rho_\mu(r) \left(\frac{d}{dr} + \frac{\kappa}{r} \right) \pi_\nu(r)\, dr, \tag{9}$$

By performing an integration by parts, we see that

$$\Pi'_{\mu\nu} = \Pi_{\nu\mu} = \left(\mathbf{\Pi}^\dagger\right)_{\mu\nu} \tag{10}$$

provided boundary terms vanish (as they will for all sensible choices). This ensures that the matrix operator in (7) is real symmetric, and hence has $2N$ real eigenvalues.

Suppose now that (10) holds and that the eigenvalues of $\mathbf{V}_N$ and $\mathbf{V}'_N$ are all greater than some number V_0, where V_0 is greater than $-2mc^2$ for all N. Then I was able to show[4] that

a) the N lowest eigenvalues $\epsilon_N^{(\nu)}, \nu = 1, \ldots, N$, of the system (7) are less than $-2mc^2$ (on our practical energy scale);

b) the N highest eigenvalues $\epsilon_N^{(\nu)}, \nu = N+1, \ldots, 2N$ are bounded below by V_0.

The details of the proof in reference 4 show that the condition on V_0 is a sufficient condition for this separation theorem to hold, though not a necessary one, a point to which we shall return below. The construction of a necessary condition has so far proved somewhat elusive.

The separation theorem is important to our analysis for two primary reasons. The first is that it gives clear guidelines for constructing basis sets for relativistic calculations that will eliminate completely the so-called "variational collapse" phenomena experienced by so many who have tried to do these calculations. The second reason is that it shows that for sufficiently weak potentials (we shall consider what is meant by this rather vague characterization below), the upper spectrum is bounded below by V_0, independent of N; because the upper spectrum is completely disjoint from the lower spectrum, its properties are now much the same as those of the nonrelativistic Schrödinger spectrum to which, as a result of (10), the spectrum converges in the nonrelativistic limit[4] for each fixed value of N. This behaviour allows us to apply many of the rigorous results from the approximation theory of the Schrödinger equation which we describe in the penultimate part of this paper.

It is perhaps worth remarking at this point that the problems mentioned in this discussion have nothing whatever to do with the so-called "Brown-Ravenhall disease[5]" which is primarily a question of handling the vacuum of QED correctly in an unquantized formalism.

CHOICE OF BASIS SETS FOR DIRAC CALCULATIONS

The requirement that the basis functions be square integrable over the range $0 < r < \infty$ leads to a natural set of boundary conditions appropriate to bound state calculations in association with the differential operator (1). It is important that these conditions be properly taken into account along with those under which the separation theorem was proved; failure to do so means that one is trying to approximate the wrong problem by the Rayleigh-Ritz method* using (6). The textbooks are usually quite vague when discussing what trial functions should be used for Rayleigh-Ritz calculations, in most cases saying only that one must use "suitable" trial functions without being more explicit. Problems of variational collapse have arisen mainly because this has not been understood, and it has been too readily assumed that any failure to satisfy boundary conditions, especially at $r = 0$, can be accounted for by adjusting the coefficients of the expansion (5).

In our work[6] with atoms with point charge nuclei, for which $V(r) = -Z/r$ as $r \to 0$, we have used what we call S-spinors (S for 'Slater') of the form

$$\pi_\nu(r) = \rho_\nu(r) = r^\gamma e^{-\zeta_\nu r}, \qquad \text{for} \quad \kappa < 0,$$

$$\pi_\nu(r) = r^\gamma e^{-\zeta_\nu r} [A + B\zeta_\nu r], \tag{11}$$

$$\rho_\nu(r) = r^\gamma e^{-\zeta_\nu r} [A' - B\zeta_\nu r], \qquad \text{for} \quad \kappa > 0,$$

in which the small components, $\rho_\nu(r)$, are related to the large components $\pi_\nu(r)$, by

$$\rho_\nu(r) = \left(\frac{d}{dr} + \frac{\kappa}{r} \right) \pi_\nu(r) \left(1 + O(1/c^2) \right). \tag{12}$$

up to terms of order $1/c^2$. The power $r^\gamma, \gamma = +\left(\kappa^2 - Z^2/c^2 \right)^{1/2}$, satisfies the indicial equation for power series solution of the Dirac equation used in textbook presentations, and the exponential ensures that the S-spinor is square integrable on $(0, \infty)$. The extra terms for κ positive are required because the coefficient of the leading power is zero in the non-relativistic limit, the dominant term being the second term of the expansion. In the specific case, $\kappa = +1$, ($\bar{p}$ states), we have

$$A = 2 - N, \qquad A' = N, \qquad B = (2N - 2)/(2\gamma_1 + 1),$$
$$\gamma_1 = +(1 - Z^2/c^2)^{1/2}, \qquad N^2 = 2 + 2\gamma_1. \tag{13}$$

Notice that the nonlinear energy parameter ζ_ν appears when κ is positive; for κ negative, the S-spinor is independent of any energy parameter.

S-spinors are discussed in more detail in reference[4] (though not under this name). The relation (12), ignoring the $O(1/c^2)$ terms, is sometimes called "kinetic balance"; since the approximation

$$2cQ(r) \approx \left(\frac{d}{dr} + \frac{\kappa}{r} \right) P(r) \tag{14}$$

---

* One has only to consider the familiar problem of small vibrations of a string with fixed endpoints at $x = 0, 1$. The equation is $x'' + \lambda^2 x = 0$, with eigenvalues $\lambda = \pi j, j = 1, 2, \ldots$. If we change the condition at $x = 1$ to allow that end of the string to move freely, $x'(1) = 0$, the eigenvalues become $\lambda = \pi(j + 1)/2, j = 0, 1, \ldots$.

is used in the Pauli nonrelativistic reduction of the Dirac equation, it is not surprising that (12) ensures that our matrix approximation reduces correctly to the matrix Schrödinger operator in the nonrelativistic (Pauli) limit[4,6,7] It is this necessary 1–1 matching that makes the dimensions of the two sums in (5) identical. It is obvious that S-spinors conform to the requirements of the separation theorem.

We have still to discuss the choice of nonlinear parameters, ζ_ν. These are often determined by costly and tedious optimization calculations. We have chosen instead to exploit the idea of "even-tempered" basis set exponents, so-called because their logarithms are equally spaced

$$\zeta_\nu = \alpha_N \beta_N^{\nu-1}, \qquad \nu = 1, 2, \ldots, N. \tag{15}$$

Even-tempered Slater basis functions are complete in $L^2(R_+)$ in the limit $N \to \infty$ provided the parameters α_N and β_N tend from above to 0 and 1, respectively, in a regular manner[8] as N tends to ∞, and it follows that the S-spinors are complete in the space $(L^2(R_+))^2$. We have found that it is not necessary to be too fussy about the choice of the α and β parameters to get good results. The most important feature is that the smallest exponent, ζ_1, must be close to the exact exponent of the tail of the mostly loosely bound orbital of interest. The larger exponents will then take care of inner loops in the wavefunctions. The S-spinors are not mutually orthogonal, so that the Gram matrices $\mathbf{S}$ and $\mathbf{S}'$ may have rather large norms when N becomes large, leading to ill-conditioning. In practice we find that this can be controlled provided β_N does not fall below about 1.3 and N is not greater than about 20.

G-spinors (G for "Gaussian") can be defined in a similar fashion; they are more appropriate to problems with finite nuclear charge density distributions[8]. The even-tempered exponent method works well here, though it is necessary to use more functions to achieve the same accuracy as S-spinors. One simplification is that it is no longer necessary to use different formulae for functions with $\kappa > 0$ in order to satisfy the requirements of the separation theorem.

A more recent development is the application of L-spinors (L for "Laguerre") which can be thought of as the analogue of the nonrelativistic "Sturmian" functions introduced by Hylleraas[9] and popularized by Rotenberg[10].* The L-spinors can be written

$$\sigma_\nu(r) = \begin{bmatrix} \pi_\nu(r) \\ \rho_\nu(r) \end{bmatrix} := \mathcal{N}_\nu r^\gamma e^{-\lambda r} \left\{ -L_{\nu-1}^{2\gamma}(2\lambda r) \pm \frac{N_\nu - \kappa}{\nu + 2\gamma} L_\nu^{2\gamma}(2\lambda r) \right\},$$

$$\nu = 0, 1, \ldots, . \tag{16}$$

---

* Rotenberg states that the term "Sturmian" was "a whim of the author ... and has no historical significance other than recognizing that the functions are a solution of one form of the Sturm-Liouville problem." For this reason, we do not consider the term "Sturmian" appropriate in our work.

where λ is an arbitrary parameter and $\mathcal{N}_\nu$ is a normalizing constant. The well-known completeness and orthonormality of associated Laguerre polynomials ensures that the Gram matrices, $\mathbf{S}_N$ and $\mathbf{S}'_N$, are tri-diagonal, the $\mathbf{\Pi}_N$ matrix is penta-diagonal, and the $\mathbf{V}_N$ and $\mathbf{V}'_N$ matrices are full. Although little work has yet been done with these functions, there are indications that L-spinors may have considerable advantages in overcoming linear dependence problems encountered with S-spinors. As the power law behaviour at $r = 0$ indicates, these functions are appropriate for calculations with point charge nuclei. There are similar expressions ("H-spinors") constructible from Hermite polynomials, but we have not yet tried these.

The "relativistic Sturmians" defined recently by Drake and Goldman[11] are rather different in character. They involve functions of the form

$$\Phi_\nu = r^{\gamma+\nu} e^{-\lambda r}, \qquad \nu = 0, 1, \ldots . \tag{17}$$

They satisfy the correct Coulomb cusp condition at $r = 0$, but do not use the pairing of equation (12). In consequence, a spurious eigenvalue appears for positive κ which is degenerate with the lowest negative κ solution; it must be identified and deleted from the calculation.

The B-splines exploited in the calculations of Johnson et al.[12] can also be regarded as finite basis sets. However, they use boundary conditions suggested by the MIT bag model, namely

$$P(0) = 0, \qquad P(R) = Q(R) = 0 \tag{18}$$

which do not fit easily into the pattern we have been examining here.

Some idea of the economy of basis set calculations made with systematic sequences of even-tempered S-spinors can be gained from the numerical results reported in reference 8. A further example, a Dirac-Fock calculation for the ground ^1S state of Argon, is shown in Table I. The GRASP calculation used the default settings listed for the radial grid parameters. The SWIRLES calculation used a systematically extended S-spinor sequence starting from the arbitrary values $\alpha_9 = 0.5$, $\beta_9 = 1.55$. The eight figure agreement seen in this calculation can be matched in calculations for all atomic numbers, as far as we can tell. The most efficient version of SWIRLES took about the same computer time to produce the tabulated data as the GRASP run on the Oxford VAX 8700/8800 cluster. This suggests that the SWIRLES approach may in future be the preferred method for some atomic applications. A detailed study is in preparation.

S-spinors have also been used for calculations in many-body perturbation theory (MBPT). Our first attempts used a toy model consisting of a hydrogenic atom, nuclear charge Z, perturbed by adding an additional nuclear charge of Z'. In the nonrelativistic case, the energy is given exactly by taking only terms of first and second order in Z' and all contributions of higher order in Z' vanish identically. The wavefunction has non-vanishing contributions in third and

Table Ia. Dirac-Hartree-Fock calculations for the ^1S ground state of Argon. Convergence of total energy computed with the SWIRLES code using a set of systematically extended even-tempered S-spinors starting from arbitrarily chosen values of α_9 and β_9. Values from the finite difference code, GRASP, using a standard radial grid $r_0 = 1.0 \times 10^{-3}$ a.u., $h = 0.0625$, $N = 220$, converged to 6×10^{-8} are given for comparison. Energies in a.u. Velocity of light, $c = 137.035\ 989\ 5$ a.u.

N	α_N	β_N	E_{tot}
9	0.500	1.550	-528.643 948 19
10	0.484	1.516	-528.683 364 15
11	0.470	1.486	-528.684 311 72
12	0.458	1.462	-528.684 289 74
13	0.447	1.440	-528.684 430 67
14	0.437	1.421	-528.684 432 46
15	0.429	1.404	-528.684 448 56
16	0.421	1.389	-528.684 449 90
17	0.413	1.376	-528.684 450 46
GRASP			-528.684 450 09

higher orders. The simplicity of the model and its characteristics make it a good testbed for evaluating computational methods in MBPT[13]. It turns out that sequences of even-tempered Slater basis functions perform very much better than traditional methods which involve continuum integrations, and one gets virtually exact energies with little computational effort. The expected cancellation of diagrams in third and higher orders is particularly impressive.[14] The relativistic generalization, which we studied in reference 15, was equally encouraging. The total energy in each order of perturbation can be obtained by expanding the Sommerfeld fine structure formula in powers of Z', though explicit expressions are no longer available for each of the energy diagrams. Because diagrams of third and higher order are relativistic corrections of order $1/c^2$, there is considerable cancellation. The contribution of negative energy states is essential for correct results in this calculation. (The results in reference 15 were obtained with an earlier version of S-spinors and would be improved at high-Z were we to use the basis sets described in this note.)

Another test of basis set quality is to evaluate expectation values of one-electron operators for checking against analytical expressions or values obtained by other methods of calculation. A favourite of this sort is the sequence of

Table Ib. Comparison of orbital expectation values, $\langle r^n \rangle$, and orbital eigenvalues: SWIRLES values above ($N = 17$), GRASP values, italicised, below.

	$\langle r^{-1} \rangle$	$\langle r \rangle$	$\langle r^2 \rangle$	ϵ
$1s$	17.702 703	0.085 624	0.009 863	-119.126 882 65
	17.702 703	*0.085 624*	*0.009 863*	*-119.126 882 65*
$2s$	3.588 960	0.409 957	0.199 112	-12.411 604 31
	3.588 960	*0.409 957*	*0.199 112*	*-12.411 604 31*
$2p_{1/2}$	3.480 113	0.373 054	0.172 429	-9.631 957 93
	3.480 113	*0.373 054*	*0.172 429*	*-9.631 957 70*
$2p_{3/2}$	3.451 497	0.375 338	0.174 404	-9.547 056 45
	3.451 497	*0.375 338*	*0.174 404*	*-9.547 056 27*
$3s$	0.967 916	1.416 157	2.331 107	-1.286 588 35
	0.967 916	*1.416 157*	*2.331 106*	*-1.286 588 33*
$3p_{1/2}$	0.818 828	1.655 632	3.282 186	-0.595 385 89
	0.818 828	*1.655 632*	*3.282 186*	*-0.595 385 84*
$3p_{3/2}$	0.813 332	1.665 395	3.321 618	-0.587 817 82
	0.813 332	*1.665 395*	*3.321 617*	*-0.587 817 78*

(nonrelativistic) dipole sum-rules given by

$$S_k = \sum_n (E_n - E_0)^k |\langle 0|\mathbf{r}|n \rangle|^2, \qquad k = 0,1,\ldots . \qquad (19)$$

In the nonrelativistic case, the (one-electron) values are given by Bethe and Salpeter

$$S_0 = \langle 0|\mathbf{r}^2|0 \rangle, \qquad S_1 = \tfrac{3}{2}, \qquad S_2 = \langle 0|\mathbf{p}^2|0 \rangle,\ldots . \qquad (20)$$

The relativistic hydrogenic sum-rules have been studied by a number of people[17] and were used by Goldman[18] and by Johnson et al.[19] to test the quality of their relativistic basis sets. The appearance of negative energy states modifies (20) so that

$$S_0^R = \langle 0|\mathbf{r}^2|0 \rangle, \qquad S_1^R = 0, \qquad S_2^R = 3c^2 \ldots . \qquad (21)$$

Some typical test results are shown in Table II using two types of basis function. In both cases, $N = 20$, $\alpha = 42/27$, $\beta = 3/2$. Set A used small components obtained from the kinetic balance condition (12) for the $\bar{p}$ states

Table II. Relativistic Thomas-Reiche-Kuhn sum rule S_1. Deviations from zero (the exact value) for the $1s$ states of hydrogenic atoms for
A. an even-tempered kinetically balanced basis set satisfying eq. (12);
B. an even-tempered S-spinor basis set.
Basis set parameters are given in the text. Notation: $a(b) = a \times 10^b$.

Basis set:	A	B
Z		
1	1.68(-7)	1.68(-7)
10	7.23(-6)	1.69(-7)
20	1.12(-4)	1.74(-7)
30	5.59(-4)	1.81(-7)
50	4.13(-3)	2.00(-7)
70	1.51(-2)	2.22(-7)
90	3.90(-2)	2.27(-7)

($\kappa = +1$). The fact that this is equivalent to the Pauli approximation means that the sum rule can only be satisfied to lowest order in the relativistic corrections, and the error in S_1 grows like $(Z/c)^4$. Set B used S-spinors and the resulting error is virtually independent of Z; it could probably have been improved had we chosen to optimize the parameters α and β for each value of Z. The results are very comparable to those obtained in references 18 and 19 though the basis set has somewhat lower dimension N.

The rate of convergence of S_1 and S_2 using a systematic sequence of even-tempered S-spinors is illustrated in Table III. It shows once again the economy obtainable with this basis set definition.

The object of these developments is to be able to carry out perturbation calculations in QED. Some steps towards defining the diagrams that must be calculated taking proper account of the vacuum and of negative energy states have been discussed by various participants in this Workshop. So far, most practical calculations have been done in the "no-pair" approximation in which the negative energy states are completely ignored. Since this is more or less equivalent to nonrelativistic MBPT done with Dirac wavefunctions, I shall refer to such calculations as MBPT calculations for the sake of convenience. Data reported by Johnson et al.[20] are the most detailed so far. The results obtained by the Chalmers group using "pair-function" methods, which do not fit into the present discussion, agree reasonably well where comparisons are possible. A detailed study with S-spinors, now in preparation, shows promise.

Table III. Convergence of relativistic dipole sum rules for a systematic sequence of even-tempered S-spinors as a function of the basis set dimension N. $\delta 1$: Absolute error in S_1. Δ_2: Relative error in S_2. Notation: $a(b) = a \times 10^b$.

N	α	β	$\delta 1$	Δ_2
9	0.3500	1.5000	5.7(-7)	2.6(- 8)
10	0.3390	1.4690	3.8(-7)	1.4(- 8)
11	0.3295	1.4430	2.7(-7)	2.0(- 8)
12	0.3210	1.4206	4.4(-8)	8.2(- 9)
13	0.3135	1.4012	9.4(-9)	1.5(- 9)
14	0.3068	1.3842	3.7(-9)	1.9(-10)
15	0.3007	1.3689	2.5(-9)	2.0(-10)

THE MATHEMATICS OF BASIS SET APPROXIMATIONS

At first sight, it is quite puzzling to find that basis set methods work so well for such things as dipole sum-rules and MBPT. We have found excellent approximations to lower bound state energies and wavefunctions but at higher energies, corresponding to the upper continuum, and also at energies in the lower continuum, the finite basis set solutions are square integrable wave packets, and can only represent the continuum in this sense. The characterization of these wave packets needs further investigation. The often made assertion that the finite matrix representation provides a good approximation to the Dirac Green's function may well be true as far as existing MBPT calculations are concerned, but has no mathematical justification. It is therefore important to see if there are any mathematically rigorous statements which can be used to define the limits of application of basis set methods. We shall see that a good deal can be done to justify their use in calculations up to the level of MBPT. At present, we cannot say much about their use in QED calculations, especially those which involve renormalization.

A central role in the mathematical theory of Schrödinger and Dirac operators[21,22] is played by norm inequalities of the form

$$\|Vf\| \le a\|f\| + b\|Tf\| \qquad f \in \mathcal{D}(T) \tag{22}$$

where $\mathcal{D}(T)$ is the domain of the positive definite operator T in the appropriate Hilbert space and a and b are positive constants; the infimum of b over $\mathcal{D}(T)$

is called the *relative bound* of the operator V relative to T. In the case of Schrödinger one-electron operators, T would be the kinetic energy operator $\mathbf{p}^2/2$, and the Hilbert space would be $L^2(R^3)$, the space of square integrable functions on 3-space (in the sense of Lebesgue). Thus $\|Vf\|$ would be defined as

$$\|Vf\|^2 = \langle Vf|Vf\rangle = \int_0^\infty |(Vf)(\mathbf{r})|^2 \, d\mathbf{r}. \tag{23}$$

It is also useful to consider quadratic form inequalities (expectation values) of the form

$$|\langle f|V|f\rangle| \le a'\langle f|f\rangle + b'\langle f|T|f\rangle. \tag{24}$$

It can be shown (reference 21, p.287) that (24) implies (22) with $a = a'$, $b = b'$ and that (22) implies (24) with

$$a'^2 = (1 + \epsilon^{-1})a^2, \qquad b'^2 = (1 + \epsilon)b^2 \tag{25}$$

where ϵ is some positive number.

Suppose that T is the nonrelativistic kinetic energy operator for an atomic or molecular system, and V is the total Coulomb interaction energy between electron pairs and between electrons and nuclear charges. We write

$$H_S = T + V \tag{26}$$

for the total Hamiltonian. Then one can show[21,22] that V satisfies (22), and that, with consequential adjustments to a, b can be taken as small as one pleases. According to the Kato-Rellich Theorem (reference 21, p.287) H_S is self-adjoint on the domain $\mathcal{D}(T)$ of T, and all its eigenfunctions lie in $\mathcal{D}(T)$. The quadratic form can be used to derive similar conclusions using (24) in which b' can be taken as small as one pleases.

Applications to Dirac Hamiltonians require much greater finesse. Let K be the free electron Dirac operator defined by

$$K := mc^2\beta + c\alpha \cdot \mathbf{p}. \tag{27}$$

which can be shown to be self-adjoint with domain $D(K) \subseteq [L^2(R^3)]^4$ where $D(K)$ is the set of all 4-spinors $u(\mathbf{r})$ with Fourier transforms $\hat{u}(\mathbf{k}) \in [L^2(R^3)]^4$ such that $(c\alpha \cdot \mathbf{k} + mc^2\beta)\hat{u}(\mathbf{k}) \in [L^2(R^3)]^4$. Then

$$K^2 = m^2c^4 + c^2\mathbf{p}^2 = m^2c^4 + 2c^2T \tag{28}$$

is clearly relatively bounded by T in the sense of (22). If we set $V = -Z/r$, the interaction between an electron and a point nucleus of charge Z, then the 'Uncertainty Principle Lemma'

$$\int \frac{1}{r^2} f^2 \, d\mathbf{r} \le 4 \int |\nabla f|^2 \, d\mathbf{r} \tag{29}$$

shows that

$$\|Vg\| \leq b\|Kg\|, \qquad g \in \mathcal{D}(K) \tag{30}$$

where $b \leq 1$ when $Z/c \leq 1/2$, or $Z \leq 68$. We require $b < 1$ in order to apply the Kato-Rellich theorem to deduce that

$$H_D = K + V \tag{31}$$

is self-adjoint on $\mathcal{D}(K)$. It is necessary to be much more subtle to improve this to $Z/c \leq \sqrt{3}/2$, (or $Z \leq 118$)[23], where we expect to encounter trouble because it is at this critical value of Z that the Coulomb exponent γ, which appears in equation (11), becomes less than $1/2$ when $|\kappa| = 1$. For higher values of Z solutions proportional to r^γ and $r^{-\gamma}$ are both square integrable in a neighbourhood of $r = 0$. (Technically, the boundary value problem is then in the limit circle case at the origin; it is always in the limit point case at infinity.) The negative exponent makes $\|Vf\|$ infinite, so that such solutions must be excluded from the domain to ensure that $\|Vf\|$ is still finite. The textbook hydrogenic eigenfunctions are well-behaved up to $Z = 137$, beyond which γ_1 is no longer real when $|\kappa| = 1$. The behaviour of hydrogenic solutions in "over-critical" potentials of point nuclei with $Z > 137$ is discussed in great detail in the book by Greiner et al.[24].

The best way to avoid these difficulties with Dirac operators in practice appears to be to replace point nuclei by distributed charges. This changes the boundary conditions at $r = 0$ so that the exponent γ is now an integer. $\|Vf\|$ is necessarily relatively bounded by $\|Kf\|$ with relative bound zero. The eigenvalues and eigenfunctions now change smoothly with Z however large it may be; the $1s$ energy falls below $-2mc^2$ for uniformly distributed spherical nuclear charges when Z increases beyond 172.[24] Such "over-critical" nuclear charges are way outside the range of Z ($0 < Z < 100$) with which we are usually concerned.

These phenomena are connected with the appearance of the constant V_0 in the matrix separation theorem and with the difficulty of formulating a necessary condition to ensure separation. We see from (31) that the expectation value of H_D for any function $f \in \mathcal{D}(K)$ satisfies

$$\langle f|H_D|f \rangle = \langle f|K|f \rangle + \langle f|V|f \rangle \tag{32}$$

The eigenspace of K consists of two closed *disjoint* subspaces

$$\mathcal{D}_\pm(K) := \{g | \widehat{Kg}(\mathbf{k}) = \pm\sqrt{(m^2c^4 + c^2\mathbf{k}^2)}\hat{g}(\mathbf{k})\}, \qquad \mathcal{D}(K) = \mathcal{D}_+(K) \cup \mathcal{D}_-(K) \tag{33}$$

where, as before, the 'hat' indicates the Fourier transform. There is a one-to-one involutory mapping (charge conjugation) connecting eigenfunctions in $\mathcal{D}_+(K)$ with those in $\mathcal{D}_-(K)$. Clearly, if we "switch on" the nuclear Coulomb attraction, a negative definite operator, the expectation of H_D given by (32) decreases by an amount which is not more than the $\inf \langle g|V|g\rangle / \langle g|g\rangle$ on $\mathcal{D}(K)$. Since finite basis set approximations to states on $\mathcal{D}(K)$ are themselves in $\mathcal{D}(K)$, we see that V_0 can be identified with $\inf \langle g|V|g\rangle / \langle g|g\rangle$ on $\mathcal{D}(K)$. Thus the separation theorem requires

$$V_0 = \inf_{\mathcal{D}(K)} \frac{\langle g|V|g\rangle}{\langle g|g\rangle} > -2mc^2. \tag{34}$$

We now see why this can only be a sufficient condition. For if g is a normalized spinor in $\mathcal{D}_+(K)$, $\langle g|H_D|g\rangle$ can only enter the lower continuum if $\langle g|K|g\rangle$ vanishes, which is not possible for non-zero g.

It is not surprising that we require a limit of the form (34). In the case of a Coulomb potential, the virial theorem[27] shows that the expectation value of $c\alpha \cdot \mathbf{p} + V$ in any eigenstate of H_D is zero. The proof (cf. reference 22, p.230) depends upon the behaviour of the operators under a dilatation, showing that V and K behave in much the same way. In order to make progress, we shall assume now that the Dirac matrix separation theorem holds, possibly with some less restrictive lower bound on $\langle g|V|g\rangle$, and that this is compatible with a quadratic form bound of the form of (23)

$$|\langle g|V|g\rangle| < a'\langle g|g\rangle + b'\langle g|(|K|)|g\rangle, \qquad g \in \mathcal{D}(K) \tag{35}$$

where $b' < 1$ for functions $g \in \mathcal{D}(K)$ and $a' > 0$.

Norm inequalities also play a key role in the theory of convergence of Rayleigh-Ritz approximations in nonrelativistic quantum mechanics. The results of Klahn and Bingel[25] can be summarized by saying that such calculations will converge to bound state eigenvalues (E-convergence) if the underlying infinite basis set is complete in the Sobolev space $W_2^{(1)}$ with norm of the form

$$\|u\|^2_{W_2^{(1)}} = \langle u|(1+T)|u\rangle = \left|(1+T)^{1/2} u\right|^2_{L^2}. \tag{36}$$

The Rayleigh-Ritz eigenfunction, ψ_N^ν then converges to the exact eigenfunction ψ^ν of the ν-th bound state in the mean whenever the corresponding eigenvalue ϵ_N^ν converges to the exact eigenvalue ϵ^ν as the dimension N of the basis set increases. Klahn and Morgan[26] extended this to study the convergence of expectation values of operators (A-convergence) which are relatively form-bounded by the nonrelativistic kinetic energy T; that is,

$$|\langle \psi|A|\psi\rangle| < a\langle \psi|\psi\rangle + b\langle \psi|T|\psi\rangle \tag{37}$$

for some fixed pair of non-negative constants a, b and for all $\psi \in \mathcal{D}(T)$. This encompasses a wide class of operators A including:

a) All bounded operators (for which we can set the relative bound $b = 0$ in (37)).

b) Coulomb potentials.

c) T itself (for which (37) is satisfied with $a = 0$, $b = 1$).

d) Components of the momentum operator, $\mathbf{p}$

e) Nonrelativistic atomic and molecular Hamiltonians, H_S,say.

Likewise T can be relatively form-bounded by $H_S + k$, where k is a positive constant chosen so that $H_S + k$ has a purely positive spectrum.

Suppose that A is a purely positive operator, self-adjoint with domain $\mathcal{D}(A)$. Then we can define a new A-norm by

$$\|\psi\|_A^2 = \langle \psi | A | \psi \rangle = \|A^{1/2}\psi\|_{L^2}^2. \tag{38}$$

In terms of the related scalar product, this induces a new Hilbert space in the A-norm. Any set of functions which is complete in the new space will be called A-complete. For example, if $A = 1 + T$, then we have constructed the Sobolev space $W_2^{(1)}$ already encountered. A necessary and sufficient condition for convergence in the A-norm can now be constructed with the aid of two inequalities for the expectation of A with respect to ψ:

$$
\begin{aligned}
|\langle A \rangle_N - \langle A \rangle| &\leq \|\psi_N - \psi\|_A^2 + 2\langle A \rangle^{1/2} \|\psi_N - \psi\|_A \\
\|\psi_N - \psi\|_A^2 &\leq |\langle A \rangle_N - \langle A \rangle| + 2\|A\psi\| \cdot \|\psi_N - \psi\|.
\end{aligned}
\tag{39}
$$

Thus $\langle A \rangle_N$ converges to $\langle A \rangle$ if and only if ψ_N converges to ψ in the A-norm. Suppose that we have an infinite set $\{\varphi_\nu\}_{\nu=1}^\infty$ in the domains of both $A^{1/2}$ and $T^{1/2}$. Then if ψ_N has the finite representation

$$\psi_N = \sum_{\nu=1}^N c_\nu \varphi_\nu, \tag{40}$$

(39) tells us that the set must be A-complete if ψ_N is to converge to ψ in the A-norm

$$\|\psi_N - \psi\|_A \xrightarrow{N \to \infty} 0.$$

Since there can be no guarantee that the Rayleigh-Ritz expansion coefficients $\{c_\nu\}$ determined from the variational principle (6) have this property, we infer that the expectation values of one operator may converge correctly, but those of another need not do so.

However, there is an important class of operators for which convergence of the Rayleigh-Ritz expectation values can be guaranteed. Suppose that the operator A is relatively form-bounded by the nonrelativistic kinetic energy operator T so that

$$|\langle \psi | A | \psi \rangle| \leq a\langle \psi | \psi \rangle + b\langle \psi | T | \psi \rangle. \tag{41}$$

Then if ψ_N is an E-convergent approximation with respect to the Sobolev norm,

$$\|\psi_N - \psi\|_{k+T}^2 = \langle \psi_N - \psi | k + T | \psi_N - \psi \rangle \xrightarrow{N \to \infty} 0, \tag{42}$$

for any positive value of k, so that

$$\|\psi_N - \psi\|_A^2 \leq \max\{\frac{a}{k}, b\}\|\psi_N - \psi\|_{k+T}^2 \tag{43}$$

and E-convergence of ψ_N implies A-convergence of ψ_N. It is for this reason that we should try to establish inequalities of the form (41) whenever possible.

The off-diagonal elements of operators converge under the same conditions as diagonal ones. Write

$$\Delta_N(ij) = \left|\langle\psi_N^{(i)}|A|\psi_N^j\rangle - \langle\psi^{(i)}|A|\psi^j\rangle\right| \tag{44}$$

One can then show that

$$\begin{aligned}\Delta_N(ij) \leq &\|\psi_N^{(i)} - \psi^{(i)}\|_A \|\psi_N^{(j)} - \psi^{(j)}\|_A \\ &+ \langle\psi^{(j)}|A|\psi^j\rangle^{1/2}\|\psi_N^{(i)} - \psi^{(i)}\|_A \\ &+ \langle\psi^{(i)}|A|\psi^i\rangle^{1/2}\|\psi_N^{(j)} - \psi^{(j)}\|_A\end{aligned} \tag{45}$$

so that $\Delta_N(ij)$ converges if both the Rayleigh-Ritz approximations to the states involved are A-convergent.

These arguments allow one to infer convergence of approximations of observable quantities, but tell us nothing about convergence rates, which must be studied afresh for each problem. Klahn and Morgan devote the rest of their paper[26] to just such investigations for hydrogenic atoms using both Slater and Gaussian basis functions. More information is needed about convergence rates in Dirac problems, although much of the behaviour in the Schrödinger case necessarily carries over provided the separation theorem holds.

We shall now try to apply these ideas to Dirac one-electron operators, assuming that the atomic potential V has a lower bound V_0 on $\mathcal{D}(K)$ given by (34) and satisfies the inequality (35). These assumptions allow us to infer the existence of two disjoint sets of matrix eigensolutions, one with N "positive" energies ($> V_0$), the other with N "negative" energies ($< -2mc^2$). It also allows us to conclude that the Dirac operator

$$H_D = K + V \tag{46}$$

is itself self-adjoint on $\mathcal{D}(K)$ and that its spectrum can also be separated in the same way, the upper part being certainly bounded below by V_0. This partitioning allows us to examine convergence of states in the upper spectrum alone. (It makes no sense to discuss convergence in the lower spectrum which is purely a continuum.)

We can define A-convergence for Dirac operators in the usual way through (38). The notion of E-convergence can be replaced by D-convergence (D for "Dirac"), that is to say convergence with respect to the D-norm

$$\|\psi\|_D = \langle\psi|K|\psi\rangle, \qquad \psi \in \mathcal{D}_+(K) \tag{47}$$

As before, we say that a basis set $\{\varphi_\nu\}_{\nu=1}^\infty$ is D-complete if $\psi_N \to \psi$ in this norm for every ψ in $\mathcal{D}_+(K)$. (Note that this is sufficient for our purposes as there is no way of constructing a Cauchy sequence in $\mathcal{D}_+(K)$ whose limit is in $\mathcal{D}_-(K)$.) We now have

A. Convergence of Rayleigh-Ritz approximations to bound state energies and eigensolutions of the one-electron Dirac operator with atomic potentials.

B. Convergence of matrix Dirac-Hartree-Fock approximations to energies and eigensolutions. Here, we note that we determine the self-consistent field by iterating with a sequence of one-electron effective potentials in which the occupied bound states generate a field screening the nuclear Coulomb attraction. At each stage of the iteration to self-consistency, the effective potential will be bounded below by the same bare nucleus value of V_0.

C. Convergence of expectation values and transition matrix elements of all operators that are relatively form-bounded by K on $\mathcal{D}_+(K)$. This is likely to include most of the operators of interest, such as:

  i) Bounded operators: constants, Dirac α- and β-matrices, and the fully-retarded radiative transition operator[28].

  ii) Powers r^λ, $\lambda \geq -1$.

  iii) Components of momentum, $\mathbf{p}$.

  iv) Pieces of the Dirac operator H_D itself such as K and V.

Of the numerical results discussed above, the only situations not covered so far have the character of sums over complete sets of eigenstates such as occur in MBPT or in the dipole sum-rules. Here the problem is that we can infer nothing about convergence in either continuum, and the arguments based on (44) and (45) do not apply to matrix elements such as $\langle 0|r|n \rangle$, where n is a continuum state. Of course, this matrix element will be finite in a basis set of the form we have been considering, but we do not expect it to converge as $N \to \infty$ to any particular number. It is not difficult to get around this in the case of sum-rules (19) where the left-hand sides reduce to expectations of bounded operators with respect to the reference state given by (20) or (21). If the reference state is one for which the Rayleigh-Ritz expectation value converges, and if the right-hand-side converges to this value as $N \to \infty$, then we can infer that the pseudo-spectrum generated by the Rayleigh-Ritz method provides an efficient set of integration points in one or both continua for this particular rule. We could draw similar conclusions were we able to find bounds for the sum-over-states of interest. However. we cannot say anything definite in other cases. In particular, we have no means of analysing sums-over-states in MBPT at this point in general.

The problem is more acute when we want to include negative energy states in QED. The present formalism takes no account of the fact that, for example, the electron self-energy (which is basically the integral over all photon energies of a second order perturbation sum involving a radiative transition from the reference state to an excited state and back) is infinite. Whilst we can in principle renormalize the self-energy to get a finite answer by standard techniques, the method of doing this using basis sets has still to be worked out.

One final question has not been explored at all. Whether MBPT or the perturbative expansion of QED has a finite sum, or even if it is merely an asymptotic series in some coupling parameter, we have no idea if a sequence of Rayleigh-Ritz approximations to the perturbation series will demonstrate the same behaviour as the theory indicates. To put it another way: Is it really possible to make realistic approximations to QED in the Furry picture using Rayleigh-Ritz eigenstates? Whilst present indications are encouraging, we have no answer to the question.

This is perhaps the ultimate challenge.

ACKNOWLEDGMENTS

I would like to thank J. D. Morgan III for useful discussions on the ideas explored in the last section, H. M. Quiney for the use of unpublished results from his SWIRLES code, and F. A. Parpia for assistance with TₑX.

1. I. P. Grant, B. J. McKenzie, P. H. Norrington, D. F. Mayers and N. C. Pyper, *Comput. Phys. Commun.* **21**, 207 (1980); B. J. McKenzie, I. P. Grant and P. H. Norrington, *Comput. Phys. Commun.* **21**, 233 (1980).
2. C. F. Fischer, *The Hartree-Fock Method for Atoms*, (Wiley, New York 1977); *Comput. Phys. Reps.* **3**, 273 (1986).
3. I. Lindgren and J. Morrison, *Atomic Many-Body Theory* (Springer- Verlag, Berlin 1983).
4. I. P. Grant, *Phys. Rev. A* **25**, 1230 (1982); *J. Phys. B* **19**, 3187 (1986).
5. G. E. Brown and D. G. Ravenhall, *Proc. Roy. Soc. London Ser. A* **208**, 552 (1951).
6. K. G. Dyall, I. P. Grant and S. Wilson, *J. Phys. B* **17**, 1201 (1984); H. M. Quiney, I. P. Grant and S. Wilson, *J. Phys. B* **20**, 1413 (1987); *Phys. Scr.* **36**, 460 (1987).
7. K. G. Dyall, I. P. Grant and S. Wilson, *J. Phys. B* **17**, 493, L45 (1984).
8. I. P. Grant and H. M. Quiney, *Adv. Atom. Molec. Phys.* **23**, 37 (1988) and references therein.
9. E. A. Hylleraas, *Z. Phys.* **48**, 469 (1928); *Adv. Quant. Chem* **1**, 1 (1964).
10. M. Rotenberg, *Adv. Atom. Molec. Phys.* **6**, 233 (1970); *Ann. Phys.* **19**, 262 (1962).
11. G. W. F. Drake and S. P. Goldman, *Adv. Atom. Molec. Phys.*, to be published.
12. W. R. Johnson, S. A. Blundell and J. Sapirstein, *Phys. Rev. A* **37**, 307 (1988).

13. R. E. Trees, *Phys. Rev.* **102**, 1553 (1956); B. F. Gray, *J. Chem. Phys.* **36**, 1801 (1962); P. J. Rossky and M. Karplus, *J.Chem. Phys.* **67**, 5419 (1977).
14. H. M. Quiney, I. P. Grant and S. Wilson, *J. Phys. B* **18**, 577 (1987).
15. H. M. Quiney, I. P. Grant and S. Wilson, *J. Phys. B* **18**, 2805 (1987).
16. H. A. Bethe and E. Salpeter, *Quantum Mechanics of One- and Two- electron atoms* (Springer-Verlag, 1957).
17. W. B. Payne and J. S. Levinger, *Phys. Rev.* **101**, 1020 (1956); I. P. Grant, *Phys. Rev.* **106**, 754 (1957); J. S. Levinger, M. L. Rustgi and K. Okamoto, *Phys. Rev.* **106** 1191 (1956); J. S. Levinger and M. L. Rustgi, *Phys. Rev.* **103**, 439 (1956).
18. G. W. F. Drake and S. P. Goldman, *Phys. Rev. A* **23**, 2093 (1981); G. W. F. Drake and S. P. Goldman, *Phys. Rev. A* **25**, 2877 (1982).
19. W. R. Johnson, S. A. Blundell and J. Sapirstein, *Phys. Rev. A* **37**, 307 (1988).
20. W. R. Johnson and J. Sapirstein, *Phys. Rev. Lett.* **57**, 1126 (1986); W. R. Johnson, M. Idrees and J. Sapirstein, *Phys. Rev. A* **35**, 3218 (1987); and other work in the press.
21. T. Kato, *Perturbation Theory for Linear Operators* (Springer- Verlag, New York 1966).
22. M. Reed and B. Simon, *Methods of Modern Mathematical Physics*, Vol IV.: *Analysis of Operators*. (Academic Press, New York 1978).
23. H. Kalf, U.-W. Schmincke, J. Walter and R. Wüst in *Springer Lect. Notes in Math.* **448**, 182 (1975).
24. W. Greiner, B. Mueller and J. Rafelski, *Quantum Electrodynamics of Strong Fields* (Springer-Verlag, Berlin 1985).
25. B. Klahn and W. A. Bingel, *Theoret. Chim. Acta (Berl.)* **44**, 9, 27 (1977).
26. B. Klahn and J. D. Morgan III, *J. Chem. Phys.* **81**, 410 (1984).
27. V. Fock, *Z. Phys.* **63**, 855 (1930); M. E. Rose and T. A. Welton, *Phys. Rev.* **86**, 432 (1952); Y. K. Kim, *Phys. Rev.* **154**, 17 (1967); and many others.
28. I. P. Grant, *J. Phys. B* **7**, 1458 (1974).

SCREENING OF RESIDUAL COULOMB INTERACTION AND STRUCTURE OF MANY-BODY PERTURBATION THEORY IN HEAVY ATOMS

V.A.Dzuba, V.V.Flambaum, P.G.Silvestrov, O.P.Sushkov[*]

Institute of Nuclear Physics
630090, Novosibirsk, U S S R

ABSTRACT

It is shown that many-body screening of a residual Coulomb interaction is a large collective effect. Taking into account this screening essentially improves the convergence of perturbation theory in heavy atoms.

Let us consider the energy levels of an atom with one unpaired electron. Standard way of calculation is to use the frozen core approximation as zero step, and then calculate corrections using the many-body perturbation theory (MBPT). It is well known that for heavy atoms this correction is of the order of 10-15%[1-3] while for light atoms just a few percent[3]. Moreover, sometimes for heavy atoms the convergence of MBPT does not look reasonable. The purpose of the present work is to reveal the structure of MBPT and by summation of special subsequences of diagrams to essentially improve the convergence of MBPT.

As an extreme example we consider the Tl atom. Why is it extreme? The Tl has configuration $1s^2...5d^{10}6s^2$ 6p. There is only one unpaired 6p-electron but the energies of 6s and 5d-electrons are close to the energy of this electron. Therefore the polarisation and

[*] Institute for Theoretical Physics Santa Barbara, CA 93106

correlation phenomena are very important.

For calculation of zero approximation we will distinguish two possibilities: V^{N-1}-approximation and V^N-approximation. In V^{N-1}-approximation we find the core wave function by usual Dirac-Hartree-Fock procedure for the closed shells. And then we generate the states of an external electron in this frozen core field. This is a very simple way. The orbitals of the external electron are automatically orthogonal to the orbitals of core electrons. However such a zero approximation does not take into account the influence of the external electron to the internal ones. Therefore one should take it into account by means of MBPT.

The second way is V^N-approximation. We mean that in calculating of core orbitals we take into account the zero multipolarity direct Coulomb field of the external electron. Therefore the self-consistent field acting on any electron looks as follows

$$V = V^{N-1} + V_o - (1 - P)V_o(1 - P) \tag{1}$$

Here V^{N-1} is the direct and exchange field of the core electrons, V_o is zero multipolarity direct field of the external $6p_{1/2}$-electron, and P is the projection operator on the core states.

$$P = \sum_{\substack{core \\ orbitals}} |i\rangle \langle i| \tag{2}$$

Of course the potential V should be calculated in a self-consistent way. It is easy to see from eq. (1) that the averaged field acting on internal electrons is equal to

$$\langle V \rangle_{int} = \langle V^{N-1} \rangle + \langle V_o \rangle , \tag{3}$$

256

and the field acting on the external electrons is

$$\langle V \rangle_{ext} = \langle V^{N-1} \rangle \qquad (4)$$

We would like to stress that the internal and external orbitals generated by Hamiltonian (1) are automatically orthogonal.

Let us discuss now the second order correlation correction to the energy level. For V^{N-1} zero approximation it is defined by the following diagrams (see e.g. ref. [1])

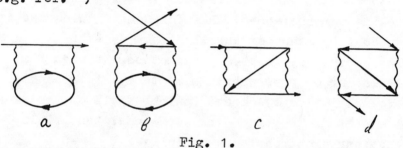

Fig. 1.

For V^N we should add the subtraction diagrams to avoid double counting. This subtraction exactly cancels the following contributions

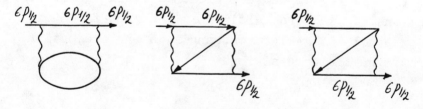

Fig. 2.

because they are absorbed by the zero order. A similar situation takes place in 3rd order. For example subtraction exactly cancels these diagrams

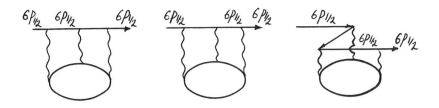

Fig. 3.

and essentially reduces the folowing contribution

Fig. 4.

The cross here denotes the subtraction potential V_0.

Experimental value of $6p_{1/2}$ state energy is
-49264 cm^{-1} [4]. Zero order calculation in V^{N-1} gives
-43909 cm^{-1}. The accuracy is -10%. Zero order in V^N
gives the value -46296 cm^{-1}. The accuracy is -6%.

Due to the all mentioned reasons we prefer to use
V^N despite of the more complicated MBPT due to the subtractions.

Thus, zero order calculation of the $6p_{1/2}$ state
energy gives the accuracy -6%. After calculation of
the second order correlation correction (fig. 1) accuracy becomes$+3.5\%$. It looks very strange. Perturbation
expansion practically does not converge despite of the
rather good accuracy of zero approximation. To understand what happens let us consider contributions of diagrams fig. 1 separately. It looks as follow (in cm^{-1})

$$
\begin{aligned}
\Sigma_{1a} &= -15383 \\
\Sigma_{1b} &= +10970 \\
\Sigma_{1c} &= +3170
\end{aligned}
\tag{5}
$$

$$\Sigma_{1d} = -3485$$

We see that the diagrams with the loop (1a,1b) are enhanced by the factor $\sim 4 - 5$ in comparison with the exchange diagrams (1c,1d). First impression is that it is usual suppression of the exchange diagrams due to the small overlapping of wave functions. But it is not so. The overlapping of 6p, 6s and 5d wave functions is not small and exchange diagrams are not supressed but direct loops are enhanced. The reason for this inhancement one can easily understand parametrically. Actually, all the excitations in the loop give contributions of the same sign. Therefore full contribution of the loop is proportional to the effective number of electrons in upper closed subshells: N_e. (This is the possible number of holes in the loop). This is the effective number since the contribution of deep electrons is suppressed by energy denominator. In Tl the upper closed subshells are $5d^{10}6s^2$ and it looks reasonable that $N_e \sim 4 - 5$.

If we have this idea of enhancement of the loops we can go to the higher orders. Let us compare for example two diagrams of the third order

$$a \qquad\qquad b$$

Fig. 5.

If denoted symbolically the dimensionless parameter of MBPT by α , then we can write

$$\Sigma_{5a} \sim \alpha^3 N_e \quad , \quad \Sigma_{5b} \sim \alpha^3 N_e^2 \qquad (6)$$

because 5a contains just a one loop and 5b contains two loops. Thus we can separate the most important higher order diagrams.

Let us consider in detail the contribution of diagram fig. 1a. In second column of table 1 we present its contribution (one loop contribution) for various multipolarities of Coulomb quanta (All values are given in cm^{-1})

Table 1

k	1 loop	1 loop +2 loops	all loops
0	-2	+246	+137
1	-9751	+2079	-4822
2	-3428	-2668	-2826
3	-1466	-1248	-1284
4 + 5 + 6	-742	-711	-712
sum	-15389	-2302	-9507

Contribution of zero multipolarity (k = 0) is small just because of subtraction. Without subtraction its contribution is about -3000 cm^{-1}. In the column three of the table 1 we present the values with 2 loops taken into account, i.e. this is the sum of diagram 1a contribution and contribution of the following diagrams

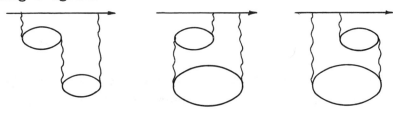

Fig. 6.

We see that situation is dramatic for dipole quantum (k = 1). Two loops contribution has opposite sign

260

and even larger than one loop contribution. In some
sense the perturbation expansion really does not converge and we should sum the whole chain of the loops.

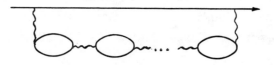

Fig. 7.

The result of this summation is presented in the
last column of the Table 1.

The similar situation takes place for the other
diagrams presented in fig. 1. The insertion of an additional loop in any Coulomb line changes the result
drastically. The conclusion is that we should use
diagrams fig. 1 just as skeleton diagrams with the
dressed Coulomb line. Namely if ⌇⌇⌇ is the bare
Coulomb line and − − − − is the dressed Coulomb line

$$- - - - = \text{⌇⌇⌇} + \text{⌇}\bigcirc\text{⌇} + \text{⌇}\bigcirc\text{⌇}\bigcirc\text{⌇} + \dots ,$$

then correlation correction to the energy is defined
by the diagrams

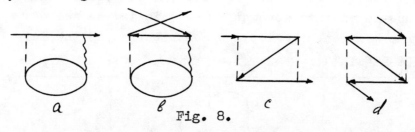

Fig. 8.

Now we would like to discuss briefly the technical
details of calculations. The usual Bueckner-Goldstone
technic is not convenient for summation of the loops
because of various energy denominators which arise in
higher orders. For example diagrams presented in fig.6

have two different combinations of denominators. Three loops diagrams contain larger number of combinations and so on. It is well known that Feynman technique is convenient for summation and we use it.

The Feynman Green's function is of the form

$$-= G_{z_1 z_2}(\varepsilon) = \sum_{n \leq 6S} \frac{\psi_n(z_1)\psi_n^+(z_2)}{\varepsilon - \varepsilon_n - i\delta} + \sum_{\gamma \geq 6p_{1/2}} \frac{\psi_\gamma(z_1)\psi_\gamma^+(z_2)}{\varepsilon - \varepsilon_\gamma + i\delta} \quad (7)$$

We remind the reader that $6s^2$ is the upper closed sub-shell in Tl, and $6p_{1/2}$ is the state of unpaired electron. The polarization operator is

$$z_1 \bigcirc z_2 = \Pi_{z_1 z_2}(\omega) = \int_{-\infty}^{+\infty} \frac{d\varepsilon}{2\pi} G_{12}(\omega+\varepsilon) G_{21}(\omega) \quad (8)$$

In the second order there are two following diagrams which contribute into the self-energy.

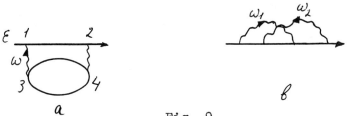

Fig. 9.

Let first consider diagram 9a

$$\sum_{9a}(\varepsilon) = i \int_{-\infty}^{+\infty} \frac{d\omega}{2\pi} G_{12}(\varepsilon+\omega) \frac{e^2}{z_{24}} \Pi_{43}(\omega) \frac{e^2}{z_{13}} \quad (9)$$

$$\varepsilon = \varepsilon_{6p_{1/2}}$$

The poles of $G(\varepsilon+\omega)$ and $\Pi(\omega)$ in complex

ω -plane are shown in fig. 10.

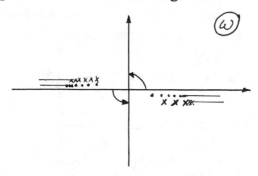

Fig. 10.

Here the points correspond to the Green's function poles and crosses to the polarization operator poles. If we integrate (9) formally then we can obtain Brueckner-Goldstone contributions corresponding to figs. 1a,1b. The diagram 1a arises from the Green's function poles in the lower half-plane and 1b arises from that in upper half-plane. Surely we do not integrate formally, but rotate the contour to the imaginary axis (see fig. 10). After rotation we can integrate over ω numerically and we do it. In this scheme there are no problem with the chaining of the loops. We just should solve the Dyson equation for the chain at complex ω

$$\sim\!\!\boxed{}\!\!\sim \;=\; \sim\!\!\bigcirc\!\!\sim \;+\; \sim\!\!\bigcirc\!\!\sim\!\!\boxed{}\!\!\sim \tag{10}$$

In similar way we work with diagram fig. 9b, but here we have double integration.

The results of calculations are presented in Table 2

Table 2 Energy levels $6p_{1/2}$, $6p_{3/2}$ (Ionization potentials) in thallium. Units are cm^{-1}.

	a	b	c	d
$6p_{1/2}$	46296	51030	50011	49264
$6p_{3/2}$	38677		42204	41471

a – zero approximation – Relativistic Hartree-Fock
(RHF) calculation in V^N potential; b – RHF+second order
correlation corrections; c – RHF+screened correlations;
d – experiment [4].

In conclusion we would like to sum the results.

1. The electron loops in Many-body perturbation theory
   expansion are stressed by the parameter N_e. N_e is
   the effective number of electrons in upper closed
   subshells. For Tl $N_e \sim 4 - 5$.

2. Due to this relatively large parameter the chaining
   of the loops is very important. It leads to the
   dressed Coulomb line.

3. Taking into account dressed Coulomb interaction instead of bare one essentially improve the conver-
   gency of many-body perturbation theory.

4. The accuracy of calculation of 6p state energy in
   Tl achieved in such a way is 1.5%. (Let us remind
   the reader that zero order gives the accuracy -6%
   and account of the second order correlation correction with the bare Coulomb interaction gives the
   accuracy +3.5%). The accuracy of fine structure calculation is 0.2%.

One of us (O.P.S.) would like to thank the Institute for Theoretical Physics, University of California,
Santa Barbara, where this manuscript was produced, for
hospitality and support (by National Science Foundation under Grant No. PHY-17853, supplemented by funds
from the National Aeronautics and Space Administration)
and for a very stimulating environment.

REFERENCES

1. V.A.Dzuba, V.V.Flambaum and O.P.Sushkov, J. Phys.
   B: At. Mol. Phys. 16, 715 (1983).

2. V.A.Dzuba, V.V.Flambaum, P.G.Silvestrov and O.P. Sushkov, J. Phys. B: At. Mol. Phys. 20, 1399 (1987).

3. W.R.Johnson, M. Idress and J.Sapirstein, Phys. Rev. A35, 3218 (1987).

4. Ch.E.Moore, Atomic Energy Levels vol.3, 1958 (Washington, DC: US Govt. Printing Office).

THE MULTICONFIGURATION DIRAC-FOCK METHOD

J.P. Desclaux

Centre d'Etudes Nucléaires de Grenoble, DRF/SPh

85X, 38041 Grenoble Cédex France

ABSTRACT

The self-consistent multiconfiguration Dirac-Fock method is first described. Then we consider the effective electron-electron interaction used in the numerical calculations and discuss why it works.

INTRODUCTION

Let us consider relativistic calculations from a pedestrian practical point of view. It is well known, in the nonrelativistic case, that a good starting point for precise calculations is provided by the solutions of the Hartree-Fock equations. Various schemes have been developed to improve this first order approximation: many body perturbation theory [1, 2], random phase approximation [3], configuration interaction [4], multiconfiguration Hartree-Fock [5] to name the most commonly used. Here we shall concentrate on the last one as applied to atomic relativistic calculations and divide the presentation in two sections: first we discuss the approximations used in this method and then try to explain why it works.

DESCRIPTION OF THE METHOD

The key approximation in the Hartree-Fock method is to replace the N electron total wave function by an antisymmetric product of one electron wave functions, i.e.:

$$\Psi(\mathbf{r}_1, \mathbf{r}_2, \ldots, \mathbf{r}_N) = \mathcal{A}\Phi(\mathbf{r}_1)\Phi(\mathbf{r}_2)\ldots\Phi(\mathbf{r}_N) \tag{1}$$

furthermore in virtually all the calculations performed up to now, the one electron wave functions Φ have been constrained to be of spherical symmetry, i.e. to be the product of a radial function times a spherical harmonic. If such a form is appropriate for closed shell systems, it already introduces a restriction for open shell ones. With this extra constraint, the one electron wave functions used in the relativistic case are eigenvectors of the total angular momemtum $\vec{j} = \vec{l} + \vec{s}$ and of its projection j_z and are of given parity. Using the principal quantum number n and the relativistic quantum number κ defined as:

$$\kappa = l \text{ if } j = l - 1/2 \quad \text{and} \quad \kappa = -(l+1) \text{ if } j = l + 1/2 \qquad (2)$$

these one electron wave functions are written as:

$$\Phi_{n\kappa m}(r,\theta,\varphi) = \frac{1}{r} \left\{ \begin{array}{l} P_{n\kappa}(r)\chi_{\kappa m}(\theta,\varphi) \\ \\ iQ_{n\kappa}(r)\chi_{-\kappa m}(\theta,\varphi) \end{array} \right\} \qquad (3)$$

P and Q are respectively the large and small radial functions and the χ's are two component Pauli spinors that are simultaneous eigenstates of l^2, s^2, j^2 and j_z with respectively $l(l+1)$, $s(s+1)$, $j(j+1)$ and m as eigenvalues. To improve the Hartree-Fock wave function given by equation 1 and restricted to a single configuration state (or in other words to consider only the Φ's of the occupied orbitals in an independent electron picture), the multiconfiguration method extends the variational space to linear combinations of configuration state functions of the form:

$$\Psi(\Pi J M) = \sum_{\nu=1}^{NCF} c_\nu |\nu \Pi J M > \qquad (4)$$

where Π is the parity of the state under consideration, J its total angular momentum of projection M and ν stands for all other values (angular momentum recoupling scheme, seniority numbers, ...) necessary to define unambiguously the νth configuration state function. To be more explicit, consider the simple case of the ground state of helium. The definition above (Eq. 4) would suggest an expansion of the type:

$$|1s^2, {}^1S_0 > = \sum_{n,l,n',l'} c_{nln'l'} |nln'l', {}^1S_0 > \qquad (5)$$

where the summation over continuum states is implicit and non relativistic orbital labels have been used for simplicity but with the understanding that all (jj) configurations that arise from a single (LS) one are included. In fact it is possible to show [5] that orthogonal transformations can be used to reduce the above expansion (Eq. 5) to:

$$|1s^2, {}^1S_0 > = \sum_{n,l} c_{nl} |(nl)^2, {}^1S_0 > \qquad (6)$$

where the new (**nl**) orbitals are linear combination of the previous ones and are known as Lowdin's natural orbitals. This transformation to the natural orbitals, that can be determined with the help of a variational calculation is the key to the efficiency of the multiconfiguration method since it is obvious from the comparison of the expansions (5) and (6) that the latter involves many less configurations, and furthermore all the configurations differ by two electrons resulting in a very small number of integrals in the expression of the electron-electron interaction.

The same expansion (Eq. 6) can be used for all the $^1S_0$ states. For example the 1s2s state will be associated to the second eigenvalue, and so on (in some cases it may be more effective to use different expansions for different states, but this will be done only for numerical efficiency). In the same way the wave functions of the first excited states can be approximated by:

$$|1s2s,^3S_1> = \sum_{l\geq0} \sum_{n\geq l+1, \Delta n=2} c_{n,l}|nl(n+1)l,^3S_1> \tag{7}$$

$$|1s2p,^{1,3}P_J> = \sum_{l\geq0}\sum_{n\geq l+1} c_{n,l}|nl(n+1)(l+1),^3P_J> \tag{8}$$

These expansions are approximate not only because they are associated with a pair-correlation wave function approach (in this respect the solution of the MCDF equations, if carried out to completess, will be equivalent to the pair equation approach discussed elsewhere in these proceedings [6]) but also because the orthogonal transformations used are specific to each (ll') subspace and, for example in the case of the P states, they will result in different p orbitals in the (01) and (12) subspaces. A more complete discussion can be found in [5]. Nevertheless it has been recently shown [7] that it is possible to recover more than 95% of the correlation energy of the He-like ions by truncating the expansion at n=4. Extension to more complex systems is straightforward at least from a theoretical point of view.

With the Dirac operator defined in atomic units (m=e=ℏ=1) by:

$$h_D = c\vec{\alpha}.\vec{p} + \beta c^2 + V_{nuc}(r) \tag{9}$$

where c is the speed of light, $\alpha_x, \alpha_y, \alpha_z$ and β are fourth order Dirac matrices and V_{nuc} is the nuclear potential. The total effective Hamiltonian is written as:

$$H = \sum_i h_D(i) + \sum_{i,j} g(i,j) \tag{10}$$

$g(i,j)$ being the electron-electron interaction, the exact form of which will be discussed in the next section. Using equation 10 to compute the expectation value of the total energy, we can apply the variational principle with respect to both the radial function P and Q and to the mixing coefficients c_ν between the configurations to obtain a double set of variational equations:

a) The mixing coefficients c_ν are the components of one of the eigenvectors of the Hamiltonian matrix depending upon the state under consideration.

b) The radial functions P and Q must satisfy a set of integrodifferential equations. For a given orbital A it reads:

$$\frac{dP_A}{dr} + \frac{\kappa_A}{r} P_A = \left[2c^2 + \frac{1}{c}(\epsilon_A - V) \right] Q_A + \sum_{B \neq A} \left(X_{AB} + \frac{\epsilon_{AB}}{c} \right) Q_B$$

$$\frac{dQ_A}{dr} - \frac{\kappa_A}{r} Q_A = -\frac{1}{c} [\epsilon_A - V] P_A - \sum_{B \neq A} \left(X_{AB} + \frac{\epsilon_{AB}}{c} \right) P_B \qquad (11)$$

with V the sum of the nuclear potential and of the direct interaction between the electrons, X_{AB} exchange type potentials and ϵ_A, ϵ_{AB} Lagrange parameters introduced to insure orthonormality between the one electron wave functions. In the above equations both V and the X's are implicit functions of the mixing coefficients c_ν, and from a practical point of view the double variational problem is solved alternatively: from a given set of initial one electron orbitals, the Hamiltonian matrix is constructed and diagonalized to obtain a first estimate of the mixing coefficients. Keeping theses values of the coefficients fixed, the integrodifferential equations for the P and Q's are solved in a self consistent way and the whole process (diagonalization of the Hamiltonian plus self consistent field) is repeated a certain number of times until a given accuracy is reached.

ELECTRON-ELECTRON INTERACTION

For the derivation of the effective electron-electron interaction of a many electron system it is usual to use Furry's bound state interaction picture of QED, then the S-matrix formulation leads to [8]:

$$S_{AB \to CD} = \alpha \int \int \Phi_C(x_1) \Phi_D(x_2) \gamma_1^\mu D_{\mu\nu}(x_1 - x_2) \gamma_2^\nu \Phi_A(x_1) \Phi_B(x_2) d^4 x_1 d^4 x_2 \qquad (12)$$

for the matrix element between the two states $|AB>$ and $|CD>$, where the Φ's are one-electron wave functions, the γ's are the standard covariant Dirac matrices and $D_{\mu\nu}$ is the photon propagator. Using Coulomb gauge and performing time integration, this matrix element can be written, in terms of the one electron wave functions as:

$$< AB|g(1,2)|CD> = < AB| \frac{1}{R} - \frac{\vec{\alpha}_1 . \vec{\alpha}_2}{R} \cos\omega R + (\vec{\alpha}_1 . \vec{\partial}_1)(\vec{\alpha}_2 . \vec{\partial}_2) \left[\frac{\cos\omega R - 1}{\omega^2 R} \right] |CD>$$
$$(13)$$

with $\omega = \omega_{AC} = \alpha|\epsilon_A - \epsilon_C| = \omega_{BD} = \alpha|\epsilon_B - \epsilon_D|$, the ϵ's being the one particle energies, α the fine structure constant and R the interelectronic distance. This form of $g(1,2)$ for the interaction is worth some comments:

a) First, to obtain it, we have made use of the Coulomb gauge, a choice that will be immaterial for one electron functions Φ that are eigenfunctions of a local potential Dirac equation, but will introduce gauge dependence in the numerical results as soon as part of the interaction between the electrons is included in the determination of the Φ's, even in an average way like in the Hartree-Fock potential [7]. The reasons for preferring the Coulomb gauge are discussed elsewhere by J. Sucher[9].

b) The apparent lack of hermiticity of g is an artifact of introducing the one particles energies ϵ's and may be restored by rewriting it explicitly in term of one particles operators, i.e. :

$$\frac{cos\omega R}{R} = \frac{e^{i\omega R} + e^{-i\omega R}}{2R}$$

then

$$\frac{e^{i\omega R}}{R} = \frac{e^{i\alpha\epsilon_A R}.e^{-i\alpha\epsilon_C R}}{R} \quad " = " \quad e^{i\alpha h_A R}\frac{1}{R}e^{-i\alpha R h_C} \qquad (14)$$

c) For matrix elements for which the energy conservation is not fullfilled $(\omega_{AC} \neq \omega_{BD})$ the symmetrization $g = [g(\omega_{AC}) + g(\omega_{BD})]/2$ has to be introduced.

Up to now all calculations have been done by using, in the self consistent process, only the leading terms of the interaction:

$$\frac{1}{R} \quad \text{or} \quad \frac{1}{R} - \frac{\vec{\alpha}_1.\vec{\alpha}_2}{R} \qquad (15)$$

because the instantaneous Coulomb repulsion is by far the most important term, followed, in decreasing order of magnitude, by the magnetic interaction and the retardation terms in various order of ω. A first order perturbation correction of the neglected terms is generally enough after self consistency has been achieved. In equation (15) no projection operators are explicitly included as needed to prevent continuum dissolution [10] and nevertheless the results obtained in the framework of the multiconfiguration Dirac-Fock method are in quite good agreement with experiment. To understand why no difficulty occurs, let us take a closer look at the self consistent field method and show that projection operators are indeed implicitly included and that they are just the ones associated to the one particle basis set used in the calculation. To see why this is indeed the case let us assume that we sandwich the interaction with a projection operator Λ_+ , then in the spirit of the self consistent field method in which the direct and exchange potential terms are calculated from the solutions at the previous iteration, we can, in principle, generate at each step a complete set of positive energy solutions and use them to construct Λ_+ . But because both the direct and exchange potentials involve only the occupied orbitals, they will remain the same whatever or not we include the projection operator just defined. So the crucial point is indeed that, at each step of the calculation, we have to be careful in selecting the *acceptable* solutions. Restricting them to be of *electron* type and not *positron* like is in fact only one more constraint since we are already imposing orthonormality requirements. These constraints just define the boundary conditions of the *acceptable* solutions.

CONCLUSION

Realizing that the effective Hamiltonian used in the Dirac-Fock method or in its multiconfiguration extension is indeed very close to the correct Hamiltanion one can derive from QED provides the answer to the question we raised in the introduction. The next step would obviously be to improve even more the agreement with experiment or at least check that the neglected terms will not destroy it. At this stage new difficulties show up and I wish to mention at least two of them.

1) First we have said that the projection operator implicitly used is the one associated to the positive energy states of the Hartree-Fock potential V_{HF}, but it has long been known [11] that this potential is only uniquely defined for the occupied orbitals. Indeed any potential $V = V_{HF} + \Omega(1 - P)$, where P is the projection operator onto the subspace of the occupied orbitals and Ω arbitrary as long as it is Hermitician, will generated the same occupied orbitals but different excited ones. This flexibility in the choice of the potential that has been used in many-body perturbation theory to speed up the convergence [1], will have to be considered when going beyond the no-pair approximation.

2) The best agreement for excitation energies is generally obtained from the difference between two self consistent calculations. So doing we don't have the same Hartree-Fock potential for the initial and final states and consequently do not refer the two states to the same vacuum. How this will show up in QED corrections remains to be investigated.

REFERENCES

1. H.P. Kelly, Advan. Chem. Phys. <u>15</u>, 129 (1969).
2. J. Morrison and I. Lindgren, *Atomic Many-Body Theory* (Springer, Berlin, 1982).
3. M. Ya. Amusia and N.A. Cherepkov, Case Studies in Atomic Physics <u>5</u>, 47 (1975).
4. J. C. Slater, *Quantum Theory of Atomic Structure* (McGraw Hill, New York, 1960).

5. C. Froese-Fischer, *The Hartree-Fock Method for Atoms* (John Wiley & Sons Inc., New York, 1977).

6. See the contribution of I. Lindgren in this volume.

7. O. Gorceix, P. Indelicato and J. P. Desclaux, J. Phys. B $\underline{20}$, 639 (1988).

8. A.I. Akhiezer and V.B. Berestetskii, *Quantum Electrodynamics* (John Whiley & Sons Inc., New York, 1965)

9. J. Sucher, to be published.

10. J. Sucher, Phys. Rev. A $\underline{22}$ 348, (1980).

11. S. Huzinaga and C. Arnau, Phys. Rev. A $\underline{1}$ 1285, (1970).

COMMENTS ON RELATIVISTIC CORRELATION

Yong-Ki Kim
National Institute of Standards and Technology
Gaithersburg, Maryland 20899

ABSTRACT

Effects of electron correlation on relativistic corrections are illustrated by calculating the fine-structure energy levels of aluminumlike ions, $3p\ ^2P_{1/2}$ and $^2P_{3/2}$, with single- and multi-configuration Dirac-Fock wave functions. While the dominant behavior of the correlation contribution to the Breit interaction for the $J = 3/2$ level varies as Z^3, as expected, that for the $J = 1/2$ level varies as Z^2. The dominant behavior of the correlation contribution to the Lamb shift for both levels varies as Z^3. We used hydrogenic values of the self energy without any screening by other bound electrons and hence the correlation effect we have seen results from the redistribution of configuration weights in the correlated wave functions. These weights themselves have a weak $Z\alpha$ dependence though they are expected to approach constant values as $Z \to \infty$.

INTRODUCTION

As is attested to by the articles in these proceedings, relativistic atomic structure theory has made impressive progress during the last decade, and we can now realistically hope that our next goal is to develop methods to predict atomic energy levels to the accuracy comparable, or at least useful, to conventional spectroscopy. To achieve such an accuracy, we must be able to predict transition energies to 0.1% or better.

Although we have seen some QED predictions that agree with experiments on one-electron atoms to 5 or more significant figures, predictions of atomic energy levels for many-electron atoms rarely attain such an accuracy. In fact, 0.5% accuracy is about the best one can do at present, except for some isolated cases where better accuracy has been achieved.

The two basic hurdles to overcome in improving current theoretical capability are electron correlation and higher-order relativistic effects that include QED corrections such as screening due to other bound electrons. The former is not new. Hylleraas[1] showed how to cope with this problem for two-electron atoms almost 60 years ago. On the other hand, serious study of the latter for many-electron atoms has been practical only since powerful computers have become readily accessible.

Unfortunately, it is difficult to generalize the Hylleraas solution to atoms with an arbitrary number of electrons. Instead, configuration mixing and many-body perturbation theory are the standard methods to account for electron correlation. Quantum chemists nowadays routinely mix tens of thousands, and sometimes millions, of configurations. Relativistic formulations based on both the configuration mixing and perturbation theory have been carried out and applied to many-electron atoms.[2,3]

There are two distinct computational approaches to incorporate configuration mixing. The first approach, which was also historically the first one developed, is to choose a set of basis functions, be it analytic or numerical, form symmetry-adapted configuration functions (i.e., eigenfunctions of L^2, J^2, etc.), and then form a Hamiltonian matrix by sandwiching the Hamiltonian between these configuration functions. In this approach, the basis functions are kept frozen and only the mixing coefficients are determined by diagonalizing the Hamiltonian matrix. Although this method is conceptually simple, it requires hundreds, if not thousands, of relativistic configuration functions to provide accurate predictions. This approach is commonly called the *configuration interaction* (CI) method.

The second approach is to allow basis functions as well as mixing coefficients to vary according to the variational principle. In practice, fewer basis functions are used in this approach than in the CI method. Usually, a set of trial basis functions is chosen and mixing coefficients are determined by diagonalizing the Hamiltonian matrix constructed from the trial basis functions. Then, the mixing coefficients are kept fixed while the basis functions are optimized, e.g., by satisfying a set of self-consistent-field (SCF) equations. The new basis functions are kept frozen and used to recalculate the Hamiltonian matrix and to determine a new set of mixing coefficients by diagonalizing the matrix. These steps are repeated until the basis functions and the resulting total energy attain preset convergence accuracy, typically one part in 10^{-5} for the basis functions and 10^{-8} for the total energy. An advantage of this approach is that it can account for a great deal of electron correlation with a small set of basis functions while still retaining the conceptual simplicity of the CI method. This approach is usually referred to as the *multiconfiguration* (MC) SCF method.

Originally the correlation energy was defined[4] as the difference between the "exact" energy, E_{exact}, and the single-configuration, nonrelativistic Hartree-Fock (HF) energy, E_{HF}:

$$E_{exact} = E_{HF} + E_{corr}. \tag{1}$$

In view of the recent advances in the relativistic atomic structure theory, this definition is inadequate and must be revised.

For example, we must include various relativistic corrections and address the fact that using relativistic basis functions introduces some gain in total energy simply from using more radial functions than an equivalent nonrelativistic case—e.g., nonrelativistic $2p$ versus relativistic $2p_{1/2}$ and $2p_{3/2}$. Moreover, the concept of a single-configuration Hartree-Fock wave function must also be revised. In the nonrelativistic formulation, single-configuration HF wave functions for the $1s2p\ ^3$P and $1s2p\ ^1$P states can be expressed independently of each other as combinations of Slater determinants. For the relativistic formulation, however, both states must be represented by a mixture of the $1s2p_{1/2}$ and $1s2p_{3/2}$ configurations; only the mixing coefficients are different for the ^{3}P$_1$ and ^{1}P$_1$ levels. The ^{3}P$_2$ level consists only of the $1s2p_{3/2}$ configuration and the ^{3}P$_0$ level consists only of the $1s2p_{1/2}$ configuration. Hence, the relativistic counterpart of E_{HF} should be the total energy produced by a relativistic Hartree-Fock (= Dirac-Fock) wave function that contains all relativistic, symmetry-adapted configurations generated from the corresponding nonrelativistic single configuration.

In this article, we have used a somewhat different definition of the correlation energy appropriate for the relativistic formulation by

$$E_{exact} = E_{DF} + E_{QED} + E_{Rcorr}, \tag{2}$$

where the Dirac-Fock energy is denoted by E_{DF}, QED corrections by E_{QED}, and the remainder is lumped into the relativistic correlation energy, E_{Rcorr}. This definition is still ambiguous because the Dirac-Fock (DF) calculation can be done with or without an extended (i.e., not point) nucleus, with or without the mass polarization correction $(\mathbf{P_i} \cdot \mathbf{P_j})$, with or without the reduced mass, the Breit interaction with or without the energy-dependent retardation, and so on. The QED correction is equally ambiguous, e.g., with or without mutual screening, with one or more virtual photon exchange, etc. These additional fine points are not important for the discussions presented in this article. We simply state that:

(1) E_{DF} was calculated with a uniformly and spherically charged nucleus and with the Breit interaction in first order *without* energy-dependent retardation. Mass polarization was ignored, while conversion to wavenumbers (in cm^{-1}) was done using the reduced mass for each ion.

(2) E_{QED} was calculated using the hydrogenic (point-nucleus) self-energy values by Mohr[5] for the $1s$, $2s$, $2p_{1/2}$, and $2p_{3/2}$ orbitals, and they were extrapolated to ns, $np_{1/2}$ and $np_{3/2}$ for $n \geq 3$ using the n^{-3} scaling. The self energy for nl for $l \geq 2$ was estimated using the $Z = 0$ values of the $F(Z\alpha)$ function, i.e., $F(Z\alpha)$ was taken to be independent of Z. The vacuum polarization was calculated by using the Uehling potential and its next term in the $Z\alpha$ expansion with the DF (not hydrogenic) orbitals.

We did not use any explicit projection operator to prevent the continuum dissolution.[6] All radial wave functions are constrained to exponentially decay at a large distance from the nucleus, and the magnitudes of small components are kept small compared to those of the large components. Their actual ratios, however, are flexible and determined by the SCF process. Negative energy states are continuum states and hence require oscillating radial wave functions at a large distance as well as a reversal of the orders of magnitude between the large and small components. All DF calculations carried out so far using numerical radial wave functions do not exhibit any explicit signs of the continuum dissolution. The computational constraints mentioned above hardly qualify as rigorous projection operators—for example, resulting orbitals for the positive energy band ($\sim mc^2$ = electron rest mass) may not be orthogonal to *all* orbitals in the negative energy band ($\sim -mc^2$). It is possible that simply the positive-energy orbitals may have negligibly small overlaps with the negative-energy orbitals and hence little effect on the total energy for the cases we have studied so far. If so, then we expect to see some indication of the continuum dissolution in the inner shells of superheavy atoms.

Past experience in nonrelativistic calculations to account for electron correlation as accurately as possible, regardless of the theoretical methods used, taught us that we rapidly reach the point of diminishing returns. Accounting for $\sim 2/3$ of the expected correlation energy can be achieved with modest effort, but recovering more than 90% of the correlation energy involves very extensive calculations. All this is true also for relativistic calculations.

Unfortunately, we must do better than getting $\sim 2/3$ of the correlation energy to achieve our goal of 0.1% or better accuracy. Moreover, comparison with experiment is done on excitation energies, which forces us to lose several significant figures in theoretical results when we take the difference between total energies, though some quantities omitted in theoretical calculations may cancel out in the subtraction.

We studied the Z dependence of correlation corrections in order to understand its systematics. In this article, we report preliminary results on the Z dependence of relativistic correlation, E_{Rcorr}, obtained by studying the fine-structure splitting between the $3p\ ^2P_{1/2}$ and $^2P_{3/2}$ levels in aluminumlike ions. The purpose of our study is not to obtain the most accurate predictions for the fine-structure splitting; rather we hope to stimulate similar studies so that we can eventually understand and develop efficient, reliable methods to deal with relativistic correlation.

COMPARISON BETWEEN CORRELATED AND UNCORRELATED RESULTS

According to the nonrelativistic Z-expansion theory,[4] the (nonrelativistic) total energy of a level can be expanded in a power series of Z^{-1}:

$$E_{tot} = (mc^2)(Z\alpha)^2(E_0 + E_1/Z + E_2/Z^2 + E_3/Z^3 + \cdots), \tag{3}$$

where α is the fine-structure constant, E_0 is simply the sum of hydrogenic energies. In an SCF calculation, E_1 represents the correlation within a complex (= all possible nl combinations for a given n), and E_2, E_3, ... represent higher-order correlation contributions.

It is known[4] that a nonrelativistic multiconfiguration Hartree-Fock (MCHF) calculation, in which all possible combinations of open-shell electron configurations within a complex are included, correctly accounts for E_1 but only parts of E_n for $n \geq 2$. To fully recover these higher-order correlation corrections, one must introduce configurations that promote occupied orbitals to virtual orbitals with different principal as well as angular quantum numbers including those in the continuum.

When relativistic wave functions are used, various orders of relativistic effects expanded in powers of $Z\alpha$ are coupled to the nonrelativistic Z dependence. For instance, we expect that a relativistic multiconfiguration result that includes all configurations within a complex should exhibit a dependence on a $(Z\alpha)^2$ in addition to the dominant Z dependence associated with E_1. We shall see shortly that this expectation is proven to be correct.

Another simple conclusion we can draw is that the Breit interaction should have $(Z\alpha)^4$ dependence because it couples large and small components and consequently the leading correlation correction on the Breit interaction should vary as $(Z\alpha)^4/Z = Z^3\alpha^4$ along an isoelectronic sequence. This is not, however, always the case.

We have calculated the two energy levels in the ground state of aluminumlike ions, $3p\,{}^2P_{1/2}$ and ${}^2P_{3/2}$, for $Z = 13$–60 using three different types of wave functions: (a) single-configuration (SC) Dirac-Fock wave functions, (b) multiconfiguration (MC) Dirac-Fock wave functions that include all three-electron combinations of M-shell configurations with odd parity and $J = 1/2$ and $3/2$, and (c) configuration-interaction (CI) Dirac-Fock wave functions that include additional configurations with one or two electrons promoted from the L shell to the M shell. All wave functions were calculated using an advanced version of the Desclaux code.[7]

The MC wave functions contain relativistic equivalents of the nonrelativistic configurations $3s^2 3p$, $3p^3$, $3s\,3p\,3d$, and $3p\,3d^2$, which exhaust all combinations of three-electron configurations with odd parity in the M shell. The MC wave function for the $J = 1/2$ level contains 13 relativistic configurations and that for the $J = 3/2$ level contains 22 relativistic configurations. These MC wave functions should account for the entire *intrashell* correlation arising from the M shell. Since the K and L shells are filled, there are no additional intrashell correlation contributions from the core.

The CI wave functions used the orbitals of the MC wave functions as frozen orbitals and only the mixing coefficients were varied. The CI wave functions contain, in addition to all the configurations used in the MC wave functions, relativistic equivalents of the nonrelativistic configurations $2p^5 3s^2 3p^2$, $2p^4 3s^2 3p^3$, and $2p^4 3s^2 3p\,3d^2$ partially representing the $L - M$ *intershell* correlation. These configurations are far from being complete; they are chosen because they have relatively significant mixing coefficients compared to other configurations such as $2p^5 3s\,3p^2 3d$ or configurations involving the promotion of $2s$ electrons to the M shell. Even with this truncated choice of configurations, the $J = 1/2$ wave function contains 109 relativistic configurations and the $J = 3/2$ wave function contains 178 relativistic configurations.

There is an important reason that we did not calculate MC wave functions that account for the L-M intershell correlation. The radial functions of virtual (i.e., unoccupied) orbitals such as the $3d$ have maximum amplitudes where they overlap the most with the maxima of the occupied orbitals. If we try to carry out a multiconfiguration calculation that correlates the $3d$ orbital with both M- and L-shell orbitals, then the numerical process becomes unstable and difficult to make converge because one radial function cannot effectively overlap with both M- and L-shell orbitals owing to their widely different peak positions. Since there are two relativistic $3d$ orbitals, $3d_{3/2}$ and $3d_{5/2}$, one might be tempted to make one of them overlap with the M-shell orbitals and the other with the L-shell orbitals.

Such a maneuver, however, turns out to be numerically unstable. It is very difficult to determine what combination of these orbitals produces the desired wave function because many of the low-lying eigenvalues are very close to each other when the Hamiltonian matrix, whose dimension is the same as the total number of configurations, is diagonalized to determine the configuration mixing coefficients.

276

Table I. Comparison of theoretical and experimental values for the fine structure splitting between $3p\ ^2P_{1/2}$ and $^2P_{3/2}$ (in cm^{-1}).

Wave function	P^{2+}	Fe^{13+}
ΔE_{SCDF} without ΔE_{QED}	571	19149
ΔE_{QED}	12	106
Nonrel. offset	−7	−20
ΔE_{SCDF} total	576	19235
ΔE_{MCDF} without ΔE_{QED}	859	18840
ΔE_{QED}	12	99
Nonrel. offset	−308	−127
ΔE_{MCDF} total	563	18812
ΔE_{CIDF} without ΔE_{QED}	914	18837
ΔE_{QED}	12	100
Nonrel. offset	−359	−115
ΔE_{CIDF} total	567	18822
Experiment	559.1±1.0[a]	18852±5[b]

[a] W. C. Martin, R. Zalubas, and A. Musgrove, J. Phys. Chem. Ref. Data **14**, 751 (1985).
[b] J. Sugar and C. Corliss, J. Phys. Chem. Ref. Data **14**, Suppl. 2 (1985).

One solution to this difficulty is to introduce nonorthogonal orbitals as Fischer did for nonrelativistic calculations.[4] For instance, we can introduce relativistic equivalents of $3s$ and $3s'$, $3p$ and $3p'$, and $3d$ and $3d'$ where the primed orbitals correlate only with L-shell orbitals and the unprimed ones only with M-shell orbitals. The introduction of nonorthogonal basis functions makes the SCF equations based on the variational principle very cumbersome and increases the total number of Slater integrals used in the SCF process by a large factor. Even with our orthonormal orbitals of truncated configurations, our CI wave function for the $J = 3/2$ level required over 35,000 Slater integrals as input that took more than 10 minutes on a Cyber 205 just to read in! Thus we chose numerical expediency over complete coverage and decided to use a limited mixture of the MC and CI methods.

In addition, we calculated nonrelativistic counterparts of SC, MC and CI Dirac-Fock wave functions by repeating the same calculations after setting the speed of light $c \to \infty$ to simulate $\alpha \to 0$.

Setting $\alpha = 0$ in the calculation of relativistic wave functions does not necessarily lead to the correct nonrelativistic limit, particularly for MC and CI wave functions. In our case, some of the relativistic configurations reduce to nonrelativistic configurations other than ^{2}P. For instance, relativistic equivalents of the $3p^3$ configuration used in the $J = 3/2$ wave function will reduce to the $^4S_{3/2}$, $^2D_{3/2}$, and $^2P_{3/2}$ levels, while those used in the $J = 1/2$ wave function will reduce only to the $^2P_{1/2}$ level. One consequence of this "unmatched" nonrelativistic limit, coupled with the fact that the total number of relativistic configurations used in the wave functions for the two J values are different, is that the two nonrelativistic limits, one for $J = 1/2$ and the other for $3/2$, are not necessarily equal, though they should be. This difference is called the

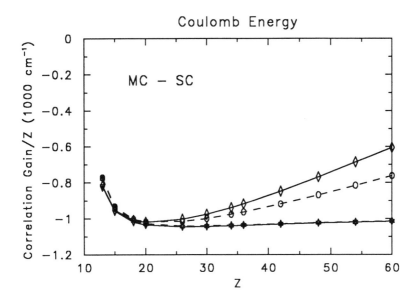

Figure 1. Gain in the Coulomb energy going from single-configuration (SC) wave functions to multiconfiguration (MC) wave functions in units of 1000 cm^{-1}. The actual ordinate plotted is the energy gain divided by the atomic number, $\delta E_{Coul}/Z$, to emphasize the Z^{-1} dependence of δE_{Coul}. Legend: $\Diamond$ = Dirac-Fock $\delta E_{Coul}(3p_{1/2})$; $\bigcirc$ = Dirac-Fock $\delta E_{Coul}(3p_{3/2})$; ∇ = nonrelativistic $\delta E_{Coul}(3p_{1/2})$; and $\triangle$ = nonrelativistic $\delta E_{Coul}(3p_{3/2})$.

nonrelativistic offset. Fine-structure splitting for ions of low charges obtained from relativistic wave functions agree well with experiment only when these nonrelativistic offsets (with appropriate signs) are subtracted from corresponding theoretical values.[8]

To illustrate the need for highly accurate theoretical results for atomic spectroscopy, we compare in Table I our results for the fine-structure splitting in P^{2+} and Fe^{13+} with experiment, since reliable values are known. The theoretical values are the sum of the Dirac one-electron energies, the Coulomb repulsion and the Breit interaction among the bound electrons, and the hydrogenic QED corrections described earlier. The hydrogenic (i.e., unscreened) QED corrections are expected to be an overestimate and hence our theoretical splitting with these QED corrections is an overestimate of the actual splitting.

For instance, a comparison of the total ΔE with the experimental value for Fe^{13+} in Table I indicates that even a theoretical accuracy of 0.1% means $\sim \pm 20$ cm^{-1}, far larger than the experimental uncertainty. The theoretical data in Table I clearly show that the CI wave functions, which include far more configurations than the MC wave functions, provide only minute changes in the *transition energies*. Hence, it is critical that we find more efficient ways to determine intershell correlation corrections.

Although an accuracy of 0.1% or better is desirable in predicting absolute transition energies, less accurate theoretical data can still be valuable in detecting erroneous experimental data by studying the Z dependence of the difference between theory and experimental data. The trend of this difference along an isoelectronic sequence is so regular that minor deviations in the difference can readily be identified and corrected.

278

Coulomb Energy

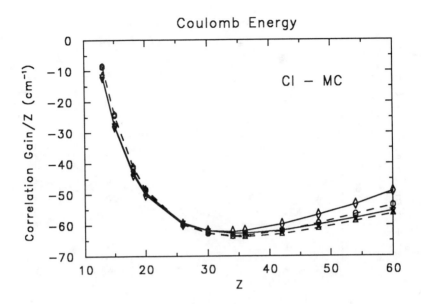

Figure 2. Gain in the Coulomb energy going from multiconfiguration (MC) wave functions to configuration interaction (CI) wave functions in cm^{-1}. The actual ordinate plotted is the energy gain divided by the atomic number, $\delta E_{Coul}/Z$, to emphasize the Z^{-1} dependence of δE_{Coul}. See the caption of Fig. 1 for the legend.

For instance, using MC wave functions, the random fluctuations in the experimental values of transition energies can easily be corrected to very high accuracy when the original experimental data are reasonably accurate.[8,9]

In Fig. 1, we present the correlation energy gained by using the MC wave functions. We used only the sum of the Dirac one-electron energies (Schrödinger energies for nonrelativistic cases) and the Coulomb repulsion. We denote this sum by E_{Coul}. The actual quantity plotted as functions of Z is E_{Coul} gained by going from SC to MC wave functions divided by Z. As expected, the nonrelativistic gains for both $J = 1/2$ and $J = 3/2$ levels become flat for $Z \sim 20$ and beyond, indicating that the E_1 in Eq. (3), which should be independent of Z, has been fully recovered. On the other hand, the relativistic gains demonstrate explicit dependence on some powers of Z, most likely to be $(Z\alpha)^2$. Another point to note is that Z-dependent trends emerge only after an ion is charged about 10 times.

In Fig. 2 we plot the gain in E_{Coul}/Z by going from MC wave functions to CI wave functions. This gain is a part of the E_2 and higher terms in Eq. (3). Here, we find that the nonrelativistic gain is very similar to that of the relativistic gain both in shape and magnitude, but neither exhibits the expected dependence of Z^{-2}. This deviation, however, is likely to be a consequence of not including enough configurations that account for intershell correlation. Relativistic gains for the $J = 1/2$ and $J = 3/2$ levels are practically the same for $Z \sim 30$ and below. This is why the gain in the transition energy, i.e., the difference between ΔE_{CIDF} and ΔE_{MCDF} in Table I, is so small, although the actual correlation energies gained by each level using the CIDF wave functions are ~ 400 cm^{-1} for P^{2+} and ~ 1600 cm^{-1} for Fe^{13+}.

A few lessons can be learned from Figs. 1 and 2. (a) One gains a significant fraction of the correlation energy by including all configurations needed to fully recover

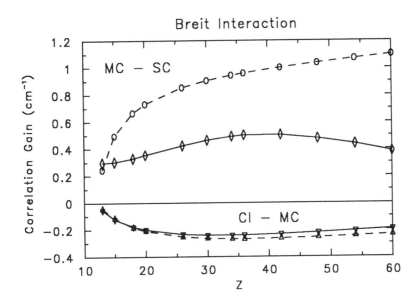

Figure 3. Correlation gain in the Breit interaction energy calculated by using single-configuration (SC), multiconfiguration (MC), and configuration mixing (CI) Dirac-Fock wave functions in units of cm^{-1}. The actual ordinate plotted is the energy gain divided by powers of Z. Legend: $\diamond = 100 \times \delta E_{Breit}(3p_{1/2})/Z^2$; $\bigcirc = 1000 \times \delta E_{Breit}(3p_{3/2})/Z^3$; $\triangledown = 100 \times \delta E_{Breit}(3p_{1/2})/Z^2$; and $\triangle = 100 \times \delta E_{Breit}(3p_{3/2})/Z^2$. Note that the gain in the $J = 3/2$ level going from SC to MC wave functions is divided by Z^3 unlike others. The actual values of δE_{Breit} in all cases are very small compared to δE_{Coul} as can be seen from Table A1 of the Appendix.

intrashell correlation in the valence shell. (b) Further effort to recover intershell correlation will be an arduous and slow process but one can expect whatever trends found in nonrelativistic calculations also to hold in corresponding relativistic calculations, e.g., choice of configurations to retain. (c) Using a nonrelativistic value of the intershell correlation instead of the actual relativistic counterpart in order to avoid numerical complexity associated with the latter[10] may be safe for $Z \sim 30$ or less but may not be so for $Z \sim 40$ and beyond.

As was mentioned earlier, we expect the gain in the Breit operator from intrashell correlation to be of the order of $(Z\alpha)^4/Z = Z^3\alpha^4$. The intrashell correlation gain in the Breit operator is shown in the top part of Fig. 3. The gain in the $J = 3/2$ level clearly exhibits the expected Z^3 dependence, but that in the $J = 1/2$ level does not. This departure from the expected Z dependence is not understood. On the other hand, the trend seen in the intershell correlation gain in the Breit operator shown in the bottom part of Fig. 3 agrees with the expected dependence, $(Z\alpha)^4/Z^2 = Z^2\alpha^4$ for both levels.

Since we used unscreened, hydrogenic values of the self energy in the Lamb shift, it appears that the self energy should not be affected by correlation. This is true only for the self energy contribution from each orbital, but the effective occupation number of each orbital changes when MC or CI wave functions are used because of the redistribution of configuration mixing coefficients. The fact that we used SCF orbitals to calculate the vacuum polarization contribution to the Lamb shift is not expected to

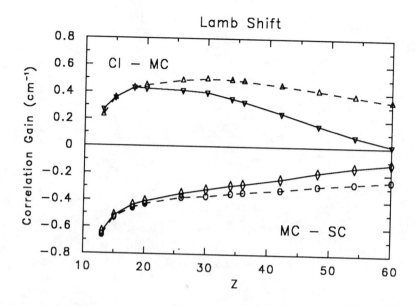

Figure 4. Correlation gain in the Lamb shift calculated by using single-configuration (SC), multiconfiguration (MC), and configuration mixing (CI) Dirac-Fock wave functions in units of cm^{-1}. The actual ordinate plotted is the energy gain divided by powers of Z. Legend: $\Diamond$ $= 100 \times \delta E_{Lamb}(3p_{1/2})/Z^3$; $\bigcirc$ $= 100 \times \delta E_{Lamb}(3p_{3/2})/Z^3$; ∇ $= 1000 \times \delta E_{Lamb}(3p_{1/2})/Z^2$; and $\triangle$ $= 1000 \times \delta E_{Lamb}(3p_{3/2})/Z^2$. Note that the gain in both levels going from MC to CI wave functions is divided by Z^2 unlike the changes in going from SC to MC wave functions. The actual values of δE_{Lamb} in all cases are very small compared to δE_{Coul} as can be seen from Table A1 of the Appendix.

significantly distort the Z dependence of the correlation gain in the Lamb shift because the vacuum polarization is a minor part of the Lamb shift for the range of Z we have studied. The intrashell correlation gain in the Lamb shift is shown in the bottom part of Fig. 4, and the intershell correlation gain in the top part. Similar to the Breit interaction, the leading Z dependence of the self energy is $\alpha(Z\alpha)^4$. According to the nonrelativistic Z-expansion theory, the mixing coefficients within a complex approach constant ratios in the limit $Z \to \infty$. This is also true in the relativistic case. Our data, however, indicate that the relativistic mixing coefficients approach their asymptotic values very slowly, exhibiting mild $Z\alpha$ dependence even at $Z = 100$. Correlation corrections to the Lamb shift, therefore, must come from a combination of this $Z\alpha$ dependence and the difference in $F(Z\alpha)$ for the individual orbitals. The intrashell gain in Fig. 4 shows a Z^3 dependence while the intershell gain exhibits a Z^2 dependence, as if the Lamb shift is modified by correlation in the same manner as the Coulomb energy is, although no two-electron operator is explicitly associated with the Lamb shift in our calculation.

Note that the actual magnitudes of the intershell correlation in the Coulomb energy are two to three orders of magnitude larger than those in the Breit interaction or in the Lamb shift. This clearly indicates that we cannot achieve the desired predictive accuracy in energy levels without solving the difficulty in recovering most of the intershell correlation, regardless of how well we can estimate the higher-order relativistic and QED corrections in many-electron atoms.

Some of the energy values used in Table I and Figs. 1–4 are presented in Table A1 of the Appendix.

CONCLUSION

Using the example of the fine-structure levels in the ground state of aluminum-like ions, we have shown that the prominent Z dependence of the intrashell and inter-shell correlation gains in the Coulomb energy (which includes the sum of the Dirac or Schrödinger one-electron energies), Breit interaction, and the Lamb shift are, in most cases, what we expect from the simple combination of the nonrelativistic Z-expansion theory and the leading $Z\alpha$ dependence of relativistic corrections. Departures from these expected behavior have been observed in the intershell correlation gain in both the relativistic and nonrelativistic Coulomb energy and the intrashell correlation gain in the Breit interaction of the $J = 1/2$ level. The former departure is likely to have resulted from the use of an insufficient number of configurations, but the latter is not understood.

For low-Z ions, we may use nonrelativistic results of the intershell correlation in the Coulomb energy instead of the relativistic results, but such a practice may not be safe for $Z \sim 40$ or higher.

The correlation gain in the Lamb shift results from the redistribution of the effective electron occupation numbers for each orbital due to configuration mixing. The mixing coefficients themselves have a mild $Z\alpha$ dependence even for very high Z unlike in the nonrelativistic case. The actual values of the intershell correlation gain in the Lamb shift are an order of magnitude smaller than those in the Breit interaction.

Finally, the uncertainty in the intershell correlation gain of the Coulomb energy far exceeds that in the Breit interaction and the Lamb shift combined. Therefore, it is important to develop practical methods to recover most of the intershell correlation in any relativistic atomic structure theory with the goal of competing in accuracy with experimental spectroscopy.

ACKNOWLEDGEMENTS

I am deeply indebted to J. P. Desclaux for his computer codes which made the present work possible. I am also grateful to numerous enlightening discussions with M. A. Ali, P. J. Mohr, and A. W. Weiss. This work was supported in part by the Office of Magnetic Fusion Energy, U.S. Department of Energy.

REFERENCES

1. E. A. Hylleraas, Z. Physik **54**, 347 (1929).
2. J. P. Desclaux, AIP Conf. Proc. **136**, 162 (1985).
3. I. Lindgren, *Many-Body Theory* in these proceedings.
4. See, for instance, C. F. Fischer, *The Hartree-Fock Method for Atoms* (Wiley, New York, 1977).
5. P. J. Mohr, Phys. Rev. A **26**, 2388 (1982).
6. J. Sucher, AIP Conf. Proc. **136**, 1 (1985).
7. J. P. Desclaux, Comp. Phys. Commun. **9**, 31 (1975).
8. Y.-K. Kim, J. Sugar, V. Kaufman, and M. A. Ali, J. Opt. Soc. Am. **B5**, 2225 (1988).
9. M. A. Ali and Y.-K. Kim, Phys. Rev. A **38**, 3992 (1988).
10. Y.-K. Kim, W. C Martin, and A. W. Weiss, J. Opt. Soc. Am. **B5**, 2215 (1988).

APPENDIX

We present in Table A1 the Coulomb energy (= one-electron energy + Coulomb repulsion), the Breit interaction, and the Lamb shift values calculated from single-configuration, multiconfiguration, and configuration interaction Dirac-Fock and corresponding nonrelativistic wave functions and used in Table I and Figs. 1–4.

Table A1. Energies calculated from SC, MC, and CI Dirac-Fock and nonrelativistic (NR) wave functions (in cm^{-1}). ($3p*=3p_{1/2}$, $3p=3p_{3/2}$)

Ion	Ca^{7+}	Zn^{17+}	Mo^{29+}	Xe^{41+}	Nd^{47+}
Z	20	30	42	54	60
Coul, 3p*,SC	-145562518	-360773352	-752710754	-1299578266	-1635460078
Breit,3p*,SC	41186.57	159198.66	477132.56	1078193.19	1519461.1
Lamb, 3p*,SC	31142.41	122512.83	377781.53	879560.11	1259217.93
NR, 3p*,SC	-144913875	-357085872	-737273259	-1254418272	-1564344764
Coul, 3p*,MC	-145582902	-360802629	-752746353	-1299615321	-1635496404
Breit,3p*,MC	41187.98	159202.83	477141.44	1078205.91	1519474.86
Lamb, 3p*,MC	31109.96	122425.77	377606.95	879318.76	1258959.45
NR, 3p*,MC	-144934586	-357117178	-737316559	-1254473360	-1564405714
Coul, 3p*,CI	-145583903	-360804484	-752748863	-1299618200	-1635499346
Breit,3p*,CI	41187.18	159200.66	477137.35	1078199.88	1519467.90
Lamb, 3p*,CI	31110.13	122426.13	377607.39	879318.99	1258959.50
NR, 3p*,CI	-144935573	-357119031	-737319154	-1254476480	-1564409040

Ion	Ca^{7+}	Zn^{17+}	Mo^{29+}	Xe^{41+}	Nd^{47+}
Coul, 3p,SC	-145557971	-360732309	-752500700	-1298897281	-1634347951
Breit,3p,SC	41021.50	158227.08	473562.66	1068965.62	1505650.33
Lamb, 3p,SC	31180.36	122696.43	378438.87	881187.35	1261535.91
NR, 3p,SC	-144913860	-357085851	-737273234	-1254418246	-1564344738
Coul, 3p,MC	-145578326	-360762299	-752539302	-1298941309	-1634393664
Breit,3p,MC	41027.35	158251.49	473636.72	1069133.92	1505887.95
Lamb, 3p,MC	31146.03	122597.68	378204.05	880757.41	1260985.32
NR, 3p,MC	-144934426	-357117055	-737316433	-1254473225	-1564405575
Coul, 3p,CI	-145579296	-360764177	-752541904	-1298944366	-1634396877
Breit,3p,CI	41026.51	158249.11	473632.06	1069126.74	1505879.48
Lamb, 3p,CI	31146.21	122598.13	378204.85	880758.51	1260986.54
NR, 3p,CI	-144935391	-357118935	-737319077	-1254476401	-1564408958

MANY-ELECTRON ATOMS: PARITY VIOLATION CALCULATIONS

MANY-BODY CALCULATIONS OF PARITY NONCONSERVATION IN CESIUM

S. A. Blundell
University of Notre Dame, Notre Dame, IN 46556

ABSTRACT

We review the atomic structure calculations necessary to interpret measurements of parity nonconservation (PNC) in cesium in terms of fundamental electron-quark couplings. The P-odd T-even interactions in atoms are summarized, and the general forms of the associated effective Hamiltonians are derived. We give a detailed review of the calculations which employ many-body perturbation theory (MBPT), which at present are among the most sophisticated; we discuss the likely theoretical uncertainty in this approach. An indication is given of the present level of agreement with the standard model, and we outline possibilities for future improvements in the atomic structure theory.

INTRODUCTION

Ever since the seminal work of the Bouchiats over a decade ago,[1] there has been considerable interest in atomic parity nonconservation, both from the experimental and the theoretical standpoints. While the initial motivation for the experiments was merely to detect weak neutral currents in electron-hadron interactions at a level consistent with the Weinberg-Salam model,[2] however, nowadays the goal is to make precision tests of this model at the level of radiative corrections. Unfortunately, the subject has been beset by difficulties on both the theoretical and experimental sides. Experimentally, there is difficulty in properly accounting for systematic errors arising from effects that mimic the very small PNC signals. On the theoretical side, there is difficulty in relating the experimental measurements to the underlying particle physics; observations have so far been made only on heavy atoms (Cs, Tl, Pb and Bi), and their interpretation involves overcoming a complex atomic many-body problem.

In general, agreement with the standard model at about the 5-10% level has already been obtained for a number of elements. Ideally, however, one would like to reduce the combined theoretical and experimental uncertainty to 1% or less in order to make a critical test of electroweak theory. The most promising element in this respect is cesium. Here, experimental measurements at the 2% level of accuracy have recently been reported, and it is possible that this uncertainty can be reduced substantially, say to the 0.2% level, in the near future.[3] In addition, of the elements of experimental interest, cesium poses the simplest atomic structure problem: it has one valence electron outside closed shells and is highly one-electron in character. Unfortunately, the present level of theoretical uncertainty in cesium, as we shall see, is about 2-5%, slightly short of the desired accuracy, and also slightly worse than the present experimental accuracy. To perform the atomic structure calculation convincingly to better than 1% remains one of the major problems facing atomic many-body theory.

In this article we shall discuss the issues involved in the atomic structure calculation and review the present status of the theory. A detailed discussion will be given of the recent MBPT calculations. First, however, we shall briefly outline the interactions

286

responsible for PNC in atoms, and discuss their relative orders of magnitude. The starting point for the atomic many-body calculations is a set of effective PNC Hamiltonians; we shall discuss the form of these Hamiltonians, and highlight the approximations involved in obtaining the standard Hamiltonians from the underlying particle physics.

The phenomena discussed here will be parity (*P*) nonconserving, but time-reversal (*T*) invariant; as such, they are all produced within the standard model. The search for an electric dipole moment of the atom, an effect which is both *P*-odd and *T*-odd, is outside the scope of the present article.

PNC ATOMIC INTERACTIONS: DOMINANT TERM

The processes which lead to PNC in an atom are summarized in Fig. 1. It turns out that the exchange of a Z^0 boson between an atomic electron and the nucleus, represented by Fig. 1(a), accounts for the majority of the observable effect, and the atomic structure theory discussed in later sections will apply exclusively to this term. We shall look at each diagram in turn, starting with Fig. 1(a).

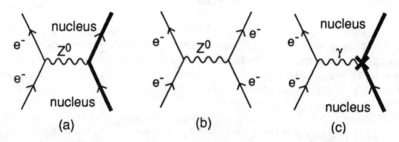

Fig. 1. Interactions responsible for PNC in atoms. Processes (a) and (b) are weak neutral current processes. In process (c), the PNC occurs within the nucleus, and is communicated electromagnetically to the atomic electrons; the black cross signifies that the nuclear current is not the usual one, but an odd-parity "anapole" current.

We shall assume the nucleus to be a stationary source of some PNC "potential", which can be used later as a perturbation in an atomic many-body calculation. To derive an expression for this potential, we shall consider the scattering amplitude for an electron scattering off a stationary nucleus. For simplicity, we shall assume a single-particle model for the nucleus, although the expressions are easily generalized to a fully correlated nuclear wavefunction, and the final result, Eqn. (9), is unchanged. We note that the momentum exchanged in atomic processes is sufficiently small that the weak interaction process in Fig. 1(a) (and Fig. 1(b)) can be represented as a contact interaction.

The PNC part of the scattering amplitude for Fig. 1(a) is conveniently treated as a sum of two terms,

$$M_{PNC}^{(a)} = M_A^{(a)} + M_V^{(a)} ,$$ (1)

$M_A^{(a)}$ arising from the axial electron current and vector nuclear current interaction,

$$M_A^{(a)} = - i \frac{G_F}{\sqrt{2}} \bar{e}\gamma_\mu \gamma_5 e \; .$$

$$\int d^3 x' e^{-i \, \mathbf{q} \cdot \mathbf{x}'} \left(C_{1n} \sum_n \bar{n}(x')\gamma^\mu n(x') + C_{1p} \sum_p \bar{p}(x')\gamma^\mu p(x') \right) , \quad (2)$$

and $M_V^{(a)}$ from the vector electron current and axial nuclear current interaction,

$$M_V^{(a)} = - i \frac{G_F}{\sqrt{2}} \bar{e}\gamma_\mu e \; .$$

$$\int d^3 x' e^{-i \, \mathbf{q} \cdot \mathbf{x}'} \left(C_{2n} \sum_n \bar{n}(x')\gamma^\mu \gamma_5 n(x') + C_{2p} \sum_p \bar{p}(x')\gamma^\mu \gamma_5 p(x') \right) . \quad (3)$$

Here G_F is the usual Fermi coupling constant, $n(x)$ and $p(x)$ are single-particle neutron and proton wavefunctions, e is a momentum-space electron spinor, $\mathbf{q}$ is the three-momentum transferred to the electron, and the sums are over all neutrons and protons in the nucleus. We have introduced coupling constants C_{1n}, C_{2n}, C_{1p} and C_{2p}; the standard model predicts values for these (which we give later), but we shall leave them as phenomenological parameters for now.

The most important term, as we shall show later, is the axial electron current term, given by Eqn. (2), and the remainder of this section will be devoted to this term. To extract an effective PNC Hamiltonian for this term, we compare Eqn. (2) with the amplitude to be expected for first-order scattering off a classical potential, and take an inverse Fourier transform with respect to $\mathbf{q}$, obtaining

$$h_1^{PNC} = \frac{G_F}{\sqrt{2}} \gamma^0 \gamma_\mu \gamma_5 \left(C_{1n} \sum_n \bar{n}(x)\gamma^\mu n(x) + C_{1p} \sum_p \bar{p}(x)\gamma^\mu p(x) \right), \quad (4)$$

where x is the electron co-ordinate. Next we take a nonrelativistic or static limit for the nucleus, in which the spatial part ($\mu = 1,2,3$) of the current-current interaction may be neglected, giving

$$h_1^{PNC} = \frac{G_F}{\sqrt{2}} (N C_{1n} + Z C_{1p}) \, \rho_1(x)\gamma_5 \; , \quad (5)$$

where N and Z are the numbers of neutrons and protons in the nucleus, and $\rho_1(x)$ is a weighted average of the neutron and proton densities, normalized to unity,

$$\rho_1(\mathbf{x}) = \frac{NC_{1n}}{NC_{1n} + ZC_{1p}} \, \rho_n(\mathbf{x}) + \frac{ZC_{1p}}{NC_{1n} + ZC_{1p}} \, \rho_p(\mathbf{x}) \ , \tag{6}$$

$$\int \rho_n(\mathbf{x}) \, d^3\mathbf{x} = \int \rho_p(\mathbf{x}) \, d^3\mathbf{x} = \int \rho_1(\mathbf{x}) \, d^3\mathbf{x} = 1 \ . \tag{7}$$

Finally, we introduce the "weak charge" defined by Bouchiat and Bouchiat,[1]

$$Q_W = 2(NC_{1n} + ZC_{1p}) \ , \tag{8}$$

and take the spherical average of the density $\rho_1(\mathbf{x})$, obtaining the final result[1]

$$h_1^{PNC} = \frac{G_F}{2\sqrt{2}} Q_W \, \rho_1(r) \gamma_5 \ . \tag{9}$$

Note that in our conventions, the Dirac matrix γ_5 has the value

$$\gamma_5 = \begin{pmatrix} 0 & I \\ I & 0 \end{pmatrix} \ , \tag{10}$$

in the Dirac-Pauli representation. There is a high degree of analogy between the derivation of Eqn. (9) from Eqn. (2), and the derivation of the usual spherically-averaged electrostatic nuclear potential $-Z(r)/r$ from single photon exchange between the nucleus and an atomic electron.

The principal motivation for studying atomic PNC is to extract a value for the weak charge from the measurements to compare with the prediction of the standard model. Now, as Q_W is the charge associated with the vector current of the nucleus appearing in Eqn. (2), which is a conserved current, Q_W does not undergo renormalization by the strong interactions. There is thus a direct relationship between Q_W and the fundamental coupling constants, with no uncertainty introduced by strong interaction effects. Similarly, the neutron and proton coupling constants are given simply in terms of the up and down quark coupling constants,

$$C_{1n} = 2C_{1d} + C_{1u}; \qquad C_{1p} = C_{1d} + 2C_{1u} \ . \tag{11}$$

This is exactly analogous to the way in which the electromagnetic charge of the neutron and proton is the sum of the charges on the valence quarks.

By using (11), (8) and (2), and the standard Feynman rules for the electroweak theory, one can easily find the tree-level predictions of the standard model,[1,4,5]

$$C_{1n} = -\frac{1}{2} \ , \tag{12}$$

$$C_{1p} = \frac{1}{2}(1 - 4 \sin^2\theta_W) \ , \tag{13}$$

$$Q_W = -N + (1 - 4\sin^2\theta_W)Z \ . \tag{14}$$

Radiative corrections to these values have been calculated by Marciano and Sirlin,[6] who obtained

$$Q_W = -0.974 N + [0.974 - 3.908 \sin^2\theta_W (M_W)]Z \ , \tag{15}$$

where $\sin^2\theta_W(M_W)$ is normalized at M_W. For cesium, the change in Q_W upon adding radiative corrections is about 2.5%, which sets the level of combined theoretical and experimental uncertainty below which significant constraints are placed on alternatives to the standard model.

A source of theoretical uncertainty that we would like to mention at this point concerns the choice for the nuclear density $\rho_1(r)$. As the present value of $\sin^2\theta_W = 0.230(5)$[7] is close to 1/4, the proton coupling constant (13) is somewhat smaller than the neutron coupling constant (12), about -10% of it. Thus the density $\rho_1(r)$ is mainly the neutron distribution. While a fair amount of empirical information is available concerning the proton (charge) distribution from electron scattering measurements, relatively little is known about the neutron distribution. In the absence of other information, we choose a Fermi distribution for $\rho_1(r)$ with parameters appropriate to the charge density.[8] For a 5% variation in the half-density radius of this distribution, the quantity of interest varies by 0.3%; thus we assign a nominal error of about 0.5% from this source, although this should probably be looked at a little more closely.

Finally, we note that the PNC Hamiltonian (9) is non-zero only within the nucleus, where the electrostatic potential for cesium is about 30 times the rest mass energy of the electron, and relativistic effects are important. It is therefore essential that the treatment of the electronic structure problem be relativistic for the level of accuracy aimed at here.

SMALLER PNC INTERACTIONS IN ATOMS

The first of the three smaller terms we shall discuss arises from the electron vector current contribution to Fig. 1(a), given in Eqn. (3). This gives rise to the following effective Hamiltonian,

$$h_2^{PNC} = \frac{G_F}{\sqrt{2}} \gamma^0 \gamma_\mu \left(C_{2n} \sum_n \bar{n}(x) \gamma^\mu \gamma_5 n(x) + C_{2p} \sum_p \bar{p}(x) \gamma^\mu \gamma_5 p(x) \right). \tag{16}$$

We take the nonrelativistic limit for the nucleus, as before, but this time the spatial part of the sum over μ remains, and the time part, $\mu = 0$, vanishes. This gives

$$h_2^{PNC} = \frac{G_F}{\sqrt{2}} \alpha \cdot \mathbf{j}(\mathbf{x}) \ , \tag{17}$$

where the current $\mathbf{j}(\mathbf{x})$ is a sort of weighted spin-density given by

$$\mathbf{j}(\mathbf{x}) = - C_{2n} \sum_n n^\dagger(\mathbf{x}) \Sigma n(\mathbf{x}) - C_{2p} \sum_p p^\dagger(\mathbf{x}) \Sigma p(\mathbf{x}) \ , \tag{18}$$

$$\Sigma = \alpha \gamma_5 = \begin{pmatrix} \sigma & 0 \\ 0 & \sigma \end{pmatrix} \tag{19}$$

(in the Dirac-Pauli representation for the γ-matrices). More generally, we can write $\mathbf{j}(\mathbf{x})$ as an expectation value over the nuclear wavefunction

$$\mathbf{j}(\mathbf{x}) = \left\langle - C_{2n} \sum_n \delta^3(\mathbf{x}_n - \mathbf{x}) \Sigma_n - C_{2p} \sum_p \delta^3(\mathbf{x}_p - \mathbf{x}) \Sigma_p \right\rangle, \tag{20}$$

where now this formula is applicable also to the more general case of a fully correlated nuclear wavefunction. The calculation of $\mathbf{j}(\mathbf{x})$ is essentially a nuclear structure problem. For atomic structure calculations, it is convenient to use the Wigner-Eckart theorem to replace $\mathbf{j}(\mathbf{x})$ by an effective operator valid for matrix elements diagonal in the nuclear spin I,

$$\mathbf{j}(\mathbf{x}) \rightarrow k_2 \mathbf{I} \rho_2(r) \ , \tag{21}$$

where k_2 is a dimensionless constant, and $\rho_2(r)$ is a suitable nuclear density normalized to unity (which we have spherically averaged). Our final result is thus

$$h_2^{PNC} = \frac{G_F}{\sqrt{2}} k_2 \rho_2(r) \alpha \cdot \mathbf{I} \ . \tag{22}$$

As the associated current is not this time conserved, the coupling constants C_{2p} and C_{2n} undergo renormalization by the strong interaction. At tree-level, the standard model gives[4]

$$C_{2p} = - C_{2n} = \tfrac{1}{2} g_A (1 - 4 \sin^2 \theta_W) \ , \tag{23}$$

$$g_A \approx 1.25 \ . \tag{24}$$

We can now see immediately why this term is much smaller than (9). Firstly, the constant k_2 is proportional to $(1 - 4\sin^2\theta_W)$, which leads to a suppression by a factor of about 0.1. In addition, in evaluating $\mathbf{j}(\mathbf{x})$, the contributions from nucleons cancel in pairs, leaving only the contribution from single unpaired nucleons. This is to be contrasted with the evaluation of Q_W, to which all nucleons contribute coherently, and leads to a further suppression factor of order $1/Z$. Thus the term is suppressed relative to (9) by a factor of order a few parts in a thousand.

By adopting a shell model of the nucleus, and assuming the contribution to come entirely from the unpaired proton, Flambaum and Khriplovich[9] obtain the following value of k_2 for the ^{133}Cs nucleus,

$$k_2 = \frac{2}{9} C_{2p} \approx 0.01 \ . \tag{25}$$

From the results in Ref. 10, we deduce that this would lead to an observable effect of about 0.2% the size of the axial electron current term, Eqn. (9).

We turn now to the neutral current electron-electron interaction in Fig. 1(b). As a scattering amplitude, the PNC part may be written

$$M_{PNC}^{(b)} = - i \frac{G_F}{\sqrt{2}} C_3 \left(\bar{e}_2 \gamma_\mu \gamma_5 e_2 \ \bar{e}_1 \gamma^\mu e_1 + \bar{e}_2 \gamma_\mu e_2 \ \bar{e}_1 \gamma^\mu \gamma_5 e_1 \right) \ . \tag{26}$$

This leads to the following two-body effective Hamiltonian, representing the interaction between electrons i and j

$$h_3^{PNC}(i, j) = \frac{G_F}{\sqrt{2}} C_3 \left((\gamma_5)_i + (\gamma_5)_j \right) \left(1 - \alpha_i \cdot \alpha_j \right) \delta^3(\mathbf{x}_i - \mathbf{x}_j), \tag{27}$$

such that the total Hamiltonian for a many-electron atom is given by

$$H_3^{PNC} = \sum_{i > j} h_3^{PNC}(i, j) \ . \tag{28}$$

In the standard model at tree level,

$$C_3 = -\frac{1}{2}(1 - 4 \sin^2 \theta_W) \ . \tag{29}$$

The effect of this term is small; Schäfer et al.[11] found that it contributed about 0.5% of the dominant term (9).

The final process we consider is depicted in Fig. 1(c). Here weak interaction processes within the nucleus endow the nucleus with an odd-parity electromagnetic moment, the so-called anapole moment,[9] and the PNC can be communicated via photon exchange to the atomic electrons. Note that the anapole moment is different from the electric dipole moment, which would require effects both P-odd and T-odd occurring within the nucleus. Unfortunately, we do not have space here to discuss the interesting nuclear processes that lead to the anapole moment.[12,9] In Ref. 9, however, it is shown that a PNC contact interaction is generated with the following form,

$$h_4^{PNC} = \frac{G_F}{\sqrt{2}} k_4 \rho_4(r) \, \alpha \cdot \mathbf{I} \, , \tag{30}$$

where k_4 is a dimensionless constant proportional to the nuclear anapole moment, and $\rho_4(r)$ is some nuclear density normalized to unity. We see that (30) is identical in form to (22); in an atomic physics experiment, one can only measure the effect of the sum of terms (30) and (22). The calculations of Flambaum et al.[13] indicate that the anapole term in fact dominates the vector electron current term (22); they find

$$k_4 = 0.063 - 0.084 \, , \tag{31}$$

which is about 6 to 8 times larger than the value for k_2, Eqn. (25). According to the calculations of Frantsuzov et al.,[10] this effect amounts to about 1 to 1.5% of that of the dominant axial electron current term, Eqn. (9).

Although the terms in Eqns. (22) and (30) are small, they can be distinguished from the dominant axial electron current term (9) by virtue of the fact that they are nuclear spin dependent, while the latter is not. Thus, measurements made on different hyperfine lines should show slight differences which can be attributed entirely to terms (22) and (30). Noecker et al.[3] report measurements on the $F=3$ to $F=4$ and $F=4$ to $F=3$ transitions in the 6s-7s Cs hyperfine multiplet, and, from the slight difference in the two results, deduce

$$k_2 + k_4 = 0.18\,(10) \, , \tag{32}$$

in good agreement with the theoretical predictions (25) and (31). The next generation of experiments should substantially reduce the error in this value.[3] Interestingly, the average value of the measurements on these two hyperfine lines is almost independent of any effect of the nuclear spin dependent terms, containing a contribution of about 0.1%.[10]

As we have seen, these additional small terms are very interesting in their own right, in particular because the anapole term gives valuable information on PNC nuclear forces. In the remainder of this review, however, we shall concentrate exclusively on the atomic theory associated with the axial electron current term, Eqn. (9), with a view to permitting an accurate extraction of Q_W from the measurements.

ATOMIC STRUCTURE THEORY: GENERAL CONSIDERATIONS

The most obvious consequence of a small PNC term in the atomic Hamiltonian is that the atomic states are no longer eigenstates of parity, but contain a small admixture of states of opposite parity. If we denote the exact (i.e. incorporating all electron correlation effects) wavefunction by $|\Psi\rangle$, then can write

$$|\Psi\rangle = |\Psi^0\rangle + |\tilde{\Psi}\rangle \, , \tag{33}$$

where $| \Psi^0 >$ is the usual atomic wavefunction, and $| \Psi >$ is the opposite parity admixture. It is necessary to treat the PNC Hamiltonian only to first order (which is furthermore consistent with its derivation above), so we have

$$| \Psi > = \sum_i \frac{| \Psi_i^0 >< \Psi_i^0| H_1^{PNC}| \Psi^0 >}{E - E_i} , \qquad (34)$$

where the sum over i extends over all exact states of opposite parity to $| \Psi^0 >$, and H_1^{PNC} is the sum over all electrons of the one-body operator h_1^{PNC}. The energy levels of the atom are unaffected to first order in the PNC Hamiltonian, since it is of odd parity.

All atomic PNC experiments to date (a recent review is given in Ref. 14) make use of the fact that the PNC Hamiltonian admixes a small electric-dipole (E1) amplitude into an otherwise magnetic-dipole (M1) transition. Let us specialize the discussion to the $6s \rightarrow 7s$ transition of interest for the cesium experiment; the amplitude for this transition for an isolated atom may be written

$$A_{6s \rightarrow 7s} = A_{M1} + A_{E1}^{PNC} . \qquad (35)$$

[In the usual phase conventions, in which the M1 amplitude is real, the PNC-E1 amplitude turns out to be pure imaginary. This may be verified directly, but is more deeply related to the fact that the effect observed is invariant under time-reversal.[1]] The experimentally observed quantity is an interference term between the PNC-E1 amplitude and a parity-conserving electromagnetic amplitude (in cesium, a Stark-induced E1 amplitude);[14] it depends linearly on the PNC-E1 amplitude, which in turn is proportional to the weak charge. If we write,

$$A_{E1}^{PNC} = a Q_W , \qquad (36)$$

then the task for the atomic theory is to calculate the constant a.[15] The remainder of this article will be devoted to this problem.

For the $6s \rightarrow 7s$ transition in Cs, we have

$$A_{E1}^{PNC} = < \Psi_{7s}| D| \Psi_{6s}^0 > + < \Psi_{7s}^0| D| \Psi_{6s} > \qquad (37)$$

$$= \sum_i \frac{< \Psi_{7s}^0| H_1^{PNC}| \Psi_i^0 >< \Psi_i^0| D| \Psi_{6s}^0 >}{E_{7s} - E_i} + $$

$$\sum_i \frac{< \Psi_{7s}^0| D| \Psi_i^0 >< \Psi_i^0| H_1^{PNC}| \Psi_{6s}^0 >}{E_{6s} - E_i} , \qquad (38)$$

where D is the dipole operator, and the intermediate state sum is over exact many-body states. One possible approach is to use Eqn. (38) directly, as has been done by Bouchiat and Piketty in their recent semi-empirical calculation.[16] In general, this direct approach involves a many-body calculation for each of a finite subset of intermediate states, and possibly some sort of extrapolation to take account of intermediate states not explicitly calculated. Experimental values are available for some of the dipole matrix elements and energies appearing in (38).

An alternative approach, which is the one to be discussed here, is the use of MBPT, an approach which has many advantages for this problem. It is completely systematic, with well-defined levels of perturbation. Moreover, it is possible to compute standard properties of the atom - energies, hyperfine splittings, allowed E1 amplitudes - at each level of perturbation as a guide to the likely accuracy of the calculation. Finally, as discussed in the previous section, it is essential for the many-body treatment to be relativistic. Terms arising in MBPT can be shown to have a natural justification in terms of quantum electrodynamics (QED);[17] in our own calculations, we use the "no pair" approximation discussed by Joe Sucher and Walter Johnson elsewhere in these proceedings. A general formalism for setting up a many-body perturbation series is discussed by Ingvar Lindgren in his contribution to these proceedings; see also Ref. 18. In the present problem, there are three perturbations: the residual Coulomb interaction (the difference between the exact Coulomb interaction of the electrons and the mean field chosen for the zeroth-order problem), the dipole operator and the PNC Hamiltonian. In principle, a complete linked diagram perturbation series can be written down using the standard graphical rules,[18] such that each diagram contains the dipole and PNC operators once each, but the residual Coulomb interaction an arbitrary number of times. The sum of all terms in this series is equal to the right-hand side of Eqn. (38).

The need to treat three perturbations makes the calculation of the PNC-E1 amplitude more complex than the corresponding calculation of some standard property, such as the hyperfine parameter, where there are only two perturbations. However, the diagrammatic book-keeping can be simplified considerably by adopting the following trick, suggested by Sandars.[19] Formally, the PNC Hamiltonian, which is a one-body operator, is regarded as being incorporated into the definition of the single-particle states $|i>$ used to build up the many-body perturbation series; thus we take (in atomic units)

$$\left(h_D + U(r) + h_1^{PNC} \right) | i > = \varepsilon_i | i > , \tag{39}$$

where

$$h_D = c\,\alpha \cdot \mathbf{p} + (\beta - 1)c^2 - \frac{Z(r)}{r} , \tag{40}$$

and $U(r)$ is the mean field. The single-particle states are now parity-mixed,

$$| i > = | i^0 > + | \tilde{i} > , \tag{41}$$

and the number of perturbations has effectively been reduced by one. In practice, Eqn. (39) is solved as the pair of equations,

$$(h_D + U(r))| i^0 > = \varepsilon_i | i^0 > , \tag{42}$$

$$(\varepsilon_i - h_D - U(r))| \tilde{i} > = h_1^{PNC} | i^0 > , \tag{43}$$

correct to first order in the PNC Hamiltonian.

We can now take an expression from MBPT for an E1 amplitude, and evaluate it with the parity-mixed single-particle states. For example, in lowest order, the PNC-E1 amplitude is given by

$$A_{E1}^{PNC} = < 7s^0 + 7\tilde{s}| D | 6s^0 + 6\tilde{s} >$$

$$= < 7s^0 | D | 6\tilde{s} > + < 7\tilde{s}| D | 6s^0 > , \tag{44}$$

to first order in the PNC Hamiltonian. In general, one takes the MBPT expression and replaces one state at a time by its PNC counterpart. For example, the following expression from second-order perturbation theory undergoes this replacement,

$$\sum_{ra} \frac{<wr| \frac{1}{r_{12}}|av > <a|D|r>}{\varepsilon_v + \varepsilon_a - \varepsilon_r - \varepsilon_w} \rightarrow$$

$$\sum_{ra} \left(\frac{<\tilde{w}r^0 \frac{1}{r_{12}}|a^0 v^0> <a^0|D|r^0>}{\varepsilon_v + \varepsilon_a - \varepsilon_r - \varepsilon_w} + \frac{<w^0 \tilde{r} \frac{1}{r_{12}}|a^0 v^0> <a^0|D|r^0>}{\varepsilon_v + \varepsilon_a - \varepsilon_r - \varepsilon_w} + \right.$$

$$\frac{<w^0 r^0 \frac{1}{r_{12}}|\tilde{a} v^0> <a^0|D|r^0>}{\varepsilon_v + \varepsilon_a - \varepsilon_r - \varepsilon_w} + \frac{<w^0 r^0 \frac{1}{r_{12}}|a^0 \tilde{v}> <a^0|D|r^0>}{\varepsilon_v + \varepsilon_a - \varepsilon_r - \varepsilon_w} +$$

$$\left. \frac{<w^0 r^0 \frac{1}{r_{12}}|a^0 v^0> <\tilde{a}|D|r^0>}{\varepsilon_v + \varepsilon_a - \varepsilon_r - \varepsilon_w} + \frac{<w^0 r^0 \frac{1}{r_{12}}|a^0 v^0> <a^0|D|\tilde{r}>}{\varepsilon_v + \varepsilon_a - \varepsilon_r - \varepsilon_w} \right). \tag{45}$$

(Here, r is an excited state, a is a core state, and v and w are the initial and final valence states respectively.) In this way, the enumeration of the MBPT diagrams is greatly simplified (although one still has to evaluate all possible opposite-parity substitutions). Later, when we compare the calculation of the PNC-E1 amplitude to the calculation of standard properties "at the same level of approximation," we shall mean this in the parity-mixed sense discussed above.

A parity-mixed single-particle basis can be used in other approaches too. Any method of calculating an allowed E1 amplitude which uses single-particle functions can be adapted to the PNC problem by writing down the parity-mixed version of the defining equations, which is obtained using the successive substitution method outlined above. This approach has been applied to the multi-configuration Dirac-Fock method,[19] the Equations-of-Motion method,[20] and the configuration-interaction method.[21]

In a MBPT calculation on cesium, it is convenient to take the zeroth-order problem to be the Dirac-Hartree-Fock (DHF) of the closed-shell core (that is, of the Cs$^+$ ion). Many MBPT diagrams vanish identically when the single-particle states are defined in this potential,[18] and the calculation of higher-order perturbation theory is greatly simplified. However, on account of the self-consistency requirement in the DHF equations, it is necessary to generalize Eqn. (39) defining the PNC counterparts. From the DHF equations,

$$\left[\varepsilon_i - h_D - V_{HF}^0 \right] | \, i^0 > \, = 0 \quad , \tag{46}$$

we construct their PNC counterpart,

$$\left[\varepsilon_i - h_D - V_{HF}^0 \right] | \, \tilde{i} > \, = \left[h_1^{PNC} + \tilde{V}_{HF} \right] | \, i^0 > \quad , \tag{47}$$

where on the right-hand side of Eqn. (47) we have introduced the PNC-DHF potential, formed by replacing one state at a time in the definition of the DHF potential by its PNC counterpart. The self-consistent solution of Eqns. (46) and (47) defines a complete parity-mixed single-particle basis, and the technique of parity-mixed perturbation theory can be employed with no further modification. The PNC-DHF equations automatically build into the PNC counterparts an infinite subset of many-body corrections analogous to the well-known "core polarization" by the hyperfine operator;[18] this point and further properties of the PNC-DHF equations are discussed in the original work by Sandars.[22]

In our calculations we use the DHF potential of the core, and perform the sums over excited states in perturbation theory by constructing a finite basis set from B-splines, as discussed by Walter Johnson elsewhere in these proceedings. The first step is to solve the DHF and PNC-DHF equations (46) and (47) for the *core* states using conventional numerical techniques. Next, we construct a finite pseudo-spectrum of states $|i^0>$ using the B-spline method.[23] For each state $|i^0>$ we now construct its PNC counterpart by using Eqn. (47) rearranged in the form

$$| \tilde{i} > \, = \sum_j \frac{| \, j^0 > < j^0 | \left(h_1^{PNC} + \tilde{V}_{HF} \right) | i^0 >}{\varepsilon_i - \varepsilon_j} \quad , \tag{48}$$

where the sum extends over all states of the pseudo-spectrum with opposite parity to $|i^0>$. Now we have a "complete" set of states with PNC partners, and can implement the procedure of parity-mixed perturbation theory by performing the summations directly. Careful tests were made to ensure that the basis set was sufficiently complete at the level of accuracy required. Full details are given in Ref. 24.

One final point concerns the expression for the dipole operator, which is commonly used in either the "length" or "velocity" form, each form being related to a different choice of gauge for the electromagnetic potential.[25] In atomic units we have for the two forms,

$$\mathbf{D}_L = -\mathbf{r} \ , \tag{49}$$

$$\mathbf{D}_V = i\,\frac{c\,\alpha}{\omega} \ , \tag{50}$$

where ω is the transition frequency. If the many-body treatment of the problem were exact, one would obtain identical results from (49) and (50). However, in any approximation scheme, the two forms may give different results. In some schemes, such as the random phase approximation, there is an identity between the two forms; in this case, the use of the two forms provides a valuable numerical cross-check on the calculation. Otherwise, one might suppose that the discrepancy arising from use of the two forms gives an indication of the degree of incompleteness of the many-body treatment. In order-by-order MBPT based on a DHF potential, the non-local exchange terms destroy length-velocity agreement at each order of perturbation theory.

MBPT RESULTS

The calculations described in this section have been performed recently by three groups: W. R. Johnson and co-workers,[8,24] O. P. Sushkov and co-workers,[26,27] and A.-M. Mårtensson-Pendrill.[28] The first calculations on cesium to employ MBPT were by B. P. Das and co-workers.[29]

In first order, using the PNC-DHF functions discussed above, one obtains

$$A_{E1}^{PNC} = <7\overline{s},m=\tfrac{1}{2}|D_z|6s^0,m=\tfrac{1}{2}> + <7s^0,m=\tfrac{1}{2}|D_z|6\overline{s},m=\tfrac{1}{2}>$$

$$= \begin{cases} 0.927 & iea_0(Q_W/-N\,)10^{-11} \quad \text{(length)} \\[2mm] 0.830 & iea_0(Q_W/-N\,)10^{-11} \quad \text{(velocity)} \end{cases} . \tag{51}$$

Note that we have factored out the sought-after quantity Q_W, divided by its approximate value $-N$. Each of the three groups cited agree on these values, to within discrepancies at the 0.4% level to be expected from slightly different choices for the nuclear density $\rho_1(r)$.

At the next order of perturbation theory, the correction to be applied is[18,30]

$$A_{E1}^{(2)} = \sum_{r\,a} \frac{\left(<7s,r|\tfrac{1}{r_{12}}|6s,a> - <r,7s|\tfrac{1}{r_{12}}|6s,a>\right)D_{ar}}{\varepsilon_{6s} + \varepsilon_a - \varepsilon_{7s} - \varepsilon_r} + c.c., \tag{52}$$

where r is an excited state, a is a core state and "c.c." denotes the complex conjugate of the previous term with 7s and 6s interchanged. The PNC amplitude is obtained by making all possible PNC substitutions, as described in the previous section. It is convenient at this stage to consider also an infinite subset of higher-order diagrams associated with the much-studied Random-Phase Approximation (RPA).[31,18] The RPA corresponds to an infinite sequence of terms, with the first- and second-order corrections, Eqns. (51) and (52), being the leading two terms. An RPA calculation has been carried out by each of the three groups above, with agreement on the total final result (first order, plus second order, plus third- and higher-order RPA corrections),[28,8,27]

$$A_{E1}^{PNC} (RPA) = 0.890 \; iea_0 (Q_W / -N) 10^{-11} \quad \text{(length and velocity)}, \quad (53)$$

at the same level as with the first-order result above. There is a well-known identity between length and velocity forms within the RPA approximation.[22] Physically, the RPA corresponds to the shielding of the electric field of the photon by the atomic electrons.[22]

Next we move on to third order, where there are many terms.[30] However, one expects on energy denominator grounds[26,32] that a subset of third-order terms dominate, namely those associated with Brueckner (or "natural") orbitals (BO). We can write this subset in the following way,

$$A_{E1}^{(3)}(BO) = < \delta (7s) | D | 6s > + < 7s | D | \delta (6s) > , \quad (54)$$

where the Brueckner orbital correction $|\delta v>$ to a valence state $|v>$ is given by

$$| \delta v > = \sum_{i \neq v} \frac{| i > < i | \Sigma (\varepsilon_v) | v >}{\varepsilon_v - \varepsilon_i} , \quad (55)$$

and $\Sigma(\varepsilon)$ is a "correlation potential",

$$< i | \Sigma (\varepsilon) | j > = \sum_{abr} \frac{(g_{abrj} - g_{barj}) g_{riba}}{\varepsilon_a + \varepsilon_b - \varepsilon - \varepsilon_r} + \sum_{ars} \frac{(g_{rsaj} - g_{rsja}) g_{airs}}{\varepsilon_a + \varepsilon - \varepsilon_r - \varepsilon_s} . \quad (56)$$

Here, a and b are core states, r and s are excited states, and i and j are general states. We have denoted Coulomb matrix elements by

$$g_{ijkl} = <ij | \frac{1}{r_{12}} | kl > . \quad (57)$$

As usual, for the PNC amplitude one must make all possible PNC substitutions in these equations. Note that the diagonal matrix element of $\Sigma(\varepsilon)$ is the second-order correction to the valence ionization energy,

$$\delta\varepsilon_v^{(2)} = < v | \Sigma (\varepsilon_v) | v > . \quad (58)$$

Physically, the BO correction corresponds to the electrostatic polarization of the core by the valence electron, and Σ is the corresponding polarization potential. Eqn. (54) gives the lowest-order BO contribution; there are BO contributions from higher orders of perturbation theory also.

The smaller remaining third-order diagrams can be classified into RPA, structural radiation (SR) and normalization contributions;[26,30] explicit expressions are given in Ref. 30. Note that we have already included the third-order RPA contributions as part of the all-order RPA calculation above.

The numerical results for third order are difficult to compare directly, because the two groups concerned have included different subsets of terms. Using the B-spline method, we have made what we believe to be a numerically accurate evaluation of the well-defined third-order BO term given in Eqn. (54). Our result is[24]

$$A_{E1}^{PNC} \text{ (3 rd order BO)} = 0.061 \; iea_0 (Q_W / -N) 10^{-11} \quad \text{(length)} , \tag{59}$$

bringing the total final result to

$$A_{E1}^{PNC} = 0.951 \; iea_0 (Q_W / -N) 10^{-11} . \tag{60}$$

By contrast, Dzuba et al.[26,27] evaluate all third-order diagrams, together with isolated higher-order contributions, namely, terms corresponding to mixtures of RPA and BO effects (a fourth-order contribution), and terms corresponding to the "chaining" of Brueckner orbitals (a fifth-order contribution). Their total final result is[27]

$$A_{E1}^{PNC} = 0.90 \; iea_0 (Q_W / -N) 10^{-11} , \tag{61}$$

to which they assign a theoretical uncertainty of 2% (discussed further in the next section).

Not all of the difference between (60) and (61) can be ascribed to the additional terms in the Russian calculation, however. From their own estimates,[27] these extra terms amount to no more than 3% of the final result, while the discrepancy to be explained is about 6%. The precise cause of this residual disagreement is unknown and is still under investigation. However, as is discussed in Ref. 24, the third-order BO result (59) arises from large cancellations between contributing terms, and is quite sensitive to basis set truncation errors. It is possible that the limited basis set and partial wave summation used by the Russian group account for the disagreement.

For reference, we give here the result of the semi-empirical calculation by Bouchiat and Piketty,[16]

$$A_{E1}^{PNC} = 0.935 (2) (3) \; iea_0 (Q_W / -N) 10^{-11} . \tag{62}$$

DISCUSSION OF MBPT RESULTS

In order to assess the likely theoretical error in these MBPT calculations arising from omitted correlation effects, it is useful to compute standard properties of the atom for comparison with their accurately known experimental values. In Table I we give results for hyperfine splitting parameters and allowed E1 matrix elements in cesium,[32] calculated at the same levels of perturbation theory as was done for the PNC-E1 amplitude above; we also give the results for the PNC-E1 amplitude from above.

Table I. Magnetic dipole hyperfine constant for the 6s state in Cs, reduced E1 transition matrix element for Cs 6s-6p$_{1/2}$, and PNC-E1 amplitude for Cs 6s-7s. To the second-order result we have added all third- and higher-order RPA corrections. The third-order result is restricted to the dominant third-order BO contributions. The values are taken from Refs. 32 and 24 and references therein.

Order	6s hfs (MHz)	E1 (length) 6s-6p$_{1/2}$ (a.u.)	PNC-E1 (length) 6s-7s ($iea_0(Q_W$-$N)10^{-11}$)
1st	1435.	5.278	0.927
2nd (RPA)	289.	- 0.303	- 0.037
3rd (BO)	642.	- 0.582	0.061
TOTAL	2366.	4.393	0.951
Experiment	2298.	4.52(1)	-

From Table I, we see that the third-order result is about twice the second-order result for all properties, showing that MBPT does not converge rapidly over the first few orders for these properties in Cs. In MBPT, although one has the general notion that the residual Coulomb interaction is "small", there is no guarantee that perturbation theory will converge rapidly over the first few orders; there is no small parameter appearing in the expansion, such as the fine structure constant in QED.

Nevertheless, it must surely be significant that both the 6s hyperfine constant and the 6s-6p$_{1/2}$ E1 amplitude end up within about 3% of their experimental values, having started in first order at -38% and +17%, respectively, from experiment. Similar results hold for other properties of cesium; after adding the RPA and third-order BO contributions, we find[32] that the 7s hyperfine constant differs from experiment by 4.8% (-28%), and the 6s-6p$_{3/2}$ E1 amplitude by -3.0% (17%). The figures in parentheses indicate the discrepancy with experiment in first order. We have performed similar calculations along the whole series of alkali atoms from lithium to cesium.[32] The same approximation gives hyperfine and allowed E1 results differing from experiment by about 0.5% for lithium, and about 1-3% for sodium and rubidium. We can also perform order-by-order MBPT for valence ionization energies in cesium, finding agreement with experiment at about the 1-2% level in second order.[32] Our conclusion is that we have identified two important classes of many-body effects, those associated with the RPA and Brueckner orbitals, which together enable one to account for matrix elements in cesium at about the 3-5% level.

The calculation of hyperfine constants and allowed E1 dipole matrix elements is particularly significant for the PNC calculation: the dipole matrix element forms half of the problem for the PNC-E1 amplitude, and tests the (implicit) quality of the wavefunction at large distances, while the s-electron hyperfine structure is sensitive to the wavefunction near the origin, which is where the PNC Hamiltonian acts. Based on this observation, and on the agreement with experiment noted above for standard properties, we assign a theoretical uncertainty of 5% to our final result given in Eqn. (60).

It should be noted, however, that the fractional effects of the RPA and BO corrections are rather smaller for the PNC calculation than for the hyperfine or allowed E1 calculations. For example, the RPA correction to the 6s hyperfine constant is a +12.5% effect, while the BO correction is an additional +28%. By contrast, the RPA correction to the PNC amplitude is a -4% effect, while the BO correction is an additional +6.4%. For this reason, the result in Eqn. (60) may be more accurate than 5%.

Turning now to the calculations by Dzuba et al.,[26,33,27] we note that they have calculated additional higher-order terms, as well as the remaining third-order diagrams, and find a shift of order only 2-3%. In addition, they make a semi-empirical adjustment to the BO calculation in order to improve the agreement of standard properties and valence ionization energies with experiment, and this changes the result in Eqn. (61) by a negligible amount. Although there is still an unexplained 3% numerical discrepancy between the lowest-order BO calculations of Refs. 27 and 24, this last step may partially compensate for any numerical error in the calculation that may be present. They assign an overall theory error of 2% to their calculation.

COMPARISON WITH STANDARD MODEL

To give some idea of the extent to which the combined experimental and theoretical results in cesium agree with the prediction of the standard model, we follow Ref. 3 in taking the average value of three recent and extensive calculations, namely those of Refs. 24 (Eqn. (60)), 27 (Eqn. (61)) and 16 (Eqn. (62)). Combining the average of these three values with their experimental result, and assuming an uncertainty of 5% in the atomic structure theory, Noecker et al.[3] obtain the following value for the weak charge,

$$Q_W = -69.4 \pm 1.5 \pm 3.8 \ , \tag{63}$$

where the first error is experimental, and the second is from the atomic structure theory.

Unfortunately, the prediction from the standard model is limited in accuracy by the present limit on the Weinberg angle, and at the level of radiative corrections also depends on the values assumed for the masses of the top quark and Higgs boson. If one takes the present world average of the Weinberg angle[7]

$$\sin^2\theta_W \equiv 1 - \frac{M_W^2}{M_Z^2} = 0.230\,(5) \ , \tag{64}$$

and makes the standard assumptions that the top quark mass is 45 GeV, and that the Higgs mass is 100 GeV, the value predicted for the weak charge is[7]

$$Q_W = -71.8\,(1.1) \ , \tag{65}$$

in good agreement with the value from atomic PNC in cesium. As an alternative, one can use the atomic data in conjunction with Eqn. (15) to obtain an independent mea surement of $\sin^2\theta_W(M_W)$. One finds[3]

$$\sin^2\theta_W\ (M_W\ ) = 0.\ 219 \pm 0.\ 007\ \pm 0.\ 018\ . \tag{66}$$

We note that, according to Langacker et al.,[34]

$$\sin^2\theta_W\ (M_W\ ) \approx 0.\ 994 \left(1 - \frac{M_w^2}{M_z^2}\right)\ . \tag{67}$$

For a discussion of how atomic PNC data can be used to explore alternatives to the standard model, for example, to place constraints on the masses of additional neutral bosons, see the review by Amaldi et al.[7]

CONCLUSIONS

The major question facing the atomic theory is how convincingly to reduce the theory error to the 1% level, or less. The uncertainty from use of third-order MBPT seems to be at about the 2-5% level. One obvious possibility is to continue the systematic application of MBPT to fourth, fifth and higher order. Unfortunately, however, the number of diagrams for even the fourth-order calculation of a matrix element in a one-valence-electron atom is prohibitively large. An alternative is to try and pick subsets of diagrams from higher order designed in such a way as to improve the results for a range standard properties. This is a highly dangerous procedure, however, because of the temptation to try those diagrams which are simplest to evaluate, and because at this level of precision it is not clear that what may happen to work for hyperfine structure and allowed E1 matrix elements will also work for a PNC amplitude. The process is not helped by the difficulty of estimating the value of a diagram with absolute certainty without resorting to an explicit evaluation.

An alternative to order-by-order MBPT is the use of an all-order method, such as the Coupled-Cluster approach which has been highly successful in quantum chemistry. In these methods, one sums systematically all one-, two-, three-, ..., etc. body interactions. Thus, the approach is as systematic as order-by-order MBPT, corresponding merely to a different ordering of perturbation theory terms. Furthermore, there is evidence from studies of small atomic and molecular systems that the sequence is a quite rapidly converging one,[18] in contrast to the order-by-order sequence which appears to be quite unstable, at least initially. These methods are briefly described by Ingvar Lindgren elsewhere in these proceedings. It seems feasible in cesium to attempt the pair approach, in which one- and two-body effects are summed to all orders. An assessment of theory error can be made by estimating the size of the next term in the series, the three-body effects, from perturbation theory; one can evaluate those MBPT diagrams which correspond to the leading order of three-body effects. It is our contention that this approach is the one most likely to lead to a reliable calculation at the 1% level, and we are presently investigating the application of spline methods to the solution of the all-order equations. See also the discussion by Ann-Marie Pendrill elsewhere in these proceedings.

The atomic theory would be strengthened, however, if a number of different approaches could be successfully applied to the problem, although each method should have a high degree of independent credibility (e.g. contain a systematic hierarchy of approximations with reliable estimates of numerical error), otherwise a disagreement would merely confuse the issue. Among the possible alternatives to the all-order approach suggested above is a large-scale configuration-interaction calculation. Also, direct use of Eqn. (38), combined with as much empirical information as possible (energies, dipole matrix elements), is a reasonable alternative to the parity-mixed technique outlined in this review.

Finally, we would like to point out that, while the present article has been confined exclusively to cesium, the case for atomic PNC would be improved overall if accurate results could be obtained on a number of different elements. Thallium, although theoretically less favorable than cesium, is structurally similar in that it has one valence electron ($6p_{1/2}$) outside closed shells. Methods that are successful on cesium are in principle applicable also to thallium. The MBPT methods discussed here have in fact been applied by Dzuba et al.[27] to the $6p_{1/2}$-$7p_{1/2}$ and $6p_{1/2}$-$6p_{3/2}$ transitions in thallium, with estimated theoretical uncertainties of 6% and 3%, respectively. Thus, if the experimental situation in thallium improves, this could form a second element in which high precision tests of the electroweak theory can be made.

REFERENCES

1. M. A. Bouchiat and C. Bouchiat, J. Phys. (Paris) **35**, 899 (1974); **36**, 493 (1975).
2. S. Weinberg, Phys. Rev. Lett. **19**, 1264 (1967); A. Salam, in *Elementary Particle Theory*, edited by N. Svartholm (Almquist and Forlag, Stockholm, 1968).
3. M. C. Noecker, B. P. Masterson and C. E. Wieman, Phys. Rev. Lett. **61**, 310 (1988).
4. E. D. Commins and P. H. Bucksbaum, *Weak Interactions of Leptons and Quarks,* (Cambridge University Press, 1983).
5. C. Quigg, *Gauge Theories of the Strong, Weak and Electromagnetic Interactions* (Benjamin/Cummings, 1983).
6. W. J. Marciano and A. Sirlin, Phys. Rev. D **27**, 552 (1983).
7. U. Amaldi et al., Phys. Rev. D **36**, 1385 (1987).
8. W. R. Johnson, D. S. Guo, M. Idrees, and J. Sapirstein, Phys. Rev. A **34**, 1043 (1986).
9. V. V. Flambaum and I. B. Khriplovich, Sov. Phys. JETP **52**(5), 835 (1980).
10. P. A. Frantsuzov and I. B. Khriplovich, Z. Phys. D **7**, 297 (1988).
11. A. Schäfer, B. Müller, and W. Greiner, Z. Phys. A **322**, 539 (1985).
12. E. M. Henley, W.-Y. P. Hwang, and G. N. Epstein, Phys. Lett. B**88**, 349 (1979).
13. V. V. Flambaum, I. B. Khriplovich, and O. P. Sushkov, Phys. Lett. B**146**, 367 (1984).
14. E. D. Commins, Physica Scripta **36**, 468 (1987).
15. More correctly, it is the ratio of the PNC-E1 amplitude to some other matrix element that is required for the interpretation. In the case of Wieman's experiment, this other matrix element is the vector transition polarizability.[35,3] However, the

atomic structure theory for this quantity is not as formidable as for the PNC-E1 amplitude, and we shall not discuss it further in this review.

16. C. Bouchiat and C. A. Piketty, Europhys. Lett. **2**, 511 (1986).
17. J. Sapirstein, Physica Scripta **36**, 801 (1987).
18. J. Morrison and I. Lindgren, *Atomic Many-Body Theory*, Vol. 13 of *Springer Series in Chemical Physics* (Springer, Berlin, 1982).
19. E. P. Plummer and I. P. Grant, J. Phys. B **18** L315 (1975).
20. C. P. Botham, S. A. Blundell, A.-M. Mårtensson-Pendrill, and P. G. H. Sandars, Physica Scripta **36**, 481 (1987).
21. B. P. Das, Physica Scripta **36**, 497 (1987).
22. P. G. H. Sandars, J. Phys. B **10**, 2983 (1977); Physica Scripta **21**, 284 (1980).
23. W.R. Johnson, S.A. Blundell, and J. Sapirstein, Phys. Rev. A **37**, 307 (1988).
24. W. R. Johnson, S. A. Blundell, Z. W. Liu, and J. Sapirstein, Phys. Rev. A **37**, 1395 (1988).
25. A. I. Akhiezer and V. B. Berestetskii, *Quantum Electrodynamics* (Interscience, New York, 1965).
26. V. A. Dzuba, V. V. Flambaum, P. G. Silvestrov and O. P. Sushkov, J. Phys. B **18** 597 (1985).
27. V. A. Dzuba, V. V. Flambaum, P. G. Silvestrov and O. P. Sushkov, J. Phys. B **20**, 3297 (1987); Physica Scripta **36**, 69 (1987).
28. A.-M. Mårtensson-Pendrill, J. Phys. (Paris) **46**, 1949 (1985).
29. B. P. Das *et al.*, Phys. Rev. Lett. **49**, 32 (1982).
30. S.A. Blundell, D.S. Guo, W.R. Johnson, and J. Sapirstein, Atomic Data and Nuclear Data Tables **37**, 103 (1987).
31. A. L. Fetter and J. D. Walecka, *Quantum Theory of Many-Particle Systems*, (McGraw-Hill, 1971).
32. W. R. Johnson, M. Idrees, and J. Sapirstein, Phys. Rev. A **35**, 3218 (1987).
33. V. A. Dzuba, V. V. Flambaum, P. G. Silvestrov, and O. P. Sushkov, J. Phys B **20**, 1399 (1987).
34. P. Langacker, *Proc. Int. Symp. on Lepton-Photon interaction (Kyoto)*, ed. M. Konuma and K. Takahashi, (1985).
35. S. L. Gilbert and C. E. Wieman, Phys. Rev. A **34**, 792 (1986).

THE NO-PAIR PAIR EQUATION -
Fundamental Problems, Numerical Solutions and Applications

Ann-Marie Mårtensson-Pendrill
Department of Physics
University of Göteborg and Chalmers University of Technology
S-412 96 Göteborg, SWEDEN   Bitnet: f3aamp@secthf51

ABSTRACT

The solution of inhomogeneous two-particle differential equations is a powerful method for treating correlation effects in the non-relativistic case and is reviewed briefly. The relativistic generalization, the so-called Dirac-Coulomb equation, is complicated by the need to distinguish between positive and negative energy states. The construction and use of projection operators onto positive energy states are presented and the application of the pair equation to other properties such as parity non-conservation is discussed.

1. INTRODUCTION

The developments in experimental techniques have provided extremely accurate results for many atomic systems. These results often contain indirect information which may be of interest outside atomic physics. E.g., the study of hyperfine structure can provide values for nuclear magnetic and electric moments[1]. Isotope shifts (i.e. small shift in energy levels between different isotopes) can tell about the differences in charge distribution[1,2]. However, in order to extract this information, a detailed understanding of the atomic structure is essential. In particular the need for accurate atomic calculations of parity non-conserving weak interaction effects in heavy atoms has stimulated the developments of methods to treat pair correlation effects in a relativistic framework. These lecture notes describe our work to generalize the solution of a non-relativistic pair equation to the relativistic case and also briefly discuss the work needed to apply the relativistic pair equation to the treatment of parity non-conservation.

The non-relativistic procedure used in our group for a long time (see e.g. Garpman *et al.*[3,4]) is based on Many-Body Perturbation Theory (MBPT), (described in detail in Ref. 5 and outlined in Lindgren's lecture notes in this volume[6]) in combination with numerical solution of partial differential equations which describe the one-and two-particle excitations from the unperturbed wave function. By iterative solution of these equations the interaction between two electrons can in principle be treated exactly and very accurate results are, indeed, obtained for two-electron systems. The MBPT provides prescriptions for generalizing the treatment to many electron systems. Although we do not explicitly include effects which involve the simultaneous excitations of more than two particles, couplings between different pair excitations are included through the iterative procedure. By using the coupled-cluster formalism, also all *products* of single and double excitations are automatically included.

The fundamental problems involved in the generalization of the non-relativistic procedure occur already for two electrons and most of the discussion below will focus on helium-like systems. Section 2 describes the non-relativistic method, the relativistic treatment is described in Sect. 3. In Section 4, we discuss alternative methods of treating additional perturbations and how the pair equation may be applied to the PNC problem.

2. THE NON-RELATIVISTIC TREATMENT
OF HELIUM-LIKE SYSTEMS

In the non-relativistic case a solution of the pair equation gives in principle the exact wave function for helium. The helium atom is therefore an important test of the equations and numerical procedures used. It is also a relatively easy example of the application of many-body perturbation theory and provides building blocks needed to treat larger systems.

One possible starting point in the treatment of helium-like systems is to include only the interaction with the nucleus in the one-body hamiltonian[*] .

$$H_0 = h_{nuc}(1) + h_{nuc}(2) \qquad h_{nuc}(i) = -\frac{\nabla_i^2}{2} - \frac{Z}{r_i} \qquad h_{nuc}|i\rangle = \varepsilon_i|i\rangle$$

and then treat the full interaction between the electrons as a perturbation:

$$V = V_{12} = \frac{1}{r_{12}}$$

The eigenvalue equation for a two-electron system may be written as

$$H\Psi = \left(h_{nuc}(1) + h_{nuc}(2) + V_{12} \right)\Psi = \left(\varepsilon_a + \varepsilon_b + \Delta E \right)\Psi = E\Psi \qquad (2.1)$$

Consider now the case of two non-equivalent electrons in orbitals a and b. These may form either a singlet or a triplet state. The spins of the electrons do not enter explicitly in the wavefunction and it is sufficient to consider only the spatial part, keeping in mind that the requirement that the total wavefunction be anti-symmetric forces the spatial function to be symmetric for singlet states and anti-symmetric for triplet states. The unperturbed spatial function can be written as

$$\Phi = \{ab\} = \frac{(|ab\rangle \pm |ba\rangle)}{\sqrt{2}}$$

where the plus (minus) sign is associated with the singlet (triplet) state and the curly brackets denote the combination with the correct symmetry properties. The projection operator P onto the model space (cf. Ref 6) including both combinations is given by

$$P = \sum_{+,-} |\Phi\rangle\langle\Phi| = \sum_{cd=ab,ba} |cd\rangle\langle cd| \qquad (2.2)$$

In the case of two equivalent electrons (i.e. $a=b$) we get instead $\Phi=\{ab\}=|ab\rangle$ and the sum in the projection operator in (2.2) contains only one term.

---

*) We use Hartree atomic units where $e=\hbar=m_e=4\pi\varepsilon_0=1$, $c=1/\alpha$.

The correction $\delta\Psi$ to this approximate wave function Φ can be expressed in terms of a pair function, ρ_{ab}, giving a total wave function

$$\Psi = \Omega\Phi = \Phi + \delta\Psi = \{ab + \rho_{ab}\} \qquad (2.3)$$

where Ω is the waveoperator, which reproduces the exact wavefunction when operating on the approximate one[5,6]. Using *intermediate normalization*, $P\Psi=\Phi$ leads to orthogonality between the correction and the unperturbed wavefunction, i.e. $<\{ab\}|\{\rho_{ab}\}>=0$, and the eigenvalue equation can be rewritten to give an equation for the correction, ρ_{ab}:

$$\left(\varepsilon_a + \varepsilon_b - h_{nuc}(1) - h_{nuc}(2)\right)\{\rho_{ab}\} = V_{12}\{ab + \rho_{ab}\} - \Delta E\{ab + \rho_{ab}\}$$

$$\Delta E = \left\langle \{ab\}\left|V_{12}\right|\{ab + \rho_{ab}\}\right\rangle = \left\langle ab\left|V_{12}\right|ab + \rho_{ab}\right\rangle \pm \left\langle ba\left|V_{12}\right|ab + \rho_{ab}\right\rangle$$

Usually we work with non-antisymmetrized pair functions, satisfying the equation

$$\left(\varepsilon_a + \varepsilon_b - h_{nuc}(1) - h_{nuc}(2)\right)\left|\rho_{ab}\right\rangle = V_{12}\left|ab + \rho_{ab}\right\rangle$$
$$- \sum_{cd=ab,ba} \left|cd + \rho_{cd}\right\rangle\left\langle cd\left|V_{12}\right|ab + \rho_{ab}\right\rangle \qquad (2.4)$$

where the term $-|cd><cd|V_{12}|ab+\rho_{ab}>$ makes the right-hand side orthogonal to cd (i.e. to ab and ba). The equation (2.4) is a special case of the Bloch equation[6,7].

$$[\Omega,H_0]\,P = (V\Omega - \Omega P V\Omega) \qquad (2.5)$$

The last term, which accounts for the energy correction, appears as backward diagrams in the diagrammatic expansion.

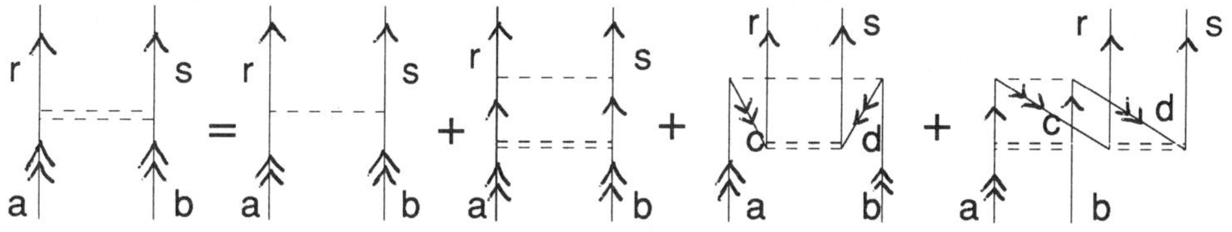

Fig. 1. Diagrammatic representation of the all-order pair equation (2.4). The last two "backward" diagrams account for the energy correction.

In effect we use a degenerate model space, which includes both the singlet and triplet state, and the orthogonalization to cd ensures orthogonality to the model space.The pair function ρ_{ab} in (2.4) will be a superposition of the singlet and triplet states and only the proper combination $\{ab+\rho_{ab}\}$ will have a physical meaning and it is easy to show that this combination satisfies the original eigenvalue equation. (If we were only interested in either the singlet or triplet state, we might move terms between the equations for ρ_{ab} and ρ_{ba} as long as the desired combination satisfies the correct equation for that state.) The correction ρ_{ab} defined in this way will contain not only double excitations but also single excitations, where one electron is left in state a or b and we have a choice of treating these together with the pair excitations or removing also the single excitations from the right-hand side and treating them separately. (The latter approach was adopted in the relativistic case, described in the next section, where the single-particle equations can be solved directly, whereas approximations are introduced in the solution of the pair equation.) The first-order pair equation is obtained by dropping all terms involving ρ_{ab} on the RHS. By using the closure relation,

$$\sum_i^{all} |i\rangle\langle i| \tag{2.6}$$

the right-hand side can be expressed in terms of a complete set of eigenfunctions, $|rs\rangle$, except the excluded combinations $|ab\rangle$ and $|ba\rangle$. The solution can then be written as

$$\rho_{ab}^{(1)} = \sum_{rs \neq ab, ba} \frac{|rs\rangle \langle rs|V_{12}|ab\rangle}{\varepsilon_a + \varepsilon_b - \varepsilon_r - \varepsilon_s} \tag{2.7}$$

This first-order wavefunction correction can then be inserted on the RHS of (2.4) to obtain the next approximation to ρ_{ab} leading to an iterative procedure, where, at each step, the best ρ_{ab} available is inserted. (If desired, it would be possible to use an order-by-order approach, but then it is necessary to store also all lower-order pair functions, rather than only the most recent.)

The exclusion of $rs=ab,ba$, from the RHS of (2.4) and (2.7) removes terms which would otherwise enter with zero energy denominator. However, the functions obtained by adding any amount of $|ab\rangle$ or $|ba\rangle$ to ρ_{ab} would still be solutions to the equation. If these homogeneous terms enter in the pair function they should be removed by orthogonalizing the solution. However, the presence of homogeneous terms may still destroy the numerical accuracy and a more stable procedure is obtained by shifting the energy denominators by adding a small correction, $\delta\varepsilon \, \rho_{ab}$ both to the left- and right-hand sides. This new equation has no homogeneous solutions and the solution to the original equation can be obtained by iteration.

In the actual calculations the pair functions are divided into several terms depending on the angular momentum of the excited states. All these terms contribute both to the energy correction and to the first term on the RHS and thus the equation for all possible l-excitations are coupled.

2.1 Results for the non-relativistic pair function

The first direct numerical solution of the non-relativistic pair equation was performed by Musher and Schulman[8] in a study of helium energies. About the same time a direct numerical solution of the eigenvalue equation for helium was performed by McKoy and Winter[9] and by Winter et al.[10]. These pioneering works were followed by applications of the pair equation to calculations to second order in the electron-electron interaction for corrections to hyperfine structure in atoms[11,3] and to correlation energies in larger systems[12]. Our own calculations rely heavily on the solutions of the pair equation. By using pair functions to create approximate Brueckner orbitals[13] and by iterative solution of the pair equation[14] we have been able to include pair correlation effects to all orders. The approach has been extended to an open-shell coupled-cluster formalism[15-16] and the procedure has proved to be capable of producing very accurate results also for atoms with many electrons (see e.g. the works on Ca and Ca[+] in Ref 17-18 and on Li[19]). An alternative approach to pair functions, which does not involve the partial wave expansion and is applicable also to molecules, is the use of "random tempering of Gaussian-type geminals" developed by Alexander *et al.*[20]

A computer program for iterative solution of the pair equation has been used in our group for over a decade. It has recently undergone major revision with intent of publication[21] and is now being tested for several problems. Among the new features included is the more accurate treatment of the cusp at $r_i = r_j$, as discussed in Ref 22. and inclusion of an energy offset term, which, together with the increased number of points, have led to the improved accuracy of the newer results for helium and beryllium, shown in Tables 1-2 and 3, respectively[23].

2.1.1 Energies for the 1s2p States of Helium Tables 1 and 2, show the results for the *1s2p* states in He as an example of the treatment of two non-equivalent electrons. Following the procedure outlined above (described in more detail in Ref 24), we solved the equations for the functions $\rho_{1s2p \rightarrow ll'}$ and $\rho_{1s2p \rightarrow l'l}$ separately, rather than for the properly (anti-)symmetrized combinations $\{\rho_{1s2p \rightarrow ll'}\}$. In the *sp*- limit we thus solve the two coupled equations for excitations into *sp* and *ps*. In the *pd*-limit these equations were coupled also to the equations for excitations to *pd and dp*. (A slight disadvantage of this approach is that the slower $(l+1/2)^{-4}$ angular convergence for the singlet space, compared to $(l+1/2)^{-6}$ for the triplet state will affect all excitations. (These relations have been strictly proven for the S states[25] where also an expression for the coefficients in the expansion has been obtained[26], but the same relations appear to hold also for the P states.) An additional disadvantage is that twice as many radial functions are coupled together, instead of treating two separate problems each with only half of the functions. For excitations from core orbitals, on the other hand, the equations are more complicated and both approaches involve couplings between all the functions and the approaches are comparable.) The results in different angular limits are compared also to the Multi-Configuration Hartree-Fock results by Froese-Fischer[27] and to the results obtained in a more recent Configuration Interaction (CI) calculation by Davis and Chung[28]

Table 1. Angular limit convergence of the energy eigenvalue for the $(1s2p)^{1,3}P$ states in Helium

2^1P	Froese-Fischer [27]	Davis and Chung [28]	Lindroth and Mårtensson-Pendrill [24]	Salomonson and Öster [23]
$sp - limit$	-2.122587	-2.1225935	-2.122592	-2.1225968
$pd - limit$	-2.123653	-2.1236858	-2.123686	-2.1236902
$df - limit$	-2.123766	-2.1237912	-2.123793	-2.1237970
$fg - limit$		-2.1238156	-2.123820	-2.1238232
$gh - limit$		-2.1238239	-2.123829	-2.1238327
$hi - limit$		-2.1238264	-2.123834	-2.1238370
$ik - limit$			-2.123835	-2.1238392
$kl - limit$				-2.1238405
$Extrapolated$				-2.123843
'Exact' [29]				-2.123843
2^3P				
$sp - limit$	-2.132367	-2.1323688	-2.132362	-2.1323714
$pd - limit$	-2.133054	-2.1331108	-2.133106	-2.1331149
$df - limit$	-2.133086	-2.1331508	-2.133147	-2.1331559
$fg - limit$		-2.1331562	-2.133153	-2.1331619
$gh - limit$		-2.1331572	-2.133154	-2.1331634
$hi - limit$			-2.133155	-2.1331638
$ik - limit$			-2.133155	-2.1331640
$kl - limit$				-2.1331641
$Extrapolated$				-2.133164
'Exact' [29]				-2.133164

Table 2. Convergence of the mass polarisation and of the normalisation integrals for the $(1s2p)^{1,3}P$ states in Helium (From Lindroth and Mårtensson-Pendrill, Ref. 24)

	$\langle p_1 \cdot p_2 \rangle$ $(10^{-2}a.u.)$		Normalisation Integral N^2	
	2^1P	2^3P	2^1P	2^3P
$sp - limit$	3.2710	-7.5157	1.9098	1.5566
$pd - limit$	4.5969	-6.4234	1.8682	1.5432
$df - limit$	4.5961	-6.4519	1.8637	1.5424
$fg - limit$	4.5994	-6.4552	1.8625	1.5422
$gh - limit$	4.6011	-6.4669	1.8621	1.5422
$hi - limit$	4.6020	-6.4562	1.8620	1.5422
$ik - limit$	4.6032	-6.4563	1.8617	1.5422
$extrapolated$	4.6043	-6.4564		
Araki et al. [a] $pd - limit$	4.6122	-6.4636		
Davis and Chung [28] $hi - limit$	4.6045	-6.4566		
'Exact' [29]	4.6045	-6.4572		

[a] G. Araki, K. Mano and M. Otha, Phys. Rev. 115, 1222 (1959)

For higher l-values, the pair function becomes more and more localized to the area close to the diagonal, $r_1 = r_2$, reflecting the increased sharpness of the cusp of $r_<^k / r_>^{k+1}$ for high k-values. For very high l-values very few grid points will be used to describe the pair function in areas where it is important and it seems that some numerical accuracy is lost, making it of little use to go beyond $l \approx 10$. To estimate the contributions of higher l-values, we perform an extrapolation using at least two powers in the expression for the angular convergence, and the final results obtained in this way are in excellent agreement with the "exact" results by Pekeris and coworkers[29] (and also with the recent, considerably more precise results by Drake and Makowski[30])

A significantly faster angular convergence (about $(l+1/2)^{-8}$ for singlet states) can be obtained by separating out one part of the pair function which takes the cusp into account, as demonstrated by Kutzelnigg[31].

2.1.2 Mass Polarization The pair functions used to evaluate the energies shown in Table 1 were also used to evaluate the expectation value of $\mathbf{p}_1 \cdot \mathbf{p}_2$ shown in Table 2, taking care to divide by the norm of the wavefunction obtained using intermediate normalization. This operator is part of the *nuclear* kinetic energy

$$\mathbf{P}^2/2M = \left(\sum_i \mathbf{p}_i\right)^2 / 2M = \sum_i \mathbf{p}_i^2 / 2M + \sum_{i \neq j} \mathbf{p}_i \cdot \mathbf{p}_j / 2M$$

where momentum conservation has been used to express the nuclear momentum $\mathbf{P}$ in terms of the electronic momenta. This expression is valid only in the non-relativistic case. In the relativistic case, the treatment of nuclear recoil is much more complicated as discussed e.g. by Yennie[32]. The first term is the familiar *'reduced mass'* effect, which is proportional to the total energy. The second term can be described as a correlation of the electronic momenta through the motion of the nucleus with its finite mass and is referred to as *'mass polarization'* Between two isotopes of an element, this gives rise to the *'specific mass shift'* (SMS). Being a two-particle operator, it is very sensitive to correlation effects and failure to treat these leads to results far from experimental values. Lack of knowledge of the SMS is, in fact, a limitation in the extraction of changes in nuclear charge radii (which give rise to the *'field'* or *'volume'* isotope shift) from experimental isotope shifts, which have been obtained for several long chains of isotopes[1] (see also e.g. Ref 33 for the case of Rb). A first calculation of correlation effects on the SMS for a many-electron atom was performed by Mårtensson and Salomonson[34]. Effects to this order have also been evaluated by Johnson *et al.*[35] for the *Na* isoelectronic sequence using relativistic wavefunctions. Although the inclusion of lowest-order correlation effects gives a significant improvement of the results, we have found that, in general, it is not sufficient to include only the lowest-order correlation contributions to the SMS (see e.g. the work on sodium in Ref 36 and on calcium in Ref 37.). The statement made in 1954 by Anne Pery[38]: *'No attempt has yet been made to calculate the Specific Shift in Calcium. The calculation would be difficult and the result of little value since the approximations in the methods hitherto employed are such that accurate results cannot be expected'* is still partially valid.

Inclusion of pair correlation effects to all orders would, in general, require the solution of pair equations including the SMS operator and we hope to develop such a procedure in the future. For two-electron systems, however, a direct evaluation of the expectation value using pair functions without the SMS operator, does give the desired results, and Table 2 shows the values for the SMS in the $1s2p$ states in helium together with the values for the norm used in the denominator. The largest contribution to the norm comes in the lowest angular limit, which includes terms which could be considered as potential corrections for the $2p$ electron due to the presence of the $1s$ electron.

2.2 Many-Electron Systems.

Prescriptions for generalizing the two-electron treatment to a many-electron system is provided by many-body perturbation theory[5,6]. The main modification of the pair equation for a system with more than two electrons is the potential and the orthogonality terms — the occupied core orbitals are not available and must be explicitly removed when the closure relation is used to obtain the sum over excited states. In the treatment of helium-like systems, discussed above, the two electrons were treated as valence electrons and single excitations were treated together with double excitations. Pair excitations from core orbitals, on the other hand, are always true two-particle excitations, S_2, and single excitations, S_1 must then be treated explicitly. The single excitations modify the orbitals to approximate Brueckner orbitals. The revised pair program[21] includes a more complete treatment of single excitation effects, with an automatic coupling between the equation for single and double excitations and it can be used also for *arbitrary one-body potentials*. For helium-like systems, the result is independent of the starting potential, which provides an additional numeric test of the program.

For larger systems, also excitations of three or more particles may occur. Part of these may be taken into account by using the coupled-cluster approach. The waveoperator is then written as an exponential of its connected one- two-.. n-body parts, S_n

$$\Omega = \{\exp(S)\} = \{\exp(S_1 + S_2 + \dots S_n)\} \approx \{\exp(S_1 + S_2)\} \tag{2.8}$$

where the curly brackets denote normal ordering[39,40]. The neglect in our work of non-factorizable excitations of three or more electrons leads to a small, but significant, dependence the starting potential, as seen from the beryllium results by Salomonson and Öster[41] shown in Table 3. A large part of the three-body effect in beryllium can be ascribed to a correction of the $1s$ orbital by the $2p^2$ admixture into $2s^2$, which is unusually large, due to the near degeneracy[42]. In a recent review, Urban *et al.*[43] compare coupled-cluster results (truncated both at S_2 and S_3 level) for a number of molecular systems to order-by-order calculations including all three-body contributions. It is clear that the relative importance of true three-body excitations depends on the system. In cases where the truncation after double excitations is not a good approximation, it is usually found that the use of the "coupled-cluster amplitudes" for the pair excitations in the evaluation of the contributions from triple excitations gives very accurate results.

Table 3. Beryllium ground state energy (a.u.)

	Correlation	Total energy		"Correlation"	Total energy
HF		-14.573 023	HFS V(N-1)		-14.544 710
CC S_2	-0.092 990				
CC S_1, S_2	-0.000 706		CC S_1, S_2	-0.121 915	
$(S_2)^2$ non fac.[a]	+0.000 028		$(S_2)^2$ non fac.[a]	+0.000 026	
	-0.093 668	-14.666 691		-0.121 889	-14.666 599
Monte Carlo[b]	-0.093 4	-14.666 4			
CI[c]	-0.094 305	-14.667 328			

[a] Eight non-factorizable $(S_2)^2$ diagrams
[b] Monte Carlo calculation by C J Umrigar, K G Wilson and J W Wilkins, Phys.Rev. Lett. **60** *1719* (1988)
[c] C F Bunge Phys. Rev. A **14** *1965* (1976), Phys. Rev. A **17** *E486* (1978)

Perturbation theory can often be applied in alternative ways. In a recent article, Heully and Salomonson study the Be-like system, Li⁻, where the weaker nuclear attraction makes correlation effects even more important than in Be. The electron affinity of Li was obtained in two different ways. In the first approach, the affinity was obtained directly by removing one electron from the closed $2s^2$ shell of Li⁻, thus studying a valence hole, using essentially the same formalism as that developed for a single valence particle, which was used in the second approach, where the energy of neutral lithium was obtained by adding a valence $2s$ electron to the $1s^2$ closed shell of Li⁺. As noted also by Hörbäck *et al.*[45] in a study of the mass shift of Ne⁺, higher-order effects are particularly important when a hole is created in a filled shell — the remaining sister electrons react more to the absence of one of them with the same spatial distribution than to an addition of an extra electron in a different region of space.

Much of the formalism developed for the non-relativistic case can be directly taken over for the relativistic case — if the pair equation can be solved. In the next section we describe our work on the generalization of the solution of the pair equation to the relativistic case.

3. RELATIVISTIC TREATMENT

The development of methods to treat relativistic correlation effects has been stimulated by the search for and observation of parity non-conserving effects in atoms[46]. Further demand for accurate calculations for heavy atomic systems come, e.g., from the studies at CERN of hyperfine structure and isotopes over long chains of radioactive isotopes[1,33]. For a heavy atom it is essential to use a relativistic treatment. The Dirac one-electron hamiltonian has both positive and negative energy eigenvalues where the negative energy states are associated with positrons. Not only the eigenvalues but also the distinction between positive and negative energy states depend on the potential V. For hydrogen-like atomic systems the potential is given by the electron-nucleus interaction and the one-electron hamiltonian becomes:

$$h_{nuc} = (\beta\, mc^2 + c\, \alpha \cdot \mathbf{p} - Z/r)$$

When the one-body Dirac equation is solved numerically for atomic states, as done in the present work, the presence of negative energy states does not normally cause a problem, since the boundary conditions can be used to ensure that only bound states are obtained[47]. The situation is, however, quite different for a many-electron equation.

Due to the existence of negative energy solutions to the one-electron equation, which correspond to positron states and are not available for electrons, the relativistic generalization of the two-electron eigenvalue equation (2.1) must be used with care to ensure that excitations *into* negative energy states are not allowed in the wave function — only excitations *from* them, corresponding to the creation of virtual electron-positron pairs, are permitted. The situation is similar to that in a large atom where the valence electrons are not allowed to be excited into the occupied core orbitals. This is accounted for[5,6] by a second-quantized representation of interactions and using Wick's theorem to get a normal ordering of creation and annihilation operators with respect to a suitably chosen "vacuum" level. In the relativistic many-electron case, failure to respect normal ordering for the negative energy states leads to the well-known problems of *"continuum dissolution"* first noted by Brown and Ravenhall[45], and brought back to attention by Sucher[46]. This problem occurs whether the electron-electron interaction V_{12} is described by the pure Coulomb interaction or includes also the effect of transverse photons, be it in Coulomb gauge or Lorentz gauge form, with or without the proper frequency dependence.

We can identify *three problems*, caused by the presence of negative energy states, which may occur if we try to solve already the first-order pair equation in the relativistic case, where the complete set of solutions making up the closure relation (2.6) includes also the negative energy states, which will thus enter, e.g., in the summations in (2.7):

> i) The right-hand sides of (2.4) and (2.7) contain a continuous infinity of states with one electron in the negative energy continuum and one electron in the high positive energy continuum ($\varepsilon \approx 2mc^2$), which are degenerate with the unperturbed initial state. This is the problem leading to the "continuum dissolution". The actual contributions from these admixtures may not be too disastrous, since for each given negative energy state, there is a strong cancellation between the contributions from the positive energy states in the other coordinate from either side of the pole[50].

> ii) Even if these terms are removed from the RHS, they may still enter when the equation is solved. These terms may be removed from the function — if we know how to project them out — but they may have destroyed the numerical accuracy by diluting the solution of interest. (This problem occurs to some extent also in the non-relativistic case, as discussed in Sect.2, but there, addition of an energy offset term can remove the degeneracy)

iii) Even if the problem of "continuum dissolution" is not likely to occur in actual calculations, where very highly excited states probably escape between the meshes of the grid, any component of negative energy states on the RHS will incorrectly be treated as an allowed electron state. This problem is particularly serious for the magnetic interactions, which mix the upper and lower components of the orbitals.

Clearly, the negative energy states should not have been included — just as occupied orbitals in a large atom should be removed. Solving the equation without projection operators would correspond to an attempt to include all the diagrams in Fig. 2 within the diagram in Fig. 2a. The problem with the relativistic generalization of the pair equation is not primarily with the description of the electron-electron interaction but in the equation itself, which does not allow for normal ordering of creation and annihilation operators.

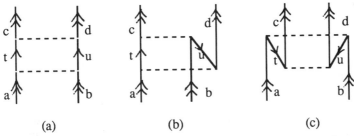

(a)   (b)   (c)

Fig. 2. Examples of second-order energy contributions for a two-electron system. Up (down) going lines correspond to positive (negative) energy states. Diagram (a) is the "no-pair" contribution, where the summation is only over excited states. Diagram (b) and (c) include effects of single and double virtual electron-positron pairs, respectively. The outgoing electron pair cd is either ab or ba.

3.1 Projection Operators

In order to treat the negative energy states correctly, it is necessary to divide the two-particle interaction into several parts by using projection operators for positive and negative energy states and to treat each part individually. The distinction between positive and negative energy states is provided by the choice of one-body hamiltonian and a positive energy state in one potential will have a small admixture of negative energy states in another basis[51].

Formally, the one-particle projection operator, λ^+, for positive energy states can easily be written down in terms of eigenstates to the Dirac equation as

$$\lambda^+ = \sum_n^{pos} |n\rangle\langle n|$$

However, if a more or less complete set of solutions to the Dirac equation is available, it may also be used directly as a basis set for a configuration interaction type calculation or in an application of perturbation theory[52] rather than to construct λ^+ explicitly. This type of approach has also been applied to calculations for heavy atoms[53-55] although only excitations into positive energy states have been considered. Such a calculation is *in principle* not much more complicated than the corresponding non-relativistic procedure, although *in practice* the demands on computing time and storage are very much larger.

3.1.1. Boundary conditions and projection operators It is sometimes claimed that the problems with the Dirac-Coulomb equation can be cured by imposing boundary conditions for electrons rather than positrons. The statement is clearly valid as long as applied to one-particle eigenvalue equations. If the eigenfunctions to the one-particle equations are used to expand the RHS of the Dirac-Coulomb equation, there is no problem — the negative energy states can easily be excluded from the summation (or treated properly), but *then, the equation has been changed!* We would also like to emphasize that the electron boundary conditions are not absolutely positron-proof. Electronic states in one potential contain admixtures of positron states in a different potential[51]. (That this is so is easily seen e.g. by the fact that the potential enters in the definition of projection operators defined below! Direct power counting would indicate that the effect is of $O(\alpha^4)$, but if the singularity at the nucleus is omitted in one potential, the energies for s electrons are changed in order $O(\alpha^3)$, as seen e.g. by Hardekopf and Sucher[56].) As an example we can consider a singly excited intermediate state in the diagram Fig 2(a), letting $t(u)=a$. The intermediate wavefunction in the other coordinate then contains part of the difference between the Dirac-Fock and the hydrogenic orbital b, caused by the direct (exchange) interaction with the other electron in state a. This difference does contain negative energy components in the hydrogenic basis and if the one-particle equation for this correction is solved, the negative energy states will automatically enter. In the case $a=b$, the diagram 2b will actually be reproduced — the same radial matrix elements enter and the angular momentum parts turn out to be the same (This procedure was used in the helium calculations by Lindroth[57].) — but this does not hold in a general case. For any given excited state u, we can obtain a one-particle equation for the summation in coordinate 1 by projecting the RHS of the two-particle equation onto the state u in coordinate 2. Solution of this equation would lead to the inclusion of a superposition of both positive and negative energy states in coordinate 1, but for the higher excited states, u, neither energy denominators nor matrix elements will be those corresponding to the diagram 2b. The higher the excitation energy, the smaller the effective denominator for the excitations to negative energy states, although for moderate excitation energies, the relative error will be $O(Z^2\alpha^2)$ in the denominator and $O(Z^4\alpha^4)$ in the contribution to the second-order energy.

We note in this context that the Multi-Configuration Dirac-Fock (MCDF) wavefunctions are approximate *no-pair* wavefunctions in a very special sense — from Brillouin's theorem follows that[58]

$$\lambda^+_{1,\mathrm{MCDF}} \, \lambda^-_{2,\mathrm{MCDF}} \, H \left| \Psi_{\mathrm{MCDF}} \right\rangle = 0$$

implying that if the projection operator onto positive energy states is approximated by the projection onto the orbitals included in the MCDF procedure, then there are no single excitations of virtual electron-positron pairs in coordinates 1 or 2 (although excitation of a virtual electron positron pair in a third coordinate may still appear). In essence, the diagram 2b has been included in the diagram 2a by the choice of one-electron orbitals.

3.1.2 Projection Operators and the Relation Between the Components In our non-relativistic work, using the inhomogeneous differential equation technique, only the occupied orbitals are obtained explicitly and the closure relation is used to provide the summation over all orbitals. The unwanted terms are then removed by performing a finite sum over occupied orbitals. In the relativistic case, there is an infinite number of unwanted negative-energy states, and we would like to find another expression for the projection operators. This can, indeed, be done by using the *relations between the upper and lower components*. The problem was investigated in some detail by Heully *et al.*59 and methods based on relations between the components have been used also in quantum chemistry to avoid the *'variational collapse'*60. It was found that operators $R_\pm$ satisfying

$$R_\pm = \pm \frac{1}{2mc^2}\left(c\sigma \cdot \mathbf{p} \mp [R_\pm, V] - R_\pm(c\sigma \cdot \mathbf{p})R_\pm\right) \qquad (3.1)$$

are such that for positive energies $G=R_+F$, whereas for negative energies we have instead $F=R_-G$. The equation (3.1) is found by direct insertion in the Dirac equation of these relations. We note also that it is impossible to find *one* operator which satisfies $G=RF$ for *all* eigenfunctions. This is most easily seen from the closure property (2.6), which can be written in component form as:

$$\sum^{all}\begin{pmatrix} |F\rangle\langle F| & |F\rangle\langle G| \\ |G\rangle\langle F| & |G\rangle\langle G| \end{pmatrix} = \begin{pmatrix} 1 & 0 \\ 0 & 1 \end{pmatrix} \qquad (3.2)$$

Applying R_+ to the relations in the left column of (3.2) with the assumption $R_+F=G$ leads to the contradiction

$$R_+ = R_+\sum^{all}|F\rangle\langle F| = \sum^{all}|G\rangle\langle F| = 0$$

This apparent contradiction is resolved by noting that the equation (3.1) for $R_\pm$ is a *second-order equation* and has *two solutions*. In the free-particle case, the two solutions for R_+ can be written as

$$R_+ = \pm \frac{c\sigma \cdot \mathbf{p}}{mc^2 \pm \sqrt{m^2c^4 + p^2c^2}} \qquad (3.3)$$

The positive square-root solution has $\sigma \cdot \mathbf{p}/2mc$ as the leading term, which is also the first term on the RHS of (3.1). This corresponds to positive energy solutions. The *negative* square-root solution of (3.3) can be rearranged as

$$R_+^{(-)} = -(c\sigma \cdot \mathbf{p})^{-1}\left(mc^2 + \sqrt{m^2c^4 + p^2c^2}\right)$$

which is minus the inverse of the solution in (3.3) with the + sign (although we have to be careful since the inverse $(c\,\sigma\cdot\mathbf{p})^{-1}$ breaks down for $p=0$). Of course, this operator should also be the inverse of the operator R_- which relates the upper component to the lower one for negative energy states, where that is a more natural relation. From the original equation (3.1) it is easy to see that

$$R_- = -R_+^\dagger$$

where R_- and R_+ refer to the 'small' solution, corresponding to the + sign in (3.3).

The relations between the components are indeed essential. That one component itself does not uniquely specify a state can be seen from the closure relation. The upper component F of any Dirac spinor (except for $p=0$ in the free-particle case) can be expanded in terms of the upper components for the other eigenfunctions:

$$F_a = \sum_i^{all} |F_i\rangle\langle F_i|F_a\rangle = |F_a\rangle(1 - |G_a\rangle\langle G_a|) + \sum_{i\neq a} |F_i\rangle\langle F_i|F_a\rangle \qquad (3.4)$$

where we have used the normalization condition for the Dirac function to get $\langle F|F\rangle=1-\langle G|G\rangle$, which, in general, differs from unity. The set of upper (or lower) components is thus linearly dependent. We conclude that *a state cannot be characterized by one of the components, but only through the relation between the components.*

Having established that the R-operators do, indeed, provide a distinction between positive and negative energy states, we can use them directly to obtain the *projection operators.* By requiring that the projection operator λ^+ leaves a positive energy state unchanged while giving zero when operating on a negative energy state, *i.e.*

$$\lambda^+ \begin{pmatrix} F \\ R_+F \end{pmatrix} = \begin{pmatrix} F \\ R_+F \end{pmatrix} \qquad \lambda^+ \begin{pmatrix} R_-G \\ G \end{pmatrix} = \lambda^+ \begin{pmatrix} -R_+^\dagger G \\ G \end{pmatrix} = 0$$

and that the sum of projection operators for positive and negative energy states is unity, we find

$$\lambda^+ = \begin{pmatrix} \Omega_+^2 & \Omega_+^2 R_+^\dagger \\ R_+\Omega_+^2 & R_+\Omega_+^2 R_+^\dagger \end{pmatrix} \qquad \lambda^- = \begin{pmatrix} R_-\Omega_-^2 R_-^\dagger & R_-\Omega_-^2 \\ \Omega_-^2 R_-^\dagger & \Omega_-^2 \end{pmatrix} \qquad (3.5)$$

where the normalization is provided by

$$\Omega_\pm = \pm\left(1 + R_+^\dagger R_+\right)^{-1/2}$$

which satisfies the relation $R_+\Omega_+ = \Omega_-R_+$. Below, we use the simplified notation $R = R_+$ and $\Omega = \Omega_+$.

3.1.3 Approximate relations So far, no approximations have been made. However, since the equation for the R operator cannot be solved for a general potential, some approximation must be made in the actual applications and we use

$$R \approx \frac{1}{2mc^2 - V} c\sigma \cdot p = \frac{B(r)}{2mc} \sigma \cdot p \equiv R_0 \qquad (3.6)$$

which differs from the exact operator in relative order α^2. This is, of course, also true if the potential is neglected in the definition of R_0, corresponding to the Pauli approximation. However, not only the order in α is important — close to the nucleus V is certainly not small. The approximate expression in (3.6) is very similar to the exact energy dependent relation between F and G obtained from the Dirac equation. For a positive energy state with $E=\varepsilon+mc^2$, we have

$$G \approx \frac{1}{2mc^2 + \varepsilon - V} c\sigma \cdot pF = \frac{B_E(r)}{2mc} \sigma \cdot pF \qquad (3.7)$$

and we find that

$$R_0F - G \approx \frac{\varepsilon}{2mc^2(1 - V/2mc^2)} G = \frac{\varepsilon}{2mc^2} B(r)G = O(\alpha^2)G$$

where $B(r)$ is given in (3.6), above. The fact that V is large close to the nucleus actually *improves* the situation since V is *negative*. To estimate the error in R_0 we note that even for the extreme case of the *1s* state in uranium, $|\varepsilon/2mc^2|\approx 0.13$. Figure 3 shows the comparison between the "exact" and approximate *B*-factors for the *1s* state in hydrogen-like cæsium and uranium, respectively. For the outer electrons of a neutral atom, the approximation is, of course, much better. On the other hand, R_0 is not adequate for a discussion of the Brown-Ravenhall disease where one electron has an energy of about $\varepsilon\approx 2mc^2$. If we had neglected V in the denominator, corresponding to the Pauli approximation, we would instead get a horizontal line and the difference would instead be $(\varepsilon-V)G/2mc^2$ which is non-negligible close to the nucleus. The advantages of using (3.6) are obvious if we consider a property such as, e.g., hyperfine structure or parity non-conservation, which involves FG/r^2 and $FG\rho_N(r)$, respectively.

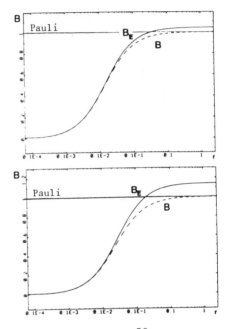

Fig. 3 Comparison[50] between B-factors for the 1s state in hydrogen-like Cs (top) and U (bottom)

To leading order in α, the projection operator in (3.5) is given by

$$\lambda^{\pm} \approx \frac{1 \pm \beta}{2} + \frac{\beta p^2}{4m^2c^2} \pm \frac{\alpha \cdot p}{2mc}$$

320

This expression, of course, coincides with the leading terms of the expansion of the Casimir free-particle projection operators in momentum space, which after slight rearrangement can be written as

$$\lambda_{\pm}^{\text{free}} = \frac{1}{2} \pm \frac{\beta/2 + \boldsymbol{\alpha} \cdot \mathbf{p}/2mc}{\sqrt{1 + \mathbf{p}^2/m^2c^2}}$$

3.2 The Relativistic Pair Equation

Equipped with projection operators, we are ready to revive the "thrice slain"[49] Dirac Coulomb equation. A two-particle Dirac function can be separated into four spatial components, each of which is a function of the coordinates for the two electrons, just like a non-relativistic pair function. For any right-hand side the two-particle equation can be written in the general form

$$
\begin{aligned}
\left(\varepsilon_a + \varepsilon_b \qquad - u_1 - u_2\right)\rho^{LL} - c(\boldsymbol{\sigma} \cdot \mathbf{p})_1 \rho^{SL} - c(\boldsymbol{\sigma} \cdot \mathbf{p})_2 \rho^{LS} &= P^{LL} \\
\left(\varepsilon_a + \varepsilon_b + 2mc^2 - u_1 - u_2\right)\rho^{LS} - c(\boldsymbol{\sigma} \cdot \mathbf{p})_1 \rho^{SS} - c(\boldsymbol{\sigma} \cdot \mathbf{p})_2 \rho^{LL} &= P^{LS} \\
\left(\varepsilon_a + \varepsilon_b + 2mc^2 - u_1 - u_2\right)\rho^{SL} - c(\boldsymbol{\sigma} \cdot \mathbf{p})_1 \rho^{LL} - c(\boldsymbol{\sigma} \cdot \mathbf{p})_2 \rho^{SS} &= P^{SL} \\
\left(\varepsilon_a + \varepsilon_b + 4mc^2 - u_1 - u_2\right)\rho^{SS} - c(\boldsymbol{\sigma} \cdot \mathbf{p})_1 \rho^{LS} - c(\boldsymbol{\sigma} \cdot \mathbf{p})_2 \rho^{SL} &= P^{SS}
\end{aligned}
\tag{3.8}
$$

where L and S denote large and small, respectively (for positive energies). The functions P appearing on the right-hand side are the respective components of $V_{12}|ab\rangle$ corrected for orthogonality terms. In the case of the Coulomb interaction we have

$$P^{LL} = \frac{1}{r_{12}}|F_a(1)F_b(2)\rangle \qquad P^{SL} = \frac{1}{r_{12}}|G_a(1)F_b(2)\rangle \qquad P^{SS} = \frac{1}{r_{12}}|G_a(1)G_b(2)\rangle$$

(where the orthogonality terms have been omitted for simplicity). In the case of magnetic interactions we have instead

$$P^{LL} = g_{12}|G_a(1)G_b(2)\rangle \qquad P^{SL} = g_{12}|F_a(1)G_b(2)\rangle \qquad P^{SS} = g_{12}|F_a(1)F_b(2)\rangle$$

with a reversed size relation between the components. Here $g_{12} = -\sigma_1 \cdot \sigma_2 / r_{12}$ in the case of the Gaunt interaction and

$$g_{12} = -\frac{\sigma_1 \cdot \sigma_2}{2r_{12}} - \frac{(\sigma_1 \cdot \mathbf{r}_{12})(\sigma_2 \cdot \mathbf{r}_{12})}{2r_{12}^3}$$

for the Breit interaction.

A possible approach to solve the set of equations in (3.8) is to eliminate the 'mixed' small-large and large-small components in the same way as the small component G can be eliminated in the one-particle case, using (3.7). This gives

$$\rho^{LS} = \frac{B_{ab}}{2mc}\left((\sigma \cdot p)_1 \rho^{SS} + (\sigma \cdot p)_2 \rho^{LL}\right) + \frac{B_{ab}}{2mc^2}P^{LS} \qquad (3.9)$$

where B_{ab} is given by

$$B_{ab} = \left(1 + \frac{\varepsilon_a + \varepsilon_b - u_1 - u_2}{2mc^2}\right)^{-1}$$

Using the expression (3.9) for ρ^{LS} and the analogous expression for ρ^{SL}, the equation for the 'large-large' component can be rewritten as

$$\left(\varepsilon_a + \varepsilon_b - u_1 - u_2 - \frac{1}{2m}\left((\sigma \cdot p)_1 B_{ab}(\sigma \cdot p)_1 + (\sigma \cdot p)_2 B_{ab}(\sigma \cdot p)_2\right)\right)\rho^{LL}$$

$$- \frac{1}{2m}\left((\sigma \cdot p)_1 B_{ab}(\sigma \cdot p)_2 + (\sigma \cdot p)_2 B_{ab}(\sigma \cdot p)_1\right)\rho^{SS} = P^{LL} + \tilde{R}_1^\dagger P^{SL} + \tilde{R}_2^\dagger P^{LS} \qquad (3.10)$$

where we have introduced the operator

$$\tilde{R}_i = \frac{B_{ab}}{2mc}(\sigma \cdot p)_i$$

which differs from the R operator in (3.1) or (3.6) used in the projection operators only in relative order α^2. The equation for ρ^{SS} is analogous to (3.10) except for an extra $4mc^2$ which normally dominates the left-hand side (although this is not the case if one electron is in a very highly excited state with $\varepsilon \approx 2mc^2$). However, instead of using this equation for ρ^{SS} we use the R operator to relate the upper and lower components of a pair function, i.e.

$$\rho^{SS} = R_1 R_2 \rho^{LL} \qquad (3.11)$$

which is found to be an active ingredient in the antidote for the Brown-Ravenhall disease —with this expression used for ρ^{SS} the equation no longer has a combination of a positive and a negative energy state as a homogeneous solution.

As an alternative to the treatment outlined here we could have used the relation

$$\rho^{LS} = R_2 \rho^{LL} \qquad (3.12)$$

rather than (3.9). Then only the term P^{LL} would enter on the right-hand side of (3.10). The left-hand side would be similar, but B rather than B_{ab} would be used in that approach and no terms involving ρ^{SS} would then appear on the left-hand side. The projection operators on the right-hand side are much more important — in effect, it leads to essentially the same terms on the RHS of (3.10), but in addition, in that case, also the normalization corrections from Ω enter in leading order to compensate for the absence of terms involving ρ^{SS}.

3.3 Application of projection operators

Although using the expression (3.11) prevents any disease of the left-hand side of the relativistic pair equation, the right-hand side may still contain illegal negative-energy components: The small upper component of a negative energy states might successfully masquerade as a large component of a positive energy state — as discussed in connection with Eq. (3.4), the upper components form a linearly dependent set. The health of the right-hand side may be ensured by applying the positive energy projection operators discussed in 3.2.

The application of projection operators in both coordinates to a right-hand side of the form in (3.8) gives

$$\overline{P}^{LL} = \Omega_1^2 \Omega_2^2 \left(P^{LL} + R_1^\dagger P^{SL} + R_2^\dagger P^{LS} + R_1^\dagger R_2^\dagger P^{SS} \right)$$

and the other components of the projected RHS can be obtained as

$$\overline{P}^{SL} = R_1 \overline{P}^{LL} \qquad \overline{P}^{LS} = R_2 \overline{P}^{LL} \qquad \overline{P}^{SS} = R_1 R_2 \overline{P}^{LL}$$

If we use the alternative approach (3.12), it would be necessary to include the normalization contributions from Ω^2, as well — in effect it is now instead included in the term involving ρ^{SS} on the LHS of (3.10). On the RHS of (3.10) we get after projection

$$\overline{P}^{LL} + R_1^\dagger \overline{P}^{SL} + R_2^\dagger \overline{P}^{LS}$$

$$= \left(1 + \tilde{R}_1^\dagger R_1 + \tilde{R}_2^\dagger R_2 \right) \Omega_1^2 \Omega_2^2 \left(P^{LL} + R_1^\dagger P^{SL} + R_2^\dagger P^{LS} + R_1^\dagger R_2^\dagger P^{SS} \right) \qquad (3.13)$$

By using the definition of Ω and the similarity between the different R operators we get

$$\left(1 + \tilde{R}_1^\dagger R_1 + \tilde{R}_2^\dagger R_2 \right) \Omega_1^2 \Omega_2^2 = 1 + O(\alpha^4)$$

and to lowest order in α^2 we need consider only the terms in the last parenthesis in (3.13). For diagonal interactions (such as the Coulomb interaction between the electrons) the terms in the last bracket are in ascending order of $(Z\alpha)^2$, with the last term of order $(Z\alpha)^4$. We thus find that the RHS in equation (3.10) is unchanged to $O(Z^2\alpha^2)$ by the application of projection operators. For non-diagonal interactions, such as the Breit interaction, the situation is very different: All four terms in the last bracket of (3.13) are of the same order and the effect of applying projection operators to the RHS enters already in the same order as the operator, itself — although the upper component of a negative energy state may be small, the contribution from excitations into *two* negative energy states are enhanced by the larger matrix elements for non-diagonal interactions

We have thus found that by solving the equation (3.10) with ρ^{SS} given by (3.11) we solve to order $O(Z^2\alpha^2)$ a *no-virtual pair pair equation*, where only excitations into the positive energy states are allowed. This equation is somewhat similar to a Schrödinger equation with relativistic corrections, although the presence of the B factors provide a welcome smoothing of the singularities close to the nucleus. The numerical procedure used to solve this equation for the large-large component is a non-trivial extension of the method presented in[14,22] and is described in detail by Lindroth[57]. Figure 4 shows the results by Lindroth for the second-order energies in the helium-like systems starting from hydrogen - like orbitals. The horizontal line corresponds to the Pauli approximation, which includes all $(Z\alpha)^2$ contributions. The discrepancy between the results obtained using the approach described here and the results by Johnson and Sapirstein[55] are of order $(Z\alpha)^4$ and higher and can be attributed to the approximations of higher order terms in α in our approach, but as seen from Figure 4, a dominating part of the correction to the Pauli approximation has been accounted for. In the study of neutral atoms, the dominating correlation effects will arise from the interaction between less relativistic electrons and we can expect the approximations, which affect only the correlation contributions, to be adequate for these cases.

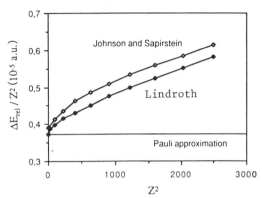

Fig. 4 Relativistic correlation contributions for helium-like systems (From Ref. 57)

3.4. The Electron-Electron Interaction

In the study of relativistic systems, a choice of the form of the electron-electron interaction must be made. For light atomic systems, it may be a reasonable approximation to use the Coulomb interaction and treat the effect of transverse photons by evaluating the expectation value of the Breit interaction, thus following the well-known advice by Bethe and Salpeter[61]. However, the problems with the frequency independent form of the Breit interaction that led to this advice is only a symptom, but not the cause, of the incorrect treatment of negative energy states, discussed above. As long as the Breit interaction is evaluated only between positive energy states, the argument no longer seems valid and the rule is being questioned — in words[62] and action[63-64] — by a number of groups. It has been found[64] that in the region $Z\approx50$ the correlation contributions involving magnetic interactions between the electrons becomes *larger* than the correlation contribution involving only the Coulomb interaction. It is then hard to justify the neglect of the Breit interaction in the wavefunctions — it seems likely that it will have significant effects not only on the energies but also on the electron orbitals and thus other properties studied. A possible approach would be to include the original energy independent Breit interaction in a self-consistent procedure. The corrections for the frequency dependence could then be treated as a perturbation (although, with its higher Z-dependence it becomes comparable to the frequency independent part of the *retardation* for Z=92, as observed in Ref 65). Treating the Breit interaction together with the Coulomb interaction has obvious computational and conceptual advantages (apart from the disadvantage that the calculations become more

324

time consuming). One avoids the need for a separate perturbation expansion — and in cases, such as the study of parity non-conserving effects, where already without the Breit interaction one is dealing with a triple (in some cases even, quadruple[66]) perturbation expansion, this aspect can hardly be overvalued. The results of such calculations including the Breit interaction will be published elsewhere[67].

The choice of gauge for the electron-electron interaction is not unimportant, as found by Gorceix and Indelicato[65] in the study of helium-like systems and the discrepancy is not cured by including more configurations in the Multi-Configuration Dirac-Fock procedure, as might have been expected[68]. After a slight rearrangement of the expressions used in Ref. 65, it is easy to show that a discrepancy to $O(Z^2\alpha^2)$ between the gauges arises within the no-pair approximation[69]. Even the original Breit interaction, which is the low-frequency limit of the transverse photon interaction in Coulomb gauge, gives more accurate results in the no-pair approximation than the frequency dependent form of the Lorentz gauge expression[70]. A complete treatment of the electron-electron interaction must take into account also the effects of excitations of virtual electron-positron pairs, including, e.g., radiative corrections, discussed elsewhere in these proceedings[71].

4. CORRELATION EFFECTS ON ADDITIONAL PERTURBATIONS

Correlation energies are usually evaluated as a natural part of the calculation of pair functions, but the pair equation can be used to study also other properties in several ways. A perturbation h can be included formally in the one-body hamiltonian (for a two-body perturbation, an effective one-body interaction analogous to the Hartree-Fock potential can be used[36]). The perturbation can also be treated together with the residual electron-electron interaction V when the corrections to wavefunctions and energies are obtained, as described by Lindgren and Morrison[5]. It is also possible to use the general expression for a matrix element of H between states f and i

$$\frac{\left\langle\phi_f\left|\Omega^\dagger h\Omega\right|\phi_i\right\rangle}{\left\langle\phi_f\left|\Omega^\dagger\Omega\right|\phi_f\right\rangle^{1/2}\left\langle\phi_i\left|\Omega^\dagger\Omega\right|\phi_i\right\rangle^{1/2}} \tag{4.1}$$

where Ω is the *waveoperator*, discussed in [5,6] and in Section 2. The formula (4.1) may be used also within an hermitian formulation of the perturbation theory, currently being developed by Lindgren and Mukherjee[72], where the intermediate normalization condition $P\Omega P=P$ is replaced by $P\Omega^\dagger\Omega P=P$

4.1 Inclusion of the Perturbation in the Hamiltonian

By inclusion of the perturbation in the one-body potential h_0, all diagrams involving single excitations are formally included already at the unperturbed level, as well as diagrams involving factorizable two-particle excitations and chains of these (sometimes referred to as 'ground-state correlation'). If, instead, the expression (4.1) were used directly, these chains would appear when several powers of two-body clusters from $\Omega^\dagger$

and Ω are connected. Although, formally, h is treated together with h_0, in practice; the calculations are usually performed so that an extra set of orbitals $|\delta a\rangle$, which include the perturbation to lowest order, is obtained from the equation

$$(\varepsilon_a - h_0)|\delta a\rangle = (h + v^h)|a\rangle - \delta\varepsilon|a\rangle \qquad (4.2)$$

where v^h is the correction to the potential due to the modification of the orbitals for the other electrons b, and $\delta\varepsilon = \langle a|h+v^h|a\rangle$ is the energy correction, making the RHS orthogonal to orbital a. Formally, the corrections to the orbitals in (4.2) include excitations into *negative energy states of the unperturbed basis* but this admixture is necessary to get the proper positive energy states in the new basis (cf. Ref. 51). An interesting, direct numerical demonstration of this was given by Quiney *et al.*[73] in a model study, where hydrogenic orbitals for one nuclear charge were used to expand hydrogenic orbitals for a different nuclear charge, which could not be reproduced unless the negative energy states were included. (The effect is particularly large in a case like this, where the boundary conditions at the nucleus are changed.)

If the Hartree-Fock or Dirac-Fock potential from the core orbitals is used, we get a correction

$$v^h = \sum_b^{core}\left(\langle b|\frac{1}{r_{12}}(1-P_{12})|\delta b\rangle + \langle\delta b|\frac{1}{r_{12}}(1-P_{12})|b\rangle\right) \qquad (4.3)$$

(using the exchange operator P_{12}, which interchanges electrons *1* and *2*). This elegant treatment was suggested by Sandars[74] in connection with the study of weak interactions and has been used by several groups to study this and other problems[54,55,75-77]. It is closely related to the *random-phase approximation (RPA)* (see e.g. Johnson *et al.*[78]) and to the method of summing all *core polarization* effects[*] to all orders described by Garpman *et al.*[4]. A wavefunction obtained in this way satisfies many identities required from the total wavefunction, and often also by orbitals obtained in a local potential, but not generally for Hartree-Fock or Dirac-Fock orbitals. E.g., the equivalence between the length and velocity forms of the dipole operator is restored by the inclusion of the correction term v^h in (4.3)[75]. Dzuba *et al.*[79] have shown that an external electric field is completely screened at the nucleus (as it must be, since the neutral atom remains stationary) by the change in electronic wavefunctions obtained in this way. In Sect. 4.3, we discuss the use of an alternative, less singular, form, originally suggested by Hiller *et al.*[80], for the weak parity non-conserving electron-nucleus interaction, which gives equivalent results only if the correction v^h is included[81].

The orbital correction i (4.2) should then be included with the unperturbed orbitals when the pair equation s are constructed and we would normally have to define a new set of pair functions i cluding the perturbation. This would also be the case in the second approach, treating h together with V, if pair correlation should be included to all orders. In lowest order this is not necessary, however, as found by Garpman *et al.*[3] who follow this second approach in deriving the expressions for all contributions to the hyperfine structure in an alkali atom which are second-order in the electron-electron interaction.

*) Note that this term is often used to refer instead to Brueckner-orbital like corrections

326

To obtain pair functions which include the perturbation appears to be the most convenient approach for studying two-particle perturbations, such as the mass polarization and the Breit interaction, where certain pair correlation effects cannot easily be included by using (4.1)37. For a non-scalar one-particle perturbation, on the other hand, it seems to be more straight-forward to use the *all-order* pair functions directly in the evaluation of matrix elements using (4.1) rather than as input to obtain new pair functions which include the perturbation.

4.2. Expectation Values in the Coupled-Cluster Approach

In the coupled cluster approach, the wave operator is expressed as an exponential of its connected parts $\Omega=\{\exp(S)\}$ (2.8). Straight-forward combinatorics leads to the expression (cf. Ref. 82)

$$\Omega^\dagger \Omega = \{\exp(S + S^\dagger + Z)\} \qquad (4.4)$$

where the curly brackets denote normal ordering and Z is the sum of all *connected* diagrams that can be formed by combining any number of cluster operators $S^\dagger$ and S. This can, in principle, lead to infinitely long chains, which may be obtained by a recursive procedure, or may be truncated at some stage. It is possible to perform a partial summation of all one-body clusters involving core orbitals using the relation (4.4). These one-body clusters include the modification of the orbitals to approximate non-normalized and non-orthogonal Brueckner orbitals. It can be shown that the summation leads to a transformation of the originally non-orthogonal Brueckner core orbitals to an ortho-normal set.

For a one-body perturbation the expression in the numerator can be obtained from that for $\Omega^\dagger \Omega$ (or any approximation thereto) by (i) adding an interaction line for h to any of the lines and (ii) by replacing any of the one-particle cluster operators S_1 by a matrix element of the interaction. If this procedure is followed, (4.1) will conserve the expectation value of the density matrix83 also for a truncated expansion. For a one-valence system this can be seen by noting that each diagram has one more line going up than down. When the interaction, $\Sigma\, a^\dagger_n\, a_n$, corresponding to the one-particle density matrix, is inserted on any of the core lines, this gives an extra minus sign which cancels a diagram with the interaction inserted on an up-going line, leaving only one uncancelled diagram, corresponding to the contribution to the normalization appearing in the denominator. The matrix element of the perturbation between different types of orbitals, of course, gives no contribution to the one-particle density matrix. A similar analysis can be performed for two-particle perturbations and density matrices. A program for evaluating expectation values between coupled-cluster wavefunctions is now being developed in our group84.

However, the expression for the expectation value between the coupled-cluster wave functions must be truncated at some stage. It may be advantageous to include long chains of factorizable two-particle excitations by using the the modified one-electron orbitals which include the perturbation as in (4.2), taking care to ensure that no double-counting occurs. If these "RPA corrections" are evaluated with Brueckner orbitals and energies, several correlation effects are included already at this level and also these terms correspond to a well-defined subset of the coupled-cluster contributions.

4.3 Parity Non-Conserving Effects

As pointed out by the Bouchiats[85], the weak interaction between the electrons and the nucleons gives rise to a non-conservation of parity in atomic states and may be observed in atomic transitions. The effect has now been observed for a number of atomic systems[46] and is in agreement with the results from theoretical calculations[86]. However, the increased experimental accuracy[87,88] provides a challenge for atomic theory. Calculations of parity non-conserving (PNC) electric dipole transition elements including correlation effects have been discussed elsewhere in this volume[89] and here we discuss only briefly the steps necessary to perform a calculation using pair functions.

The calculation of the PNC transition elements requires the inclusion of two additional perturbations, the weak PNC interaction and the electric dipole transition operator. The weak interaction, being a pseudo-scalar, can be conveniently included in the pair functions. The inclusion of the dipole transition operator in the orbitals leads to the RPA in the one-particle case, but its inclusion in the pair functions leads to an order of magnitude larger number of excitations and it seems more convenient to use (4.1) with the pair functions already obtained. The expression (4.1) also has the advantage of giving a more symmetric treatment of the initial and final states, i and f, in the evaluation of transition elements. The most viable procedure for including pair correlation to all orders seems to be to combine both approaches, calculating PNC pair functions and using these in the evaluation of PNC transition elements using (4.1)

4.3.1 The Parity Non-Conserving Weak Interaction
The spin-independent part of the parity non-conserving electron-nucleon interaction can be written as

$$h^{PNC} = \frac{G_F Q_W}{2\sqrt{2}} \gamma_5 \rho_N(r)$$

where $G_F = G_\mu(\hbar c)^3 = 89.6 eV/fm^3 = 2.22 \cdot 10^{-14}$ a.u. and Q_W is the quantity to be determined from a comparison between atomic theory and experiment and ρ_N is a normalized nuclear density. To include this interaction in the pair equation is not completely trivial, since the pair equation is normally solved only for points outside the nucleus. However, it is possible to use an alternative operator. If $\rho_N(r)$ has the same shape as the charge density giving rise to the nuclear potential, $V_N(r)$, it can be written as

$$\rho_N(r) = \nabla^2 V_N(r)/4\pi Z$$

Generalizing the point nucleus treatment by Hiller et al.[80] gives

$$4\pi Z \gamma_5 \rho_N(r) = \left[\alpha \cdot \mathbf{p}, [\sigma \cdot \mathbf{p}, V_N(r)]\right] + 2\left(\frac{1}{r}\frac{dV_N}{dr}\right)\alpha \cdot \mathbf{l}$$

$$= \frac{1}{c}\left[h_0, [\sigma \cdot \mathbf{p}, V_N(r)]\right] + 2\left(\frac{1}{r}\frac{dV_N}{dr}\right)\alpha \cdot \mathbf{l}$$

where h_0 is a Dirac one-electron hamiltonian. Consider first the commutator term. The inner commutator

$$C = \frac{1}{c}\left[\sigma \cdot \mathbf{p}, V_N(r)\right] = \frac{1}{c}\sigma \cdot \mathbf{r}\frac{1}{r}\frac{dV_N(r)}{dr}$$

is a local, antihermitian one-electron operator. The antihermiticity, $C^\dagger = -C$, is necessary to make the perturbation $[h_0, C]$, itself, hermitian. Since C is local, it commutes with any local potential, which can thus be added to h_0. The $\alpha \cdot \mathbf{p}$ terms for the other electrons commute with C and it is thus also possible to use the full hamiltonian as long as the electron-electron interaction is described by a local operator. The commutator term can then be written as

$$\left[\alpha \cdot \mathbf{p}, [\sigma \cdot \mathbf{p}, V_N(r)]\right] = [H, C]$$

Were it not for the problems with negative energy states, the unperturbed wavefunction could have been written as a solution to $H\Psi = E\Psi$ and the correction would then be given directly by $\delta\Psi = -C\Psi$. However, as we shall see below, this relation still holds.

For a local potential, with one-electron orbitals given by

$$(\varepsilon_a - h_0)|a_0\rangle = 0$$

the lowest-order correction $|\delta a\rangle$ introduced by the perturbation $[h_0, C]$ is a solution to the equation (analogous to (4.2))

$$(\varepsilon_a - h_0)|\delta a\rangle = [h_0, C]|a_0\rangle$$

and it is seen directly that $|\delta a\rangle = -C|a_0\rangle$ is a solution to this equation. The first-order energy correction, $\delta\varepsilon = \langle a_0|[h_0, C]|a_0\rangle$, vanishes identically.

For the non-local Dirac-Fock potential, the perturbation can be written as

$$\left[\alpha \cdot \mathbf{p}, [\sigma \cdot \mathbf{p}, V_N(r)]\right] = [h_{DF} - v_{DF}, C]$$

It is, however, easily seen that $|\delta a\rangle = -C|a_0\rangle$ is a solution also in this case — provided the correction to the Dirac-Fock potential caused by the orbital corrections is treated together with the perturbation (cf. (4.3)). The equation for the correction then becomes

$$(\varepsilon_a - h_{DF})|\delta a\rangle = [h_{DF}, C]|a_0\rangle - [v_{DF}, C]|a_0\rangle + \delta v_{DF}|a_0\rangle$$

Direct insertion of the orbital corrections $|\delta b\rangle = -C|b_0\rangle$ in the expression for the correction to the Dirac-Fock potential gives

$$\left(-[v_{DF}, C] + \delta v_{DF}\right)_1 = -\sum_b^{occ}\left\{\left[\langle b_2|V_{12}(1 - P_{12})|b_2\rangle, C_1\right] + \langle b_2|[V_{12}(1 - P_{12}), C_2]|b_2\rangle\right\}$$

$$= -\sum_b^{occ}\left\{\langle b_2|[V_{12}, C_1] - V_{12}P_{12}C_1 + C_1V_{12}P_{12}|b_2\rangle\right.$$
$$\left. + \langle b_2|[V_{12}, C_2] - V_{12}P_{12}C_2 + C_2V_{12}P_{12}|b_2\rangle\right\}$$

The direct parts of both terms vanish, since C commutes with the electron-electron interaction, V_{12}. Between the exchange terms there is a cross-wise cancellation — the relation $P_{12}C_1 = C_2P_{12}$ shows that the second term in the first matrix element cancels the last term in the second matrix element and vice versa. Thus, also for the Dirac-Fock potential the only term surviving on the right-hand side is the commutator, giving

$$(\varepsilon_a - h_{DF})|\delta a\rangle = [h_{DF}, C]|a_0\rangle$$

with the solution |δa>=-C|a>. Since this holds for each orbital, it holds also for the total wavefunction, as stated above.

To get the cancellation between $[v^{DF}, C]$ and δv^{DF}, it is necessary that all states —also those with negative energy —are allowed in the summation on the RHS, since only C, but not $\lambda^+ C \lambda^+$, commutes with V_{12}. The matrix elements of any operator commuting with C will be unchanged by using the transformed PNC interaction

$$\tilde{h}^{PNC} = \frac{G_F Q_w}{8\sqrt{2}\,\pi Z}\left(\frac{1}{r}\frac{dV_N}{dr}\right)\alpha \cdot 1 \tag{4.5}$$

This holds for the length form of the dipole operator, whereas for the velocity form, a correction term

$$\langle\Psi_f|[C, d_v]|\Psi_i\rangle = \left\langle\Psi_f\left|\frac{1}{r}\frac{dV_N}{dr}\frac{\alpha \times r}{\omega}\right|\Psi_i\right\rangle$$

must be added[81].

We expect that the alternative, less singular, form (4.5) of h^{PNC} can be used in the pair equation without particular complications. The milder singularity of this operator may also improve the numerical accuracy of other approaches, but it has not yet been used.

4.3.2 Other Parity Non-Conserving Effects The interaction discussed above is independent of the nuclear spin. However, the increased experimental accuracy makes it possible to observe also a spin-dependence in the parity non-conservation[89], which may arise from the spin-dependent part of the weak electron-nucleon interaction, but also from a nuclear anapole moment as discussed by Khriplovich and coworkers[90]. Its calculation is somewhat more complicated than for the spin-independent part, since both external perturbation are non-scalars, leading to a larger number of excitations.

Closely related to the calculations of PNC weak interaction is the study of atomic Electric Dipole Moments (EDM)[86]. If found, an atomic EDM would imply a violation of symmetry under both parity and time reversal, and may arise from a P and T violating interaction or from an intrinsic EDM of an electron or nucleon, through several mechanisms. When correlation effects are to be included, these calculations do, however, get more complicated that the corresponding PNC calculations, due to the two-body character of the interaction with the electric field from the other electrons. Further, since the effect vanishes in the non-relativistic limit, effects of transverse photons enter in the same order in α as the interaction itself, giving rise to additional operators[91].

330

5. CONCLUSION

Many-body perturbation theory is a powerful technique. Much of the development in the non-relativistic case can be taken over to the relativistic case. However, the existence of negative energy states makes it necessary to find ways of distinguishing them from positive energy ones. Further, a choice of gauge must be made for the electron-electron interaction. In a "complete" treatment, effects involving negative energy states must be included. Radiative corrections give important contributions for heavy atoms. The treatment of relativistic pair correlation effects poses new challenging problems, many of which have not been discussed here, but have been treated in other lectures in this volume.

ACKNOWLEDGEMENTS

Financial support by the Swedish Natural Science Research Council (NFR) is gratefully acknowledged.

Most of the work presented here has been carried out in Ingvar Lindgren's research group in Göteborg and I would like thank him and other past and present coworkers in the group, John Morrison, Johannes Lindgren, Sten Salomonson, Jean-Louis Heully, Eva Lindroth, Per Öster and Anders Ynnerman, for happy and fruitful collaboration. Ernest Henley and Larry Wilets have provided kind guidance both into parity non-conserving phenomena and into the problems connected with negative energy states. Further, I would like to express my appreciation of the invitation to participate in this workshop at the ITP, which made possible several clarifying discussions, in particular with Joe Sucher, Marvin Mittlemann, Jim Vary, Walter Johnson, Jonathan Sapirstein and Steve Blundell. I would also like to thank Ian Grant, Harry Quiney and Steve Wilson for stimulating correspondence and Pat Sandars for never-failing support and interest in the work.

REFERENCES

1. E W.Otten, *Nuclear Radii and Moments of Unstable Nulcei,* Treatise on Heavy Ion Physics (ed .D A Bromley) Vol. **8**, Plenum Press, New York (1987)
2. W H King, *Isotope Shifts in Atomic Spectra,* Plenum Press, New York (1984)
3. S Garpman, I Lindgren, J Lindgren and J Morrison, Phys. Rev. A**11** *758* (1975)
4. S Garpman, I Lindgren, J Lindgren and J Morrison, Z. Physik A**276** *167* (1976)
5. I Lindgren and J Morrison, *Atomic Many-Body Theory* Springer Series in Chemical Physics, vol **13** (1982)
   S Wilson, *Electron Correlation in Molecules* (Clarendon Press, Oxford, 1984) and Comp. Phys. Rep. **2** *389* (1985)
6. I Lindgren, *Many-Body Theory* (this volume)
7. C Bloch, Nucl. Phys. **6** *329* (1958)
8. J I Musher and J M Schulman, Phys. Rev. **173** *93* (1968)
9. V McKoy and N W Winter, J.Chem. Phys. **48** *5514* (1968)
10. N W Winter, A Laferriere and V McKoy, Phys. Rev. A**2** *49* (1970)
11. J M Schulman and W S Lee, Phys. Rev. A**5** *13* (1972)

12. J C Morrison, J. Phys. **B6** *2205* (1973)
13. I Lindgren, J Lindgren and A-M Mårtensson, Z. Physik **A279** *113* (1976)
14. A-M Mårtensson, J. Phys. **B12** *3995* (1979)
15. I Lindgren and S Salomonson, Physica Scripta **21** *335* (1980)
16. J C Morrison and S Salomonson, Physica Scripta **21** *343* (1980)
    S Salomonson, I Lindgren and A-M Mårtensson , Physica Scripta **21** *351* (1980)
17. S Salomonson, Z. Physik **A316** *135* (1984)
18. A-M Mårtensson-Pendrill and S Salomonson, Phys. Rev. **A30** *712* (1984)
19. I Lindgren, Phys. Rev. **A31** *1273* (1985)
20. S A Alexander, H J Monkhorst and K Szalewicz,
    J. Chem. Phys. **85** 5821 (1986), *ibid.* **87** *3976* (1987), **89** *355* (1988)
21. S Salomonson, P Öster, I Lindgren and A-M Mårtensson-Pendrill,to be published
    S Salomonson and P Öster, Abstract VIII-3, 11th ICAP conference,
    Paris, July 4-8, 1988 (Ed. C Fabre and D Delande)
22. A-M Mårtensson-Pendrill, *Numerical Determination of Non-relativistic and
    Relativistic Pair Correlation in* Proceedings of NATO-ARW on *"Numerical
    Determination of the Electronic Structure of Atoms, Diatomic and Polyuatomic
    Molecules"* Versailles, 17-22 April 1988 (D. Reidel, Dordrecht, ed. J Delhalle
    and M Defrancheschi)
23. S Salomonson and P. Öster, private communication
24. E Lindroth and A-M Mårtensson-Pendrill, Z. Physik **A316** *265* (1984)
25. C Schwartz, Phys. Rev. **126** *1015* (1962)
26. R N Hill, J. Chem. Phys. **83** *1173* (1985)
27. C Froese-Fischer, J. Comput. Phys. **13** *502* (1973)
28. B F Davis and K T Chung, Phys. Rev. **A25** *1328* (1982)
29. Y Accad, C L Pekeris and B Schiff B, Phys. Rev. **A4** *516* (1971)
    B Schiff, H Lifson, C L Pekeris and P Rabinowitz,
    Phys. Rev. **140** *A1104* (1965)
30. G W F Drake and A J Makowksi, J. Opt. Soc. **B**, October (1988), see also
    G W F Drake, Nucl. Inst. Meth. **B31** *7* (1988) and
    Phys. Rev. Lett. **59** *1549* (1987)
31. W Kutzelnigg, Theor. Chim. Acta **68** *445* (1985) and
    Chem. Phys. Lett. **134** *17* (1987)
32. H Grotch and D R Yennie, Rev. Mod. Phys. **41** *350* (1969) and
    D R Yennie, Lecture notes in this volume
33. C Thibault *et al.* Phys. Rev. **C23** *2720* (1981)
34. A-M Mårtensson and S Salomonson, J. Phys. **B15** *2115* (1982)
35. W R Johnson, S A Blundell and J Sapirstein, Phys. Rev. **A38** *2699* (1988)
36. E Lindroth and A-M Mårtensson-Pendrill, Z. Physik **A309** *277* (1983)
37. E Lindroth E, A-M Mårtensson-Pendrill and S Salomonson
    Phys. Rev. **A30** *712* (1985)
38. A Pery, Proc. Phys. Soc. **A67** *181* (1954)
39. I Lindgren, Int. J. Quant. Chem **S12** *33* (1978)
40. J Čížek, J. Chem. Phys. **45** *4256* (1966), Adv. Chem. Phys. **14** *35* (1969)
41. S Salomonson and P Öster, to be published
42. C Froese-Fischer and K M S Saxena, Phys. Rev. **A9** *1498* (1974)
43. M. Urban, I Černušák, V Kellö and J Noga, *Electron Correlation in Molecules*
    in *Methods in Computational Chemistry* , Vol 1 (Ed. S Wilson, Plenum,
    New York, 1987)
44. J-L Heully and S Salomonson, Physica Scripta, **38** *143* (1988)

332

45. S Hörbäck, A-M Mårtensson-Pendrill, S Salomonson and U Österberg,
    Physica Scripta **28** *469* (1983)
46. M A Bouchiat and L Pottier, Science **234** *1203* (1986), Atomic Physics **9** *246*
    (Ed. E N Fortson and R S van Dyck, World Publishing, Singapore, 1984)
    E D Commins, Physica Scripta **35** *468* (1987)
47. I P Grant, Phys. Rev. **A25** *1230* (1982)
48. G E Brown and D G Ravenhall, Proc. Roy. Soc. **A208** *552* (1951)
49. J Sucher, Phys. Rev. **A22** *348* (1980),
    Int. J. Quant. Chem. **25** *3* (1984) and lecture notes in this volume
50. E Lindroth, Physica Scripta, **36** *485* (1987)
51. J-L Heully, I Lindgren, E Lindroth and A-M Mårtensson-Pendrill
    Phys. Rev. A33 *4426* (1986)
52. H M Quiney, I P Grant and S Wilson J. Phys. B18 *2805* (1985),
    Physica Scripta **36** *460* (1987) and
    H M Quiney, *Relativistic Many-Body Perturbation Theory* in *Methods in
    Computational Chemistry,* Vol 2 (Ed. S Wilson, Plenum, New York, 1987)
53. M Vajed-Samii, S N Ray, T P Das and J Andriessen,
    Phys. Rev. **A20** *1787* (1979)
54. V A Dzuba, V V Flambaum, P G Silvestrov and O P Sushkov,
    J. Phys. **B17** *1953* (1985) Physica Scripta **35** *69* (1987),
    J. Phys. **B20,** *1399* and *3297* (1987)
55. W R Johnson and J Sapirstein, Phys. Rev. Lett. **57** *1126* (1986)
    W R Johnson, S A Blundell and J Sapirstein,
    Phys. Rev. **A37** *307, 1395* (1988)
    W R Johnson, S A Blundell, Z W Liu and J Sapirstein,
    Phys. Rev. **A37** *2764* (1988)
56. G Hardekopf and J Sucher, Phys. Rev. **A30** *703* (1984)
57. E Lindroth, Phys. Rev **A37** *316* (1988) and
    Thesis, University of Göteborg, (1987)
58. C Froese Fischer *The Hartree-Fock Method for Atoms - a Numerical Approach,*
    (Wiley, 1976, in particular Chapter 3.6)
59. J-L Heully, I Lindgren, E Lindroth, S Lundqvist and A-M Mårtensson-Pendrill,
    J. Phys. **B19** *2799* (1986)
60. W Kutzelnigg, Int. J. Quantum Chem. **25** *107* (1984)
    W H E Schwarz and H Wallmeier, Mol. Phys. **46** *1045* (1982)
61. H A Bethe and E E Salpeter,*Quantum Mechanics of Two-Electron Atoms,*
    (Springer, 1957)
62. L Armstrong, *Relativistic Effects in Many-Body Systems,* Atomic Physics **8** *129*
    (Plenum, ed I Lindgren, A Rosén and S Svanberg, 1983)
    M H Mittlemann, Phys. Rev. A5, *2395* (1972)
63. H M Quiney, I P Grant and S Wilson, J. Phys. B **20,** 1413 (1987)
    W R Johnson, *Relativistic Many-Body Calculations*, Atomic Physics **8** *149*
    (Plenum, ed I Lindgren, A Rosén and S Svanberg, 1983)
64. O. Gorceix, P Indelicato and J-P Desclaux, J. Phys. B **20** *639* (1987)
    P Indelicato, O. Gorceix and J-P Desclaux, J. Phys. B **20** *651* (1987)
65. O. Gorceix and P Indelicato, Phys. Rev. A37 *1087* (1988)
66. A-M Mårtensson-Pendrill and P Öster, Physica Scripta **36** *444* (1988)
67. E Lindroth, A-M Mårtensson-Pendrill, P Öster and A Ynnerman, to be published
68. I P Grant, J. Phys. B20 *L735* (1987)

69. E Lindroth and A-M Mårtensson-Pendrill, *Comment on the Complete Breit Interaction*, Phys. Rev. A Submitted
70. J Sucher, J. Phys. B **21** *L585* (1988)
71. See, e.g., the lecture notes by P Mohr and J. Sapirstein in this volume.
72. I Lindgren and D Mukherjee, to be published
73. H M Quiney, I P Grant and S Wilson, J. Phys. B**18** *577* and *2305* (1985)
74. P G H Sandars, J. Phys. B**10** *2983* (1977)
75. A-M Mårtensson-Pendrill, J. de Physique (Paris) **46** *1949* (1985)
76. A-M Mårtensson-Pendrill, Phys. Rev. Lett. **54** *1153* (1985),
    A-M Mårtensson, E M Henley and L Wilets, Phys. Rev. A**24** *308* (1981)
77. W R Johnson, D S Guo, M Idrees and J Sapirstein,
    Phys. Rev. A**32** *2093* (1985), *ibid.* A**34** *1043* (1986)
78. W R Johnson, C D Lin, K T Cheng and C M Lee, Physica Scripta **21** *409* (1980)
79. V A Dzuba, V V Flambaum, P G Silvestrov and O P Sushkov,
    Phys. Lett. A**118** *177* (1987)
80. J Hiller, J Sucher and G Feinberg, Phys. Rev. A**18** 2399 (1978)
81. A-M Mårtensson-Pendrill, E Lindroth and P Öster,
    Physica Scripta T**22** *300* (1988)
82. V Kvasnička, V Laurinc and S Biskupič, Phys. Rep. **90** *159* (1982) (Eq 3.33)
83. H Kümmel, Int. J. Quant. Chem. **24** *79* (1983)
    see also J Vary, lecture notes, this volume.
84. Anders Ynnerman and A-M Mårtensson-Pendrill, to be published
85. C and M-A Bouchiat, J. de Physique **35** *111* (1974)
86. P G H Sandars, Physica Scripta **36** *904* (1987)
87. M-A Bouchiat, J Guéna, L Pottier and L Hunter,
    J. de Physique, **46** *1897* (1985), **47** *1175* and *1709* (1986)
88. S L Gilbert, M C Noecker, B P Masterson and J Cooper,
    Phys. Rev. Lett. **58** *1738* (1987)
    M C Noecker, B P Masterson and C E Wieman, Phys. Rev. Lett (submitted)
89. See the lecture notes by S A Blundell and by O P Sushkov in this volume
90. V V Flambaum, I B Khriplovich and O P Sushkov,
    Phys. Lett. B**146** *367* (1984)
    P A Frantsuzov and I B Khriplovich, Z. Phys. D**7** *297* (1988)
    see also E R Boston and P G H Sandars, Abstract I-1,
    11th ICAP conference, Paris, July 4-8, 1988 (Ed. C Fabre and D Delande)
91. E Lindroth, B W Lynn and P G H Sandars, *Order α^2 theory of Atomic Electric Dipole Moment due to an Electric Dipole Moment of the Electron*, submitted to J. Phys. B

MANY-ELECTRON ATOMS: FORMAL ASPECTS

POTENTIALS FROM FIELD THEORY: NON-UNIQUENESS, GAUGE DEPENDENCE AND ALL THAT

J. Sucher
Department of Physics, University of Maryland,
College Park, MD 20741

ABSTRACT

I explore the field-theoretic background and attendant ambiguities of configuration-space operators, "potentials", used to study relativistic effects in many-body systems. Topics such as non-locality, energy-transfer dependence, the role played by particle identity and the difference between bosons and fermions in this regard are discussed and the construction of Hilbert-space operators which take energy transfer into account is described. Particular attention is paid to the so-called "gauge dependence" of effective potentials in gauge theories; an analysis is given which reveals the misleading nature of this terminology.

I. INTRODUCTION

A. Motivation

These notes contain "truths" about the topics in the title, which I believe have a Lincoln-type property. All the people know some of them, some people know all of them, but all the people do not know all of them. In any case, there does not seem to be available a simple, self-contained discussion of these matters.

The discussion is motivated by the existence of a variety of "potentials", used in the study of level shifts in atomic systems where relativistic effects are important, either because Z is large, or because the experimental accuracy is high, or both. Chief among these are two "competing" potentials associated with exchange of a photon between two electrons. One of these is the somewhat messy-looking Coulomb-Breit potential V_I, given by

$$V_I(\mathbf{r}) = \left[1 - \frac{1}{2}(\boldsymbol{\alpha}_A \cdot \boldsymbol{\alpha}_B + \boldsymbol{\alpha}_A \cdot \hat{r}\boldsymbol{\alpha}_B \cdot \hat{r})\right] U_C(r) \qquad (1.1)$$

where $U_C(r)$ is the Coulomb potential

$$U_C(r) = \left(\frac{e_A e_B}{4\pi}\right)\frac{1}{r}. \qquad (1.2)$$

[For later purposes, I distinguish between the particles, referring to them as "A" and "B"; for two electrons, $e_A = e_B = -e$ and, with $\hbar = c = 1$, $e^2/4\pi = \alpha = 1/137$.] The other is the neat-looking Coulomb-Gaunt-Møller potential V_{II}, given by

$$V_{II}(\mathbf{r}) = (1 - \boldsymbol{\alpha}_A \cdot \boldsymbol{\alpha}_B)U_C(r). \qquad (1.3)$$

The potential V_{II} arises "naturally" if one uses the Feynman gauge for the photon-propagator in the computation of the amplitude for one-photon exchange, viz.

$$\tilde{D}_{F\mu\nu}(k) = -ig_{\mu\nu}/(k^2 + i\epsilon),\qquad(1.4)$$

whereas V_I arises "naturally" if one uses the Coulomb-gauge, also called the radiation gauge, for the photon propagator, viz.

$$\tilde{D}_{C\mu\nu}(k) = \begin{cases} i/\mathbf{k}^2 \;;\; \mu = \nu = 0 \\ i(\delta_{ij} - \hat{k}_i\hat{k}_j)/(k^2 + i\epsilon) \;;\; \mu = i \neq 0 \,,\, \nu = j \neq 0 \\ 0 \;;\; \mu = 0\,, \nu \neq 0 \text{ or } \mu \neq 0 \,,\, \nu = 0 \end{cases} \qquad (1.5)$$

It is therefore universally said that the difference between V_I and V_{II} arises from a different choice of gauge and is just an example of gauge-dependence of the potential. I will argue later that this statement is misleading because it ignores some subtleties which arise in getting from a scattering amplitude to a potential, in the framework of a Dirac spinor description of wave functions for spin-1/2 particles. This is why I have put quotes on the word "naturally".

To complicate the situation still further, there are modified versions $V_I(\mathbf{r}; \Delta)$ and $V_{II}(\mathbf{r}; \Delta)$ of both $V_I(\mathbf{r})$ and $V_{II}(\mathbf{r})$ which are associated with an "energy-transfer" Δ and which come equipped with instructions for their use. The one associated with $V_{II}(r)$, given long ago by Gerry Brown,[1] has the relatively simple form

$$V_{II}(\mathbf{r}; \Delta) = V_I(\mathbf{r}) \cos(\Delta r),$$
$$= \frac{e_A e_B}{4\pi r} (1 - \boldsymbol{\alpha}_A \cdot \boldsymbol{\alpha}_B) \cos \Delta r \qquad (1.6)$$

The definition of $V_I(r; \Delta)$ is considerably more complicated,

$$V_I(\mathbf{r}; \Delta) = \frac{e_A e_B}{4\pi} \left[\frac{(1 - \boldsymbol{\alpha}_A \cdot \boldsymbol{\alpha}_B \cos \Delta r)}{r} - (\boldsymbol{\alpha}_A \cdot \nabla)(\boldsymbol{\alpha}_B \cdot \nabla) \left(\frac{\cos(\Delta r) - 1}{\Delta^2 r} \right) \right].$$
$$(1.7)$$

Here $\nabla \equiv \partial/\partial \mathbf{r}$ acts only on the factor to the right, not on the wave function. Note that $V_i(\mathbf{r}; \Delta)$ reduces to $V_i(\mathbf{r})$ for $\Delta \to 0 : V_i(\mathbf{r}; 0) = V_i(\mathbf{r})$.

To paraphrase the Bard: Who is Δ, what is she, that all our swains commend her? As you might agree, it is unlikely that Δ is holy, and considering the triangular shape it is unlikely that many will consider her fair. The real question is, is she wise? If so, I for one am ready to admire her.

B. Some Significant Issues

Let us now turn to a study of these and related questions. I will proceed by example, starting with the simplest possible circumstances. My main purpose is to show that a number of different and *independent* issues come into play here and that it is useful to distinguish among them. They include

(i) the role of relativistic quantum field theory (RQFT),

(ii) the desire to have a "potential" description of particle interactions, for use in practical calculations,

(iii) the role of particle identity,

(iv) the role of gauge choices in gauge theories

and

(v) the role of spin, especially the Dirac description of spin-1/2 particles.

With regard to items (i) and (ii) on this list, of course in nonrelativistic quantum theory (NRQT) the theoretical study of N-particle systems requires "only" an analysis of solution of the Schrödinger equation

$$\frac{i\partial}{\partial t}\psi = H_{nr}\psi \tag{1.8a}$$

or, for stationary states,

$$E\psi = H_{nr}\psi, \tag{1.8b}$$

where the nonrelativistic Hamiltonian H_{nr} is an operator in the N-particle configuration space (CS), viz.

$$H_{nr} = \sum_i \left(\mathbf{p}_i^2/2m_i + V_i^{ext}\right) + \sum_{i<j} V_{ij}. \tag{1.8c}$$

Here not only the interactions V_i^{ext} with an external field, if any, are given *a priori*, but also the interactions V_{ij} between the particles are given, *e.g.*, $V_{ij} = e^2/4\pi r_{ij}$ for N-electron systems.

In RQFT not only is no potential given but the very concept of an N-particle system requires discussion – even the concept of a 1-particle system is non-trivial. What is given instead is an interaction involving products of fields $\phi_i(x)$, whose quantized form involves the creation and destruction of particles. So one is a long-way from the familiar Schrödinger equation and it is not surprising that funny things can happen on the way to a CS description of the dynamics of N-particles systems. To the extent that such a description is possible, one has to derive a CS-type Hamiltonian from the theory, including of course approximate forms of the potentials.

II. SIMPLE YUKAWA-TYPE THEORIES

We will study first the simplest non-gauge theories imaginable, with the minimum complications from spin, and inquire about the lowest-order potentials that arise "naturally" and of what use they might be.

We consider distinct particles A and B interacting with each other via the exchange of a spin-0 meson of mass μ. For A and B spinless and not the same as their antiparticles, the simplest Lagrangian density describing such an interaction is given by

$$\mathcal{L}_1 = -G_A \phi_A^\dagger(x)\phi_A(x)\varphi(x) - (A \to B), \tag{2.1}$$

where ϕ_A and ϕ_B are complex spin-0 fields associated with A and B respectively and ϕ is a real spin-0 field associated with the meson. If A and B have spin-1/2, the analogue of (2.1) is

$$\mathcal{L}_2 = -g'_A \overline{\psi}_A(x)\psi_A(x)\varphi(x) - (A \to B), \qquad (2.2)$$

where ψ_A and ψ_B are Dirac spinor fields, with $\overline{\psi} = \psi^\dagger \gamma^0$. In addition we will contemplate the presence of additional interactions with a static external (scalar) source $\varphi_{ext}(\mathbf{x})$, viz.

$$\mathcal{L}_1^{ext} = -G_A \phi_A^\dagger \phi_A \varphi_{ext}(\mathbf{x}) - (A \to B), \qquad (2.3)$$

and, similarly

$$\mathcal{L}_2^{ext} = -g'_A \overline{\psi}_A \psi_A \varphi_{ext}(\mathbf{x}) - (A \to B). \qquad (2.4)$$

A. Lowest-order Scattering Amplitudes And Level-shifts

1. Scattering amplitudes

In any RQFT, the S-matrix element for a transition $|i\rangle \to |f\rangle$ has the form

$$S_{fi} = \delta(f,i) - (2\pi)^4 i\delta(P_f - P_i)T_{fi}, \qquad (2.5)$$

where $\delta(f,i)$ is a generalized Kronecker-delta [Dirac delta function in continuous labels], P_i and P_f are total initial and final four-momenta, and T_{fi} is the transition amplitude. This amplitude is related to the invariant Feynman amplitude M_{fi} by

$$T_{fi} = N_f M_{fi} N_i, \qquad (2.6)$$

where the N's are kinematical factors determined by the normalization of one-particle states. For a scattering process

$$A + B \to A + B$$

with initial and final four-momenta p_A, p_B and p'_A, p'_B respectively, the conventional choice for spin-0 particles, viz. plane waves normalized to $\langle \mathbf{p}'|\mathbf{p}\rangle = (2\pi)^3 \delta(\mathbf{p}' - \mathbf{p})$, leads to

$$N_i = (\sqrt{2E_A}\sqrt{2E_B})^{-1} \ , N_f = (\sqrt{2E'_A}\sqrt{2E'_B})^{-1}. \qquad (2.7a)$$

These also hold for spin-1/2 particles, provided the Dirac spinors are normalized to

$$\bar{u}(\mathbf{p})u(\mathbf{p}) = 2m, \qquad (2.7b)$$

rather than $\bar{u}u = 1$.

(a) Distinct spin-0 bosons

There is only one Feynman diagram in lowest order, shown in Fig. 1. The meson propagator is $\tilde{D}_F(k) = i/(k^2 - \mu^2)$ and the Feynman rules, from the Lagrangian (2.1), give a factor $(-iG)$ at each vertex in the calculation of $-iM$. Thus

$$-iM^{(2)} = (-iG_A)(-iG_B)(i/(k^2 - \mu^2)).$$

With $k = Q$, the four-momentum transfer,

$$Q \equiv p_A - p'_A = -(p_B - p'_B), \qquad (2.8)$$

we have

$$M^{(2)} = (G_A G_B)/(t - \mu^2), \qquad (2.9)$$

where t is the invariant squared momentum transfer:

$$t = Q^2. \qquad (2.10)$$

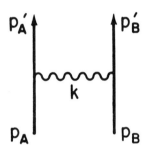

Figure 1: Feynman diagram describing the one-meson exchange contribution to the scattering amplitude for distinct particles A and B.

For later comparison with the level-shift calculation let us also do this in the "old-fashioned way." The interaction Hamiltonian is

$$H_I = \int \mathcal{H}_I dx = -\int \mathcal{L}_I dx, \qquad (2.11)$$

with $\mathcal{L}_I = \mathcal{L}_1$. Working in the Schrödinger picture, one finds that the matrix element for emission of a meson of three-momentum $\mathbf{k}$ (and energy $w \equiv \sqrt{k^2 + \mu^2}$) is given by

$$\langle \mathbf{p}'_A, \mathbf{k} | H_I | \mathbf{p}_A \rangle = G_A \langle \mathbf{p}'_A | \frac{e^{-i\mathbf{k} \cdot \mathbf{r}_A}}{(2\pi)^{3/2}\sqrt{2w}} | \mathbf{p}_A \rangle \frac{1}{\sqrt{4E'_A E_A}} \frac{1}{(2\pi)^3}$$

$$= G_A \delta(\mathbf{p}'_A + \mathbf{k} - \mathbf{p}_A)(8w E'_A E_A)^{-1/2}(2\pi)^{-3/2}, \quad (2.12)$$

with $\mathbf{k} \to -\mathbf{k}$ for absorption. The matrix element t_{fi} for the transition $|i\rangle = |\mathbf{p}_A, \mathbf{p}_B\rangle \to |f\rangle = |\mathbf{p}'_A, \mathbf{p}'_B\rangle$ is therefore given, in lowest order, by

$$t_{fi}^{(2)} = \sum_{int} \frac{\langle f|H_I|\Psi_{int}\rangle\langle\Psi_{int}|H_I|i\rangle}{E_i - E_{int}} \tag{2.13}$$

$$= G_A G_B N_f N_i \int \frac{d\mathbf{k}}{(2\pi)^3} \left[\frac{\langle\mathbf{p}'_B|\frac{e^{i\mathbf{k}\cdot\mathbf{r}_B}}{\sqrt{2w}}|\mathbf{p}_B\rangle\langle\mathbf{p}'_A|\frac{e^{-i\mathbf{k}\cdot\mathbf{r}_A}}{\sqrt{2w}}|\mathbf{p}_A\rangle}{E_i - (E'_A + E_B + w)} + (A \to B) \right].$$

The two terms in (2.13) correspond to the two *time-ordered* diagrams shown in Fig. 2. Since $E_i = E_A + E_B = E'_A + E'_B$, (2.13) reduces to

$$t_{fi}^{(2)} = N_f N_i (2\pi)^3 \delta(\mathbf{P}_f - \mathbf{P}_i) \frac{G_A G_B}{2w} \left[\frac{1}{E'_B - E_B - w} + \frac{1}{E'_A - E_A - w} \right]. \tag{2.14}$$

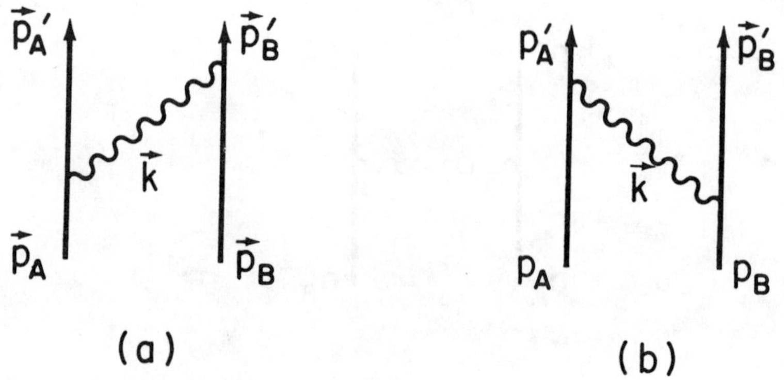

Figure 2: The two time-ordered diagrams corresponding to Fig. 1.

Since further $E'_A - E_A = -(E'_B - E_B) = -Q_0$, the square bracket in (2.14) reduces to $(-2w)/(-Q_0^2 + w^2) = 2w/(t - \mu^2)$ and we see that

$$t_{fi}^{(2)} = (2\pi)^3 \delta(\mathbf{P}_f - \mathbf{P}_i) T_{fi}^{(2)} \tag{2.15}$$

where

$$T_{fi}^{(2)} = N_f M^{(2)} N_i, \tag{2.16}$$

with $M^{(2)}$ given by (2.9), as required for agreement with equation (2.5) and (2.6).

(b) Distinct spin-1/2 fermions:

The Feynman diagram is the same as for Fig. 1 and with the convention (2.7b), we get

$$M^{(2)} = (\bar{u}'_A u_A)(\bar{u}'_B u_B) g'_A g'_B / (t - \mu^2),$$
(2.17)

i.e., the same result as in (2.9), apart from the spinor factors. The calculation in terms of old-fashioned perturbation theory is analogous to that above and yields, as required,

$$t^{(2)}_{fi} = (2\pi)^3 \delta(\mathbf{P}_f - \mathbf{P}_i) N_f N_i (\bar{u}'_A u_A)(\bar{u}'_B u_B) g'_A g'_B / (t - \mu^2).$$
(2.18)

2. Effect of external field: level shifts

Although we could study the scattering of A and B in the presence of a static external field, it is of much greater physical interest to consider A and B *bound* to a common core, acting as the source of the external field, and to compute the lowest-order level shift arising from the interaction of A and B. We can avoid making any approximation with regard to the external field by using the Furry picture, which involves expansion of the field operators in terms of eigenfunctions of the associated external-field Klein-Gordon or Dirac equation.

While a discussion of the expansion for spin-1/2 particles may be found in textbooks, this does not appear to be so for spin-0 particles. In any case, these expansions are described in Appendix A, with special attention to the kinematical differences between the two cases.

Consider a total Hamiltonian of the form

$$H = H_0 + H_I$$
(2.19)

and an eigenvector Ψ of H_0 with eigenvalue E_0 such that $\langle \Psi | H_I | \Psi \rangle = 0$. The part of the second-order level shift arising from H_I which is of interest for us here is given by

$$\Delta E^{(2)} = \langle \Psi | H_I \frac{1}{E_0 - H_0} H_I | \Psi \rangle'$$
(2.20)

where the prime indicates that self-energy parts are to be omitted. For H_0 we take the sum of the Hamiltonians for the fields associated with particles A and B, the interaction with the external field included, together with the free Hamiltonian for the meson field. In the Furry picture the state then has the form

$$\Psi = a^\dagger_A(n_a) a^\dagger_B(n_b) |vac\rangle$$
(2.21a)

and

$$E_0 = E_{n_a} + E_{n_b}.$$
(2.21b)

For H_I we take (2.11) with $\mathcal{L}_I$ given by either (2.1) or (2.2). On insertion of a complete set of intermediate states (only three-particle states can contribute) and use of the matrix elements computed in Appendix A, one finds the following results:

(a) Distinct spin-0 bosons

For the sake of comparison with the spin-1/2 case it is convenient to introduce dimensionless coupling constants g_A and g_B related to G_A and G_B via

$$G_A = 2m_A g_A \quad , \quad G_A = 2m_A g_B. \tag{2.22}$$

On use of Eq. (A11) of Appendix A, one then gets

$$\Delta E^{(2)} = g_A g_B \, N^2 \int \frac{d\mathbf{k}}{(2\pi)^3} \frac{1}{2w} \left[\frac{\langle n_b | e^{i\mathbf{k}\cdot\mathbf{r}_B} | n_b \rangle \langle n_a | e^{-i\mathbf{k}\cdot\mathbf{r}_A} | n_a \rangle}{-w} + (A \to B) \right] \tag{2.23}$$

where

$$N = (m_A m_B / E_{n_a} E_{n_b})^{\frac{1}{2}}. \tag{2.24}$$

(b) Distinct spin-1/2 fermions

On use of Eq. (A20) of Appendix A one finds, for the counterpart of (2.23),

$$\Delta E^{(2)} = g'_A g'_B \int \frac{d\mathbf{k}}{(2\pi)^3} \frac{1}{2w} \left[\frac{\langle n_b | \beta_B e^{i\mathbf{k}\cdot\mathbf{r}_B} | n_b \rangle \langle n_a | \beta_A e^{-i\mathbf{k}\cdot\mathbf{r}_A} | n_a \rangle}{-w} + (A \leftrightarrow B) \right]. \tag{2.25}$$

In each case the energy denominator $-w$ arises as the difference between the energy $E_0 = E_{n_a} + E_{n_b}$ of Ψ and the intermediate-state energy which is just $E_0 + w$. The two terms in (2.23) or (2.25) may be represented by the graphs shown in Fig. 3, which are the bound-state counterparts of the time-ordered scattering graphs shown in Fig. 2. From these graphs one sees directly that $E_{int} = E_0 + w$, since the energy of each particle remains unchanged in the intermediate state. (Note that in the corresponding scattering problem this is only true in the c.m. system.)

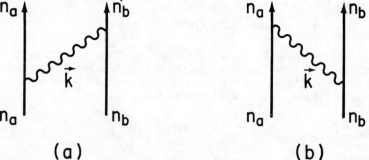

(a)   (b)

Figure 3: The two time-ordered diagrams associated with the level shift arising from one-meson exchange between distinct particles A and B, bound to a common core.

B. Reinterpretation In Terms Of "Potentials"

It is natural to try to re-express both the scattering amplitudes and the level shifts as matrix elements of operators which act directly in configuration space. After all, this is how such quantities would be expressed in nonrelativistic quantum mechanics. I will refer to such operators as "effective potentials" or "effective potential operators". However, as we shall see, the concept of effective potential is fraught with ambiguity – and here I mean actual ambiguity, not just potential ambiguity. Let us first consider what happens if one simply follows one's nose, or rather the noses of our forefathers in this matter, among whom are Breit, Yukawa, *et al.* Because the effective "level-shift potentials" are somewhat simpler to deal with I start with these.

1. Level-shift potentials

The time-honored procedure for obtaining an effective potential from an expression such as (2.23) or (2.25) for a level shift is simply to reverse the orders of integration over the configuration space variables $\mathbf{r}_A$ and $\mathbf{r}_B$ and the momentum $\mathbf{k}$ of the virtual meson.

(a) Spin-0

From (2.23) one gets

$$\Delta E^{(2)} = N^2 \langle n_a, n_b | U_Y(\mathbf{r}) | n_a, n_b \rangle \qquad (2.26a)$$

where, with $\mathbf{r} = \mathbf{r}_A - \mathbf{r}_B$,

$$U_Y(\mathbf{r}) \equiv -g_A g_B \int \frac{d\mathbf{k}}{(2\pi)^3} \frac{e^{-i\mathbf{k}\cdot\mathbf{r}} + e^{i\mathbf{k}\cdot\mathbf{r}}}{2w^2}. \qquad (2.26b)$$

Integration over $\mathbf{k}$ yields

$$U_Y(\mathbf{r}) = \frac{-g_A g_B}{4\pi} \frac{e^{-\mu r}}{r}, \qquad (2.26c)$$

which is a Yukawa-type potential.

(b) Spin-1/2

Following the same procedure for (2.25) one gets a slightly different result,

$$\Delta E^{(2)} = \langle n_a, n_b | \beta_A \beta_B U_Y'(\mathbf{r}) | n_a, n_b \rangle \qquad (2.27a)$$

where U_Y' differs from U_Y only by replacement of the g's by their primed counterparts,

$$U_Y'(\mathbf{r}) = \frac{-g_A' g_B'}{4\pi} \frac{e^{-\mu r}}{r}. \tag{2.27b}$$

Let us compare the result (2.26) for the spin-0 case with the result (2.27) for the spin-1/2 case. Note that in the latter case the level shift is already expressed as the expectation value of an operator, which in this case happens to be a local potential when considered as a mapping of the product Dirac configuration space of A and B, viz.

$$V^{(2)} = \beta_A \beta_B \, U_Y'(r). \tag{2.28}$$

This is not yet true of (2.26a), because of the presence of the state-dependent factor N^2. Can this be fixed up somehow? It will not do to simply incorporate this factor in the wave functions, because these would then no longer be normalized to unity. Note that because of the factor N^2 the result (2.26a) differs from that found by Yukawa; this is not a contradiction because Yukawa considered only infinitely massive point sources for the meson field. Of course, in the weak binding limit $[(E_n - m)/m \ll 1]$ the factor N approaches unity.

One way to obtain an operator for the spin-0 case is to introduce the positive square root, $\mathcal{E}_{op}$, of the Klein-Gordon operator D_{op} defined in Appendix A (Eq. (A3)),

$$\mathcal{E}_{op} = (D_{op})^{\frac{1}{2}} = (\mathbf{p}_{op}^2 + m^2 + G \, \varphi_{ext})^{\frac{1}{2}}. \tag{2.29}$$

Let us denote the corresponding operators for A and B by $\mathcal{E}_A^{op}$ and $\mathcal{E}_B^{op}$; the wave functions $\phi_{n_a}(\mathbf{x})$ and $\phi_{n_b}(\mathbf{x})$ are then eigenfunctions of these operators with eigenvalue E_{n_a} and E_{n_b}, respectively. Now consider the operator $V^{(2)}$ defined by

$$V^{(2)} = y_{op}^{ext} \, U_Y \, y_{op}^{ext} \tag{2.30}$$

where y_{op}^{ext} is defined by

$$y_{op}^{ext} = (m_A m_B / \mathcal{E}_A^{op} \mathcal{E}_B^{op})^{\frac{1}{2}}. \tag{2.31}$$

It is easy to see that (2.26a) can now be written in the form

$$\Delta E^{(2)} = \langle n_a, n_b | V^{(2)} | n_a, n_b \rangle \tag{2.32}$$

so that we have achieved our goal. However, although the operator defined by (2.30) is hermitian, like the one defined by (2.28), unlike the latter it is not local and in fact not even translation invariant. *Mais, c'est la vie* – it does the job.

It may be worth noting that the difference between the spin-0 and spin-1/2 cases is not as large as may appear from the foregoing. The reason is that the use of the Dirac formalism for the spin-1/2 particles hides some of the non-locality lurking in the background. To be precise, if one uses the external-field Dirac equation to express, say in the standard rep for the Dirac matrices, the lower components of the bound-state wave functions in terms of the upper components and writes the level shift as an expectation value involving only the upper components, *i.e.*, in terms of Pauli wave functions, the resulting effective operator would also be both non-local and non-invariant under translations.

2. Scattering potentials

Let us now consider the related problem of expressing the lowest-order scattering amplitudes in terms of effective potentials. The time-honored procedure in this case is simply that of inverse Fourier transformation; this is because the initial and final states are just product plane waves.

To simplify matters I will consider the problem here only from the view point of the c.m. system, *i.e.*, the system in which the total initial and final three-momenta are zero. In this system the initial and final four-momenta have the form

$$p_A = (W_A, \mathbf{p}) \quad , \qquad p_B = (W_B, -\mathbf{p}) \tag{2.33a}$$
$$p'_A = (W_A, \mathbf{p}') \quad , \qquad p'_B = (W_B, -\mathbf{p}') \tag{2.33b}$$

where W_A and W_B are the energies of A and B, respectively, both before and after the scattering,

$$W_A = E_A(\mathbf{p}) \quad , \qquad W_B = E_B(\mathbf{p}) \tag{2.33c}$$

and

$$|\mathbf{p}'| = |\mathbf{p}|. \tag{2.33d}$$

The four-momentum transfer Q defined by (2.8) then takes the form $Q = (0, \mathbf{q})$ where

$$\mathbf{q} = \mathbf{p} - \mathbf{p}', \tag{2.33e}$$

and the invariant t may be written as

$$t = -\mathbf{q}^2. \tag{2.33f}$$

In the c.m. frame the relation (2.6) between the frame-dependent transition amplitude T and the invariant Feynman amplitude M then takes the form, in view of the choice (2.7a),

$$T = M/4W_A W_B. \tag{2.34}$$

We now consider the spin-0 and spin-1/2 cases in turn.

(a) Spin-0

In terms of the reduced coupling constants defined by (2.22) the second-order transition amplitude $T^{(2)}$ is given, in view of (2.9), (2.34) and (2.33f), by

$$T^{(2)} = -g_A g_B (m_A m_B / W_A W_B) \, (q^2 + \mu^2)^{-1}. \tag{2.35}$$

We now introduce a configuration space S_{rel}, spanned by "relative plane waves" $|\mathbf{k}\rangle$, and look for a linear operator $V^{(2)}$ in this space which has the property

$$\langle \mathbf{k}' | V^{(2)} | \mathbf{k} \rangle = T^{(2)} \tag{2.36a}$$

whenever

$$\mathbf{k} = \mathbf{p} \quad , \qquad \mathbf{k}' = \mathbf{p}'. \tag{2.36b}$$

With $\mathbf{p}_{op}$ defined so that $\mathbf{p}_{op} |\mathbf{k}\rangle = \mathbf{k} |\mathbf{k}\rangle$ and

$$W_A^{op} = E_A(\mathbf{p}_{op}) \quad , \qquad W_B^{op} = E_B(\mathbf{p}_{op}) \tag{2.37}$$

we may eliminate the square root factors in (2.36a) by writing, in analogy with (2.30) and (2.31),

$$V^{(2)} = y_{op} U^{(2)} y_{op} \tag{2.38a}$$

where

$$y_{op} = \left[m_A m_B / W_A^{op} W_B^{op} \right]^{\frac{1}{2}}. \tag{2.38b}$$

Substitution of (2.28a) into (2.36a) yields the condition that

$$\langle \mathbf{k}' | U^{(2)} | \mathbf{k} \rangle = -g_A g_B / (q^2 + \mu^2) \tag{2.39}$$

whenever (2.36b) is satisfied. The *simplest* solution of (2.39) is obtained by assuming that in the r-space conjugate to the k-space, with $\langle \mathbf{r} | \mathbf{k} \rangle = \exp(i \mathbf{k} \cdot \mathbf{r})$, the operator $U^{(2)}$ is strictly local,

$$\langle \mathbf{r}' | U^{(2)} | \mathbf{r} \rangle = U^{(2)}(\mathbf{r}) \delta(\mathbf{r}' - \mathbf{r}). \tag{2.40}$$

Then (2.39) reduces to

$$\int d\mathbf{r} \, \exp(i \mathbf{K} \cdot \mathbf{r}) U^{(2)}(\mathbf{r}) = -g_A g_B / (\mathbf{K}^2 + \mu^2) \tag{2.41}$$

where $\mathbf{K} \equiv \mathbf{k} - \mathbf{k}' = \mathbf{q}$. Since $\mathbf{q} = \mathbf{p} - \mathbf{p}'$ can assume all real vector values, even though $|\mathbf{p}| = |\mathbf{p}'|$, so can $\mathbf{K}$. We can therefore invert the relation (2.40) to find

$$U^{(2)}(\mathbf{r}) = U_Y(\mathbf{r}). \tag{2.42}$$

Thus in this case the scattering potential is

$$V^{(2)} = y_{op} U_Y y_{op}. \tag{2.43}$$

On comparison with the level shift potential (2.30) we see that the two are not the same, contrary to what one might have guessed. Some light is shed on this difference by a consideration of the spin-1/2 case.

(b) Spin-1/2

This case can be disposed of rather quickly by using the results for spin-0. In view of (2.17) and (2.34), the second-order transition amplitude is now given in the c.m. system by

$$T^{(2)} = -g'_A g'_B (S/4W_A W_B) \, (q^2 + \mu^2)^{-1} \tag{2.44}$$

where S is a spinor factor,

$$S = (\bar{u}'_A u_A) \, (\bar{u}'_B u_B). \tag{2.45}$$

and the u's are normalized acording to (2.7b). However, in ordinary three-dimensional calculations the free Dirac spinors are normalized to unity via $u^\dagger u = 1$ and correspondingly, the normalization $\int u_n^\dagger(x) u_n(x) = 1$ is used for bound states, as in the Furry expansion used in computing the level-shift. To define a scattering potential $V^{(2)}$ which may be reasonably compared with the level-shift potential (2.28), let us first introduce conventionally normalized spinors $\hat{u}$, with

$$\hat{u}^\dagger \hat{u} = 1. \tag{2.46}$$

We now require that $V^{(2)}$, acting in the space spanned by multi-Dirac spinor wave functions of the form $u_A(\mathbf{k}) u_B(-\mathbf{k}) \exp(i\mathbf{k} \cdot \mathbf{r})$, satisfy, for $\mathbf{k} = \mathbf{p}$, $\mathbf{k}' = \mathbf{p}'$,

$$\langle \mathbf{k}' | \left(\hat{u}'_A{}^\dagger \hat{u}'_B{}^\dagger V^{(2)} \hat{u}_A \hat{u}_B \right) | \mathbf{k} \rangle = T^{(2)} \tag{2.47}$$

where $T^{(2)}$ is given by (2.44). Since $\hat{u}$ is related to u by

$$\hat{u}(\mathbf{k}) = u(\mathbf{k})/2E(\mathbf{k}) \tag{2.48}$$

it follows that we may take $V^{(2)}$ to have the r-space form

$$V^{(2)} = \beta_A \beta_B \, U'_Y(r). \tag{2.49}$$

Thus in this case we have been able to choose a scattering potential which coincides with the level-shift potential (2.28). This was not possible for the spin-0 case.

C. Interim Summary And Discussion

We have seen that for the simple theories considered the effect of one-meson exchange can be represented by effective potentials both for the calculation of the lowest-order c.m. system scattering amplitude and for the lowest-order level shift. When A and B are both spin-1/2 Dirac particles the scattering and level-shift potential can be the same and strictly local. When A and B are both spin-0 Klein-Gordon particles the scattering and level-shift potentials we found were not the same and neither of them was local. In the spin-0 scattering case it is easy to get rid of the non-locality, at the expense of having a potential which depends on a parameter, namely the invariant squared energy s defined by

$$s = (p_A + p_B)^2 = W^2 \tag{2.50}$$

where W is the total energy in the c.m. system,

$$W = E_A(\mathbf{p}) + E_B(\mathbf{p}), \tag{2.51}$$

or, equivalently, on the magnitude $\mathbf{p} = |\mathbf{p}| = |\mathbf{p}'|$ of the c.m. system three-momentum. To this end it is only necessary to replace the operator $\mathbf{p}_{op}$ in the definition of y_{op} in (2.38b) by $\mathbf{p}$ and to define an s-dependent potential $V^{(2)}[s]$ via

$$V^{(2)}[s] = y^2 U_Y \tag{2.52a}$$

where $y = y[s]$ is defined by

$$y = [m_A m_B / E_A(\mathbf{p}) E_B(\mathbf{p})]^{\frac{1}{2}}. \tag{2.52b}$$

This example is illustrative of the fact that effective potentials are far from unique and leads to questions which I have merrily by-passed up to now.

i) What good are effective potentials?

ii) If they are any good, are some better than others?

With regard to the first question, they certainly have psychological value, because of their "anschaulichkeit" and the fact that potentials are familiar from NRQM. Moreover, for the case of the level shift the form involving the potential is likely to be easier to evaluate than the original form which emerges from field theory, so that in this sense they may offer some calculational advantage. However, all this is pretty small potatoes. Their main value lies in the fact that they can be used in Schrödinger-type equations to take into account some higher-order effects, in situations where brute force perturbation theory is inapplicable. With regard to the second question, there is no general answer; the main point to be made is that this question cannot even be addressed until the equation in which the potential is to be used is completely specified. I will return to these issues later on. Let us now study the modifications which are needed when A and B are identical.

D. Effect Of Particle Identity

1. Scattering potentials

The only modification to the preceding analysis when A and B represent identical particles is that there is now a new diagram, the exchange diagram shown in Fig. 4. The corresponding amplitude, denoted by $M_{ex}^{(2)}$, can be obtained from the amplitude $M^{(2)}$, which I will now call the "direct" amplitude and denote by $M_{dir}^{(2)}$, simply by interchanging the labels A and B on the final four-momenta and spins and supplying a factor -1 for fermions. The four-momentum k of the virtual boson is now $p_A - p_B'$ so that t is replaced by the squared crossed-momentum transfer u, defined by

$$u = (p_A - p_B')^2 = (p_B - p_A')^2. \tag{2.53}$$

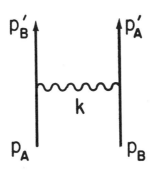

Figure 4: Feynman diagram describing the exchange scattering amplitude arising from one-meson transfer between identical particles A and B.

(a) Identical spin-0 bosons

On putting $G_A = G_B = G$ in (2.9) we get

$$M_{dir}^{(2)} = \frac{G^2}{t - \mu^2} \quad , \qquad M_{ex}^{(2)} = \frac{G^2}{u - \mu^2} . \tag{2.54}$$

The total second-order amplitude is then

$$M_{tot}^{(2)} = M_{dir}^{(2)} + M_{ex}^{(2)}. \tag{2.55}$$

We now ask (i) Can we include the exchange term by adding a new operator to the original potential given by (2.43), which I now denote by $V_{dir}^{(2)}$? (ii) *Must* we add a new operator to $V_{dir}^{(2)}$ in order to include the exchange term? The answer is: YES and NO!

(i) YES: Note that in the c.m. system, for $m_A = m_B = m$ the quantity u is obtainable from t by the substitution $\mathbf{p} \to -\mathbf{p}$ or $\mathbf{p}' \to -\mathbf{p}'$:

$$t = -(\mathbf{p} - \mathbf{p}')^2 \quad , \qquad u = -(\mathbf{p} + \mathbf{p}')^2. \tag{2.56}$$

Let us introduce the space-exchange operator P_{ex}, defined by

$$P_{ex}f(\mathbf{r}_A, \mathbf{r}_B) = f(\mathbf{r}_B, \mathbf{r}_A),\qquad (2.57a)$$

and note that, for $\mathbf{r} = \mathbf{r}_A - \mathbf{r}_B$,

$$P_{ex}e^{i\mathbf{p}\cdot\mathbf{r}} = e^{-i\mathbf{p}\cdot\mathbf{r}}.\qquad (2.57b)$$

From (2.54) and (2.56) we see that with

$$V_{dir}^{(2)} = y_{op}U_Y y_{op}\qquad (2.58)$$

evaluated for $G_A = G_B, m_A = m_B$, the operator $V_{dir}^{(2)}P_{ex}$ will reproduce the exchange term. Thus the choice

$$V_{tot}^{(2)} = V_{dir}^{(2)}(1 + P_{ex})\qquad (2.59)$$

does the job. Note that because P_{ex} is a non-local operator so is $V_{tot}^{(2)}$. So there *appears* to be an exchange force present, and non-locality is present even apart from that associated with the y_{op} factors.

(ii) NO: The concept of "exchange force" is *not* one which is properly associated with identical particles, but, rather one in which two distinct particles can change their identity, *e.g.*, $p \to n$ and $n \to p$ via $\pi^{\pm}$ exchange. The reason is simple: All such apparent effects can be incorporated by using properly symmetrized or antisymmetrized wave functions, *i.e.*, in the boson case

$$|\mathbf{p}\rangle_{sym} = (e^{i\mathbf{p}\cdot\mathbf{r}} + e^{-i\mathbf{p}\cdot\mathbf{r}})/\sqrt{2}.\qquad (2.60)$$

Historically, one does speak of "exchange force" even for electrons, but this is really an abuse of language – in all such cases one is dealing simply with an exchange *integral*. For the case at hand we may write

$$T^{(2)} = {}_{sym}\langle\mathbf{p}'|V_{dir}^{(2)}|\mathbf{p}\rangle_{sym}\qquad (2.61)$$

so that the original potential need not be changed.

(b) Identical spin-1/2 fermions

Similarly, for fermions, we put $g'_A = g'_B = g'$ and incorporate the exchange term by using an antisymmetrical wave function, with spinors included, viz.

$$|\mathbf{p};\lambda_1\lambda_2\rangle_{a.sym} \equiv \frac{1}{\sqrt{2}}\left[u_A(\mathbf{p}, \lambda_1)u_B(-\mathbf{p}, \lambda_2)e^{i\mathbf{p}\cdot\mathbf{r}} - u_A(-\mathbf{p}, \lambda_2)u_B(\mathbf{p}, \lambda_1)e^{-i\mathbf{p}\cdot\mathbf{r}}\right].$$
$$(2.62)$$

Thus

$$T^{(2)} = {}_{a.sym}\langle p';\lambda'_1\lambda'_2|V_{dir}^{(2)}|\mathbf{p};\lambda_1\lambda_2\rangle_{a.sym}\qquad (2.63)$$

where

$$V_{dir}^{(2)} = \frac{-g'^2}{4\pi} \frac{e^{-\mu r}}{r} \beta_A \beta_B.$$ (2.64)

2. Level-shift potentials

Let us now consider the analogous problem for the level shift. On using old-fashioned perturbation theory and repeating the analysis carried out previously for distinct particles, we find that the total second-order level-shift $\Delta E^{(2)}$ is given as a sum of a direct term $\Delta E_{dir}^{(2)}$ and an exchange term $\Delta E_{ex}^{(2)}$:

$$\Delta E^{(2)} = \Delta E_{dir}^{(2)} + \Delta E_{ex}^{(2)}.$$ (2.65a)

The direct term has the same form as in the case of distinct particles, viz. (2.27) or (2.32), specialized to equal coupling constants and masses. In particular for the fermion case we have

$$\Delta E_{dir}^{(2)} = \langle n_a, n_b | V_{dir}^{(2)} n_a, n_b \rangle$$ (2.65b)

where $V_{dir}^{(2)}$ is given by (2.64). The exchange contribution is given in the fermion case, by

$$\Delta E_{ex}^{(2)} = -g'^2 \, \mathcal{R}e \int \frac{dk}{(2\pi)^3} \frac{1}{2w} \left[\frac{\langle n_b | \beta_B e^{i\mathbf{k}\cdot\mathbf{r}_B} | n_b \rangle \langle n_b | \beta_A e^{-i\mathbf{k}\cdot\mathbf{r}_A} | n_a \rangle}{D_a} \right. $$
$$\left. + \frac{\langle n_b | \beta_A e^{i\mathbf{k}\cdot\mathbf{r}_A} | n_a \rangle \langle n_a | \beta_B e^{-i\mathbf{k}\cdot\mathbf{r}_B} | n_b \rangle}{D_b} \right]$$ (2.65c)

where the D's are energy denominators and "$\mathcal{R}e$" means "real part". The two terms in (2.64) correspond to the graphs shown in Fig. 5. From these we see that

$$D_a = (E_{n_a} + E_{n_b}) - (E_{n_b} + E_{n_b} + w) = \Delta_{ab} - w,$$
$$D_b = (E_{n_a} + E_{n_b}) - (E_{n_a} + E_{n_a} - w) = -\Delta_{ab} - w,$$

where Δ_{ab} is a virtual energy-transfer defined by

$$\Delta_{ab} \equiv E_{n_a} - E_{n_b}.$$ (2.66)

354

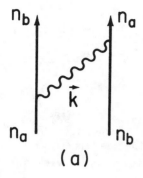 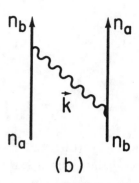

(a)   (b)

Figure 5: The two time-ordered graphs associated with the exchange contribution to the level-shift arising from one-meson transfer between identical particles A and B, bound to a common core.

Thus (2.64) reduces to

$$\Delta E_{ex}^{(2)} = -\langle n_b n_a | \beta_A \beta_B U_Y'(r; \Delta_{ab}) | n_a n_b \rangle \qquad (2.67a)$$

where

$$U_Y'(r; \Delta) = g'^2 \, P \int \frac{dk}{(2\pi)^3} \frac{1}{2w} \left(\frac{e^{-i k \cdot r}}{\Delta - w} + \frac{e^{i k \cdot r}}{-\Delta - w} \right). \qquad (2.67b)$$

and P denotes a principal value integration. On letting $k \to -k$ in the second term the denominators can be combined to give a factor $(k^2 + \mu^2 - \Delta^2)^{-1}$ and the resulting integral can be done exactly:

$$U_Y'(r; \Delta) = \frac{g'^2}{4\pi} \frac{e^{-\sqrt{\mu^2 - \Delta^2} r}}{r} \qquad (\mu^2 > \Delta^2), \qquad (2.68a)$$

and

$$U_Y'(r; \Delta) = \frac{g'^2}{4\pi} \frac{\cos(\sqrt{\Delta^2 - \mu^2} r)}{r} \qquad (\mu^2 < \Delta^2). \qquad (2.68b)$$

We see that for $\mu^2 > \Delta^2$, the function U_Y' is of the Yukawa-type, i.e., a decaying exponential with decay length $(\mu^2 - \Delta^2)^{-\frac{1}{2}}$, whereas for $\mu^2 < \Delta^2$ the function U_Y' oscillates, with oscillation length $(\Delta^2 - \mu^2)^{-\frac{1}{2}}$. In particular, if $\mu = 0$, as for QED, only the second alternative is possible and

$$U_Y'(r; \Delta) \to U'(r; \Delta) = \frac{-g'^2}{4\pi} \frac{\cos \Delta r}{r}. \qquad (2.68c)$$

Thus we have finally arrived at an object similar to the Δ-dependent operators $V_I(\mathbf{r}; \Delta)$ and $V_{II}(\mathbf{r}; \Delta)$ mentioned in Sec. IA. We see now that such dependence arises only in the case of identical particles.

Note that we have not yet succeeded in expressing the exchange level-shift ΔE_{ex} as the expectation value of an explicitly given, state-independent operator – one must supply the order: "When $U'_Y(\mathbf{r}; \Delta)$ acts on the product state $|n_a, n_b\rangle$, replace Δ by Δ_{ab}." This is what I meant by saying that the $V(r; \Delta)$ come equipped with instructions for their use. We are dealing with a family of operators or, equivalently, with an operator which depends on a parameter, whose value is determined only for product states. This is an awkward state of affairs. One way to proceed is to introduce another tool, that of "ordered exponential".

3. Ordered exponentials

Consider any two operators A and B, and an arbitrary product C of powers of these operators, $e.g.$, $C = BA^3B^2AB^5$. For any such product C we define an associated ordered product $C_{A,B}$ by moving all A's to the left and all B's to the right, $e.g.$,

$$(BA^3B^2AB^5)_{A,B} = A^4B^8.$$

It's just like the normal-ordering process in field theory, except that A and B are treated symmetrically, $e.g.$, $(AB)_{B,A} = BA$. With $(\,)_{A,B}$ defined to be distributive over sums, we then have

$$(e^{iAB})_{A,B} \equiv \sum_{n=0}^{\infty} \frac{(iAB)^n_{A,B}}{n!} = \sum_{n=0}^{\infty} \frac{i^n A^n B^n}{n!}. \tag{2.69}$$

Instead of $(e^{iAB})_{A,B}$ I will just write e^{iAB}_{ord}, with the understanding that whatever factor stands to the *left* in the exponent to be moved to the left in the ordering process.

(a) Application to level shift.

With the help of ordered exponentials it is relatively straightforward to express an expectation value such as (2.66) in terms of *bona fide* operators. For simplicitly, let us concentrate on the case $\mu = 0$. Ignore the matrix factors for the moment. Since

$$\frac{\cos(\Delta r)}{r} = \frac{1}{2}\left(\frac{e^{i\Delta r}}{r} + \frac{e^{-i\Delta r}}{r}\right)$$

our problem reduces to expressing the expectation value of functions such as $\exp(i\Delta_{ab}r)/r$ in terms of *bona fide* operators. Note that the labels n_a and n_b now run through the same set of values, and with

$$h_A \equiv \boldsymbol{\alpha}_A \cdot \mathbf{p}_A + \beta_A m + g'\varphi_{ext}(\mathbf{r}_A) \quad , \quad h_B = \boldsymbol{\alpha}_B \cdot \mathbf{p}_B + \beta_B m + g'\varphi_{ext}(\mathbf{r}_B)$$

we have

$$h_A \, u_{n_a}(\mathbf{r}_A) = E_{n_a} u_{n_a}(\mathbf{r}_a) \quad , \quad h_A u_{n_b}(\mathbf{r}_A) = E_{n_b} u_{n_b}(\mathbf{r}_A)$$

and a similar pair of equations with $A \to B$. Since

$$\frac{e^{i\Delta_{ab}r}}{r} = e^{-iE_{n_b}r}\frac{1}{r}e^{iE_{n_a}r}$$

it follows that in terms of ordered exponentials we have

$$\langle n_b, n_a| \frac{e^{i\Delta_{ab}r}}{r} |n_a, n_b\rangle = \langle n_b, n_a| e_{ord}^{-ih_A r} \frac{1}{r} e_{ord}^{irh_A} |n_a, n_b\rangle \tag{2.70a}$$

and, similarly

$$\langle n_b, n_a| \frac{e^{-i\Delta_{ab}r}}{r} |n_a, n_b\rangle = \langle n_b, n_a| e_{ord}^{-ih_B r} \frac{1}{r} e_{ord}^{irh_B} |n_a, n_b\rangle. \tag{2.70b}$$

On reinstating the factors β_A and β_B we see that $\Delta E_{ex}^{(2)}$ may be written in the form

$$\Delta E_{ex}^{(2)} = -\langle n_b, n_a| V^{(2)} |n_a, n_b\rangle \tag{2.71a}$$

where

$$V^{(2)} = \frac{1}{2}\left[e_{ord}^{-ih_A r} V_{dir}^{(2)} e_{ord}^{irh_A} + e_{ord}^{-ih_B r} V_{dir}^{(2)} e_{ord}^{irh_B} \right] \tag{2.71b}$$

and $V_{dir}^{(2)}$ is defined by (2.64). It is easy to verify that

$$\langle n_a, n_b| V^{(2)} |n_a, n_b\rangle = \langle n_a, n_b| V_{dir}^{(2)} |n_a, n_b\rangle \tag{2.72}$$

so that finally we get

$$\Delta E^{(2)} = \langle n_a, n_b| (1 - P_{ex}) V^{(2)} |n_a, n_b\rangle \tag{2.73}$$

or, equivalently,

$$\Delta E^{(2)} = {}_{a.sym.}\langle n_a, n_b| V^{(2)} |n_a, n_b\rangle_{a.sym.} \tag{2.74}$$

Thus we have achieved our goal of writing the level shift in terms of a *bona fide* operator. Of course, this is not the only way to do it, but it appears to be one of the simplest.

In an entirely similar manner one can show that the total level shift in the two-boson case can be written in the form

$$\Delta E^{(2)} = {}_{sym}\langle n_a, n_b| V^{(2)} |n_a, n_b\rangle_{sym} \tag{2.75}$$

where now $V^{(2)}$ is defined by

$$V^{(2)} = \frac{1}{2}\left[e_{ord}^{-i\epsilon_A^{op} r} V_{dir}^{(2)} e_{ord}^{ir\epsilon_A^{op}} + e_{ord}^{-i\epsilon_B^{op} r} V_{dir}^{(2)} e_{ord}^{ir\epsilon_B^{op}} \right] \tag{2.76}$$

with $V_{dir}^{(2)}$ defined by setting $G_A = G_B$ and $m_A = m_B$ in (2.30) and (2.31).

Let us take stock: On the minus side, the effective level-shift potentials we have found are fairly complicated, because of the presence of the ordered exponentials. On the plus side, one can imagine using them in a relativistic many-body calculation, as part of the interaction terms in a configuration space Hamiltonian (with projection operators included in the fermion case), without encountering any ambiguity: The matrix elements of these operators are not defined just between product states. The energy-levels of

$$H = \sum_i h_{op}^{ext}(i) + \sum_{i<j} V^{(2)}(i,j) \tag{2.77}$$

are guaranteed to be exact to all orders in the external field and to leading order in the strength of one-meson exchange.

Another point to note is that the potentials as constructed are formally hermitian, since, for hermitian A and B,

$$(e_{ord}^{iAB})^\dagger = e_{ord}^{-iBA}. \tag{2.78}$$

Note finally that when the particles are identical, the effective potential is now nonlocal even in the fermion case.

III. QUANTUM ELECTRODYNAMICS

We are now ready to consider some of the other issues raised in Sec. IIA, especially that of gauge-dependence – let's plunge right in and deal with the case of greatest physical interest for the purpose of this workshop, spinor QED.

A. Scattering potentials

The amplitude $M^{(2)}$ for one-photon exchange between spin-1/2 particles A and B, treated as distinct for the while, is given by

$$-iM^{(2)} = \bar{u}_A'(-ie_A\gamma^\mu)u_A \; \bar{u}_B'(-ie_B\gamma^\mu)u_B \tilde{D}_{\mu\nu} \tag{3.1a}$$

where $\tilde{D}_{\mu\nu}$ is the photon propagator. This may be rewritten in the form

$$M^{(2)} = \bar{u}_A'\bar{u}_B' \; F \; u_A u_B \tag{3.2a}$$

where F is defined by

$$F = -i \, e_A e_B \, \gamma_A^\mu \gamma_B^\nu \tilde{D}_{\mu\nu}. \tag{3.2b}$$

If now, on the one hand, we use Feynman propagator (1.4), F takes the form

$$F_{II} = \frac{-e_A e_B \gamma_A \cdot \gamma_B}{t}, \tag{3.3}$$

where, as before, $t = Q^2$ and $Q = p_A - p'_A$ is the four-momentum transfer. We now repeat the procedure used in Sec. IIB.2: We evaluate F_{II} in the c.m. system, where $t = -\mathbf{q}^2$ with $\mathbf{q} = \mathbf{p} - \mathbf{p}'$ the three-momentum transfer,

$$F_{II} \to F_{II}^{cm} = e_A e_B \, \gamma_A \cdot \gamma_B / \mathbf{q}^2 \qquad (3.4a)$$

and take the Fourier transform of $\beta_A \beta_B F_{II}$ with respect to $\mathbf{q}$ to define V_{II}:

$$V_{II}(\mathbf{r}) = (2\pi)^{-3} \int d\mathbf{q} \, e^{i\mathbf{q}\cdot\mathbf{r}} \beta_A \beta_B F_{II}^{cm} \qquad (3.4b)$$

This yields

$$V_{II}(\mathbf{r}) = (1 - \boldsymbol{\alpha}_A \cdot \boldsymbol{\alpha}_B) U_C(r) \qquad (3.5)$$

where U_C is the Coulomb potential defined by (1.2). If, on the other hand, we use the Coulomb-gauge propagator (1.5) we get instead

$$F_I = e_A e_B \left[\frac{\beta_A \beta_B}{Q^2} + \frac{\gamma_A \gamma_B - \gamma_A \cdot \hat{Q} \gamma_B \cdot \hat{Q}}{Q^2} \right] \qquad (3.6)$$

an object quite different in appearance from F_{II}. In the c.m. system it takes the form

$$F_I \to F_I^{c.m.} = e_A e_B \left[\beta_A \beta_B - (\gamma_A \cdot \gamma_B - \gamma_A \cdot \hat{q} \gamma_B \cdot \hat{q}) \right] / \mathbf{q}^2 \qquad (3.7a)$$

and the associated potential is defined by

$$V_I = (2\pi)^{-3} \int d\mathbf{q} e^{i\mathbf{q}\cdot\mathbf{r}} \beta_A \beta_B F_I^{cm}. \qquad (3.7b)$$

This yields

$$V_I(\mathbf{r}) = \left[1 - \frac{1}{2}(\boldsymbol{\alpha}_A \cdot \boldsymbol{\alpha}_B + \boldsymbol{\alpha}_A \hat{\cdot} r \boldsymbol{\alpha}_B \cdot \hat{r}) \right] U_C(r). \qquad (3.8)$$

The potentials V_I and V_{II} coincide, respectively, with the Coulomb-Breit and Coulomb-Moller-Gaunt potential discussed in Sec. IA; for the case of identical fermions, exchange can be taken into account by using antisymmetrized states, as before.

We see that the difference between V_I and V_{II} indeed appears to arise from the difference between the two photon propagators. We have, it seems, encountered our first example of a *gauge dependence of the potential*. Let us postpone discussion of this point and consider next the associated level-shift problem.

B. Level-shift potentials

Let us deal with the case of identical fermions right from the start, with $e_A = e_B = -e$, the electron charge and $m_A = m_B = m$, the electron mass, and consider two electrons bound to a common core. To be specific, let us think of a helium-like ion with $Z \gg 1$. We use the Furry picture and treat all electron-electron interaction as a perturbation. To lessen the profusion of subscripts, let $a, b, \ldots n$, denote quantum numbers of one-electron states and let $C^\dagger(n)$ denote the operator which creates an electron with bound-state Dirac wave function $u_n(\mathbf{x})$, an eigenfunction with eigenvalue E_n of

$$h_{ext} = \boldsymbol{\alpha} \cdot \mathbf{p}_{op} + \beta m + V_{ext}. \tag{3.9}$$

The level shift ΔE in a state $\Psi_{ab}^{(0)} = C^\dagger(a)C^\dagger(b)|vac\rangle$, arising from one-photon exchange between the electrons, is then given, for the case (II) of covariant quantization of the electromagnetic field (indefinite metric and Feynman gauge), on use of stationary-state perturbation theory by

$$\Delta E_{II} = \Delta E_{II}^{dir} + \Delta E_{II}^{ex}, \tag{3.10}$$

where, with $w = |\mathbf{k}|$,

$$\Delta E_{II}^{dir} = e^2 \int \frac{d\mathbf{k}}{(2\pi)^3 2w} \left(\frac{\Gamma_{bb}^\nu(-\mathbf{k})\Gamma_{aa}^\mu(\mathbf{k})}{-w} + \frac{\Gamma_{aa}^\mu(-\mathbf{k})\Gamma_{bb}^\nu(\mathbf{k})}{-w} \right) (-g_{\mu\nu}) \tag{3.11a}$$

is the direct term and, with a principal-value integration understood,

$$\Delta E_{II}^{ex} = -e^2 \int \frac{d\mathbf{k}}{(2\pi)^3 2w} \left(\frac{\Gamma_{ab}^\nu(-\mathbf{k})\Gamma_{ba}^\mu(\mathbf{k})}{E_a - E_b - w} + \frac{\Gamma_{ba}^\mu(-\mathbf{k})\Gamma_{ab}^\nu(\mathbf{k})}{E_b - E_a - w} \right) (-g_{\mu\nu}) \tag{3.11b}$$

is the exchange term. Here the Γ's are bound-state vertex functions defined by

$$\Gamma_{dc}^\mu(\mathbf{k}) = \langle d|\alpha^\mu e^{-i\mathbf{k}\cdot\mathbf{x}}|c\rangle \equiv \int d\mathbf{x} u_d^\dagger(\mathbf{x})\alpha^\mu e^{-i\mathbf{k}\cdot\mathbf{x}} u_c(\mathbf{x}) \tag{3.12}$$

with $(\alpha^\mu) = (1, \boldsymbol{\alpha})$. For the case (I) of Coulomb gauge, we have instead

$$\Delta E_I = \Delta E_I^{dir} + \Delta E_I^{ex}, \tag{3.13}$$

where now

$$\Delta E_I^{dir} = e^2 \int \frac{d\mathbf{k}}{(2\pi)^3} \left[\frac{\Gamma_{bb}^0(-\mathbf{k})\Gamma_{ab}^0(\mathbf{k})}{\mathbf{k}^2} \right.$$
$$\left. + \frac{\tilde{\delta}_{ij}}{2w} \left(\frac{\Gamma_{bb}^j(-\mathbf{k})\Gamma_{aa}^i(\mathbf{k})}{-w} + \frac{\Gamma_{aa}^i(-\mathbf{k})\Gamma_{bb}^j(\mathbf{k})}{-w} \right) \right] \tag{3.14a}$$

is the direct term and

$$\Delta E_I^{ex} = -e^2 \int \frac{d\mathbf{k}}{(2\pi)^3} \left[\frac{\Gamma_{ab}^0(-\mathbf{k})\Gamma_{ba}^0(\mathbf{k})}{\mathbf{k}^2} \right.$$
$$\left. + \frac{\tilde{\delta}_{ij}}{2w} \left(\frac{\Gamma_{ab}^j(-\mathbf{k})\Gamma_{ba}^i(\mathbf{k})}{E_a - E_b - w} + \frac{\Gamma_{ba}^i(-\mathbf{k})\Gamma_{ab}^j(\mathbf{k})}{E_b - E_a - w} \right) \right] \tag{3.14b}$$

is the exchange term, with $\tilde{\delta}_{ij} = \delta_{ij} - \hat{k}_i \hat{k}_j$. The first line in either (3.14a) or (3.14b) represents the effect of the Coulomb interaction and the second line the effect of transverse-photon exchange. From the definition (3.12) it follows that, *for V_{ext} a local potential,*

$$\mathbf{k} \cdot \mathbf{\Gamma}_{dc} = (E_d - E_c)\Gamma^0_{dc}. \tag{3.15}$$

Using (3.15) in (3.11a) and (3.11b), one can show straightforwardly that

$$\Delta E_I^{dir} = \Delta E_{II}^{dir} \quad , \quad \Delta E_I^{ex} = \Delta E_{II}^{ex}, \tag{3.16}$$

as is required by gauge invariance. Thus, in this context the choice of gauge is irrelevant.

The two well-known forms for an effective electron-electron interaction in configuration space may be obtained from either of the (equal!) expressions ΔE_I and ΔE_{II} by reversing the order of integration over $\mathbf{k}$ and electron coordinates, in (3.14a,b) and (3.11a,b). With

$$\Delta_{ab} = E_a - E_b$$

this traditional procedure yields

$$\Delta E_I^{dir} = \langle ab | V_I(\mathbf{r}) | ab \rangle \tag{3.17a}$$
$$\Delta E_I^{ex} = -\langle ba | V_I(\mathbf{r}; \Delta_{ab}) | ab \rangle \tag{3.17b}$$

where $V_I(\mathbf{r})$ and $V_I(\mathbf{r}; \Delta)$ are defined by (1.1) and (1.7) respectively, and

$$\Delta E_{II}^{dir} = \langle ab | V_{II}(\mathbf{r}) | ab \rangle, \tag{3.18a}$$
$$\Delta E_{II}^{ex} = -\langle ba | V_{II}(\mathbf{r}; \Delta_{ab}) | ab \rangle, \tag{3.18b}$$

where $V_{II}(\mathbf{r})$ and $V_{II}(\mathbf{r}; \Delta)$ are defined by (1.3) and (1.6a) respectively, all with $e_A = e_B = -e$. We have thus recovered the "competing potentials" of Sec. IA, together with the required instruction for the use of $V_I(\mathbf{r}; \Delta)$ and $V_{II}(\mathbf{r}; \Delta)$ implicit in (3.17b) and (3.18b).

Note that for $\Delta \to 0$, $V_i(\mathbf{r}; \Delta)$ tends to $V_i(\mathbf{r})$:

$$V_I(\mathbf{r}; 0) = V_I(\mathbf{r}) \quad , \quad V_{II}(\mathbf{r}; 0) = V_{II}(\mathbf{r}). \tag{3.19}$$

Further, the $V_i(\mathbf{r}; \Delta)$ are related by the identity

$$V_{II}(\mathbf{r}; \Delta) - V_I(\mathbf{r}; \Delta) = A + \Delta^{-2} [h_{ext}(1), [h_{ext}(2), A]] \tag{3.20a}$$

where

$$A = \frac{e^2 [\cos(\Delta r) - 1]}{r}, \tag{3.20b}$$

and the $\Delta = 0$ limit of this relation has the form

$$V_{II}(\mathbf{r}) - V_I(\mathbf{r}) = \frac{-e^2}{2} [h_{ext}(1), [h_{ext}(2), r]]. \tag{3.21}$$

As a check, one can verify first equality in (3.16) by using (3.21) and the second equality in (3.16) by use of (3.20a).

Note also that the $V_i(r; \Delta)$ have made their appearance only in connection with identical particles and only for the level shift, as in Sec. IID. Because of the presence of the parameter Δ they cannot, as they stand, be put into a Schrödinger-type of equation to account for some higher-order effects. By use of the methods described in Sec. IID they can be extended into *bona fide* operators. Such an extension is briefly described in Appendix B.

IV. DISCUSSION

So much for brute force calculation. Let us now discuss and try to clarify some of the issues which have been left hanging.

A. Gauge dependence of the potentials

Let us consider, for simplicity, distinguishable fermions A and B, and examine more closely the source of the difference between the scattering potentials $V_{II}(\mathbf{r})$ and $V_I(\mathbf{r})$, defined by (3.5) and (3.8), respectively. Let us verify first that M_I, the Feynman amplitude as computed in the Coulomb gauge is both (i) Lorentz invariant and (ii) equal to the amplitude M_{II}, which is computed in the Feynman gauge and is manifestly Lorentz invariant. To this end note that the initial and final Dirac spinors u satisfy

$$\not{p}_A u_A = m_A u_A \,, \quad \not{p}'_A u'_A = m_A u'_A \tag{4.1a}$$

$$\not{p}_B u_B = m_B u_B \,, \quad \not{p}'_B u'_B = m_B u'_{n_B} \tag{4.1b}$$

with $\not{p} = \gamma \cdot p$, so that

$$\bar{u}'_A \gamma \cdot Q u_A = Q_0 \bar{u}'_A \beta_A u_A \,, \quad \bar{u}'_B \gamma \cdot Q u_B = Q_0 \bar{u}'_B \beta_B \bar{u}_B. \tag{4.2}$$

It follows that when sandwiched between the spinor products $u_A u_B$ and $\bar{u}'_A \bar{u}'_B$ the matrix F_I defined by (3.6) may be replaced by

$$F'_I = e_A e_B \left[\frac{\beta_A \beta_B}{Q^2} - \frac{\gamma_A \cdot \gamma_B}{Q^2} - \left(\frac{Q_0^2}{Q^2} \right) \frac{\beta_A \beta_B}{Q^2} \right]$$

Since the coefficient of $\beta_A \beta_B$ inside the square bracket is

$$(1/Q^2) \left(1 - \frac{Q_0^2}{Q^2} \right) = -\frac{1}{Q^2}$$

we see that

$$F'_I = -\frac{e_A e_B \gamma_A \cdot \gamma_B}{Q^2} = F_{II}. \tag{4.3}$$

This shows that M_I is equal to M_{II} and, *a fortiori*, that M_I is Lorentz invariant.

The difference between $V_{II}(\mathbf{r})$ and $V_I(\mathbf{r})$ therefore arises solely from the fact that the matrices F_{II}^{cm} and F_I^{cm}, while both yielding the same on-shell scattering amplitudes, are different matrix functions of the c.m. system three-momentum transfer $\mathbf{q}$. The reason that more than one such matrix function can exist is not that one can use different gauges for the calculation of M – we have just seen that both of the gauges in question give rise to the same M – but rather that the Dirac spinors u_A, $u'_A, \ldots$ are not free four-component objects, being constrained by the relation $\not{p}u = mu$. Thus the difference arises basically from the fact that the matrix function F is not uniquely determined by the physical amplitude M and the decision to use one or another choice for F is what determines V.

To see this aspect of the ambiguity in the potential as clearly as possible, consider the reduction of the F matrices to Pauli form. An efficient way to carry this out is to note that with projection matrices $\beta_\pm$ defined by

$$\beta_\pm = \frac{1 \pm \beta}{2} \tag{4.4a}$$

the relation $\not{p}u = mu$ allows one to write the spinor u in terms of the projected spinor $u^{(+)}$, defined by

$$u^{(+)} = \beta_+ u, \tag{4.4b}$$

in the form

$$u = S(\mathbf{p})u^{(+)}. \tag{4.4c}$$

Here $S(\mathbf{p})$ is defined by

$$S(\mathbf{p}) = N(\mathbf{p})\left(1 + \frac{\boldsymbol{\alpha} \cdot \mathbf{p}}{m + E(\mathbf{p})}\right)\beta_+, \quad N(\mathbf{p}) = \left(\frac{(E(\mathbf{p}) + m)}{2E(\mathbf{p})}\right)^{\frac{1}{2}} \tag{4.4d}$$

and is pseudo-unitary, $S^\dagger(\mathbf{p})S(\mathbf{p}) = \beta_+$. In the standard representation for the Dirac matrices $u^{(+)}$ then has the form

$$u^{(+)} = \begin{pmatrix} \chi \\ 0 \end{pmatrix} \tag{4.4e}$$

where χ is a Pauli spinor, normalized to unity.

When expressed in terms of Pauli spinors the two versions of the amplitude $M^{(2)}$ then take the form, in an obvious notation,

$$M_i^{(2)} = (\chi'_A \chi'_B)^\dagger F_i^{red} \chi_A \chi_B \qquad (i = I, II) \tag{4.5a}$$

where the reduced F's are defined by

$$F_i^{red} = \text{``}(S'_A S'_B)^\dagger F_i^{c.m.} S_A S_B.\text{''} \tag{4.5b}$$

363

Here the quotes on the right-hand side mean that after the $\boldsymbol{\alpha}$ matrices are eliminated, by use of the relation

$$\boldsymbol{\alpha} \cdot \mathbf{a}\boldsymbol{\alpha} \cdot \mathbf{b} = \mathbf{a} \cdot \mathbf{b} + i\boldsymbol{\sigma}^D \cdot \mathbf{a} \times \mathbf{b}$$

where $\boldsymbol{\sigma}^D$ is the double of the Pauli matrix vector $\boldsymbol{\sigma}$, and the β's are replaced by unity, $\boldsymbol{\sigma}^D$ is to be replaced by $\boldsymbol{\sigma}$. The F_i^{red} can thus be expressed solely in terms of $\mathbf{p}, \mathbf{p}'$ and the Pauli matrices $\boldsymbol{\sigma}_A$ and $\boldsymbol{\sigma}_B$: $F_i^{red} = F_i^{red}[\mathbf{p}, \mathbf{p}', \boldsymbol{\sigma}_A, \boldsymbol{\sigma}_B]$. You can now verify that, on the energy shell, *i.e.*, for $|\mathbf{p}| = |\mathbf{p}'|$, we have

$$F_I^{red} = F_{II}^{red}. \tag{4.6}$$

Thus in the Pauli framework there is only one matrix function which yields the amplitude $M^{(2)}$ when sandwiched between Pauli spinors. *Any freedom in defining the second-order potential then has nothing to do with the original choice of gauge but rather with how one chooses to handle the dependence on $\mathbf{p}$ and $\mathbf{p}'$ – the so-called gauge dependence of the potential has disappeared.*

Yet another way to emphasize this point is to consider the case of one-photon exchange in scalar QED. Here the second-order amplitude $M^{(2)}$ is given in the Feynman gauge, by

$$M_{II}^{(2)} = \frac{-e_A e_B P_A \cdot P_B}{Q^2} \quad (P_A = p_A + p_A', \ P_B = p_B + p_B') \tag{4.7a}$$

and in the Coulomb gauge by

$$M_I^{(2)} = e_A e_B \left[\frac{P_A^0 P_B^0}{Q^2} + \frac{(\mathbf{P}_A \cdot \mathbf{P}_B - \mathbf{P}_A \cdot \hat{Q}\mathbf{P}_B \cdot \hat{Q})}{Q^2} \right] \tag{4.7b}$$

Since, as is easily seen, $M_I^{(2)} = M_{II}^{(2)}$ on the energy shell, as is indeed required by gauge invariance, the question of gauge dependence does not arise at all, if one defines the potential by starting with the on-shell amplitude.

This last example highlights the fact that the real source of the difference between the potentials V_I and V_{II} is that they correspond to different ways of going off the energy shell. In spinor QED this feature is masked by the "hidden" constraints on the Dirac spinors. A systematic exposition of these matters is given in a recent paper by G. Feinberg and myself, the first outlining a gauge-independent approach to the relativistic two-body problem[2] and the second dealing with the definition and calculation of the two-photon exchange potential for scalar QED.[3] A third paper will deal with the extension of this work to spinor QED.[4]

B. Which potential is "better"?

However one interprets this question, it can only be answered by using the potentials in a nontrivial way, that is, not just by computing the second-order scattering amplitude, where by construction they give the same result. As already discussed, the real purpose of computing potentials is that they can be used in Schrödinger-type equations to take into account, at least partially, higher-order effects. Let us therefore compare the "no-pair" two-body equations[5]

$$h_I\phi = E\phi, \quad h_I = \left(\boldsymbol{\alpha}_A \cdot \mathbf{p}_{op} + \beta_A m_A\right) + \left(-\boldsymbol{\alpha}_B \cdot \mathbf{p}_{op} + \beta_B m_B\right) + \wedge_{++} V_I \wedge_{++}$$
$$(4.8a)$$

and

$$h_{II}\phi = E\phi, \quad h_{II} = \left(\boldsymbol{\alpha}_A \cdot \mathbf{p}_{op} + \beta_A m_A\right) + \left(-\boldsymbol{\alpha}_B \cdot \mathbf{p}_{op} + \beta_B m_B\right) + \wedge_{++} V_{II} \wedge_{++}$$
$$(4.8b)$$

where $\wedge_{++}$ is the product of Casimir-type positive energy projection operators. The nonrelativistic limit of both these equations is of course just

$$h_{nr}\varphi = W\varphi, \quad h_{nr} = \mathbf{p}_{op}^2/2m_{red} + U_C(r) \qquad (4.9)$$

where φ is a Pauli-type wave function, and m_{red} is the reduced mass.

It is convenient and indeed of greater interest for us to consider the application of (4.8a) and (4.8b) to the bound state problem rather than the scattering problem. Thus we take $e_A = -e_B = e$ and think of A as, say, either a positron or a proton and of B as an electron. The energy levels W are then proportional to $\alpha^2 m_{red}$ with m_{red} of order m_e. One can now show that the eigenvalues E obtained from (4.8a) are correct to order $\alpha^4 m_e$ but that the ones from (4.8b) are already wrong in this order. This is not a contradiction, since in the bound-state problem matrix elements of the potential between virtual plane-wave states of different energy enter (implicitly) and these matrix elements will in general not be the same for the two potentials. Thus, from this point of view the Coulomb-Breit potential V_I is "better" than V_{II}.

In a short paper, whose writing was prompted by our discussions in this workshop, I emphasize that the same conclusion holds with regard to the preferred choice of potential in the context of bound-states of many-electron atoms.[6]

A quantitative understanding of the fact that V_{II} gives incorrect results at the order in question can be obtained by computing $V^{(4)}$, the potential associated with the exchange of two photons. As discussed in Ref. 2, this potential depends on the choice of one-photon exchange potential $V^{(2)}$. It turns out that when V_{II} is used, $V^{(4)}$ has a leading long-range part $V_{LR}^{(4)}$ given by

$$V_{LR}^{(4)} = e_A e_B/2(m_A + m_B)r^2 \qquad (4.10)$$

and the expectation value of this potential with the nonrelativistic two-body wave function φ gives a shift which precisely accounts for the difference between V_I and V_{II} in order $\alpha^4 m_e$.

C. Concluding remarks

We have seen that the concept of potential in quantum field theory is subtle, fraught with ambiguity. In this discussion I have tried to emphasize the difference between what might be called "actual potential ambiguities", associated with the choices that are made in going off the energy shell as well as with the choice of lowest-order potential, and what might be called "phantom potential ambiguities", apparently connected with the choice of gauge in gauge theories. The choice of gauge plays a role only if one, so to speak, "lets it", by using it in giving a prescription for going off shell. But there is no need to "let it".

In the work with Gary Feinberg mentioned above it is shown that these ambiguities can be sharply limited by exploiting the fact that on-shell scattering amplitudes are both gauge invariant and analytic functions of t for fixed s. To be precise, once the form of the relativistic two-body equation which is to be used in describing a two-body system has been specified and a lowest-order potential has been chosen, the higher-order terms in the potential are uniquely determined by the requirement that the potential reproduce the amplitude given by field theory.

I am grateful to many members of this workshop for stimulating discussions of some of these topics, especially to Jean-Paul Desclaux, Walter Johnson, and Ingvar Lindgren.

APPENDIX A: FURRY PICTURE FOR SPIN-0 AND SPIN-1/2 FIELDS

In the text we need the Furry picture for a spin-0 field. Although a straightforward extension of the spin-1/2 case originally treated by Furry, it does not seem to be discussed in text books.

Let $\phi(x)$ denote a quantized spin-0 field, interacting only with an external scalar field $\varphi_{ext}(x)$, the interaction with other quantized fields regarded as switched off. The Heisenberg equation of motion for $\phi(x)$ is

$$\left(\Box + m^2 + G\varphi_{ext}\right)\phi(x) = 0, \tag{A1}$$

or equivalently

$$-\partial^2\phi/\partial t^2 = D_{op}\phi(x) \tag{A2}$$

where

$$D_{op} = \mathbf{p}_{op}^2 + m^2 + G\varphi_{ext}. \tag{A3}$$

For a not too strong static field $\varphi_{ext} = \varphi_{ext}(\mathbf{x})$, the operator D_{op} will be hermitian and have only positive eigenvalues, some of which may belong to a discrete spectrum if $G\varphi_{ext}$ is sufficiently attractive. The eigenfunctions $\phi_n(\mathbf{x})$ of D_{op} will then form a complete orthogonal set and with the normalization

$$\langle n|n'\rangle = \int d\mathbf{x}\; \phi_n^*(\mathbf{x})\phi_{n'}(\mathbf{x}) = \delta(n, n') \tag{A4a}$$

will satisfy the completeness relation

$$\sum_n \phi_n(\mathbf{x})\phi_n^*(\mathbf{x}') = \delta(\mathbf{x} - \mathbf{x}'). \tag{A4b}$$

Here the sigma symbol denotes a sum over discrete states and an integration over continuum states; the symbol $\delta(n, n')$ denotes a generalized Kronecker delta, equal to $\delta_{n,n'}$ for discrete values of the labels. The stationary-state solutions of (A2) may then be taken in the form

$$\phi_n^{(\pm)}(x) = \phi_n(\mathbf{x})\exp(\mp iE_n t) \quad (E_n > 0). \tag{A5}$$

Note that the simplicity of the relation between the positive- and negative-energy solutions is a consequence of the assumed scalar nature of the external field; for an external vector field there are two distinct sets of spatial wave functions.

The field $\phi(x)$ may be expanded in the form

$$\phi(x) = \sum_n (2E_n)^{-\frac{1}{2}}\left\{a(n)\phi_n^{(+)}(x) + b^\dagger(n)\phi_n^{(-)}(x)\right\}. \tag{A6}$$

The canonical conjugate momentum density $\pi(x)$, related to $\phi(x)$ via $\pi(x) = \partial\phi^\dagger(x)/\partial t$, is required to satisfy the equal-time commutation relation

$$[\phi(x), \pi(x')] = i\delta(\mathbf{x} - \mathbf{x}') \quad (t = t'). \tag{A7a}$$

This is assured if

$$[a(n), a^\dagger(n')] = \delta(n, n'), \quad [b(n), b^\dagger(n')] = \delta(n, n'), \qquad (A7b)$$

with other commutators vanishing. The a's, b's and their hermitian conjugates may be identified as annihilation and creation operators, with, *e.g.*,

$$\Psi_n = a^\dagger(n)|vac\rangle \qquad (A8)$$

representing a bound state if the associated wave function $\phi_n(\mathbf{x})$ is normalizable.

Let us now introduce an additional interaction with a neutral quantized scalar field $\varphi(x)$, like that given by (2.1), viz.

$$H' = G \int d\mathbf{x} \; \phi^\dagger(x)\phi(x)\varphi(x) \qquad (A9)$$

where $\varphi(x)$ is taken to have the familiar free-field expansion

$$\varphi(x) = (2\pi)^{-3/2} \int (d\mathbf{k}/2w) \{a(\mathbf{k}) \exp(-ik \cdot x) + h.c.\} \qquad (A10)$$

with $k \cdot x = wt - \mathbf{k} \cdot \mathbf{x}$. In the Furry-Schrödinger picture H' is evaluated for $t = 0$ and the matrix element for emission of a meson of momentum $\mathbf{k}$ while the particle makes a transition from a state Ψ_n to a state $\Psi_{n'}$ is given by

$$\langle a^\dagger(\mathbf{k})\Psi_{n'}|H'|\Psi_n \rangle = \left(G/(2\pi)^{3/2}\right) \langle n'| \exp(-i\mathbf{k} \cdot \mathbf{x})|n\rangle(8wE_{n'}E_n)^{-\frac{1}{2}}. \qquad (A11)$$

This is the analog of the expression (2.12) for emission by a free particle.

To establish the notation used in the text, I also review here the more familiar spin-1/2 case. Let $\psi(x)$ denote a quantized Dirac field, interacting only with the external scalar field $\varphi_{ext}(x)$ considered above, with Lagrangian density

$$-g\bar{\psi}(x)\psi(x)\varphi_{ext}(x). \qquad (A12)$$

The Heisenberg equation of motion is now

$$(i\partial/\partial t)\psi(x) = h_{ext}\psi(x) \qquad (A13)$$

where h_{ext} is the external-field Dirac Hamiltonian, defined by

$$h_{ext} = h_{op} + g\varphi_{ext}, \qquad (A14)$$

and h_{op} is the free Dirac Hamiltonian

$$h_{op} = \boldsymbol{\alpha} \cdot \mathbf{p}_{op} + \beta m. \qquad (A15)$$

For a not too strong, attractive external field there will be a positive-energy spectrum with perhaps some bound states in the region $0 < E < m$, with associated eigenfunctions $u_n(x)$, and a negative energy spectrum in the region

$E < -m$, with associated eigenfunctions $v_m(x)$. With normalization conventions analogous to (A4a), viz.

$$\langle n|n'\rangle = \int d\mathbf{x}\ u_n^\dagger(\mathbf{x})u_{n'}(\mathbf{x}) = \delta(n, n'),\qquad (A16a)$$

the completeness relation takes the form

$$\sum_n u_n(\mathbf{x})u_n^\dagger(\mathbf{x}') + \sum_m v_m(\mathbf{x})v_m^\dagger(\mathbf{x}') = \delta(\mathbf{x} - \mathbf{x}').\qquad (A16b)$$

The expansion of $\psi(x)$ in terms of the u's and v's is then

$$\psi(x) = \sum_n a(n)u_n(\mathbf{x})\exp(-iE_n t) + \sum_m b^\dagger(m)v_m(\mathbf{x})\exp(-iE_m t).\qquad (A17)$$

The momentum density conjugate to $\psi(x)$ is now $i\psi^\dagger(x)$ and is required to satisfy the anti-commutation relation

$$\left\{\psi(x), i\psi^\dagger(x')\right\} = i\delta(\mathbf{x} - \mathbf{x}')\qquad (t = t').\qquad (A18a)$$

This is assured if the a's and b's satisfy the A.C.R.

$$\left\{a(n), a^\dagger(n')\right\} = \delta(n, n')\ ,\qquad \left\{b(n), b^\dagger(n')\right\} = \delta(n, n').\qquad (A18b)$$

The additional interaction with the quantized scalar field now has the form

$$H'' = g\int d\mathbf{x}\ \bar{\psi}(x)\psi(x)\varphi_{ext}(x)\qquad (A19)$$

and the matrix element anologous to (A11) for emission of a meson of momentum $\mathbf{k}$ is now given by

$$\langle a^+(\mathbf{k})\Psi_{n'}|H''|\Psi_n\rangle = \left(g/(2\pi)^{3/2}\right)\langle n'|\beta\ \exp(-i\mathbf{k}\cdot\mathbf{x})|n\rangle(2\omega)^{-\frac{1}{2}}\qquad (A20)$$

where

$$\langle n'|\beta\ \exp(-i\mathbf{k}\cdot\mathbf{x})|n\rangle = \int d\mathbf{x}\ u_{n'}^\dagger(x)\beta\ \exp(-i\mathbf{k}\cdot\mathbf{x})u_n(\mathbf{x}).\qquad (A21)$$

Note that because there are no square-root of energy factors in the expansion (A17), in contrast to (A6), none associated with the emitting particle appear in (A20), in contrast to (A11). This is the reason that the one-meson exchange potential is local in the case of the Dirac particles but not for two Klein-Gordon particles.

APPENDIX B: OPERATORS FOR $V_I(\mathbf{r}; \Delta)$ and $V_{II}(\mathbf{r}; \Delta)$

The extension of the Δ-dependent "potential" $V_{II}(\mathbf{r}; \Delta)$ defined by (1.6a) into an ordinary operator V_{II} is immediate, since the Δ-dependence just involves a factor $\cos(\Delta r)$, as in the case treated in Sec. II.D.3. Thus we need only replace $V_{dir}^{(2)}$ in (2.71) by the Coulomb-Gaunt-Moller potential $V_{II}(\mathbf{r})$, given by (1.3), and define V_{II} by

$$V_{II} = (1/2) \left\{ e_{ord}^{-ih_A r} V_{II}(\mathbf{r}) e_{ord}^{irh_A} + e_{ord}^{-ih_B r} V_{II}(\mathbf{r}) e_{ord}^{irh_B} \right\}. \tag{B1}$$

For $V_I(\mathbf{r}; \Delta)$, defined by (1.7), such an extension is also feasible but is more intricate because of the relatively complicated dependence of $V_I(\mathbf{r}; \Delta)$ on the quantity Δ. One way to proceed is to first carry out the indicated differentiations in the second term of (1.7). This yields

$$V_I(\mathbf{r}; \Delta) = U_c 9r) \left[1 + a(\Delta r)\alpha_A \cdot \alpha_B + b(\Delta r)\alpha_A \cdot \hat{r}\alpha_B \cdot \hat{r} \right] \tag{B2}$$

where

$$a(x) = -\cos x + \sin x / x + (\cos x - 1)/x^2 \tag{B3}$$

$$b(x) = \cos x - 3 \sin x / x - 3(\cos x - 1)/x^2. \tag{B4}$$

As a check note that $a(0) = b(0) = -1/2$, so that $V_I(\mathbf{r}; 0) = V_I(\mathbf{r})$, the Coulomb-Breit potential defined by (1.1). Terms of the form $f(\mathbf{r}) \cos x$ term, with $x = \Delta_{ab} r$ and $f(\mathbf{r})$ a matrix function of $\mathbf{r}$ can be dealt with as before. To deal with the other terms note that

$$\sin x / x = \int_0^1 dy \; \cos(xy) \tag{B5}$$

and

$$(\cos x - 1)/x^2 = \int_0^1 dy(y - 1) \cos(xy). \tag{B6}$$

It follows that

$$a(x) = -\cos x + \int_0^1 dy \; y \cos(xy) \tag{B7}$$

and

$$b(x) = \cos x - 3 \int_0^1 dy \; y \cos(xy). \tag{B8}$$

The integration over the variable y can now be reserved and terms of the form $f(\mathbf{r}) \cos(xy)$ can now be handled like the $f(\mathbf{r}) \cos x$ terms, i.e.,

$$f(\mathbf{r}) \cos(xy) \rightarrow (1/2) \left\{ e_{ord} f(\mathbf{r}) e_{ord} + (A \rightarrow B) \right\}. \tag{B9}$$

In this way one arrives at an operator replacement for $V_I(\mathbf{r}; \Delta)$,

$$V_I^{op} = U_c + (1/2) \left\{ e_{ord}^{-ih_A r} \left(\alpha_A \cdot \hat{r}\alpha_B \cdot \hat{r} - \alpha_A \cdot \alpha_B \right) U_c e_{ord}^{irh_A} + (A \rightarrow B) \right\} \tag{B10}$$

$$- (1/2) \int_0^1 dy \; y \left\{ e_{ord}^{ih_A ry} \left(3\alpha_A \cdot \hat{r}\alpha_B \cdot \hat{r}\alpha_B \cdot \hat{r} - \alpha_A \cdot \alpha_B \right) U_c e_{ord}^{irh_A y} + (A \rightarrow B) \right\},$$

370

with the desired property that

$$\langle a, b | V_I^{op} | a, b \rangle = \langle a, b | V_I(\mathbf{r}) | a, b \rangle, \tag{B11}$$

and

$$\langle b, a | V_I^{op} | a, b \rangle = \langle b, a | V_I(r; \Delta_{ab} | a, b \rangle. \tag{B12}$$

Note that V_I^{op} is hermitian as well as symmetric under interchange of A and B.

REFERENCES

1. G.E. Brown, *Phil. Mag.* **43**, 467 (1952). For further discussion and references see H.A. Bethe and E.E. Salpeter, "Quantum Mechanics of One- and Two-Electron Systems," (Plenum, New York, 1977) pp. 198-200.

2. G. Feinberg and J. Sucher, University of Maryland TR#88-97.

3. G. Feinberg and J. Sucher, NSF-ITP preprint, to appear in *Phys. Rev. D*.

4. G. Feinberg and J. Sucher, in preparation.

5. For numerical studies of Eqs. (4.8a) and (4.8b) see G. Hardekopf and J. Sucher, *Phys. Rev.* **A30**, 703 (1984); ibid **31**, 2020 (1985).

6. J. Sucher, NSF-ITP preprint, to appear in *J. Phys. B* (Letters).

Effective Potentials in Relativistic Many-Body Theory

Ingvar Lindgren
Department of Physics, Chalmers University of Technology/
University of Gothenburg, S-412 96 Göteborg, Sweden

Abstract

The purpose of this study is to investigate the possibilities of transforming the non-relativistic many-body perturbation scheme (MBPT) to a corresponding relativistic one. Non-relativistic MBPT has proven to be quite successful for atomic and molecular calculations, particularly in the all-order or coupled-cluster formulations, where important classes of higher-order contributions are generated in a systematic way by means of iterations /1,2/. The general idea of extending the non-relativistic scheme to the relativistic regime will here be to express the electron-electron interaction and the radiative corrections in terms of *effective potentials* in order to be able to maintain the iterative character of the scheme. The ultimate aim is to be able to include the relativistic and QED effects successively into the many-body scheme in a systematic and rigorous manner, and this work may serve as a step towards that goal.

In the present notes the exchange of one and two photons between electrons is first analysed, and the corresponding effective potentials are derived. The standard S-matrix formulation is used, which yields directly the energy shift or diagonal part of such potentials. By a modification of the procedure it is possible to evaluate also the evolution operator (or wave operator), i.e., the *non-diagonal* part of the interaction potentials. This is done only in the one-photon case.

First- and second-order radiative corrections are studied in a similar fashion, and the results are expressed in terms of one- and two-body potentials. The effect of renormalization is not studied in any detail here and requires further study. The effect of nuclear recoil is also left out in this study.

The main derivations are made in the covariant Feynman gauge, but the results can be transferred to the Coulomb gauge, which is more suitable when combined with the non-relativistic theory, where the Coulomb interaction is treated to all orders.

In a separate study the construction of a many-body scheme of coupled-cluster type is discussed, based on the effective potentials derived in the present work /3/.

I. Single-photon exchange

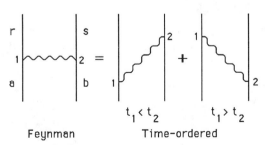

Feynman   Time-ordered

Figure 1. The Feynman representation of the exchange of a single, virtual photon between two electrons.

I.A. Scattering amplitude

We consider first the exchange of a single, covariant photon with four polarizations between two electrons in arbitrary states, as illustrated by the Feynman diagram in Fig. 1. The scattering amplitude is given by the standard rules of QED to be

$$S^{(2)}_{rs,ab} = \langle rs | S^{(2)} | ab \rangle \tag{I.1}$$

where $S^{(2)}$ is

$$S^{(2)} = \frac{(-i)^2}{2!} \int_{-\infty}^{\infty}\int_{-\infty}^{\infty} dt_1\, dt_2\ T\,[\,H_{int}(t_2)\,H_{int}(t_1)\,]\ e^{-\alpha_1|t_1|-\alpha_2|t_2|}$$

$$= -\frac{1}{2} \iint d^4x_1\, d^4x_2\ T\,[\,(\,\overline{\Psi}\,\gamma\cdot A\,\Psi\,)_2\,(\,\overline{\Psi}\,\gamma\cdot A\,\Psi\,)_1\,]\ e^{-\alpha_1|t_1|-\alpha_2|t_2|} \tag{I.2}$$

$$= -\frac{c^2}{2} \iint dt_1\, d^3x_1\, dt_2\, d^3x_2\ T\,[\,(\,\Psi^\dagger\,\bar{\gamma}\cdot A\,\Psi\,)_2\,(\,\Psi^\dagger\,\bar{\gamma}\cdot A\,\Psi\,)_1\,]\ e^{-\alpha_1|t_1|-\alpha_2|t_2|}$$

T is the time-ordering operator and H_{int} the interaction operator

$$H_{int}(t) = -\int d^3x\ \overline{\Psi}\ c\ \gamma\cdot A\ \Psi = -c\int d^3x\ \Psi^\dagger\ \bar{\gamma}\cdot A\ \Psi \tag{I.3}$$

Hartree atomic units are used ($e = m = \hbar = 4\pi\varepsilon_0 = 1$; $c = 1/\alpha$) and the following notations (see, for example, ref. 4, p 5)

$$x = (x_0,\, \mathbf{x}),\ x_0 = ct;\ \gamma = (\gamma_0,\, \boldsymbol{\gamma}),\ \gamma_0 = \beta,\ \boldsymbol{\gamma} = \gamma_0\,\boldsymbol{\alpha},\ (\,\boldsymbol{\alpha},\beta\ \text{Dirac operators})$$

$$a\cdot b = a^\mu b_\mu = g_{\mu\mu}\, a_\mu b_\mu = a_0 b_0 - \mathbf{a}\cdot\mathbf{b}$$

The field operators are given in a *Furry-type of representation*

$$\Psi = \sum b_i(t)\,\phi_i;\quad \Psi^\dagger = \sum b_i^\dagger(t)\,\phi_i^*;\quad \overline{\Psi} = \Psi^\dagger \gamma_0;\quad \bar{\gamma} = \gamma_0\,\gamma \tag{I.4}$$

where $\{\phi_i\}$ is a complete set of single-electron orbitals - with positive as well as negative energy - generated by the Dirac hamiltonian with an additional potential U

$$h_{Dirac} = c\,\boldsymbol{\alpha}\cdot\mathbf{p} + \beta\,c^2 - Z/r + U \tag{I.5}$$

The functions $|ab\rangle$ and $|rs\rangle$ in (I.1) are direct - *not antisymmetrized* - products of such orbitals. $b_i(t)$, $b_i^\dagger(t)$ are the time-dependent destruction/creation operators in the interaction picture with the time-dependence

$$b_i(t) = b_i(0)\,e^{-i\varepsilon_i t},\quad b_i^\dagger(t) = b_i^\dagger(0)\,e^{i\varepsilon_i t} \tag{I.6}$$

where ε_i is the orbital energy. The external potential, U, is so far unspecified, but it will in the many-body application be assumed to be built up by the remaining electrons in some self-consistent manner.

A is the vector potential for the radiation in the covariant Feynman gauge

$$A = \sum_{\mathbf{k}m} \sqrt{\frac{\hbar}{2\,\varepsilon_0\,\omega V}}\ \{\,a_{\mathbf{k}m}\,\varepsilon^{(m)}\,e^{-ikx} + a_{\mathbf{k}m}^\dagger\,\varepsilon^{(m)*}\,e^{ikx}\,\} \tag{I.7}$$

$k = (k_0,\, \mathbf{k})$; $k_0 = \omega/c = |\mathbf{k}|$; $k\cdot x = \omega t - \mathbf{k}\cdot\mathbf{x}$ and $\hbar / \varepsilon_0 = 4\pi$ in atomic units.

V is the volume of the box considered. a_{km} , $a_{km}^\dagger$ are the destruction/ creation operators for photons, which satisfy the commutation relation

$$[a_{km}, a_{km}^\dagger] = - g_{mm} \qquad (I.8)$$

This relation is required in order to satisfy the canonical equal-time commutation relation (ref 4, p 107,128)

$$[A_\mu(t,\mathbf{x}), \pi_\nu(t,\mathbf{x}')] = i \, \hbar \, g_{\mu\nu} \, \delta^3 (\mathbf{x-x'}) \text{ or } [\dot{A}_\mu(t,\mathbf{x}), A_\nu(t,\mathbf{x}')] = i \, \hbar \, \varepsilon_0 \, g_{\mu\nu} \, \delta^3 (\mathbf{x-x'}) \quad (I.9)$$

$\varepsilon^{(m)}$ are the polarization vectors (m=0,1,2,3), which satisfy the relation (ref 4, p 129)

$$\sum_{m=0}^3 g_{mm} \; \varepsilon_\mu^{(m)} \varepsilon_\nu^{(m)*} = g_{\mu\nu} \qquad (I.10)$$

Considering only distinct *pairs* (a,b), with the sums over r and s running independently, the amplitude (I.1) can be expressed

$$S_{rs,ab}^{(2)} = - \langle rs| \iint dt_1 \, dt_2 \; e^{i(\varepsilon_r - \varepsilon_a)t_1} e^{-\alpha_1|t_1|} e^{i(\varepsilon_s - \varepsilon_b)t_2} e^{-\alpha_2|t_2|}$$

$$\times c^2 (\bar{\gamma}_1 \cdot \bar{\gamma}_2) \, D_F(1,2) | ab \rangle \qquad (I.11)$$

Here, D_F is the *photon propagator* defined by means of the vacuum expectation value

$$\langle 0 | T[A_\mu(x) A_\nu(x')] | 0 \rangle = \langle 0 | \Theta(t-t') A_\mu(x) A_\nu(x') + \Theta(t'-t) A_\nu(x') A_\mu(x) | 0 \rangle$$

$$= g_{\mu\nu} \, D_F(x-x') \qquad (I.12)$$

With $\hbar / \varepsilon_0 = 4\pi$ in atomic units this gives

$$D_F(x-x') = - \sum_{\mathbf{k}} \frac{4\pi}{2\omega V} \left\{ \Theta(t-t') \, e^{-i \, k \cdot (x-x')} + \Theta(t'-t) \, e^{i \, k \cdot (x-x')} \right\} \qquad (I.13)$$

The expression in the curly brackets can be replaced by the integral

$$- \frac{2\omega}{2\pi i} \int_{-\infty}^{\infty} dz \frac{e^{iz(t-t')}}{z^2 - \omega^2 + i\delta} \; e^{i \, \mathbf{k} \cdot (\mathbf{x-x'})} \qquad (\omega^2 = c^2 \, \mathbf{k}^2)$$

and transforming the k sum into an integral

$$\frac{1}{V} \sum_{\mathbf{k}} \to (2\pi)^{-3} \int d^3 k$$

we get the following form of the photon propagator

$$D_F(x-x') = \frac{1}{2\pi^2} \int d^3 k \, e^{i \, \mathbf{k} \cdot (\mathbf{x-x'})} \frac{1}{2\pi i} \int_{-\infty}^{\infty} dz \frac{e^{iz(t-t')}}{z^2 - \omega^2 + i\delta} \qquad (I.14)$$

With $z = ck_0$, $\omega^2 = c^2 \, \mathbf{k}^2$, $k^2 = k_0^2 - \mathbf{k}^2$ this can be written as a four-dimensional integral, in agreement with the standard definition (ref 4, p 133).

Performing the time integrations of (I.11) gives

$$S_{rs,ab}^{(2)} = \langle rs| \, 2\pi i \, (\bar{\gamma}_1 \cdot \bar{\gamma}_2) \int_{-\infty}^{\infty} dz \, \Delta_{\alpha_1}(z+\varepsilon_r-\varepsilon_a) \, \Delta_{\alpha_2}(z-\varepsilon_s+\varepsilon_b)$$

$$\times \int d^3k \; \frac{e^{i\mathbf{k}\cdot\mathbf{r}_{12}}}{z^2 - \omega^2 + i\delta} \; |ab\rangle \qquad (\mathbf{r}_{12} = \mathbf{x}_1 - \mathbf{x}_2)$$

The Δ function is defined

$$\Delta_\alpha(x) = \frac{1}{\pi} \frac{\alpha}{x^2 + \alpha^2} \tag{I.15}$$

and has the following properties

$$\lim_{\alpha \to 0} \Delta_\alpha(x) = \delta(x) \; ; \; \lim_{\alpha \to 0} \pi\,\alpha\,\Delta_\alpha(x) = \delta_{x,0}$$

$$\int_{-\infty}^{\infty} dx\; \Delta_\alpha(x-a)\, \Delta_\beta(x-b) = \Delta_{\alpha+\beta}(a-b) \tag{I.16}$$

Carrying out the Fourier transform $(\omega^2 = c^2\,\mathbf{k}^2)$

$$\frac{c^2}{2\,\pi^2} \int d^3k \; \frac{e^{i\mathbf{k}\cdot\mathbf{r}_{12}}}{z^2 - \omega^2 + i\delta} = - \frac{e^{i\,|z|\,r_{12}/c}}{r_{12}} \tag{I.17}$$

then gives the single-photon scattering amplitude

$$S^{(2)}_{rs,ab} = -2\pi i\, \Delta_{\alpha_1+\alpha_2}(\varepsilon_r+\varepsilon_s-\varepsilon_a-\varepsilon_b)\, \langle rs|\; \frac{\bar{\gamma}_1\cdot\bar{\gamma}_2}{r_{12}} \; e^{i\,|q|\,r_{12}} \,|ab\rangle \tag{I.18}$$

where $\bar{\gamma}_1\cdot\bar{\gamma}_2 = 1 - \boldsymbol{\alpha}_1\cdot\boldsymbol{\alpha}_2$, $q = (\varepsilon_a-\varepsilon_r)/c = (\varepsilon_s-\varepsilon_b)/c$ and $\langle rs| = \langle ab|$ or $\langle ba|$.

I.B. Equivalent potential

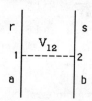

Figure 2. The single-photon exchange is compared with the scattering of a potential, V_{12}

The results obtained above can formally be expressed by means of a *potential*, V_{12}, in analogy with potential scattering (see Fig. 2)

$$S^{(2)}_{rs,ab} = -2\pi i\, \Delta_{\alpha_1+\alpha_2}(\varepsilon_r+\varepsilon_s-\varepsilon_a-\varepsilon_b)\, \langle rs|V_{12}|ab\rangle \tag{I.19}$$

where the potential is defined by means of its matrix elements

$$\langle rs|V_{12}|ab\rangle = \langle rs|\; \frac{\bar{\gamma}_1\cdot\bar{\gamma}_2}{r_{12}} \; e^{i\,|q|\,r_{12}}\,|ab\rangle = \langle rs|\,(1 - \boldsymbol{\alpha}_1\cdot\boldsymbol{\alpha}_2)\; \frac{e^{i\,|q|\,r_{12}}}{r_{12}} \;|ab\rangle \tag{I.20}$$

(Note that the two-electron functions are straight product states - not anti-symmetrized). This result is obtained with the *Feynman gauge*. The corresponding results in the Coulomb gauge can be obtained by means of a simple transformation.

The quantity $\bar{\gamma}_1 \cdot \bar{\gamma}_2 = 1 - \boldsymbol{\alpha}_1 \cdot \boldsymbol{\alpha}_2$, appearing in the Feynman gauge, as a result of the summation over all four polarizations, has in the Coulomb gauge to be replaced by the corresponding sum over the two *transverse* polarizations, $-\boldsymbol{\alpha}_1 \cdot \boldsymbol{\alpha}_2 + (\boldsymbol{\alpha}_1 \cdot \hat{\mathbf{k}})(\boldsymbol{\alpha}_2 \cdot \hat{\mathbf{k}})$. This implies that the Fourier transform (I.17),

$\bar{\gamma}_1 \cdot \bar{\gamma}_2 \, e^{i|q|r_{12}}/r_{12}$, is replaced by the Coulomb potential and

$$B_{12}(q) = -\boldsymbol{\alpha}_1 \cdot \boldsymbol{\alpha}_2 \frac{e^{i|q|r_{12}}}{r_{12}} + \left[\boldsymbol{\alpha}_1 \cdot \nabla_1 , \left[(\boldsymbol{\alpha}_2 \cdot \nabla_2) , \frac{e^{i|q|r_{12}} - 1}{q^2 r_{12}} \right] \right] \qquad (I.21)$$

which is the (diagonal) potential corresponding to the exchange of one virtual transverse photon, i.e., the *(retarded) Breit interaction*. When $q \to 0$, using the relation

$$(\boldsymbol{\alpha}_1 \cdot \nabla_1)(\boldsymbol{\alpha}_2 \cdot \nabla_2) \, r_{12} = - [\boldsymbol{\alpha}_1 \cdot \boldsymbol{\alpha}_2 - (\boldsymbol{\alpha}_1 \cdot \hat{\mathbf{r}}_{12})(\boldsymbol{\alpha}_2 \cdot \hat{\mathbf{r}}_{12})]/r_{12} \qquad (I.22)$$

we retrieve the standard (unretarded) Breit interaction

$$B_{12} = -1/2 \, [\boldsymbol{\alpha}_1 \cdot \boldsymbol{\alpha}_2 + (\boldsymbol{\alpha}_1 \cdot \hat{\mathbf{r}}_{12})(\boldsymbol{\alpha}_2 \cdot \hat{\mathbf{r}}_{12})]/r_{12} \qquad (I.23)$$

I.C. Calculation of the first-order energy shift.

In calculating the first-order energy shift, corresponding to the exchange of a single photon, we employ the Gell-Mann-Low-Sucher prescription /5,6/ and set all $\alpha_i = \alpha$. Then the energy shift in the (closed-shell) state Φ is given by

$$\delta E = \lim_{\alpha \to 0} \frac{i\alpha}{2} \frac{\sum n \langle \Phi | S^{(n)} | \Phi \rangle}{\langle \Phi | S | \Phi \rangle} \qquad (I.24)$$

which in lowest order becomes

$$\delta E = \lim_{\alpha \to 0} i\alpha \langle \Phi | S^{(2)} | \Phi \rangle \qquad (I.25)$$

Using (I.16), we get from (I.19)

$$\lim_{\alpha_1 + \alpha_2 \to 0} \frac{i(\alpha_1 + \alpha_2)}{2} S^{(2)}_{rs,ab} = \delta_{\varepsilon_r + \varepsilon_s, \, \varepsilon_a + \varepsilon_b} \langle rs | V_{12} | ab \rangle \qquad (I.26)$$

and the first-order energy shift becomes

$$\delta E = \langle \Phi | V_{12} | \Phi \rangle = \langle ab | V_{12} | ab \rangle - \langle ba | V_{12} | ab \rangle \qquad (I.27)$$

where Φ is the *antisymmetrized*, unperturbed state $|\{ab\}\rangle$. This result is consistent with the interpretation of V_{12} as an equivalent perturbing potential.

Due to the energy conservation of the scattering process, only *diagonal* (on-the-energy shell) matrix elements are obtained by this procedure. The evaluation of *non-diagonal* (off-the-energy shell) elements is discussed in section IV.

II. Two-photon exchange
II.A. Non-crossing photons

Next we consider the exchange of two "non-crossing" photons between the electrons, according to the Feynman diagram in Fig. 3. The scattering amplitude is given by

$$S^{(4)}_{rs,ab} = \langle rs | S^{(4)} | ab \rangle \qquad (II.1)$$

where $S^{(4)}$ is

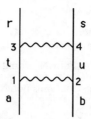

Figure 3. Exchange of two non-crossing photons

$$S^{(4)} = \frac{(-i)^4}{4!} \iiiint dt_1 \, dt_2 \, dt_3 \, dt_4 \, T \, [\, H_{int}(t_4) \, H_{int}(t_3) \, H_{int}(t_2) \, H_{int}(t_1) \,]$$

$$\times e^{-\alpha_1 |t_1| - \alpha_2 |t_2| - \alpha_3 |t_3| - \alpha_4 |t_4|} \qquad (II.2)$$

and H_{int} is the same as before (I.3), which gives

$$S^{(4)}_{rs,ab} = \frac{c^4}{4!} \iiiint dt_1 \, d^3 x_1 \, dt_2 \, d^3 x_2 \, dt_3 \, d^3 x_3 \, dt_4 \, d^3 x_4$$

$$\times T \, [\, (\, \Psi^\dagger \, \bar{\gamma} \cdot A \, \Psi \,)_4 \, (\, \Psi^\dagger \, \bar{\gamma} \cdot A \, \Psi \,)_3 \, (\, \Psi^\dagger \, \bar{\gamma} \cdot A \, \Psi \,)_2 \, (\, \Psi^\dagger \, \bar{\gamma} \cdot A \, \Psi \,)_1 \,] \qquad (II.3)$$

The vertices can be connected by the photon propagators in 6 distinct ways and by the electron propagators in two distinct ways, corresponding to the non-crossing diagram. Considering only one of these combinations, yields a weight of 1/2, when summed over abrstu independently. Considering only distinct *pairs* (a,b) as in the one-photon case (with the sums over rstu still running independently), gives the final weight of unity, or

$$S^{(4)}_{rs,ab} = \langle rs | \, \iiiint dt_1 \, dt_2 \, dt_3 \, dt_4 \, e^{i(\epsilon_r t_3 + \epsilon_s t_4)} \, e^{-i(\epsilon_a t_1 + \epsilon_b t_2)}$$

$$\times e^{-\alpha_1 |t_1| - \alpha_2 |t_2| - \alpha_3 |t_3| - \alpha_4 |t_4|}$$

$$\times c^2 \, (\, \bar{\gamma}_1 \cdot \bar{\gamma}_2 \,) \, c^2 \, (\, \bar{\gamma}_3 \cdot \bar{\gamma}_4 \,) \, D_F(1,2) \, D_F(3,4) \, S_F(3,1) \, S_F(4,2) | ab \rangle \qquad (II.4)$$

where the photon propagators are given by (I.14)

$$D_F(1,2) = \frac{1}{2 \pi^2} \int d^3 k \, e^{i \, \mathbf{k}_1 \cdot \mathbf{r}_{12}} \, \frac{1}{2\pi i} \int dz_1 \frac{e^{i z_1 (t_1 - t_2)}}{z_1^2 - \omega_1^2 + i\delta_1}$$

$$D_F(3,4) = \frac{1}{2 \pi^2} \int d^3 k \, e^{i \, \mathbf{k}_2 \cdot \mathbf{r}_{34}} \, \frac{1}{2\pi i} \int dz_2 \frac{e^{i z_2 (t_3 - t_4)}}{z_2^2 - \omega_2^2 + i\delta_2} \qquad (II.5)$$

The *electron propagator* is here defined as the vacuum expectation

$$S_F(x-x') = \langle 0 | T[\Psi(x) \, \Psi^*(x')] | 0 \rangle$$

$$= \langle 0 | \Theta(t-t') \, \Psi(x) \, \Psi^*(x') - \Theta(t'-t) \, \Psi^*(x') \, \Psi(x) | 0 \rangle \qquad (II.6)$$

which with the representation (I.4) can be expressed

$$S_F(x-x') = \frac{1}{2\pi i} \int dz \sum_j \frac{|\phi_j(\mathbf{x})\rangle\langle\phi_j(\mathbf{x}')|}{\epsilon_j - z - i\delta \epsilon_j} \, e^{i z (t' - t)} \qquad (II.7)$$

(This differs from the standard definition /6/ by a factor of γ_0, which is here included in $\bar{\gamma}$). In particular, this gives

$$S_F(3,1) = \frac{1}{2\pi i} \int dz_3 \sum_t \frac{|\phi_t(3)\rangle\langle\phi_t(1)|}{\varepsilon_t - z_3 - i\delta_3\varepsilon_t} e^{iz_3(t_1-t_3)}$$

$$S_F(4,2) = \frac{1}{2\pi i} \int dz_4 \sum_u \frac{|\phi_u(4)\rangle\langle\phi_u(2)|}{\varepsilon_u - z_4 - i\delta_4\varepsilon_u} e^{iz_4(t_2-t_4)} \tag{II.8}$$

After the time integrations the scattering amplitude (II.4) becomes

$$S^{(4)}_{rs,ab} = \langle rs | (\bar{\gamma}_1 \cdot \bar{\gamma}_2)(\bar{\gamma}_3 \cdot \bar{\gamma}_4) \int\!\!\int\!\!\int\!\!\int dz_1\, dz_2\, dz_3\, dz_4$$

$$\times \Delta_{\alpha_1}(z_1+z_3-\varepsilon_a)\,\Delta_{\alpha_2}(-z_1+z_4-\varepsilon_b)\,\Delta_{\alpha_3}(z_2-z_3+\varepsilon_r)\,\Delta_{\alpha_4}(-z_2-z_4+\varepsilon_s)$$

$$\times \left(\frac{c^2}{2\pi^2}\right)^2 \int\!\!\int d^3k_1\, d^3k_2\, \frac{e^{i\mathbf{k}_1\cdot\mathbf{r}_{12}}}{z_1^2 - \omega_1^2 + i\delta_1}\, \frac{e^{i\mathbf{k}_2\cdot\mathbf{r}_{34}}}{z_2^2 - \omega_2^2 + i\delta_2}$$

$$\times \sum_{tu} \frac{|\phi_t(3)\rangle\langle\phi_t(1)|}{\varepsilon_t - z_3 - i\delta_3\varepsilon_t}\, \frac{|\phi_u(4)\rangle\langle\phi_u(2)|}{\varepsilon_u - z_4 - i\delta_4\varepsilon_u} |ab\rangle \tag{II.9}$$

and after the Fourier transform (I.17)

$$S^{(4)}_{rs,ab} = \langle rs | \frac{\bar{\gamma}_1\cdot\bar{\gamma}_2}{r_{12}}\, \frac{\bar{\gamma}_3\cdot\bar{\gamma}_4}{r_{34}} \int\!\!\int\!\!\int\!\!\int dz_1\, dz_2\, dz_3\, dz_4$$

$$\times \Delta_{\alpha_1}(z_1+z_3-\varepsilon_a)\,\Delta_{\alpha_2}(-z_1+z_4-\varepsilon_b)\,\Delta_{\alpha_3}(z_2-z_3+\varepsilon_r)\,\Delta_{\alpha_4}(-z_2-z_4+\varepsilon_s)$$

$$\times e^{i|z_1|r_{12}/c}\, e^{i|z_2|r_{34}/c} \sum_{tu} \frac{|\phi_t(3)\rangle\langle\phi_t(1)|}{\varepsilon_t - z_3 - i\delta_3\varepsilon_t}\, \frac{|\phi_u(4)\rangle\langle\phi_u(2)|}{\varepsilon_u - z_4 - i\delta_4\varepsilon_u} |ab\rangle \tag{II.10}$$

Integral evaluation

For the integral evaluation we expand the denominators in (II.10) according to

$$\frac{1}{\varepsilon_t - z_3 - i\delta_3\varepsilon_t}\, \frac{1}{\varepsilon_u - z_4 - i\delta_4\varepsilon_u} =$$

$$\left[\frac{1}{\varepsilon_t - z_3 - i\delta_3\varepsilon_t} + \frac{1}{\varepsilon_u - z_4 - i\delta_4\varepsilon_u}\right] \frac{1}{\varepsilon_t + \varepsilon_u - z_3 - z_4 - i(\delta_3\varepsilon_t+\delta_4\varepsilon_u)} \tag{II.11}$$

or in the limit where the δ's go to zero

$$\left[\frac{P}{\varepsilon_t - z_3} + i\pi\, \mathrm{sign}(\varepsilon_t)\,\delta(\varepsilon_t - z_3) + \frac{P}{\varepsilon_u - z_4} + i\pi\, \mathrm{sign}(\varepsilon_u)\,\delta(\varepsilon_u - z_4)\right]$$

$$\times \left[\frac{P}{\varepsilon_t + \varepsilon_u - z_3 - z_4} + i\pi\, \mathrm{sign}(\delta_3\varepsilon_t+\delta_4\varepsilon_u)\,\delta(\varepsilon_t+\varepsilon_u-z_3-z_4)\right] \tag{II.12}$$

where P stands for the principal value. The integral can then be separated into four parts, which will be referred to as the δ-δ part, the P-δ part, the δ-P part and the

P-P part, respectively. The δ function in the second factor will contribute only "<u>on the energy shell</u>", i.e., when the intermediate state has the same energy as the initial state, and the P-part of the second factor only "<u>off the shell</u>", when the intermediate state has an energy different from that of the initial state. That the δ function only contributes on the shell is obvious for the <u>δ-δ part</u>, which contains essentially only products of δ functions. Furthermore, the <u>P-δ part</u>, containing the principal values from the first factor and the δ function from the second one, will not contribute at all, since the two principal terms will exactly cancel each other after integration over z_3 or z_4.

Off-shell contribution

Off the energy shell we need not consider the imaginary part of the second factor in (II.11). Performing the integrations over z_2, z_3 and z_4 in (II.10), we then get $(z = z_1)$

$$\Delta_{\alpha_1+\alpha_2+\alpha_3+\alpha_4}(\varepsilon_a+\varepsilon_b-\varepsilon_r-\varepsilon_s) \int dz \, e^{i|z|r_{12}} e^{i|q_{ar}-z|r_{34}}$$

$$\times \left[\frac{1}{z - q_{at} - i\delta_3\varepsilon_t} - \frac{1}{z + q_{bu} + i\delta_3\varepsilon_t} \right] / \Delta E \qquad (II.13)$$

where $\Delta E = \varepsilon_a+\varepsilon_b-\varepsilon_t-\varepsilon_u$ and $q_{ij} = (\varepsilon_i-\varepsilon_j)/c$. This can be separated into a <u>δ-part</u>

$$-i\pi \, \Delta_{\alpha_1+\alpha_2+\alpha_3+\alpha_4}(\varepsilon_r+\varepsilon_s-\varepsilon_a-\varepsilon_b)$$

$$\times \left[\text{sign}(\varepsilon_t) \, e^{i|q_{at}|r_{12}} e^{i|q_{tr}|r_{34}} + \text{sign}(\varepsilon_u) \, e^{i|q_{ub}|r_{12}} e^{i|q_{su}|r_{34}} \right] / \Delta E \qquad (II.14)$$

and a <u>principal (P) part</u>

$$\Delta_{\alpha_1+\alpha_2+\alpha_3+\alpha_4}(\varepsilon_a+\varepsilon_b-\varepsilon_r-\varepsilon_s) \int dz \, e^{i|z|r_{12}} e^{i|q_{ar}-z|r_{34}}$$

$$\times \left[\frac{P}{z - q_{at}} - \frac{P}{z + q_{bu}} \right] / \Delta E \qquad (II.15)$$

The off-shell contributions can also be evaluated by considering (II.9) *before* the Fourier transformation is performed (which is closely related to the Greens-function treatment of Don Yennie). Then it is found that the δ part above corresponds to the *average* of the residues of the poles of the *electron* propagators, evaluated above and below the real axes, respectively, while the P part corresponds to the similar average of the poles of the *photon* propagators.

The integrations above can also be performed essentially along the imaginary axis, as will be discussed further below. For the present analysis, though, we shall here use the separated form given above.

On-shell contribution

On the energy shell, i.e., when the intermediate state has the same energy as the initial one, we use the expression (II.12). We then have to consider the δ-P and the P-P parts as well as the δ-δ part, and we shall first show that the first two terms do not contribute. In order to evaluate the contribution of the <u>δ-P part</u> on the shell, we make the substitution $z_4 \Leftrightarrow \varepsilon_t + \varepsilon_u - z_3$ in the second term. Then, assuming $\varepsilon_t+\varepsilon_u = \varepsilon_a+\varepsilon_b$, we see that this term becomes identical to the first one with opposite sign.

To estimate the on-shell contribution of the P-P part, we express the principal parts more accurately as

$$\left[\ \frac{\varepsilon_t - z_3}{(\varepsilon_t - z_3)^2 + \delta^2} + \frac{\varepsilon_u - z_4}{(\varepsilon_u - z_4)^2 + \delta^2}\ \right]\ \frac{\varepsilon_t + \varepsilon_u - z_3 - z_4}{(\varepsilon_t + \varepsilon_u - z_3 - z_4)^2 + 4\,\delta^2}$$

(II.16)

When the α's go to zero, only $z_3 + z_4 \approx \varepsilon_a + \varepsilon_b$ contributes to the integral. But on the shell we also have $z_3 + z_4 \approx \varepsilon_t + \varepsilon_u$ or $\varepsilon_t - z_3 \approx -(\varepsilon_u - z_4)$ and the two terms in the bracket tend to cancel for each value of δ. It can then be shown that the integral vanishes when α and δ go to zero. Therefore, there is no P-P contribution on the shell.

Finally, we have to consider the the δ-δ part of the integral (II.10,12). This has a **double singularity** and therefore has to be evaluated more carefully. After integrations over z_3 and z_4 this part becomes

$$(i\pi)^2 \sum_{tu} [\text{sign}(\varepsilon_t) + \text{sign}(\varepsilon_u)]\ \text{sign}(\delta_3 \varepsilon_t + \delta_4 \varepsilon_u)$$

$$\times \iint dz_1\, dz_2\, \Delta_{\alpha_1}(z_1 + \varepsilon_t - \varepsilon_a)\, \Delta_{\alpha_2}(-z_1 + \varepsilon_u - \varepsilon_b)\, e^{i|z_1|r_{12}/c}$$

$$\times \Delta_{\alpha_3}(z_2 - \varepsilon_t + \varepsilon_r)\, \Delta_{\alpha_4}(-z_2 - \varepsilon_u + \varepsilon_s)\, e^{i|z_2|r_{34}/c}$$

(II.17)

The sign factors give a factor of two if the signs of ε_t and ε_u are equal and zero otherwise. Assuming the signs to be positive, we can write the corresponding contribution to the scattering amplitude (II.10,12) in the following way

$$S^{(4)}_{rs,ab}(\delta\text{-}\delta) = 2 \sum_{tu} \langle rs\,|\, i\pi\, \frac{\bar{\gamma}_3 \cdot \bar{\gamma}_4}{r_{34}} \int dz_2\, \Delta_{\alpha_3}(z_2 - \varepsilon_t + \varepsilon_r)\, \Delta_{\alpha_4}(-z_2 - \varepsilon_u + \varepsilon_s)\, e^{i|z_2|r_{34}/c}\,|\,tu\rangle$$

$$\times \langle tu\,|\, i\pi\, \frac{\bar{\gamma}_1 \cdot \bar{\gamma}_2}{r_{12}} \int dz_1\, \Delta_{\alpha_1}(z_1 + \varepsilon_t - \varepsilon_a)\, \Delta_{\alpha_2}(-z_1 + \varepsilon_u - \varepsilon_b)\, e^{i|z_1|r_{12}/c}\,|\,ab\rangle$$

(II.18)

By comparing with the single-photon exchange (I.19), we see that the result above can be **exactly** expressed in terms of single-photon scattering elements

$$S^{(4)}_{rs,ab}(\delta\text{-}\delta) = \frac{1}{2}\, S^{(2)}_{rs,tu} \times S^{(2)}_{tu,ab}$$

(II.19)

Obviously, this part contributes only on the shell, where $\varepsilon_t + \varepsilon_u = \varepsilon_a + \varepsilon_b = \varepsilon_r + \varepsilon_s$. When the α's go to zero, this part has a double singularity, due to the two S matrix elements involved. This term, however, is *exactly* eliminated, when the energy shift is evaluated (see III.2 below). Consequently, **there is no on-shell contribution to the energy.**

It should be noted that starting directly by separating (II.10) into principal and δ parts, the products of the δ factors would only give one half of (II.19), with the other part of the singularity hidden in the products of the principal parts. For the ground-state of He-like systems the elimination of the on-shell contribution can be shown in a more direct and simpler way by shifting the vacuum level to that of the closed $1s^2$ shell. Then the part with vanishing energy denominator is represented by a disconnected diagram, which is eliminated by the normalization.

II.B. Two-photon exchange. Crossing photons

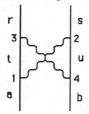

Figure 4. Exchange of two crossing photons

For the exchange of two "crossing" photons between the electrons according to Fig. 4, the scattering amplitude is given by (II.1-4) with

$$S_{rs,ab}^{(4)} = \langle rs| \iiiint dt_1\, dt_2\, dt_3\, dt_4\; e^{i(\varepsilon_r t_3 + \varepsilon_s t_2)}\; e^{-i(\varepsilon_a t_1 + \varepsilon_b t_4)}$$

$$\times e^{-\alpha_1|t_1| - \alpha_2|t_2| - \alpha_3|t_3| - \alpha_4|t_4|}$$

$$\times c^2\,(\,\bar{\gamma}_1 \cdot \bar{\gamma}_2\,)\, c^2\,(\,\bar{\gamma}_3 \cdot \bar{\gamma}_4\,)\, D_F(1,2)\, D_F(3,4)\, S_F(3,1)\, S_F(2,4)|ab\rangle \tag{II.20}$$

where the photon and electron propagators are the same as before (II.5,8), except

$$S_F(2,4) = \frac{1}{2\pi i} \int dz_4 \sum_u \frac{|\phi_u(2)\rangle\langle\phi_u(4)|}{\varepsilon_u - z_4 - i\delta_4\varepsilon_u}\, e^{i z_4 (t_4 - t_2)} \tag{II.21}$$

After the time integrations and the Fourier transforms the scattering amplitude becomes in analogy with (II.10)

$$S_{rs,ab}^{(4)} = \langle rs| \frac{\bar{\gamma}_1 \cdot \bar{\gamma}_2}{r_{12}}\, \frac{\bar{\gamma}_3 \cdot \bar{\gamma}_4}{r_{34}} \iiiint dz_1\, dz_2\, dz_3\, dz_4$$

$$\times \Delta_{\alpha_1}(z_1 + z_3 - \varepsilon_a)\, \Delta_{\alpha_2}(-z_1 - z_4 + \varepsilon_s)\, \Delta_{\alpha_3}(z_2 - z_3 + \varepsilon_r)\, \Delta_{\alpha_4}(-z_2 + z_4 - \varepsilon_b)$$

$$\times e^{i|z_1|r_{12}/c}\, e^{i|z_2|r_{34}/c} \sum_{tu} \frac{|\phi_t(3)\rangle\langle\phi_t(1)|}{\varepsilon_t - z_3 - i\delta_3\varepsilon_t}\, \frac{|\phi_u(2)\rangle\langle\phi_u(4)|}{\varepsilon_u - z_4 - i\delta_4\varepsilon_u}|ab\rangle \tag{II.22}$$

By comparing the results for crossing and non-crossing photons, we see that the results can be transformed into each other in the following way

1. interchange b and s
2. interchange creation/absorption for b, u and s (incl. change of sign of
   ε_b, ε_u and ε_s) \hfill (II.23)
3. change overall sign.

Therefore, we can directly get the results for crossed diagram from the result above of the uncrossed one.

On the energy shell we find from (II.17) after changing the sign of ε_u, that _there is no singularity in the crossed-photon diagram for intermediate states with positive energy._ In other words, there is no bound intermediate state which leads to zero energy denominator, and therefore this does not require any special consideration.

The effects of the two-photon exchange on the energy and the equivalent potential is discussed in the next section. The results here are obtained in the Feynman gauge, but they can be converted to the Coulomb gauge, as in the one-photon case, although this is more involved in the present case.

III. Calculation of the second-order energy shift. Equivalent potential

To evaluate the energy shift corresponding to the two-photon exchange we use again the formula (I.24), setting all $\alpha_i = \alpha$. For the one- and two-photon exchange we then get - with the limit $\alpha \to 0$ being understood -

$$\delta E = \frac{i\,\alpha}{2} \, \langle \Phi | 2\, S^{(2)} + 4\, S^{(4)} | \Phi \rangle / \langle \Phi | 1 + S^{(2)} + S^{(4)} | \Phi \rangle$$

$$\approx \frac{i\,\alpha}{2} \left[\langle \Phi | 2\, S^{(2)} + 4\, S^{(4)} | \Phi \rangle - 2 \langle \Phi | S^{(2)} | \Phi \rangle^2 \right] \qquad (III.1)$$

By means of the relation (II.19) it is then found that the last term *exactly* cancels the on-shell (δ-δ) part of $S^{(4)}$. The result to second order can then be expressed in terms of the off-shell parts

$$\delta E = \frac{i\,\alpha}{2} \, \langle \Phi | 2\, S^{(2)} + 4\, S^{(4)} \, (\delta\text{-}P) + 4\, S^{(4)} \, (P\text{-}P) | \Phi \rangle \qquad (III.2)$$

Formally, we can express the scattering amplitude in terms of a "potential" in analogy with the single-photon exchange (I.19)

$$S^{(4)}_{rs,ab} = -\, 2\pi i\, \Delta_{\alpha_1+\alpha_2+\alpha_3+\alpha_4} (\varepsilon_r + \varepsilon_s - \varepsilon_a - \varepsilon_b)\, \langle rs | V_{12,34} | ab \rangle \qquad (III.3)$$

Using (I.16), with all α's equal, we can also write this relation as

$$\lim_{\alpha \to 0} \frac{i}{2}\, 4\, \alpha\, S^{(4)}_{rs,ab} = \delta_{\varepsilon_r + \varepsilon_s,\, \varepsilon_a + \varepsilon_b}\, \langle rs | V_{12,34} | ab \rangle \qquad (III.4)$$

The two-photon contribution to the energy shift is then according to (III.2) given by the expectation values of the off-shell parts of this potential

$$\delta E^{(2)} = \langle \Phi | V_{12,34}\,(\delta) + V_{12,34}\,(P) | \Phi \rangle \qquad (III.5)$$

The potentials for the uncrossed diagram (Fig. 3) are obtained from (II.14,15)

$$\langle rs | V_{12,34}(\delta) | ab \rangle_{uncrossed} = 1/2 \sum_{tu} [\text{sign}(\varepsilon_t)\, \langle rs | V_{tr} | tu \rangle \langle tu | V_{at} | ab \rangle$$

$$+\, \text{sign}(\varepsilon_u)\, \langle rs | V_{su} | tu \rangle \langle tu | V_{ub} | ab \rangle] / \Delta E \qquad (III.6)$$

$$\langle rs | V_{12,34}(P) | ab \rangle_{uncrossed} = \langle r_3\, s_4 | \frac{\bar{\gamma}_3 \cdot \bar{\gamma}_4}{r_{34}} \sum_{tu} \langle t_1\, u_2 | \frac{\bar{\gamma}_1 \cdot \bar{\gamma}_2}{r_{12}}$$

$$\times\, \frac{1}{2\pi i} \int dz\, e^{i|z|r_{12}}\, e^{i|q_{ar}-z|r_{34}} \left[\frac{P}{z - q_{at}} - \frac{P}{z + q_{bu}} \right] |t_3\, u_4 \rangle |a_1\, b_2 \rangle / \Delta E \qquad (III.7)$$

where $V_{ij} = \bar{\gamma}_1 \cdot \bar{\gamma}_2\, e^{i|q_{ij}|r_{12}} / r_{12}$, $q_{ij} = (\varepsilon_i - \varepsilon_j)/c$, $\Delta E = \varepsilon_a + \varepsilon_b - \varepsilon_t - \varepsilon_u$ and where r_3 stands for $\phi_r(3)$ etc. For the crossed diagram (Fig. 4) we get the corresponding results by means of the substitution (II.23)

$$\langle rs | V_{12,34}(\delta) | ab \rangle_{crossed} = -1/2 \sum_{tu} [\text{sign}(\varepsilon_t) \langle ru | V_{tr} | tb \rangle \langle ts | V_{at} | au \rangle$$

$$- \text{sign}(\varepsilon_u) \langle ru | V_{bu} | tb \rangle \langle ts | V_{us} | au \rangle] / \Delta E \qquad \text{(III.8)}$$

$$\langle rs | V_{12,34}(P) | ab \rangle_{crossed} = - \sum_{tu} \langle r_3 u_4 | \frac{\bar{\gamma}_3 \cdot \bar{\gamma}_4}{r_{34}} \langle t_1 s_2 | \frac{\bar{\gamma}_1 \cdot \bar{\gamma}_2}{r_{12}}$$

$$\times \frac{1}{2\pi i} \int dz\, e^{i|z|r_{12}}\, e^{i|q_{ar}-z|r_{34}} \left[\frac{P}{z - q_{at}} - \frac{P}{z - q_{su}} \right] |t_3 b_4\rangle |a_1 u_2\rangle / \Delta E \qquad \text{(III.9)}$$

where $\Delta E = \varepsilon_a - \varepsilon_s - \varepsilon_t + \varepsilon_u$.

Notice the difference in the matrix elements and the energy denominators between the uncrossed and crossed diagrams, which are the same as in the non-relativistic MBPT diagrams

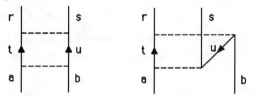

Evaluation of the principal parts

The integrals appearing in the principal parts above can be evaluated analytically. For $q_{ar} = 0$ (and $q_{bs} = 0$), i.e., the *direct* contribution, the uncrossed diagram gives

$$\langle rs | V_{12,34}(P) | ab \rangle_{uncr, dir} = \langle r_3 s_4 | \frac{\bar{\gamma}_3 \cdot \bar{\gamma}_4}{r_{34}} \sum_{tu} \langle t_1 u_2 | \frac{\bar{\gamma}_1 \cdot \bar{\gamma}_2}{r_{12}}$$

$$\times 1/\pi \left[\cos(q_{at} R)\, \text{Si}(q_{at} R) - \sin(q_{at} R)\, \text{Ci}(q_{at} R) \right.$$
$$+ \cos(q_{bu} R)\, \text{Si}(q_{bu} R) - \sin(q_{bu} R)\, \text{Ci}(q_{bu} R) \left. \right] |t_3 u_4\rangle |a_1 b_2\rangle / \Delta E \qquad \text{(III.10)}$$

where $R = r_{12} + r_{34}$ and the Si and Ci functions are defined /7/

$$\text{Si}(x) = \int_0^x \frac{\sin t}{t}\, dt = x - \frac{x^3}{3 \cdot 3!} + \ldots \qquad \text{(III.11)}$$

$$\text{Ci}(x) = - \int_x^\infty \frac{\cos t}{t}\, dt = \gamma + \ln |x| + \int_0^x \frac{\cos t - 1}{t}\, dt = \gamma + \ln |x| - \frac{x^2}{2 \cdot 2!} + \frac{x^4}{4 \cdot 4!} - \cdots$$

and γ is Euler's constant. The corresponding crossed-photon contribution is obtained by the substitution (II.23). The exchange part can be expressed by means of the same functions but is more complicated.

As mentioned previously, the expressions given here are not very useful for numerical work, but they will be used for the analysis below. In the case of $q_{ar} = 0$ the integration can be performed over the imaginary axis, as demonstrated by Blundell et al. /8/ in their study of the He ground state. This leads to the following expression for the entire potential (δ and P part) of the uncrossed diagram

$$\langle rs|V_{12,34}|ab\rangle_{\text{uncr, dir}} = \langle r_3\, s_4| \frac{\bar{\gamma}_3\cdot\bar{\gamma}_4}{r_{34}} \sum_{tu} \langle t_1\, u_2| \frac{\bar{\gamma}_1\cdot\bar{\gamma}_2}{r_{12}}$$

$$\times \left\{ -\frac{1}{\pi} \int_0^\infty dy\, e^{-yR} \left[\frac{q_{at}}{y^2+q_{at}^2} + \frac{q_{bu}}{y^2+q_{bu}^2} \right] + e^{i\,q_{at}R} + e^{i\,q_{bu}R} \right\} |t_3\, u_4\rangle |a_1\, b_2\rangle \qquad \text{(III.12)}$$

where the last two terms appear only for $0 < \varepsilon_t < \varepsilon_a$ and $0 < \varepsilon_u < \varepsilon_b$, respectively.
(For $q_{ar} \neq 0$ the integration path has to be slightly modified). By means of the
relation (ref. 7, p 312)

$$\int_0^\infty dy\, e^{-yR} \frac{q}{y^2+q^2} = \pi/2\, \text{sign}(q)\, \cos(qR) + \sin(qR)\, \text{Ci}(qR) - \cos(qR)\, \text{Si}(qR) \qquad \text{(III.13)}$$

it is easy to show that the results given above (III.6,10) are exactly reproduced. The
first term in (III.13) generates the δ part and the last two the P part. The
corresponding expression for the crossed diagram is obtained with the substitution
(II.23)

$$\langle rs|V_{12,34}|ab\rangle_{\text{crossed, dir}} = - \langle r_3\, u_4| \frac{\bar{\gamma}_3\cdot\bar{\gamma}_4}{r_{34}} \sum_{tu} \langle t_1\, s_2| \frac{\bar{\gamma}_1\cdot\bar{\gamma}_2}{r_{12}}$$

$$\times \left\{ -\frac{1}{\pi} \int_0^\infty dy\, e^{-yR} \left[\frac{q_{at}}{y^2+q_{at}^2} - \frac{q_{su}}{y^2+q_{su}^2} \right] + e^{i\,q_{at}R} + e^{i\,q_{su}R} \right\} |t_3\, b_4\rangle |a_1\, u_2\rangle \qquad \text{(III.14)}$$

Power expansion

We shall now analyse the four contributions (III.6-9) to the two-photon exchange
we obtained in the Feynman gauge, i.e., the δ and the P parts of the uncrossed
diagram (Fig. 3) and of the crossed diagram (Fig. 4), respectively. For energy
transfers of the order of Rydbergs we then find that the δ part will have even and
the P part odd powers of α. The leading contribution is the δ part of the uncrossed
diagram, which contains the unretarded term

$$\sum_{tu} \langle rs|V_{12}|tu\rangle \langle tu|V_{12}|ab\rangle / \Delta E$$

and the corresponding energy contribution

$$\delta E = \sum_{tu} \langle \Phi|V_{12}|tu\rangle \langle tu|V_{12}|\Phi\rangle] / \Delta E \qquad \text{(III.15)}$$

where V_{12} is the unretarded potential, $V_{12} = \bar{\gamma}_1\cdot\bar{\gamma}_2 / r_{12} = (1 - \boldsymbol{\alpha}_1\cdot\boldsymbol{\alpha}_2) / r_{12}$.

It is obvious, though, that this potential does not yield all contributions to $O(\alpha^2\, \text{Ry})$.
With the excitation energies of the order of Rydbergs, the retardation part of (III.6)
is also of $O(\alpha^2\, \text{Ry})$. The δ part of the crossed diagram does not have any unretarded
contribution when the intermediate energies are positive, due to the sign factors,
but also here the retarded part is of $O(\alpha^2\, \text{Ry})$. The contribution of the negative
energy states is for both diagrams at most of $O(\alpha^3\, \text{Ry})$.

The leading P parts of the individual diagrams are of O(α Ry), but the first term is cancelled when the two diagrams are considered together. Therefore, the leading P contribution of the *combined* diagrams is of O(α^3 Ry). (It will be shown in section IV that the situation is quite different when the Coulomb gauge is used.)

The major part of the two-photon exchange given above is **reducible** in the sense that it can be represented by a potential interaction, applied twice. This is obviously the case for the unretarded part, but, as will be shown in the next section, also part of the retarded interaction can be included in that way. The remaining, **irreducible** part can be included by means of an additional potential.

IV. Evaluation of the wave operator. Non-diagonal potential

For bound-state problems the standard S-matrix approach is restricted to *diagonal* elements of the interaction, i.e., to energy shifts and transition probabilities, due to the energy conservation implicit in the scattering problem. It is possible, though, to use a similar technique to evaluate *non-diagonal* contributions, i.e., contributions to the evolution operator or the wave operator by *applying electron propagators to the outgoing lines and setting the final time equal to zero.*

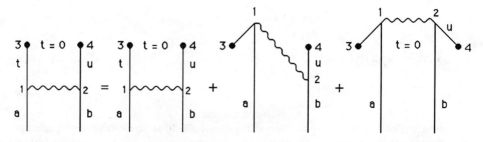

Figure 5. Diagrams for evaluating the evolution operator or the non-diagonal one-photon potential. The Feynman diagram to the left corresponds to the time-ordered diagrams to the right. The dots represent time t=0.

We consider the Feynman diagram in Fig. 5 (left), with the exchange of a single photon between two electrons and with electron propagators on the outgoing lines (t and u). The evaluation is very much the same as in the two-photon exchange, the only difference being that there is only one photon propagator and that the final times (t_3 and t_4) are zero. This represents the *evolution operator*, U(0,- ∞), which in first order is identical to the *wave operator*, used in MBPT /1/. The evolution operator is related to the S matrix by

$$S = U(\infty,- \infty) = U(\infty,0)\, U(0,- \infty) = U^{\dagger}(0,- \infty)\, U(0,- \infty) \qquad (IV.1)$$

In analogy with the two-photon case (II.4), we get

$$\langle tu \,|\, U(0,- \infty) \,|\, ab \rangle = - \langle tu | \iint dt_1\, dt_2\, e^{-i\,\varepsilon_a t_1}\, e^{-i\,\varepsilon_b t_2}\, e^{-\alpha_1 |t_1| - \alpha_2 |t_2|}$$

$$\times c^2\, (\, \bar{\gamma}_1 \cdot \bar{\gamma}_2\,)\, D_F(1,2)\, S_F(3,1)\, S_F(4,2)\,|\, ab \rangle \qquad (IV.2)$$

where the propagators are

$$D_F(1,2) = \frac{1}{2\pi^2} \int d^3k \, e^{i\mathbf{k}_1 \cdot \mathbf{r}_{12}} \frac{1}{2\pi i} \int dz_1 \frac{e^{iz_1(t_1-t_2)}}{z_1^2 - \omega_1^2 + i\delta_1}$$

$$S_F(3,1) = \frac{1}{2\pi i} \int dz_3 \sum_t \frac{|\phi_t(3)\rangle\langle\phi_t(1)|}{\varepsilon_t - z_3 - i\delta_3\varepsilon_t} e^{iz_3 t_1}$$

$$S_F(4,2) = \frac{1}{2\pi i} \int dz_4 \sum_u \frac{|\phi_u(4)\rangle\langle\phi_u(2)|}{\varepsilon_u - z_4 - i\delta_4\varepsilon_u} e^{iz_4 t_2} \qquad \text{(IV.3)}$$

Note that there are no time factors on the *outgoing* lines. These have been replaced by the propagators. For ε_t and ε_u being positive these propagators give $e^{i\varepsilon_t t_1}$ and $e^{i\varepsilon_u t_2}$, respectively, provided t_1 and t_2 are negative. This corresponds to the first time-ordered diagram in Fig. 5. When the orbital energies are negative, we get the same factors with opposite signs provided t_1 and t_2 are positive, corresponding to the last two time-ordered diagrams.

After time integrations, Fourier transform and integration over z_1 we get

$$\langle tu | U(0,-\infty) | ab \rangle = - \langle tu | \frac{\bar{\gamma}_1 \cdot \bar{\gamma}_2}{r_{12}} \frac{1}{2\pi i} \iint dz_3 \, dz_4 \, \Delta_{\alpha_1 + \alpha_2}(z_3 + z_4 - \varepsilon_a - \varepsilon_b)$$

$$\times e^{i|\varepsilon_a - z_3|r_{12}/c} \frac{1}{\varepsilon_t - z_3 - i\delta_3\varepsilon_t} \frac{1}{\varepsilon_u - z_4 - i\delta_4\varepsilon_u} | ab \rangle \qquad \text{(IV.4)}$$

The denominators are expanded as before (II.12)

$$\left[\frac{P}{\varepsilon_t - z_3} + i\pi \, \text{sign}(\varepsilon_t) \, \delta(\varepsilon_t - z_3) + \frac{P}{\varepsilon_u - z_4} + i\pi \, \text{sign}(\varepsilon_u) \, \delta(\varepsilon_u - z_4) \right]$$

$$\times \left[\frac{P}{\varepsilon_t + \varepsilon_u - z_3 - z_4} + i\pi \, \text{sign}(\delta_3\varepsilon_t + \delta_4\varepsilon_u) \, \delta(\varepsilon_t + \varepsilon_u - z_3 - z_4) \right] \qquad \text{(IV.5)}$$

As in the two-photon case, the δ-δ part contributes only on the energy shell and the δ-P and P-P parts only off the shell. The P-δ part vanishes also as before. We are here primarily interested in the off-shell part, but as a corroborate we shall evaluate also the on-shell part.

On-shell part

Assuming the energies of the outgoing lines to be positive, we get after integration

$$\langle tu | U(\delta - \delta) | ab \rangle = - i\pi \, \Delta_{\alpha_1 + \alpha_2}(\varepsilon_t + \varepsilon_u - \varepsilon_a - \varepsilon_b) \langle tu | \frac{\bar{\gamma}_1 \cdot \bar{\gamma}_2}{r_{12}} e^{i|\varepsilon_a - \varepsilon_t|r_{12}/c} | ab \rangle \qquad \text{(IV.6)}$$

This agrees with the one-photon result (I,18), apart from a factor of two, which is due to the fact that we are evaluating $U(0,-\infty)$ rather than $S = U(\infty,-\infty)$. In the present case the energy shift is given by

$$\delta E = \lim_{\alpha \to 0} 2 i \alpha \langle \Phi | U(0,-\infty) | \Phi \rangle \qquad \text{(IV.7)}$$

which gives the same result as (I.22).

Off the shell

The off-shell parts of the evolution operator become in analogy with the two-photon case (III.6) and (III.7)

$$\langle tu|U(\delta)|ab\rangle = \langle tu|\frac{\bar\gamma_1\cdot\bar\gamma_2}{r_{12}} 1/2\,[\,\mathrm{sign}(\varepsilon_t)\,e^{i\,|q_{at}|\,r_{12}} + \mathrm{sign}(\varepsilon_u)\,e^{i\,|q_{ub}|\,r_{12}}]|ab\rangle/\Delta E \quad \text{(IV.8)}$$

$$\langle tu|U(P)|ab\rangle = \langle tu|\frac{\bar\gamma_1\cdot\bar\gamma_2}{r_{12}}\frac{1}{2\pi i}\int dz\,e^{i\,|z|\,r_{12}}\,[\,\frac{P}{z-q_{at}} - \frac{P}{z+q_{bu}}\,]|ab\rangle/\Delta E \quad \text{(IV.9)}$$

The matrix elements above represent a *potential*

$$\langle tu|V_{12}|ab\rangle = \langle tu|\frac{\bar\gamma_1\cdot\bar\gamma_2}{r_{12}} 1/2\,[\,\mathrm{sign}(\varepsilon_t)\,e^{i\,|q_{at}|\,r_{12}} + \mathrm{sign}(\varepsilon_u)\,e^{i\,|q_{ub}|\,r_{12}}]$$

$$+ \frac{\bar\gamma_1\cdot\bar\gamma_2}{r_{12}}\frac{1}{2\pi i}\int dz\,e^{i\,|z|\,r_{12}}\,[\,\frac{P}{z-q_{at}} - \frac{P}{z+q_{bu}}\,]|ab\rangle \quad \text{(IV.10)}$$

The first (δ) part is essentially the potential of Mittleman /9/. The additional term is due to the principal part of the electron propagators.

The *imaginary* part of the potential is

$$\langle tu|V_{12}|ab\rangle_I = \langle tu|\frac{\bar\gamma_1\cdot\bar\gamma_2}{r_{12}} 1/2\,[\mathrm{sign}(\varepsilon_t)\sin(|q_{at}|\,r_{12}) + \mathrm{sign}(\varepsilon_u)\sin(|q_{bu}|\,r_{12})]$$

$$- \frac{\bar\gamma_1\cdot\bar\gamma_2}{r_{12}}\frac{1}{2\pi}\int dz\,\cos(z\,r_{12})\,[\,\frac{P}{z-q_{at}} - \frac{P}{z+q_{bu}}\,]|ab\rangle$$

$$= 1/2\,\langle tu|\frac{\bar\gamma_1\cdot\bar\gamma_2}{r_{12}}\,[\mathrm{sign}(\varepsilon_t)\sin(|q_{at}|\,r_{12}) + \mathrm{sign}(\varepsilon_u)\sin(|q_{bu}|\,r_{12})$$

$$+ \sin(q_{at}\,r_{12}) + \sin(q_{bu}\,r_{12})]|ab\rangle \quad \text{(IV.11)}$$

Note that there are no absolute signs in the last terms. The q parameters are defined $q_{ij} = (\varepsilon_i-\varepsilon_j)/c$, which means that if ab represents the ground state and t and u are *positive* energy states, then q_{at} and q_{bu} are always *negative*. It then follows that the imaginary part vanishes, as it should. If, on the other hand, ab is *not* the ground state, then some q's are positive, and the imaginary part does *not* vanish, corresponding to possible decay channels.

The potential above can also be expressed in analogy with (III.12)

$$\langle tu|V_{12}|ab\rangle =$$

$$\langle tu|\frac{\bar\gamma_1\cdot\bar\gamma_2}{r_{12}}\{-\frac{1}{\pi}\int_0^\infty dy\,e^{-y\,r_{12}}\,[\,\frac{q_{at}}{y^2+q_{at}^2} + \frac{q_{bu}}{y^2+q_{bu}^2}\,] + e^{i\,q_{at}\,r_{12}} + e^{i\,q_{bu}\,r_{12}}\}|ab\rangle \quad \text{(IV.12)}$$

Here, the imaginary part is entirely due to the last two terms, which appear only if $0 < \varepsilon_t < \varepsilon_a$ and $0 < \varepsilon_u < \varepsilon_b$, respectively.

Applying the potential obtained here twice in accordance with (IV.1), yields the *reducible* part of the two-photon scattering amplitude. The remaining *irreducible* part corresponds to the situations, where there are two photons present at time t=0, as illustrated in Fig. 6. In principle, these parts could be included in the evolution operator, although this leads to quite a complicated expression. A better alternative seems to be to include these parts - when needed - by means of an "*irreducible two-photon potential*", as discussed below.

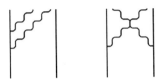

Figure 6. Evolution-operator diagrams corresponding to the irreducible part of the two-photon scattering amplitude.

It should be noted that *with the Feynman gauge it is not sufficient to use the frequency dependent potential (IV.10) or (IV.12) in order to get all contributions of $O(\alpha^2 R_y)$* - also the irreducible part has to be considered /10/. As we shall see below, the situation is different in the Coulomb gauge.

In order to convert the results above to the Coulomb gauge, we apply the transformation used in the one-photon case (I.21). The potential (IV.10) then yields

$$\langle tu | B_{12} | ab \rangle = \langle tu | 1/2 \, [\text{sign}(\varepsilon_t) \, B_{12} \, (q_{at}) + \text{sign}(\varepsilon_u) \, B_{12} \, (q_{ub})]$$

$$+ \frac{1}{2\pi i} \int dz \, B_{12} \, (z) \left[\frac{P}{z - q_{at}} - \frac{P}{z + q_{bu}} \right] | ab \rangle \qquad (IV.13)$$

where B_{12} (z) is defined in (I.21). The first part of this expression can be regarded as the <u>*non-diagonal Breit interaction in the Coulomb gauge*</u>. It is the analogue of the Mittleman potential in the Feynman gauge (IV.10). When the q's go to zero, for ε_t and ε_u being positive, this part goes over into the standard Breit interaction (I.23). This verifies that *this potential is justified also in evaluating non-diagonal elements, as long as the intermediate states have positive energy (no-virtual-pair approximation).* An alternative, and for numerical work more convenient, expression is obtained by applying an analogous transformation to (IV.12).

The integral appearing in (IV.13) is similar to that appearing in the corresponding Feynman-gauge expression. Due to the $\alpha_1 \cdot \alpha_2$ operators of the B operator, though, there will in the present case be no contribution of $O(\alpha \, Ry)$. The leading contribution from this integral is therefore of $O(\alpha^3 \, Ry)$.

Applying the potential (IV.13) twice together with the Coulomb potential, corresponds to the **reducible** part of the two-photon exchange in the Coulomb gauge. That will yield all contributions of $O(\alpha^2 \, Ry)$ and important parts of those due to the two-photon exchange up to $O(\alpha^4 \, Ry)$. The **irreducible** part, which can be evaluated as the Feynman case, will then have most of the remaining parts up to that order. This situation should be contrasted to that with the Feynman gauge, where also the irreducible parts were needed in order to get all contributions to $O(\alpha^2 \, Ry)$.

V. Radiative corrections

In this section we shall consider the first- and second-order radiative corrections (self-energy interaction and vacuum polarization). The technique applied in the previous sections can be generalized into the following set of Feynman rules, modified for evaluating diagrams with electron orbitals generated in an external potential

1. $\overline{\Psi}$ (Ψ) for each free outgoing (incoming) line
2. $-2\pi i\,\gamma^\mu$ at each vertex
3. Electron propagator $S_F(2,1,z)$ for each internal orbital line with the arrow going from 1 to 2

$$S_F(2,1,z) = \frac{1}{2\pi i}\sum_j \frac{|\phi_j(2)\rangle\langle\overline{\phi}_j(1)|}{\varepsilon_j - z - i\delta\varepsilon_j}$$

where $\{\phi_j\}$ are the single-electron orbitals, generated in some potential, U, and ε_j the corresponding orbital energy.

$\overline{\Psi}, \Psi, \gamma_\mu, S_F$ are read from right to left in the direction of the orbital line.
4. Photon propagator $g_{\mu\nu}\,D_F(2,1,z)$ for each internal photon line

$$D_F(2,1,z) = -\frac{1}{2\pi i}\frac{e^{\,i|z|r_{12}/c}}{r_{12}}$$

5. Energy conservation at each vertex, considering z as the energy of the internal lines
6. Integration over all z's from $-\infty$ to $+\infty$
7. Trace symbol and a factor of -1 for each closed orbital loop
8. Weight factor depending on the symmetry of the diagram (equal to one for an unsymmetric diagram and one half for a diagram with reflection symmetry). Note that the labeling of the lines may destroy the symmetry.

It is assumed here that in the normal-ordering positive energy states are treated as particles and negative energy states as holes. If the vacuum level is shifted to some closed-shell state, as normally used in non-relativistic MBPT, then the core states should be treated as holes, i.e., with a *positive* imaginary part in the electron propagator.

First-order corrections

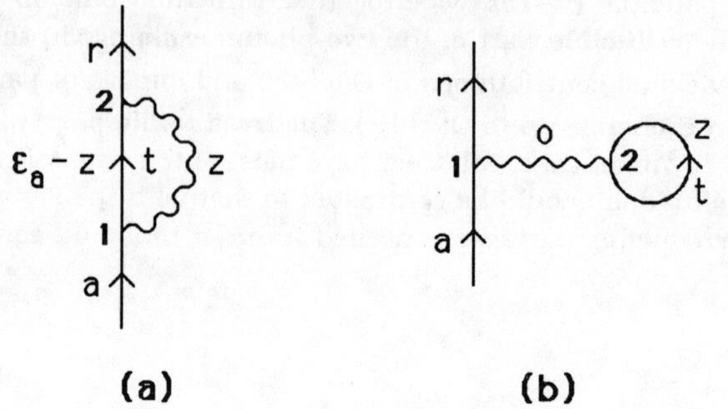

(a) **(b)**

Figure 7. First-order radiative corrections to the S matrix.

For the first-order corrections to the S matrix due to electron self-interaction and the vacuum polarization we get with the notations of Fig. 7

$$\bar{\Psi}\,(2)\,\gamma^\alpha\,S_F(2,1,\varepsilon_a\text{-}z)\,\gamma^\beta\,\Psi(1)\,g_{\alpha\beta}\,D_F(2,1,z)\qquad\text{(V.1a)}$$

$$-\,\bar{\Psi}\,(1)\,\gamma^\alpha\,\Psi(1)\,g_{\alpha\beta}\,D_F(1,2,0)\,\mathrm{Tr}[\gamma^\beta\,S_F(2,2,z)]\qquad\text{(V.1b)}$$

which leads to the S-matrix elements

$$\langle r\,|\,S^{(2)}(a)\,|\,a\rangle = -\sum_t \langle r\,t\,|\,\frac{\bar{\gamma}_1\cdot\bar{\gamma}_2}{r_{12}}\int dz\,\frac{e^{i|z|r_{12}}}{z - q_{at} - i\delta\varepsilon_t}\,|\,t\,a\rangle\qquad\text{(V.2a)}$$

$$\langle r\,|\,S^{(2)}(b)\,|\,a\rangle = \sum_t \langle r\,t\,|\,\frac{\bar{\gamma}_1\cdot\bar{\gamma}_2}{r_{12}}\int dz\,\frac{1}{\varepsilon_t - z - i\delta\varepsilon_t}\,|\,a\,t\rangle$$

$$= i\pi\sum_t \mathrm{sign}(\varepsilon_t)\,\langle r\,t\,|\,\frac{\bar{\gamma}_1\cdot\bar{\gamma}_2}{r_{12}}\,|\,a\,t\rangle\qquad\text{(V.2b)}$$

where $\bar{\gamma} = \gamma_o\gamma$ and $q_{at} = \varepsilon_a - \varepsilon_t$. (For convenience we let here ε denote the orbital energy divided by c). The corresponding potentials (on the energy shell) are obtained from the potential-scattering relation

$$\langle r\,|\,S\,|\,a\rangle = -2\pi i\,\delta(\varepsilon_r - \varepsilon_a)\,\langle rs\,|\,V_{12}\,|\,ab\rangle\qquad\text{(V.3)}$$

or, in other words, by dividing the S-matrix elements by $-2\pi i$. Off-the-energy-shell matrix elements can be obtained by replacing the outgoing lines by electron propagators, setting the final times equal to zero (see Fig. 8). The Feynman rules then yield the evolution operator, $U(0,-\infty)$, which in first order is identical to the wave operator, Ω. The expressions (V.1) are

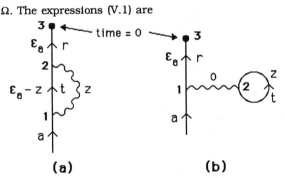

Figure 8. First-order radiative corrections to the evolution operator.

then replaced by

$$S_F(3,2,\varepsilon_a)\,\gamma^\alpha\,S_F(2,1,\varepsilon_a\text{-}z)\,\gamma^\beta\,\Psi(1)\,g_{\alpha\beta}\,D_F(2,1,z)\qquad\text{(V.4a)}$$

$$-\,S_F(3,2,\varepsilon_a)\,\gamma^\alpha\,\Psi(1)\,g_{\alpha\beta}\,D_F(1,2,0)\,\mathrm{Tr}[\gamma^\beta\,S_F(2,2,z)]\qquad\text{(V.4b)}$$

which leads to the evolution operators

$$\langle r\,|\,U^{(2)}(a)\,|\,a\rangle = \frac{1}{2\pi i}\sum_t \langle r\,t\,|\,\frac{\bar{\gamma}_1\cdot\bar{\gamma}_2}{r_{12}}\int dz\,\frac{e^{i|z|r_{12}}}{z - q_{at} - i\delta\varepsilon_t}\,|\,t\,a\rangle/(\varepsilon_a - \varepsilon_r)\qquad\text{(V.5a)}$$

$$\langle r | U^{(2)}(b) | a \rangle = -\frac{1}{2} \sum_t \text{sign}(\varepsilon_t) \langle r\, t | \frac{\bar{\gamma}_1 \cdot \bar{\gamma}_2}{r_{12}} | a\, t \rangle / (\varepsilon_a - \varepsilon_r) \qquad \text{(V.5b)}$$

The numerators are here the matrix elements of the equivalent potentials, which have the same form as the on-shell elements obtained from the S matrix

$$\langle r | V^{(2)}(a) | a \rangle = \frac{1}{2\pi i} \sum_t \langle r\, t | \frac{\bar{\gamma}_1 \cdot \bar{\gamma}_2}{r_{12}} \int dz \frac{e^{i|z|r_{12}}}{z - q_{at} - i\delta\varepsilon_t} | t\, a \rangle \qquad \text{(V.6a)}$$

$$\langle r | V^{(2)}(b) | a \rangle = -\frac{1}{2} \sum_t \text{sign}(\varepsilon_t) \langle r\, t | \frac{\bar{\gamma}_1 \cdot \bar{\gamma}_2}{r_{12}} | a\, t \rangle \qquad \text{(V.6b)}$$

The integral in the first expression can be evaluated by integration over the imaginary axis

$$\langle r | V^{(2)}(a) | a \rangle = -\frac{1}{\pi} \sum_t \langle r\, t | \frac{\bar{\gamma}_1 \cdot \bar{\gamma}_2}{r_{12}} \int_0^\infty dy\, e^{-y\, r_{12}} \frac{q_{at}}{y^2 + q_{at}^2} | t\, a \rangle \qquad \text{(V.7)}$$

By shifting the vacuum level, used to separate particles and holes in the second-quantized formalism, the first-order radiative corrections can be separated into the Hartree-Fock potentials of the core (states with positive energy below the vacuum level) and pure radiative corrections, as discussed in a separate paper /3/.

Most of the expressions above are divergent as the basis set gets infinitely large and therefore have to be *renormalized*, essentially by taking out the free-electron part. The technique of doing this in the present case requires further study.

Second-order corrections

In second order we consider the four diagrams shown in Fig. 9 for which the rules given above lead to the following expressions

$$\bar{\Psi}(4)\, \gamma^\delta S_F(4,2,\varepsilon_r\text{-}z')\, \gamma^\beta S_F(2,3,\varepsilon_a\text{-}z\text{-}z')\, \gamma^\gamma S_F(3,1,\varepsilon_a\text{-}z)\, \gamma^\alpha \Psi(1)\, g_{\alpha\beta}\, D_F(1,2,z)$$
$$\times g_{\gamma\delta}\, D_F(3,4,z') \qquad \text{(V.8a)}$$

$$-\bar{\Psi}(4)\, \gamma^\delta S_F(4,1,\varepsilon_a\text{-}z)\, \gamma^\alpha \Psi(1)\, \text{Tr}[\gamma^\gamma S_F(3,2,z')\, \gamma^\beta S_F(2,3,z'\text{-}z)]\, g_{\alpha\beta}\, D_F(1,2,z)$$
$$\times g_{\gamma\delta}\, D_F(3,4,z) \qquad \text{(V.8b)}$$

$$\bar{\Psi}(4)\, \gamma^\alpha S_F(4,1,\varepsilon_r\text{-}z)\, \gamma^\gamma S_F(1,3,\varepsilon_a\text{-}z)\, \gamma^\beta \Psi(3)\, g_{\alpha\beta}\, D_F(3,4,z)\, g_{\gamma\delta}\, D_F(1,2,\varepsilon_s\text{-}\varepsilon_b)$$
$$\times g\, \bar{\Psi}(2)\, \gamma^\delta \Psi(2) \qquad \text{(V.8c)}$$

$$-\bar{\Psi}(4)\, \gamma^\delta \Psi(4)\, g_{\gamma\delta}\, D_F(3,4,q_{sb})\, \text{Tr}[\gamma^\gamma S_F(3,2,z)\, \gamma^\beta S_F(2,3,z\text{-}q_{ar})]\, g_{\alpha\beta}\, D_F(1,2,q_{ar})$$
$$\times g\, \bar{\Psi}(1)\, \gamma^\alpha \Psi(1) \qquad \text{(V.8d)}$$

This leads to the following S-matrix elements

$$\langle rs | S^{(4)}(a) | ab \rangle = \frac{1}{2\pi i} \sum_{tuv} \langle u_3\, r_4 | \frac{\bar{\gamma}_3 \cdot \bar{\gamma}_4}{r_{34}} \langle t_1\, v_2 | \frac{\bar{\gamma}_1 \cdot \bar{\gamma}_2}{r_{12}} \times$$

$$\iint dz\, dz' \frac{e^{i|z|r_{12}}\, e^{i|z'|r_{34}}}{(z - q_{at} - i\delta\varepsilon_t)(z' - q_{rv} - i\delta\varepsilon_v)(z + z' - q_{au} - i\delta\varepsilon_u)} | t_3\, v_4 \rangle | a_1\, u_2 \rangle \qquad \text{(V.9a)}$$

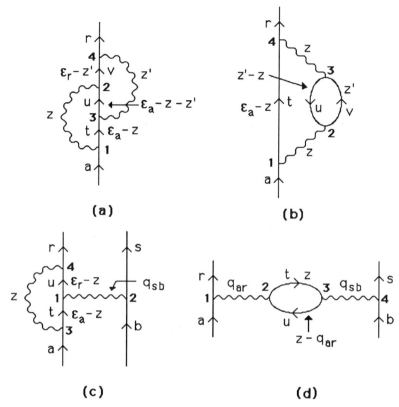

(a)   **(b)**

(c)   **(d)**

Figure 9. Second-order radiative corrections.

$$\langle r\,s\,|\,S^{(4)}(b)\,|\,ab \rangle = -\frac{1}{2\pi i}\sum_{tuv}\langle u_3 r_4|\;\frac{\bar{\gamma}_3\cdot\bar{\gamma}_4}{r_{34}}\;\langle t_1 v_2\,|\;\frac{\bar{\gamma}_1\cdot\bar{\gamma}_2}{r_{12}}\times$$

$$\iint dz\,dz'\;\frac{e^{i\,|z|r_{12}}\;e^{i\,|z'|r_{34}}}{(z - q_{at} - i\delta\,\varepsilon_t)\,(z' - \varepsilon_v - i\delta\,\varepsilon_v)\,(z + z' - \varepsilon_u - i\delta\varepsilon_u)}\;|v_3 t_4\rangle|a_1 u_2\rangle \quad\text{(V.9b)}$$

$$\langle rs\,|\,S^{(4)}(c)\,|\,ab\rangle = \sum_{tu}\langle t_3 r_4|\;\frac{\bar{\gamma}_3\cdot\bar{\gamma}_4}{r_{34}}\;\langle u_1 s_2\,|\;\frac{\bar{\gamma}_1\cdot\bar{\gamma}_2}{r_{12}}\;e^{i\,|q_{sb}|r_{12}}$$

$$\times \int dz\;\frac{e^{i\,|z|r_{34}}}{z - q_{ru} - i\delta\varepsilon_u}\;\frac{1}{z - q_{at} - i\delta\varepsilon_t}\;|a_3 u_4\rangle|t_1 b_2\rangle \quad\text{(V.9c)}$$

$$\langle rs\,|\,S^{(4)}(d)\,|\,ab\rangle = \sum_{tu}\langle u_3 s_4|\;\frac{\bar{\gamma}_3\cdot\bar{\gamma}_4}{r_{34}}\;\langle r_1 t_2\,|\;\frac{\bar{\gamma}_1\cdot\bar{\gamma}_2}{r_{12}}\;e^{i\,|q_{ar}|r_{12}}\;e^{i\,|q_{sb}|r_{34}}$$

$$\times \int dz\;\frac{1}{z - \varepsilon_t + i\delta\varepsilon_t}\;\frac{1}{z - \varepsilon_u - q_{ar} + i\delta\varepsilon_u}\;|t_3 b_4\rangle|a_1 u_2\rangle \quad\text{(V.9d)}$$

Also these expressions can be separated into standard many-body contributions and remaining radiative corrections by shifting the vacuum level, as in the first-order case, although this is less straightforward in the higher orders.

Also here, of course, the renormalization question has to be investigated further.

The construction of a many-body scheme of coupled-cluster type, based on the potentials derived in this work, is discussed in a separate paper /3/.

References

1. I. Lindgren and J. Morrison, *Atomic Many-Body Theory*, Second edition, Springer Series on Atoms and Plasmas, Vol. 3, Springer-Verlag, 1986
2. I.Lindgren, lecture notes in this volume
3. I.Lindgren, *A Relativistic Coupled-Cluster Procedure with Radiative Corrections*, Proceedings of the Symposium on Many-Body Methods in Quantum Chemistry, Tel Aviv 28-31 August 1988 (Ed. U.Kaldor) Springer-Verlag
4. C.Itzykson and J.-B.Zuber, *Quantum-Field Theory*, McGraw-Hill (1965)
5. M.Gell-Mann and F.Low, Phys.Rev. **84**, 350 (1951)
   J.Sucher, Phys. Rev. **107**, 1448 (1957)
6. P.Mohr, lecture notes in this volume
7. I.S.Gradshteyn and J.M.Ryzhik, *Tables of Integrals, Series and Products*, Translation from the Russian, Fourth edition, Academic Press (1965)
8. S.Blundell, W.Johnson, P.Mohr and J.Sapirstein, preprint
9. M.Mittleman, Phys.Rev. **A5**, 2395 (1971)
10. J.Sucher, preprint

A PERTURBATIVE APPROACH TO MANY ELECTRON ENERGY LEVELS

D. R. Yennie
Laboratory of Nuclear Studies,
Cornell University, Ithaca, NY 14853

ABSTRACT

A method which has proved successful for the study of the two-body problem and radiative level shifts is extended to the many electron problem. The aim is to identify perturbatively the relativistic and radiative corrections with the expectation that once they are properly understood it may prove possible to treat them non-perturbatively. I think that the methods are basically correct and may provide a viable approach to the proper identification of such corrections.

Introduction

We assume that the electrons are moving in a time-independent binding potential. Our aim is to formulate the problem of calculating their energy levels by studying a (time) Fourier transform of a multi-electron Green's function. The energy levels appear as poles of this Fourier transform as a function of the total energy. This method has proved useful in other contexts, such as the formulation of the Lamb shift which is alluded to below and the formulation of the two-body problem which is described in a separate paper in this volume. In brief, we study

$$-i \int_{-\infty}^{+\infty} dt \, e^{iEt} S^{(n)}(\vec{x}_i, t; \vec{y}_j, 0) \tag{1a}$$

where $S^{(n)}$ is the n-particle Green's function in the Heisenberg picture with all orders of photon interactions. It may be expressed as

$$S^{(n)} = \frac{1}{n!} < \text{vac}|T[\prod_i \psi_H(\vec{x}_i, t) \prod_j \bar{\psi}_H(\vec{y}_j, 0)]|\text{vac} > \tag{1b}$$

where $\psi_H, \bar{\psi}_H$ are Heisenberg picture operators. Note that all initial times are set equal to 0 and all final times to t. The energy levels show up as poles in the variable E. To see this, consider the time ordering $t > 0$ and insert a complete set of states $|n >$ between the products of operators. The expectation value then contains a factor $e^{-iE_n t}$, which gives a pole at $E = E_n$ when the integration from $t = 0$ to $t = \infty$ is evaluated. E_n is the energy of a state containing n electrons; in addition to the electrons, it may contain photons and electron-positron pairs.[1,2] To obtain energies of states involving positrons, one could introduce factors of ψ_H at $t = 0$ and factors of $\bar{\psi}_H$ at t. If we supply exponential damping as $t \to \infty$, we see that the pole lies below the real axis. A similar argument can be made for negative times with the result that the pole lies above the axis at $E = -E_n$. These general results are confirmed in the example of non-interacting particles described below.

To carry out a perturbative treatment, we may switch to the Furry interaction picture where

$$S^{(n)} = \frac{1}{n!} < \text{vac}|T[\prod_i \psi(\vec{x}_i, t) S(\infty, -\infty) \prod_j \bar{\psi}(\vec{y}_j, 0)]|\text{vac} > \ . \tag{1c}$$

Now ψ and $\bar{\psi}$ are second quantized electron operators and $S(\infty, -\infty)$ is the time ordered transition operator; both are expressed in the Furry representation interaction picture where S has a perturbation expansion. If all photon corrections are ignored, the n-particle Green's function takes the form

$$\prod_i \otimes S_F(\vec{x}_i, t; \vec{y}_i, 0). \tag{2a}$$

In general we may write

$$S_F(x, y) = \int_{-\infty}^{+\infty} \frac{d\xi}{-2\pi i} \sum_n \frac{u_n(\vec{x}) \bar{u}_n(\vec{y}) e^{-i\xi(x_0 - y_0)}}{\xi - E_n(1 - i\delta)} \tag{2b}$$

where the u_n's are eigenfunctions of the c-number Dirac equation in the external potential. They have the non-covariant normalization ($\int u^\dagger u d^3 x = 1$).

An expected advantage of this approach is that since it is a field theoretical rather than a Hamiltonian formulation, difficulties such as "continuum dissolution," which has been described by Joe Sucher,[3] will not occur. An indication of how this difficulty is avoided is given below. A complication is that it makes the additivity of the energies of non-interacting particles a little less direct. Beyond that lies the question of how one can economically find correct expressions for the energy shifts caused by radiative corrections and interactions between electrons. That depends on the experience of actually trying different methods and comparing their relative ease of sorting out various contributions. While I am convinced that the method to be described is a viable one and will lead to the right answers, it is not yet clear whether it is more convenient than some others which are described in this volume. Probably one can gain valuable insights by comparing the various methods. One method that fascinated me was the ϵ-limiting procedure of Peter Mohr,[4] based on earlier work of Joe Sucher.[5] If that is to be carried very far, it entails the possibly delicate compensation of higher powers of $1/\epsilon$ in order to extract the coefficient of the first power. The counterpart of that in the present discussion is that the pole in E is found by identifying a geometric series in $1/(\epsilon + i\delta)$, where now ϵ is the difference between E and the unperturbed energy. The energy shift is expressible in terms of the coefficient of the term with the factor $1/(\epsilon + i\delta)^2$. One must be persuaded that this works, and I try to be convincing but cannot present a general proof here. If one accepts this argument, it turns out to be relatively straightforward to derive perturbative expressions for energy shifts. I have made no attempt to fit this together with a Hartree-Fock or similar type analysis. The main idea is to find a method which treats relativity and radiative corrections conceptually correctly.

It is clear on physical grounds that to obtain a non-vanishing contribution to (1a) from the non-interacting particle term (2a), all sums must consist of either all positive energies or all negative energies (for $t > 0$ only positive energies contribute; while for $t < 0$ only negative energies contribute). As an example, suppose we consider the case $n = 3$ and take all energies positive. We suppress the spinor dependence and examine the denominator structure. After the t-integration, we find (from now on, a set of integrations extending from $-\infty$ to ∞ is represented by a single integral sign)

$$\int \frac{d\xi_1}{-2\pi i} \frac{d\xi_2}{-2\pi i} \frac{d\xi_3}{-2\pi i} \frac{-2\pi i \delta(E - \xi_1 - \xi_2 - \xi_3)}{(\xi_1 - E_{n_1} + i\delta)(\xi_2 - E_{n_2} + i\delta)(\xi_3 - E_{n_3} + i\delta)}.$$

After carrying out one of the integrations using the δ-function, the others may be done by using contours; and we find the result

$$\frac{1}{E - E_{n_1} - E_{n_2} - E_{n_3} + i\delta}. \tag{3}$$

The energies of non-interacting particles are additive, as anticipated. This confirms the expectations described below (1b). It is easy to see explicitly how a combination of positive and negative energy values gives a zero contribution. For example, suppose that only E_{n_3} were negative. Then the sign of $i\delta$ would be reversed in the third denominator; and after the δ-function is used, one finds it possible to close one of the remaining contours so as to avoid enclosing any poles. If all E_n are negative, one finds a result similar to (3) except that the sign of δ is reversed. As anticipated from the discussion following (1b), E has a pole at the negative of the energy of a possible state.

Simply stated, the non-interacting term contains a pole for each combination of energies of the bound states in the external potential. Usually, we wish to find a particular energy as perturbed by the interactions, rather than deal with the whole spectrum at once. To perturb around a particular state (labeled by the one electron states occupied as i, j, ...), we simply replace the operators ψ and $\bar{\psi}$ in (1c) by (Furry) interaction picture operators as follows

$$\bar{\psi}(\vec{y}_j, 0) \to b_i^\dagger = \int \bar{\psi}(\vec{y}, 0)\gamma_0 u_i(\vec{y})d^3y$$

and

$$\psi(\vec{x}_i, t) \to b_i(t) = \int \bar{u}_i(\vec{y})\gamma_0 \psi(\vec{y}, t)d^3y \ .$$

With this replacement, only one particular pole occurs in (3) and the numerator spinors have disappeared.

To make the subsequent discussion a little more compact, define

$$E \equiv \epsilon + E_i + E_j + \dots \ .$$

Without any interactions, we have a single pole $1/(\epsilon + i\delta)$. The effect of the interactions is to shift the position of this pole. I assert that when the interactions are taken into account, the result will take the form

$$C(\epsilon) + F(\epsilon)\left\{\frac{1}{\epsilon + i\delta} + \frac{\Delta(\epsilon)}{(\epsilon + i\delta)^2} + \frac{\Delta(\epsilon)^2}{(\epsilon + i\delta)^3} + \dots\right\} \tag{4a}$$

$$= C(\epsilon) + F(\epsilon)\frac{1}{\epsilon - \Delta(\epsilon) + i\delta} \ .$$

I.e., some of the perturbations "cancel" the original denominator leaving a term (C) regular at $\epsilon = 0$. The remaining terms produce a geometric series which may be summed, together with a factor F which affects the residue of the pole, but not its position. The function Δ generally depends on ϵ, so the position of the pole is given by expanding it about $\epsilon = 0$:

$$\delta E \approx \Delta(0) + \Delta(0)\Delta'(0) + \dots \tag{4b}$$

where the prime denotes differentiation with respect to ϵ.

In practice, I think this method may be more straightforward than some other ones for finding perturbative expressions for the energy shifts. The goal is to find the function Δ as the coefficient of the $1/(\epsilon + i\delta)^2$ term in the expansion. The procedure would be the following. One starts with simple perturbations in (1c) and proceeds to more complicated ones. At each level, F and Δ will have been worked out to some accuracy as a series of terms; but higher powers of Δ will be incomplete, and some terms in F discovered in the first term of the expansion may not appear in the higher terms. New terms may contribute to C, or fill out the previously missing terms, or be genuinely new contributions to Δ. It is necessary to identify successively these various terms in order to verify the form of (4a) and find the new contributions to Δ. Since Δ is a sum of terms, higher powers must contain products of these terms with the right weight according to a multinomial expansion. While all of this may sound complicated, it is quite straightforward and obvious in practice. One starts with the simplest interaction diagrams. These may have terms with zero, one, or two denominators ($\epsilon + i\delta$). The zero denominator term, if any, is a contribution to C. The one denominator term is a contribution to F. The two denominator term gives an energy shift. In the next level of complication, we discover terms which may have up to three denominators. The zero and one denominator terms give further modifications of C and F. The two denominator term may either be absorbed into a previously discovered term in F multiplying the previously discovered contribution to Δ, which must appear at some point, or it gives an additional contribution to the energy shift. Finally, three denominator terms should be contributions from the square of the previously discovered terms in the energy shift. It is of course important to verify that these terms occur exactly, i.e., with the right combinatorial factors, etc. I have satisfied myself that this works, but haven't tried to write out a general proof. Once this is established (or believed), one can forget these higher terms and simply study the coefficient with the denominator squared. Here it is important to keep tabs on changes in F which show up first in the one denominator term so that they are not misidentified as energy shifts. They seem to have a characteristic structure and should not cause great difficulty.

The reader should note that the definitions of F and Δ are not individually unique. For example, suppose the functions are altered slightly by the transformation

$$F \to F(1 + g), \quad \Delta \to \Delta(1 + f) - \epsilon f . \tag{4c}$$

This leaves the pole term unchanged; accordingly, it can not affect the value of the energy shift. Such modifications could occur in various ways. For example, under a gauge change, the definition of F will certainly be altered. Also, there is a certain amount of flexibility in treating various perturbations and it may be possible to take advantage of this property to make things simpler.

Example (a): External Potential

We'll try to build up some technical expertize by considering a succession of more and more complicated examples. Suppose we have a single electron in a time-independent external potential given by the perturbation Hamiltonian

$$\mathcal{V} = \int \bar{\psi} V \psi d^3 x . \tag{5a}$$

We perturb about a state a. All time integrations have an integrand of the form $\exp[-i(\xi_i - \xi_j)x_0]$. Including the factor $-i$ from the perturbation expansion, the integration produces a factor $-2\pi i \delta(\xi_i - \xi_j)$. Each free integration that remains has a denominator $-2\pi i$. In the present case, all ξ-integrations are eliminated and we obtain the series

$$\frac{1}{E - E_a + i\delta} + \frac{V_{aa}}{(E - E_a + i\delta)^2} + \sum_n \frac{V_{an}V_{na}}{(E - E_a + i\delta)^2(E - E_n(1 - i\delta))}$$

$$+ \sum_{n,n'} \frac{V_{an}V_{nn'}V_{n'a}}{(E - E_a + i\delta)^2(E - E_n(1 - i\delta))(E - E_{n'}(1 - i\delta))} + \dots \tag{5b}$$

where

$$V_{nn'} \equiv \int \bar{u}_n V u_{n'} d^3 x \; .$$

In (5b), it is important to separate out the terms in the sum where $n(n') = a$ from the others. Then one can identify Δ in this case to be

$$\Delta_a = V_{aa} + \sum_{n \neq a} \frac{V_{an} V_{na}}{E - E_n(a - i\delta)} + \sum_{n,n' \neq a} \frac{V_{an} V_{nn'} V_{n'a}}{(E - E_n(1 - i\delta))(E - E_{n'}(1 - i\delta))} + \dots . \quad (5c)$$

The index a refers to the fact that we are perturbing about the state a. To be persuaded of this result, one should check that the expansion works out correctly. For example, the terms with $n = n' = \dots = a$ in the original expansion correspond to the geometrical series associated with the first term of (5c). If we take either n or n' to be a in the double sum of (5b), we find the cross term between the first two terms of Δ_a in two different ways. One may proceed in a similar way to identify higher terms in the expansion and see that they produce the correct higher powers of Δ_a with the correct combinatorial factor. This is a particularly simple example, but it does illustrate the idea. Observations: I have ignored possible complications requiring degenerate perturbation theory; also note that negative energy states are properly included.

Our one electron result is

$$\frac{1}{\epsilon_a - \Delta_a(\epsilon_a) + i\delta} \quad (6a)$$

where $\epsilon_a = E - E_a$. This is the same result which would be obtained by using the resolvent kernel approach. There is no C in this case. For later application, it is necessary to note briefly the analytic behavior of this as a function of ϵ_a. A more complete argument than we can give here shows that denominator zeros can occur only near the real axis. The denominator has a zero for ϵ_a close to zero, and also at values corresponding to all those states which are coupled to the state a through the interaction. The easiest way to see this is to plot the denominator as a function of ϵ_a and note the effect of the denominators in (5c). Examination shows that the additional poles produced by the sum occur *below* the axis for positive energies (just like the one near E_a) while negative energy poles lie above the axis. This agrees with the discussion following (1b). The complete expression, which includes all the poles is

$$\sum_j \frac{R_j}{E - \tilde{E}_j(1 - i\delta)} \quad (6b)$$

where R_i is the residue of the ith pole. Here $\tilde{E}_i$ is the exact energy of one of the states. Since we are interested only in the pole near E_a, we separate out that term

$$\frac{R_a}{E - \tilde{E}_a + i\delta} \; . \quad (7)$$

We don't care about the value of the residue, which comes from expanding Δ about the point $\tilde{\epsilon}_a$. These forms will be convenient when we consider multi-electron contributions.

A generalization of this is to study what happens when there are one or more "bystander" electrons in positive energy states while the interaction with a single electron takes place. For each bystander, we need to introduce a ξ integration. Each bystander electron then provides a factor $1/(\xi_i - E_i + i\delta)$, while a term $-\xi_i$ is added to all the denominators of (5b). We find the result (from now on, we simply ignore terms contributing to C)

$$\frac{1}{E - E_a - \sum_{i>1} \xi_i - \Delta_a(E - E_a - \sum_{i>1} \xi_i) + i\delta} \times \prod_{i>1} \frac{1}{\xi_i - E_i + i\delta} \ . \qquad (8)$$

We aim to carry out the ξ-integration by the method of contours. In general, we can use (6b) for the first factor in (8) by inserting $-\sum \xi_i$ in the denominator. For negative energy states in the first factor, we can close the each contour above and find zero. For positive energy states, we close each contour below and keep only the terms from the poles at $\xi_i = E_i$. For the general result, one includes $-\sum_{i>1} E_i$ in the denominator of (6b). Thus we obtain the reasonable result that the energy of an individual electron is unaffected by the presence of the others as long as they don't interact. For the contribution with small ϵ_a we simply use (7) to obtain the term of interest.

The next level of complication is to consider interactions on several electron lines. We sum the interactions along each line separately and find

$$\frac{1}{E - E_a - \sum \xi_i - \Delta_a(E - E_a - \sum \xi_i) + i\delta} \times \prod_{i>1} \frac{1}{\xi_i - E_i - \Delta_i(\xi_i - E_i) + i\delta} \ . \qquad (9a)$$

Although each factor here has many poles, the ones we want correspond to using (7) for each factor for the relevant poles. Thus we replace (9a) by

$$\frac{R_a}{E - \tilde{E}_a - \sum \xi_i + i\delta} \times \prod_{i>1} \frac{R_i}{\xi_i - \tilde{E}_i + i\delta} \ . \qquad (9b)$$

Next we carry out the ξ_i integrations by closing the contours below and find

$$\frac{\text{residue}}{E - \sum_{i=1}^{n} \tilde{E}_i + i\delta}$$

where $E_1 \equiv E_a$. Thus the energy is given by

$$E = \sum_{i=1}^{n} \tilde{E}_i \ . \qquad (9c)$$

The energies are still additive, as they must be by common sense.

How about the poles we have ignored in (9a)? I assert that the ignored poles will all produce different energy eigenvalues. If we use (6b) for the individual factors, we can get a non-zero result only if all energies are negative or all energies are positive (just like the free case). If all energies are positive, the pole is just below the axis at the point where E is the total energy. If all are negative, it is above the axis at the point where E is the negative of the energy.

Example (b): Individual line radiative corrections

Only a sketch of the formal method of treating the Lamb shift is given here[6]. The electron propagator in an external field may be written formally as

$$\frac{1}{\slashed{A} - m + i\delta} \qquad (10a)$$

where $\Pi^\mu = p^\mu - eA^\mu$, with $p^0 \equiv \epsilon + E_1$. The components of Π^μ are non-commuting because $\vec{p}$ is the differential operator and A^μ is the space-dependent external potential. With radiative corrections, this becomes

$$\frac{1}{\slashed{\Pi} - m + [\Sigma(\Pi) - \delta m] + i\delta} \tag{10b}$$

where Σ is the electron self energy in the Furry representation. A renormalization analysis show that it may be written

$$[\Sigma(\Pi) - \delta m] = B[\slashed{\Pi} - m] + (1 - B)\Sigma_c . \tag{10c}$$

Here δm and B have the same definitions as in the usual renormalization and Σ_c is the finite part of the electron self-energy. Here we encounter one difference from the free particle case. If Π were a c-number, we would define B so as to require that $\Sigma_c \sim (\slashed{\Pi} - m)^2$. Now, however, the components of Π^μ are non-commuting; in fact, the commutators are just components of the electromagnetic field strengths. This leads to additional terms in Σ_c which need not have factors of $\slashed{\Pi} - m$. In fact, it is these additional terms which are responsible for the anomalous magnetic moment and the Lamb shift. Now we see that the propagator with radiative corrections is simply

$$\frac{1}{1 - B} \frac{1}{\slashed{\Pi} - m - \Sigma_c + i\delta} . \tag{10d}$$

For purposes of identifying the level (i.e., Lamb) shift, we may ignore the overall factor and analyze the (formal) expression

$$\bar{u}_a \gamma_0 \frac{1}{\slashed{\Pi} - m - \Sigma_c + i\delta} \gamma_0 u_a . \tag{11a}$$

The next step is to expand up Σ_c and examine the resulting series using the fact that in coordinate space $1/(\slashed{\Pi} - m)$ corresponds to (2b) to introduce intermediate sums over states

$$\frac{1}{\epsilon + i\delta} + \frac{\Sigma_{aa}}{(\epsilon + i\delta)^2} + \sum_n \frac{\Sigma_{an}\Sigma_{na}}{(\epsilon + i\delta)^2(\epsilon + E_a - E_n + i\delta)} + \dots \tag{11b}$$

where we define $\Sigma_{nn'} = \bar{u}_n \Sigma_c u_{n'}$. This rearranges into a geometric series whose sum is

$$\frac{1}{\epsilon - \sigma(\epsilon) + i\delta}$$

where

$$\sigma(\epsilon) \equiv \Sigma_{aa} + \sum_{n \neq a} \frac{\Sigma_{an}\Sigma_{na}}{\epsilon + E_a - E_n(1 - i\delta)} + \dots . \tag{11c}$$

This fits the scheme defined by (4a) and yields the usual formal expression for the Lamb shift, including derivative terms if carried to high enough order. Details of the actual calculation are quite lengthy and are not relevant to our present discussion. The purpose for mentioning the Lamb shift here is to indicate the utility of the analysis of the geometrical series in various contexts.

The extension to the case of several electrons with radiative corrections and possible additional external potentials follows through in exactly the same manner as at the end of Example (a). It is even possible to carry through the renormalization analysis prior to treating the interaction between electrons or the recoil of the nucleus. This is very helpful in avoiding spurious terms of lower order than the ones ultimately of interest.

Example (c): Interactions between the electrons

Consider the contributions arising from photon exchanges (treated covariantly) between the electrons. In the Feynman gauge, each photon propagator has a factor

$$D_{F\mu\nu}(x - y) = -g_{\mu\nu} \int \frac{d^4k}{-(2\pi)^4 i} \frac{e^{-ik\cdot(x-y)}}{k^2 + i\delta} \,. \tag{12}$$

Alternatively, we could use the Coulomb gauge; but since our aim here is primarily to examine the general structure, we do not do so. Thus, at any vertex there will be a time-dependent factor like

$$e^{i(\xi_1 - \xi_2 + k_0)x_0},$$

whose time integral produces a factor $-2\pi i\delta(\xi_1 - \xi_2 + k_0)$ which we use either to express k_0 in terms of the ξ's or to relate different ξ's. Thus, after taking these δ-functions into account, the direct term with one photon exchanged produces an integrand

$$\frac{\bar{u}_a \gamma_\mu e^{i\vec{k}\cdot\vec{x}} u_a \bar{u}_b \gamma^\mu e^{-i\vec{k}\cdot\vec{x}} u_b}{(\xi_1 + i\delta)(\xi_2 + i\delta)(\epsilon - \xi_1 + i\delta)(\epsilon - \xi_2 + i\delta)[(\xi_1 - \xi_2)^2 - \vec{k}^2 + i\delta]} \,. \tag{13a}$$

Here, $\epsilon = E - E_a - E_b$. As usual, we attempt to carry out the ξ-integrations by closing the contours below. The most important contributions are those arising from the electron poles at $\xi_{1,2} = 0$; they give the result

$$\frac{\bar{u}_a \gamma_\mu e^{i\vec{k}\cdot\vec{x}} u_a \bar{u}_b \gamma^\mu e^{-i\vec{k}\cdot\vec{x}} u_b}{(\epsilon + i\delta)^2 [-\vec{k}^2 + i\delta]} \tag{13b}$$

which corresponds to an obvious contribution to the energy. Note that the $\mu = 0$ term is the Coulomb interaction between the electrons even though we started with a covariant photon propagator.

There are also contributions from the photon poles (in this case, either ξ_1 or ξ_2, but not both, has such a contribution). The photon pole contributions have the form

$$\frac{\bar{u}_a \gamma_\mu e^{i\vec{k}\cdot\vec{x}} u_a \bar{u}_b \gamma^\mu e^{-i\vec{k}\cdot\vec{x}} u_b}{2\vec{k}^2(\epsilon - |\vec{k}| + i\delta)(\epsilon + i\delta)} \,. \tag{13c}$$

Since this has only one denominator $\epsilon + i\delta$, it is identified as a modification of the overall factor F, rather than as a contribution to the energy shift. I believe that in a more complete discussion it could be identified as a wave-function modification. In any case, we need not calculate it seriously, but we should be prepared to find similar coefficients in higher terms of the expansion. The two equal contributions to F found here can be regarded as arising from the initial or final electron pairs. When the analysis is carried further, it will appear that F factorizes; i.e., $F = F_f F_i$, with

$$F_i = 1 - 4\pi\alpha \int \frac{<a|\gamma_\mu e^{i\vec{k}\cdot\vec{x}}|a><b|\gamma^\mu e^{-i\vec{k}\cdot\vec{x}}|b>}{2\vec{k}^2(\epsilon - |\vec{k}| + i\delta)} \frac{d^3k}{(2\pi)^3} + \dots \,. \tag{13d}$$

It is mildly disconcerting that the modification of F apparently has an infrared divergence giving $\ln \epsilon$. However, I do not think this is a real problem; it should be interesting to see whether this infrared behavior exponentiates.[7]

The electron exchange contribution has an integrand similar to (13a)

$$-\frac{\bar{u}_a\gamma_\mu e^{i\vec{k}\cdot\vec{x}}u_b\bar{u}_b\gamma^\mu e^{-i\vec{k}\cdot\vec{x}}u_a}{(\xi_1+i\delta)(\xi_2+E_a-E_b+i\delta)(\epsilon-\xi_1+i\delta)(\epsilon-E_a+E_b-\xi_2+i\delta)[(\xi_1-\xi_2)^2-\vec{k}^2+i\delta]}. \tag{14a}$$

The most important contribution again arises from the electron poles. It is

$$-\frac{\bar{u}_a\gamma_\mu e^{i\vec{k}\cdot\vec{x}}u_b\bar{u}_b\gamma^\mu e^{-i\vec{k}\cdot\vec{x}}u_a}{(\epsilon+i\delta)^2[(E_a-E_b)^2-\vec{k}^2+i\delta]}. \tag{14b}$$

Obviously, this produces an exchange contribution to the energy. Again, the photon pole terms give a modification to the overall factor F which we do not record here explicitly. It is worth mentioning that similar terms occur in the Lamb shifts of the two particles and would exactly cancel these terms if they were added together. This is a manifestation of the exclusion principle; it eliminates a spurious imaginary part which would otherwise occur if $|\vec{k}|$ runs through the energy difference. However, it is better not to manifest the cancellation of the real parts since the Lamb shift is best calculated by keeping all intermediate states (e.g., they are involved in a correct treatment of mass renormalization).

Next consider two-photon exchange in the ladder configuration. The contribution in which there is no overall exchange of the electrons is

$$\frac{\bar{u}_a\gamma_\mu e^{i\vec{k}\cdot\vec{x}}u_n\bar{u}_n\gamma_\nu e^{i\vec{k}'\cdot\vec{x}}u_a}{(\xi_1+i\delta)(\xi_2+E_a-E_n[1-i\delta])(\xi_3+i\delta)}$$
$$\times\frac{\bar{u}_b\gamma^\mu e^{-i\vec{k}\cdot\vec{x}}u_{n'}\bar{u}_{n'}\gamma^\nu e^{-i\vec{k}'\cdot\vec{x}}u_b}{(\epsilon-\xi_1+i\delta)(\epsilon-\xi_2+E_b-E_{n'}[1-i\delta])(\epsilon-\xi_3+i\delta)} \tag{15a}$$
$$\times\frac{1}{[(\xi_1-\xi_2)^2-\vec{k}^2+i\delta][(\xi_2-\xi_3)^2-\vec{k}'^2+i\delta]}$$

where the intermediate states n, n' are of course to be summed over later. We carry out the $\xi_{1,3}$ integrations first, ignoring the photon pole terms (to be discussed later) and find the result

$$\frac{\bar{u}_a\gamma_\mu e^{i\vec{k}\cdot\vec{x}}u_n\bar{u}_n\gamma_\nu e^{i\vec{k}'\cdot\vec{x}}u_a\bar{u}_b\gamma^\mu e^{-i\vec{k}\cdot\vec{x}}u_{n'}\bar{u}_{n'}\gamma^\nu e^{-i\vec{k}'\cdot\vec{x}}u_b}{(\epsilon+i\delta)^2(\xi_2+E_a-E_n[1-i\delta])(\epsilon-\xi_2+E_b-E_{n'}[1-i\delta])}. \tag{15b}$$
$$\times\frac{1}{[\xi_2^2-\vec{k}^2+i\delta][\xi_2^2-\vec{k}'^2+i\delta]}$$

Now there are a large number of possibilities to consider. Suppose at first that the pair $\{n,n'\}\neq\{a,b\}$. Then (15b) corresponds to an energy shift and does not contain higher terms in the geometric series because the ξ_2-integration is unable to produce an additional denominator $(\epsilon+i\delta)$. Our aim here is to identify terms rather than discuss them in detail, so we limit ourselves to a few remarks. For some situations, the ξ-integrations are done in detail in an Appendix. The integration over ξ_2 depends on whether the energies $E_{n,n'}$ are positive or negative. If both are positive and we close the contour below the axis, ignoring the photon poles, the result is

$$\frac{\bar{u}_a\gamma_\mu e^{i\vec{k}\cdot\vec{x}}u_n\bar{u}_n\gamma_\nu e^{i\vec{k}'\cdot\vec{x}}u_a\bar{u}_b\gamma^\mu e^{-i\vec{k}\cdot\vec{x}}u_{n'}\bar{u}_{n'}\gamma^\nu e^{-i\vec{k}'\cdot\vec{x}}u_b}{(\epsilon+i\delta)^2(\epsilon+E_a+E_b-E_n-E_{n'}+i\delta)} \tag{15c}$$
$$\times\frac{1}{[(E_a-E_n)^2-\vec{k}^2+i\delta]}\frac{1}{[(E_a-E_n)^2-\vec{k}'^2+i\delta]}.$$

For reasons of symmetry, Ingvar Lindgren likes to average this with the similar term obtained by closing the contour above and using the other electron pole. The apparent asymmetry of either of these contributions taken alone is removed when the photon poles are taken into account (see Appendix). The photon poles produce a characteristically smaller contribution than the electron poles because they have $\xi_2 = k, k'$ and produce larger denominators (at least for light nuclei where $k \sim Z\alpha m_e$ and $E_{a,b,n} \sim (Z\alpha)^2 m_e$). (In the case of Coulomb interactions, they of course give no contributions and the two electron poles give the same contribution.) More importantly, the photon pole residues eliminate the apparent poles at $k, k' = |E_a - E_n|$. This changes the character of the functions obtained by integrating out the $\vec{k}, \vec{k}'$ dependencies. Instead of producing an oscillating function of the coordinate difference, it produces one with higher inverse powers of the distance between the electrons at large separation.

If one of the energies is positive and the other is negative, we obtain a contribution only from a photon pole because we may close the contour away from both electron poles. I will not record it here. This eliminates the continuum dissolution problem which would lead to an expression similar to (15c) independently of the signs of the energies, permitting the denominator to become zero. If both are negative and we avoid the photon poles, we obtain a result similar to (15c) except for the following two changes: the sign of δ is reversed, the overall sign of the expression is reversed. Of course, in this case the photon poles should be no less important that the electron poles.

In the special case where $n = a, n' = b$, (15c) gives the term with the direct contribution squared, as expected from the geometric series expansion. The other case where $n = b, n' = a$ corresponds to the square of the exchange contribution. In addition to these terms, the residue from the photon pole cannot be ignored in evaluating (15b). With ξ_2 set equal to k or k', we are left with a $1/(\epsilon + i\delta)^2$ term representing an energy shift. Note that there is not an actual pole at $k = k'$ in spite of appearances. This case is discussed further below. Of course with the Coulomb interaction in the Coulomb gauge, this contribution to the energy does not occur. This does not mean that the energies are gauge dependent. Rather, it is the calculation that is gauge dependent; but when carried out consistently to appropriate order in any gauge, the results should agree.

To show the presence of the square of the complete one photon contribution to the energy shift, we must consider the contribution where there is overall an exchange of the two electrons and the intermediate states are taken to be the pair $\{a, b\}$. When we evaluate the leading electron poles, we find the expected cross term between the direct and exchange one-photon contributions. It occurs in two ways and has the right sign, as expected. The cases where the intermediate states differ from the initial pair do produce additional contributions to the energy.

Now we have to identify the contributions arising from various contributions of the $\xi_{1,3}$ photon poles. We can understand this best by rearranging denominators on line 1 of (15a)

$$\frac{1}{\xi_{1,3} + i\delta} = -2\pi i\delta(\xi_{1,3}) - \frac{1}{-\xi_{1,3} + i\delta} \, . \qquad (16)$$

The contributions just described correspond to taking the product of the two δ-functions. Any combination of the other (remainder) terms must correspond to some photon pole residues. Consider first the cases where one δ-function is used with the remainder from the other. The situation then depends on whether $\{n, n'\} = \{a, b\}$ or not. With $\{n, n'\} \neq \{a, b\}$, we have now only one $\epsilon + i\delta$ denominator; and hence we find a new modification of the overall coefficient F_i or F_j, depending on which remainder is used.

With $\{n, n'\} = \{a, b\}$, we find a correction to the $1/(\epsilon + i\delta)^2$ term with the same coefficient that occurred in a previous $1/(\epsilon + i\delta)$ term. I.e., this supplies an expected term so that $F_i F_f$ indeed multiplies every term in the series. If both remainders are used in the $\{n, n'\} = \{a, b\}$ case, we find a single denominator term which is the product of one photon contributions in F_i and F_f. This shows that the product of these factors is being developed in perturbation theory.

For the case $n = a, n' = b$, the electron pole of (15b) is removed by a rearrangement similar to (16) so that its photon pole contribution is given by

$$-\frac{\bar{u}_a \gamma_\mu e^{i\vec{k}\cdot\vec{x}} u_a \bar{u}_a \gamma_\nu e^{i\vec{k}'\cdot\vec{x}} u_a \bar{u}_b \gamma^\mu e^{-i\vec{k}\cdot\vec{x}} u_b \bar{u}_b \gamma^\nu e^{-i\vec{k}'\cdot\vec{x}} u_b}{(\epsilon + i\delta)^2(-\xi_2 + i\delta)(\epsilon - \xi_2 + i\delta)} \frac{1}{(\xi_2^2 - \vec{k}^2 + i\delta)(\xi_2^2 - \vec{k}'^2 + i\delta)}$$

$$(17a)$$

and, with intermediate states exchanged, by

$$-\frac{\bar{u}_a \gamma_\mu e^{i\vec{k}\cdot\vec{x}} u_b \bar{u}_b \gamma_\nu e^{i\vec{k}'\cdot\vec{x}} u_a \bar{u}_b \gamma^\mu e^{-i\vec{k}\cdot\vec{x}} u_a \bar{u}_a \gamma^\nu e^{-i\vec{k}'\cdot\vec{x}} u_b}{(\epsilon + i\delta)^2(-\xi_2 - E_a + E_b + i\delta)(\epsilon - \xi_2 + E_b - E_a + i\delta)} \frac{1}{(\xi_2^2 - \vec{k}^2 + i\delta)(\xi_2^2 - \vec{k}'^2 + i\delta)}.$$

$$(17b)$$

They are displayed this way rather than after integration so that it may more easily be seen that these terms are partially canceled by the crossed photon contribution.

Next we write down the expression for the crossed-photon integrand with no overall exchange

$$\frac{\bar{u}_a \gamma_\mu e^{i\vec{k}\cdot\vec{x}} u_n \bar{u}_n \gamma_\nu e^{i\vec{k}'\cdot\vec{x}} u_a \bar{u}_b \gamma^\nu e^{-i\vec{k}'\cdot\vec{x}} u_{n'} \bar{u}_{n'} \gamma^\mu e^{-i\vec{k}\cdot\vec{x}} u_b}{(\xi_1 + i\delta)(\xi_2 + E_a - E_n[1 - i\delta])(\epsilon - \xi_1 + i\delta)(\epsilon + \xi_2 - \xi_1 - \xi_3 + E_b - E_{n'}[1 - i\delta])}$$

$$\times \frac{1}{(\xi_3 + i\delta)(\epsilon - \xi_3 + i\delta)} \frac{1}{[(\xi_1 - \xi_2)^2 - \vec{k}^2 + i\delta][(\xi_2 - \xi_3)^2 - \vec{k}'^2 + i\delta]}.$$

$$(18a)$$

To obtain a contribution to the energy shift, we must use the first terms of (16); the remainders give a correction to F_i or F_f. At this stage, we have

$$\frac{\bar{u}_a \gamma_\mu e^{i\vec{k}\cdot\vec{x}} u_n \bar{u}_n \gamma_\nu e^{i\vec{k}'\cdot\vec{x}} u_a \bar{u}_b \gamma^\nu e^{-i\vec{k}'\cdot\vec{x}} u_{n'} \bar{u}_{n'} \gamma^\mu e^{-i\vec{k}\cdot\vec{x}} u_b}{(\epsilon + i\delta)^2(\xi_2 + E_a - E_n[1 - i\delta])(\epsilon + \xi_2 + E_b - E_{n'}[1 - i\delta])}$$

$$\times \frac{1}{(\xi_2^2 - \vec{k}^2 + i\delta)(\xi_2^2 - \vec{k}'^2 + i\delta)}.$$

$$(18b)$$

The case $n = a, n' = b$ is especially interesting. If we reverse the sign of ξ_2 in the integration, we see that it exactly cancels one of the terms remaining from the pole contribution to the ladder graph, namely (17a). While either of these terms contribute individually, we need not calculate them.

On the other hand, if we take $n = b, n' = a$ in (18b) and reverse the sign of ξ_2, we do not obtain a term which cancels any photon ladder terms. Even if $E_a = E_b$, it is not possible to get the matrix elements to match. Of course, there may be special circumstances where the dominant parts do cancel, leaving a small result.

At this point, the general pattern is beginning to emerge, and we can now justify the form (4a), with $F = F_i F_f$, for two interacting electrons. Any graph can be broken up into a succession of two-electron irreducible kernels separated by two electron lines. We first have to identify how the different numbers of inverse powers of $(\epsilon + i\delta)$ can arise.

If a two-electron state combination between kernels is the same as that of the external electrons, integration over the associated ξ produces such a factor from the electron poles, but not from the photon poles. The same is true for the external pair of electron lines. Photon pole contributions and contributions from other intermediate two-electron states produce a new type of kernel which may be treated together with original two-electron irreducible kernels. Now the general structure of a given contribution is easily described. From the external lines, there is either a factor of $1/(\epsilon + i\delta)$ or not. If not, some structure occurs in combination with the external line before a factor of $1/(\epsilon + i\delta)$ is reached. Together, these produce a contribution to F_i or F_f external to the first such denominator. This factor is universal; i.e., it is independent of the details of the remainder of the graph. Now consider the remaining structure after these factors are identified. It consists of one or more factors of $1/(\epsilon + i\delta)$ separated by our generalized kernels. Each generalized kernel has a structure independent of every other one, and when that structure is summed over all possibilies, it produces a contribution to $\Delta(\epsilon)$. Knowing that this structure works out generally, it should be easy to find the terms of interest.

So far, we have discussed the interactions between two electrons at a time. Let me give a very brief indication of how the argument may be extended to more than two electrons. As a starting point, consider the contribution to (1c) where electrons #1 and #2 exchange a photon and then electrons #2 and #3 exchange a photon. One of the many contributions will involve the following sequences of transitions for the individual electrons

$$
\begin{aligned}
\#1: \quad & a \to a \\
\#2: \quad & b \to n; \quad n \to b \ . \\
\#3: \quad & c \to c
\end{aligned}
\tag{19}
$$

If $n \neq b$ or a photon pole contribution occurs, this will yield a new contribution to Δ; it is an intrinsic three electron contribution. On the other hand, if $n = b$ and only electron pole contributions are considered, this gives a contribution to the $\Delta^2/(\epsilon + i\delta)^3$ term in (4a). While this term involves only pairwise interactions of electrons, it cannot be obtained by considering only two electrons at a time in (1c). It does not yet have the correct coefficient of 2, but another sequence gives an identical contribution, so the geometric series is confirmed. One can also easily pick up contributions in this term of the series in which electrons are exchanged in the separate factors of Δ^2.

Another example is a disconnected graph in which electrons #1 and #2 exchange a photon and electrons #3 and #4 exchange an electron (of course, one must be sure not to double count because the electrons are identical). One has a product of factors for $\{a,b\} \to \{a,b\}$ and $\{c,d\} \to \{c,d\}$. Now the combinatorial factor 2 which is needed for the geometrical expansion does not appear to occur at first sight. However, when the contour integrations are worked out using Cauchy's theorem, it is restored. This indicates how in general disconnected graphs produce higher terms in the geometric series rather than new contributions to the energy.

I hope these brief remarks indicate how one could, in principle, obtain perturbative expressions for the energy shift due to interactions between the electrons by systematically finding kernels involving individual electrons (Lamb shift type terms), then kernels involving pairs of the electrons (including radiative corrections, if desired), then kernels involving three electrons, and so on. Disconnected kernels can be ignored.

Acknowledgements: In addition to support received from the program through ITP, this work was supported in part at Cornell by the National Science Foundation. While this work was not presented formally at the program, it was inspired by other presentations and benefited from conversations with several other members of the program. I particularly wish to thank Stanley Brodsky, Walter Johnson, Ingvar Lindgren, Peter Mohr, Jonathan Sapirstein, and Joe Sucher for their interest.

Appendix

There is a procedure for dealing with the many poles in the integral which may be easier than brute force contour integration. Various factors, or combinations of factors may first be rearranged using the method of partial fractions. When done in a careful manner, this can help avoid the presence of spurious poles. As an example, consider the ξ-dependence of the denominator combination of (15a). We may rearrange pairs of denominators and find (for positive energy intermediate states) the integrand

$$
\frac{-1}{4kk'(\epsilon + i\delta)^2(\epsilon + E_a + E_b - E_n - E_{n'} + i\delta)}
$$
$$
\times \left[\frac{1}{\xi_1 + i\delta} + \frac{1}{\epsilon - \xi_1 + i\delta}\right]\left[\frac{1}{\xi_2 - \xi_1 - k + i\delta} - \frac{1}{\xi_2 - \xi_1 + k - i\delta}\right]
$$
$$
\times \left[\frac{1}{\xi_2 + E_a - E_n + i\delta} - \frac{1}{\xi_2 - \epsilon - E_b + E_{n'} - i\delta}\right]
$$
$$
\times \left[\frac{1}{\xi_2 - \xi_3 - k' + i\delta} - \frac{1}{\xi_2 - \xi_3 + k' - i\delta}\right]\left[\frac{1}{\xi_3 + i\delta} + \frac{1}{\epsilon - \xi_3 + i\delta}\right] . \tag{20}
$$

Before the ξ-integrations are carried out, there are no actual poles at $\epsilon = 0$, but the introduction of $+i\delta$ in the denominators permits us to separate terms in the remaining factor before doing the integrations.

The $\xi_{1,3}$-integrations are now easily carried out. For example, in the product of the first two factors in [...]'s, the terms in which all the ξ_1-poles are on the same side of the axis give zero. As a result, after integration the first [...] disappears and the ξ_1 in the first term of the next factor is replaced by zero; while that of the second term is replaced by ϵ. A similar result holds for the ξ_3-integration. The remaining ξ_2-integrand is[8]

$$
\frac{-1}{4kk'(\epsilon + i\delta)^2(\epsilon + E_a + E_b - E_n - E_{n'} + i\delta)}
$$
$$
\times \left[\frac{1}{\xi_2 - k + i\delta} - \frac{1}{\xi_2 - \epsilon + k - i\delta}\right]\left[\frac{1}{\xi_2 + E_a - E_n + i\delta} - \frac{1}{\xi_2 - \epsilon - E_b + E_{n'} - i\delta}\right]
$$
$$
\times \left[\frac{1}{\xi_2 - k' + i\delta} - \frac{1}{\xi_2 - \epsilon + k' - i\delta}\right] . \tag{21}
$$

The ξ_2-integration has eight terms altogether, but two give zero because the poles are all on the same side of the axis. Each of the remaining terms has two poles on one side of the axis and one on the other; it is of course easiest to close the contour on the side with only one pole. The result of this integration is

$$
\frac{-1}{4kk'(\epsilon + i\delta)^2(\epsilon + \lambda_{an} + \lambda_{bn'} + i\delta)}
$$

$$
\times \left[\frac{1}{(k - \epsilon - \lambda_{bn'} - i\delta)(-\epsilon + k + k' - i\delta)} + \frac{1}{(-\epsilon - \lambda_{an} + k - i\delta)(-\epsilon - \lambda_{an} + k' - i\delta)} \right.
$$

$$
+ \frac{1}{(k' - \epsilon - \lambda_{bn'} - i\delta)(-\epsilon + k + k' - i\delta)} + \frac{1}{(\epsilon - k + \lambda_{an} + i\delta)(\epsilon - k - k' + i\delta)}
$$

$$
\left. + \frac{1}{(\epsilon + \lambda_{bn'} - k + i\delta)(\epsilon + \lambda_{bn'} - k' + i\delta)} + \frac{1}{(\epsilon - k' + \lambda_{an} + i\delta)(\epsilon - k - k' + i\delta)} \right]
$$

$$(22)$$

where we have introduced the notation $\lambda_{ab} = E_a - E_b$, etc.

When the numerator factors are incorporated, this gives a contribution to $\Delta(\epsilon)$. The leading contribution to the integrand of the energy shift is obtained by removing the factor $1/(\epsilon + i\delta)^2$ and taking $\epsilon = 0$

$$
\frac{-\bar{u}_a\gamma_\mu e^{i\vec{k}\cdot\vec{x}}u_n \, \bar{u}_n\gamma_\nu e^{i\vec{k}'\cdot\vec{x}}u_a \, \bar{u}_b\gamma^\mu e^{-i\vec{k}\cdot\vec{x}}u_{n'} \, \bar{u}_{n'}\gamma^\nu e^{-i\vec{k}'\cdot\vec{x}}u_b}{4kk'(E_a + E_b - E_n - E_{n'} + i\delta)}
$$

$$
\times \left\{ \frac{1}{(\lambda_{an} - k + i\delta)(\lambda_{an} - k' + i\delta)} + \frac{1}{(\lambda_{bn'} - k + i\delta)(\lambda_{bn'} - k' + i\delta)} \right.
$$

$$
\left. + \frac{1}{-k - k' + i\delta} \left[\frac{1}{\lambda_{bn'} - k' + i\delta} + \frac{1}{\lambda_{an} - k + i\delta} + \frac{1}{\lambda_{bn'} - k + i\delta} + \frac{1}{\lambda_{an} - k' + i\delta} \right] \right\}
$$

$$
= \frac{-\bar{u}_a\gamma_\mu e^{i\vec{k}\cdot\vec{x}}u_n \, \bar{u}_n\gamma_\nu e^{i\vec{k}'\cdot\vec{x}}u_a \, \bar{u}_b\gamma^\mu e^{-i\vec{k}\cdot\vec{x}}u_{n'} \, \bar{u}_{n'}\gamma^\nu e^{-i\vec{k}'\cdot\vec{x}}u_b}{4kk'}
$$

$$
\times \left\{ \frac{1}{E_a + E_b - E_n - E_{n'} + i\delta} \left[\frac{1}{\lambda_{an} - k + i\delta} + \frac{1}{\lambda_{bn'} - k + i\delta} \right] \right.
$$

$$
\times \left[\frac{1}{\lambda_{an} - k' + i\delta} + \frac{1}{\lambda_{bn'} - k' + i\delta} \right]
$$

$$
\left. + \frac{1}{-k - k' + i\delta} \left[\frac{1}{(\lambda_{an} - k + i\delta)(\lambda_{bn'} - k' + i\delta)} + \frac{1}{(\lambda_{bn'} - k + i\delta)(\lambda_{an} - k' + i\delta)} \right] \right\} \, .
$$

This last form has a structure which could easily have been derived from old-fashioned perturbation theory.

It is instructive to identify contributions to F arising from (22). If $\{n, n'\} \neq \{a, b\}$, any ϵ may be expanded up from the denominator to produce such a correction. If $n = a$, $n' = b$, we can obtain expressions like (13d) by rearranging appropriate denominators. It appears that we should *not* rearrange the denominators $\epsilon - k - k' + i\delta$ so as to produce a factor of ϵ in the numerator. This ϵ-dependence should remain in the definition of Δ.

REFERENCES

1. When there are continuum states, the sum over n may lead to a branch cut in the variable E, but we shall be concerned primarily with discrete bound states.

2. Since this work is exploratory and the author is relatively unfamiliar with the field, no strenuous effort has been made to give complete references.

3. J. Sucher, Phys. Rev. Lett. **55**, 1033 (1985).

4. P. Mohr, described in this volume.

5. J. Sucher, Phys. Rev. **107**, 1448 (1957).

6. The approach outlined here is presented in J. Fox and D. R. Yennie, Ann. Physics (N.Y.) **81**, 483 (1973). Some of the methods for working out the Lamb shift along these lines is given in G. W. Erickson and D. R. Yennie, Ann. Physics (N.Y.) **35**, 271 and 447 (1965).

7. This mild blemish on the formalism is actually a consequence of the choice of the Feynman gauge. It does not occur in the Coulomb gauge because the Coulomb potential has no poles in the ξ-integrations and the transverse photons couple less strongly near $\vec{k} = 0$. One may note that the integral actually produces a branch cut in the variable E starting at $E = E_a + E_b$. Physically this corresponds to the presence of a real photon in addition to the electrons.

8. Observation: one can easily pick out the original contribution (15b) by setting to 0 the ϵ's which appear in the photon denominators. The remainder in which an ϵ is expanded up gives contributions to F.

QED POTENTIALS IN MANY-ELECTRON ATOMS

Bernard Zygelman
Center for Astrophysics, Harvard University
60 Garden Street, Cambridge Ma. 02138

ABSTRACT

A many-electron configuration space equation is obtained using a combination of the no-pair approximation and a series of decoupling transformations in Fock space. The contribution of QED three-body potentials to the structure of heavy atoms is discussed. Some remarks concerning the non-uniqueness, or gauge dependence, of the two-electron potential are made.

INTRODUCTION

Accurate atomic structure calculation of heavy atoms and ions must include relativistic effects derived in a systematic way from quantum electrodynamics. Because most calculations are done using a many-electron configuration space wave equation as the starting point a necessary intermediate step is the extraction of a first-quantized wave equation from the field theory. Keeping in mind that a first-quantized Hamiltonian cannot describe all field-theoretic effects, our goal is somewhat limited. Our intent is to incorporate at least the more gross field-theoretic features in the configuration space equation.

In section I we review the transition from Fock to configuration space in a non-relativistic many-body theory. For a relativistic theory the electron number is not conserved and the transition to configuration space is no longer trivial. The no-pair approximation[1,2,3] is used to derive an approximate relativistic wave equation for N-electrons in the field of an infinitely heavy nucleus. In section II we allow for the exchange of virtual transverse photons among the electrons. In section III we consider two-photon exchange among triplets of electrons. A class of these diagrams manifest as three-electron potentials[2] in configuration space. In section IV we discuss the non-relativistic limit of these three-body potentials and recover the, classical, Primakoff-Holstein potential[4]. QED three-body potentials contribute to the hyperfine structure of atoms[5], and serve as a prototype for the phenomenological three-nucleon potentials that are believed to contribute to the binding energy of three-nucleon systems.[6] In section V we calculate the contribution of the three-electron potentials to the binding energy of high-Z lithium like ions. Our results suggest that three-electron potentials play a minor role in the gross structure of heavy atoms. Throughout, we set (unless otherwise stated) $\hbar = c = 1$.

I. CONFIGURATION SPACE EQUATIONS FROM FOCK SPACE

A. Nonrelativistic case

Consider a system of nonrelativistic charged particles interacting via an arbitrary two-body potential, $v(x_{12})$. In Fock space the Hamiltonian is given by,

$$H = \int dx \psi^\dagger(x)h(x)\psi(x) + \frac{1}{2}\int dx_1 \int dx_2 \psi^\dagger(x_1)\psi^\dagger(x_2)v(x_{12})\psi(x_2)\psi(x_1) \qquad (1.1)$$

where $\psi(x)$, $\psi^\dagger(x)$ are field destruction and creation operators respectively and $h(x)$ is a configuration space Hamiltonian for the noninteracting particles. The field operators obey the equal time commutation relations

$$[\psi(x), \psi^\dagger(x')] = \delta^3(x - x'), \quad [\psi(x), \psi(x')] = 0, \quad [\psi^\dagger(x), \psi^\dagger(x')] = 0, \qquad (1.2)$$

and can be expanded in terms of a complete set of eigenstates, $\phi_n(x)$, of $h(x)$

$$\psi(x) = \sum_n a_n \phi_n(x)$$

$$\psi^\dagger(x) = \sum_n a_n^\dagger \phi_n^\dagger(x). \qquad (1.3)$$

In order that the wave operators satisfy (1.2) the expansion coefficients are chosen so that,

$$[a_n, a_{n'}^\dagger] = \delta_{n,n'}, \quad [a_n, a_{n'}] = 0, \quad [a_n^\dagger, a_{n'}^\dagger] = 0. \qquad (1.4)$$

We define a vacuum state, $|0>$, so that $a_n|0>= 0$ for all n, and introduce the number operator,

$$N = \int dx \psi^\dagger(x)\psi(x) = \sum_s a_s^\dagger a_s = \sum_s N_s. \qquad (1.5)$$

N_s counts the number of particles in mode s and N counts the total number of particles. An N-body state $|\Psi_N >$ is constructed by the repeated application of $\psi^\dagger(x)$, on the vacuum state, and summing over all configurations via an amplitude $\chi(x_1 x_2 ... x_N)$, i.e.

$$|\Psi_N >= \int dx_1 dx_2 ... dx_N \chi(x_1 x_2 ... x_N)\psi^\dagger(x_1)\psi^\dagger(x_2)...\psi^\dagger(x_N)|0 > . \qquad (1.6)$$

410

Because N commutes with the total Hamiltonian χ uniquely specifies this N-body eigenstate. An eigenvalue equation for χ is obtained by requiring,

$$\delta\{< \Psi_N|H|\Psi_N > -E < \Psi_N|\Psi_N| >\} = 0 \qquad (1.7)$$

where the variation δ is taken with respect to the amplitudes $\chi^\dagger$ and χ. Using (1.6) in (1.7) and proceeding with the variation with respect to $\chi^\dagger$ and χ we get

$$\sum_i^N h(i)\chi(x_1 x_2 ... x_N) + \sum_{i<j}^N v(x_{ij})\chi(x_1 x_2 ... x_N) = E\chi(x_1 x_2 ... x_N) \qquad (1.8)$$

and it's conjugate equation respectively.

B. Relativistic case

We generalize this procedure for the relativistic case. First, we consider a system of fermions interacting via an instantaneous two-fermion interaction. In Fock space the Hamiltonian is given by

$$H = \int dx \sum_{\alpha\beta} \psi_\alpha^\dagger(x) h_\beta^\alpha(x) \psi^\beta(x)$$

$$+\frac{1}{2}\sum_{\alpha\beta}\sum_{\mu\nu}\int dx_1 \int dx_2 : \psi_\alpha^\dagger(x_1)\psi_\beta^\dagger(x_2)v_{\mu\nu}^{\beta\alpha}(x_{12})\psi^\nu(x_2)\psi^\mu(x_1) : . \qquad (1.9)$$

The wave operators $\psi^\alpha(x)$, $\psi_\alpha^\dagger(x)$ are now four component spinor, and conjugate spinor operators, and the noninteracting Hamiltonian, $h_\alpha^\beta(x)$, is a Dirac - central field spinor operator. At this stage we do not specify the exact form for the central field but we do assume that the nucleus is infinitely heavy. The fermion wave operators obey anticommutator relations,

$$\{\psi_\alpha^\dagger(x), \psi^\beta(x')\}_+ = \delta_\alpha^\beta(x - x'), \quad \{\psi_\alpha^\dagger(x), \psi_\beta^\dagger(x')\}_+ = 0, \quad \{\psi^\alpha(x), \psi^\beta(x')\}_+ = 0 \qquad (1.10)$$

where we have explicitly included all spinor indicies α, β. Below we introduce a notation that treats all spinor indicies implictly, we avoid confusion by treating $\psi(x)$ as a four component column matrix and $\psi^\dagger(x)$ as a row matrix. Products of operators are positioned so that matrix multiplication is implied i.e., $\psi^\dagger(x)\psi(x)$ is a scalar whereas $\psi(x)\psi^\dagger(x)$ is a 4 × 4 square matrix (spinor) operator. First, let us expand the wave operators

$$\psi(x) = \sum_n a_n\phi_n(x), \quad \psi^\dagger(x) = \sum_n a_n^\dagger\phi_n^\dagger(x) \qquad (1.11)$$

where the $\phi_n(x)$ are eigenspinors of $h(x)$, and $a_n^\dagger, a_n$ obey anticommutation relations so that (1.10) is satisfied. A vacuum is defined so that $a_n|0>=0$ for all n and $N_s \equiv \sum_s a_s^\dagger a_s$ is a number operator for the mode s. Using (1.9) and expansion (1.11) we get for the non-interacting Fock space Hamiltonian,

$$H_0 = \int dx \psi^\dagger(x) h(x) \psi(x) =$$

$$\sum_n^+ |E_n| a_n^\dagger a_n - \sum_n^- |E_n| a_n^\dagger a_n. \tag{1.12}$$

The sum $\sum^+$ is over the positive energy eigenvalues of $h(x)$, and $\sum^-$ represents the sum over the negative eigenvalues. With this choice for the vacuum the the non-interacting Hamiltonian does not possess a lower bound and the theory is unsatisfactory. We can construct a meaningful field theory if we redefine our vacuum, hence we expand the field operators

$$\psi(x) = \sum_n^+ b_n \phi_n(x) + \sum_n^- a_n^\dagger \phi_n^\dagger(x)$$

$$\psi^\dagger(x) = \sum_n^+ b_n^\dagger \phi_n^\dagger(x) + \sum_n^- a_n \phi_n(x)$$

where

$$\{a_n^\dagger, a_{n'}\}_+ = \delta_{nn'}, \quad \{b_n^\dagger, b_{n'}\}_+ = \delta_{nn'} \tag{1.13}$$

and all other anticommutators vanish. One can verify that the wave operators in this representation satisfy relations (1.10). Inserting (1.13) into the free-fermion Hamiltonian, we get

$$H_0 = \sum_n^+ |E_n| b_n^\dagger b_n + \sum_n^- |E_n| a_n^\dagger a_n - \sum_n^- |E_n|. \tag{1.14}$$

We now define the fermion vacuum so that $a_n|0>=0$, $b_n|0>=0$ for all n; thus $N_n^{e-} = a_n^\dagger a_n$, and $N_n^{e+} = b_n^\dagger b_n$ are number operators each with the spectrum $0, 1$ for each n. We can then rewrite (1.14)

$$H_0 = \sum_n^+ |E_n| N_n^{e-} + \sum_n^- |E_n| N_n^{e+} - \sum_n^- |E_n| \tag{1.15}$$

Hamiltonian (1.15) is bounded from below. Using the equations of motion for the field operators we obtain the conservation law

$$\frac{d}{dt}(e : \psi^\dagger(x,t)\psi(x,t) :) = e\vec{\nabla} \cdot (: \psi^\dagger(x,t)\vec{\alpha}\psi(x,t) :) \tag{1.16}$$

and identify $Q = -|e| \int dx : \psi(x)^\dagger \psi(x) :$ as the conserved charge operator. Using 1.13 we get

$$Q = -|e| \sum_n^+ N_n^{e-} + |e| \sum_s^- N_s^{e+}. \tag{1.17}$$

The spectrum of Q is $q = (0, \pm 1, \pm 2, \ldots \pm n)|e|$, therefore, we identify N^{-e} as the number operator for electrons and N^{+e} the number operator for positrons. The total charge of the system is equal to the total number of electrons and positrons. Although Q is conserved, the number operator for the total fermion number operator $N = N^{e-} + N^{e+}$ is not since $[H, N] \neq 0$. The noncommutivity of the number operator with the total Hamiltonian is due to the fact that the fermion-fermion interaction operator can create or destroy fermion (electron-positron) pairs. An eigenstate of H with charge $-|e|N$ (non relativistically we call this an N-electron system) is expressed by the infinite sum

$$|\Psi_N> = |\Psi_0> + |\Psi_1> + |\Psi_2> + \ldots \tag{1.18}$$

where $|\Psi_0>$ is a charge eigenstate that contains N electrons, $|\Psi_1>$ contains N electrons plus a pair, $|\Psi_2>$ contains N electrons plus two pairs, etc. In the configuration space representation $|\Psi_N>$ is determined by the set of amplitudes $\chi^i(x_1, x_2, \ldots x_N; x_1', x_2' \ldots x_{2i})$ defined so that

$$|\Psi_0> = \int dx_1 dx_2 \ldots dx_N \psi^\dagger(x_1)\psi^\dagger(x_2) \ldots \psi^\dagger(x_N)\chi^0(x_1, x_2, \ldots x_N)|0>$$

$$|\Psi_1> = \int dx_1 dx_2 \ldots dx_N \int dx_1' dx_2'$$

$$\psi^\dagger(x_1)\psi^\dagger(x_2) \ldots \psi^\dagger(x_N)\psi^\dagger(x_1')\psi(x_2')\chi^1(x_1, x_2, \ldots x_N, x_1', x_2')|0> \tag{1.19}$$

and so on. The set of kets in (1.18) span the complete Hilbert space of states with charge $-|e|N$. In order to specify the amplitudes χ^i we require

$$\delta\{< \Psi_N|H|\Psi_N> -E < \Psi_N|\Psi_N >\} = 0 \tag{1.20}$$

where the variation is taken with respect to the complete set of χ^i. Proceeding with the variation one obtains an infinite set of coupled equations for the amplitudes. This is not the best procedure for practical calculation. Instead, we make an approximation by truncating the series (1.18) after the first term and using $\chi^0 \equiv \chi$ as the variational parameter. This approximation includes all the effects where pairs are not present, we call it the no-pair approximation.[3] Using the no-pair ansatz we construct a configuration space equation by requiring

$$\delta_\chi \{< \Psi|H|\Psi > - E < \Psi|\Psi >\} = 0. \tag{1.21}$$

Contracting the fermion operators, we get

$$\delta_\chi \Bigg\{ \int dx'_1...dx'_N \int dx_1....dx_N 2\chi^\dagger(x'_1...x'_N) \times$$

$$[\sum_i \int dx \Lambda_+(x'_i, x)h(x)\Lambda_+(x, x_i) + \sum_{i<j}^N \int dz_1 \int dz_2 \times$$

$$\Lambda_+(x_i, z_1)\Lambda_+(x_j, z_2)v(z_{12})\Lambda_+(z_2, x_j)\Lambda_+(z_1, x_i)$$

$$- E\Lambda_+(x'_1, x_1)...\Lambda_+(x'_N, x_N)]\chi(x_1...x_N) \Bigg\} = 0 \tag{1.22}$$

where the positive energy projection operators are given by,

$$\Lambda_+(x, x') \equiv < 0|\psi(x)\psi^+(x')|0 > = \sum_s^+ \phi_s(x)\phi_s^\dagger(x'). \tag{1.23}$$

Carrying out the variation we get,

$$[\sum_i^N \Lambda_+(i)h(i) + \sum_{i<j}^N \Lambda_+(i)\Lambda_+(j)v(i,j)\Lambda_+(j)\Lambda_+(i)]\omega(x_1...x_N) = E\omega(x_1...x_N) \tag{1.24}$$

where,

$$\omega(x_1...x_N) = \int dx'_1...dx'_N \Lambda_+(x_1, x'_1)...\Lambda_+(x_N, x'_N)\chi(x'_1...x'_N), \tag{1.25}$$

and

$$\Lambda_+(i)f(i) \equiv \int dx'_i \Lambda_+(x_i, x'_i)f(x'_i). \tag{1.26}$$

The appearance of the projection operators in the configuration space equation (1.24) is a direct result of using the no-pair approximation. In the single-time reduction of the, two-electron, Bethe-Salpeter equation[7] some pair effects are taken into account. In this theory the projection operators surrounding $v(x_{12})$ are replaced by

$$\Lambda_+(i)\Lambda_+(j) \rightarrow \Lambda_+(i)\Lambda_+(j) - \Lambda_-(i)\Lambda_-(j)$$

where Λ_- is a negative energy projection operator. Pair effects which induce three-electron potentials in configuration space are discussed in Section III.

The exact form of the projection operator depends on the choice of the Furry representation. For a many-electron atom Mittleman[8] has suggested that a Hartree-Fock representation is optimal. Throughout, we assume a Furry representation defined in a local potential (mean field).

II. TRANSVERSE PHOTON EXCHANGE

The configuration space equation (1.24) is a possible starting point for structure calculations of relativistic multi-electron atoms. It includes all the kinematic effects of relativity and incorporates the leading instantaneous electron-electron Coulomb interactions. However, it does not include the effects where virtual transverse photons are exchanged between the electrons. In this section we investigate the modifications that are necessary when single-transverse photon exchange is included.

In the Coulomb gauge, the QED Hamiltonian is given by,

$$H_{QED} = H_0^R + H_0^M + H_C + H_I$$

where H_0^R and H_0^M are the non-interacting Hamiltonians for transverse photons, and fermions (electrons, positron) respectively. H_C is the Coulomb interaction term given by,

$$H_C = \frac{1}{2} \int dx_1 \int dx_2 : \psi^\dagger(x_1)\psi^\dagger(x_2)v_C(x_{12})\psi(x_2)\psi(x_1) : \qquad (2.1)$$

where $v_C(x_{12}) \equiv e^2/|x_1 - x_2|$. The fermion-transverse photon interaction is

$$H_I = - \int dx \vec{J}(x) \cdot \vec{A}(x)$$

where,

$$\vec{J}(x) = -e : \psi(x)^\dagger \vec{\alpha} \psi(x) : \qquad (2.2)$$

and $\vec{A}(x)$ is the transverse photon field operator. The fermion field operators satisfy the anticommutation relation (1.10) and the photon field obeys the transversality condition $\vec{\nabla} \cdot \vec{A} = 0$. We decompose the photon field in a plane wave representation,

$$\vec{A}(x) = \sqrt{\frac{2\hbar c}{Vk}} \sum_{\kappa,\lambda}(a_{\kappa\lambda}^\dagger \hat{\epsilon}_{\kappa\lambda}e^{i\kappa x} + a_{\kappa\lambda}\hat{\epsilon}_{\kappa\lambda}e^{-i\kappa x})$$

$$\hat{\kappa} \cdot \hat{\epsilon}_{\kappa\lambda} = 0 \quad \lambda = 1,2 \quad k \equiv |\vec{\kappa}|. \qquad (2.3)$$

The operators $a_{\kappa\lambda}$ obey commutation relations,

$$[a_{\kappa\lambda}^{\dagger}, a_{\kappa'\lambda'}] = \delta_{\kappa,\kappa'}\delta_{\lambda,\lambda'}, \quad [a_{\kappa\lambda}^{\dagger}, a_{\kappa'\lambda'}^{\dagger}] = 0, \quad [a_{\kappa\lambda}, a_{\kappa'\lambda'}] = 0, \quad (2.4)$$

and the photon vacuum, $|0>$, is defined so that $a_{\kappa\lambda}|0>$ for all κ and λ. We introduce the dressing transformation[1,2]

$$H \rightarrow H' = e^{i\sigma} H e^{-i\sigma}$$

where the generator σ is given by,

$$\sigma = -\frac{1}{2}\int_{-\infty}^{\infty} d\tau \epsilon(\tau) H_I(\tau) e^{-\eta|\tau|}, \qquad H_I(\tau) \equiv -\int dx \vec{J}(x,\tau)\cdot\vec{A}(x,\tau)$$

$$\vec{J}(x,\tau) \equiv e^{iH_0\tau}\vec{J}(x)e^{-iH_0\tau}, \qquad \vec{A}(x,\tau) \equiv e^{iH_0\tau}\vec{A}(x)e^{-iH_0\tau} \qquad (2.5)$$

$H_0 = H_0^R + H_0^M$, $\epsilon(\tau) = \theta(\tau) - \theta(-\tau)$, $\theta(\tau)$ is the Heaviside function, $e^{-\eta|\tau|}$ is a convergence factor and the $\lim \eta \rightarrow 0$ is implied after all integrations are done. Using the identity

$$e^{i\sigma}Oe^{-i\sigma} = \sum_{n=0}^{\infty} \frac{i^n}{n!}C_n[\sigma, O]$$

$$C_0[\sigma, O] = 1, \quad C_1[\sigma, O] = [\sigma, O], \quad C_2[\sigma, O] = [\sigma, [\sigma, O]]... \qquad (2.6)$$

we get,

$$H' = \sum_{n=0}^{\infty} \frac{i^n}{n!}(C_n[\sigma, H_0] + C_n[\sigma, H_I] + C_n[\sigma, H_C])$$

$$= H_0 + i[\sigma, H_0] + \frac{i^2}{2}[\sigma, [\sigma, H_0]] + H_I + H_C + i[\sigma, H_I] + O(e^3) + ... \qquad (2.7)$$

where the infinite series is truncated to include terms that are second order in e. We now evaluate the commutator $i[\sigma, H_0]$. Noting

$$i[\sigma, H_0] = -\frac{1}{2}\int dx \int_{-\infty}^{\infty} d\tau \epsilon(\tau)\frac{\partial}{\partial \tau}(\vec{J}(x,\tau)\cdot\vec{A}(x,\tau))e^{-|\eta|\tau}, \qquad (2.8)$$

and integrating the parts we get,

$$i[\sigma, H_0] = -H_I + \lim_{\eta \rightarrow 0}\frac{\eta}{2}\int_{-\infty}^{\infty} d\tau H_I(\tau)e^{-\eta|\tau|} \qquad (2.9)$$

416

where we have used, $\partial \epsilon(\tau)/\partial \tau = 2\delta(\tau)$ and $\partial e^{-|\tau|\eta}/\partial \tau = -\eta e^{-|\tau|\eta}\epsilon(\tau)$. The second term in (2.9) vanishes in the limit as $\eta \to 0$ unless the integral is proportional to $1/\eta$. This occurs for energy on-shell matrix elements, i.e., real photon transitions. Because we are interested in virtual photon exchange processes only, we can neglect this term in our discussion, thus

$$i[\sigma, H_0] = -H_I. \tag{2.10}$$

Using (2.10) in (2.7) we get,

$$H' = H_0 + \frac{i}{2}[\sigma, H_I] + H_C + O(e^3) \tag{2.11}$$

which is an order e^2 approximation to H_{QED} and is diagonal with respect to the photon number. The effects due to the exchange of virtual photons manifest themselves in the term $\frac{i}{2}[\sigma, H_I]$. This term is a sum of one-body and two-body operators. The one-body terms are due to processes where a fermion can emit and absorb its own photon. These terms must be treated with care via a renormalization program in order to eliminate unwanted infinite contributions. We collectively call the one-body contributions (both finite and infinite) as self-energy terms. The two-body terms describe processes where different fermions exchange virtual photons. We expand the commutator to get,

$$H_B \equiv \frac{i}{2}[\sigma, H_I] =$$

$$-\frac{i}{4} \int dx \int dx' \int_{-\infty}^{\infty} d\tau \epsilon(\tau) e^{-\eta|\tau|} J^a(x,\tau) J^b(x',\tau)[A_a(x,\tau), A_b(x',\tau)] \tag{2.12}$$

where we neglected current commutator terms since these give the aforementioned one-body terms, and used the summation convention $V_a V^a \equiv \vec{V} \cdot \vec{V}$. The commutator for the photon field is a c-number given by,

$$D_{ab}(x - x', \tau - \tau') \equiv [A_a(x,\tau), A_b(x',\tau')] =$$

$$-\frac{i\hbar c}{2\pi^2} \int \frac{d^3\kappa}{k} e^{i\kappa(x-x)'} sin[ck(\tau - \tau')](\delta_{ab} - \hat{\kappa}_a\hat{\kappa}_b). \tag{2.13}$$

It is convenient to express the current operator in terms of the wave operators $\psi(x)$, $\psi^\dagger(x)$, we get

$$\vec{J}(x,\tau) = -e : (\psi^\dagger(x)e^{ih(x)\tau})\vec{\alpha}(e^{-ih(x)\tau}\psi(x)) : \tag{2.14}$$

where $h(x)$ is the single particle Dirac-central field Hamiltonian. The brackets restrict the exponential operators to operate on the $\psi^\dagger$ and ψ fields to the left and the right respectively. Inserting (2.14) into (2.12) we get

$$H_B = \frac{1}{2} \int dx_1 \int dx_2 : \psi^\dagger(x_1)\psi^\dagger(x_2)v_B(x_{12})\psi(x_2)\psi(x_1) :$$

$$v_B(x_{12}) = \frac{-ie^2}{4} \int_{-\infty}^{\infty} d\tau \epsilon(\tau)e^{ih(x_1)\tau}\alpha_1^a D_{ab}(x_{12},\tau)\alpha_2^b e^{-ih(x_1)\tau} + (1 \leftrightarrow 2) \qquad (2.15)$$

where $1 \leftrightarrow 2$ represents symmetrization with respect to the particle coordinates. The above form is similar to the two-body potential operator, given by (2.1) except that the kernel v_B is now a non-local operator. The transition to configuration space is carried out in an identical manner to that described in the previous section. Using the no pair approximation for H and invoking (1.20) we obtain a configuration space equation for an N-electron atomic system given by,

$$\left[\sum_i^N \Lambda_+(i)h(i) + \sum_{i<j}^N \Lambda_+(i)\Lambda_+(j)\big(v_C(x_{ij}) + v_B(x_{ij})\big)\Lambda_+(j)\Lambda_+(i)\right]\omega(x_1...x_N) =$$

$$E\omega(x_1...x_N). \qquad (2.16)$$

It is useful to express the non-local potential $v_B(x_{12})$ in the occupation representation, i.e. its' matrix elements with respect , ϕ_n, the eigenstates of $h(x)$. We define

$$\eta_{nl,n'l'} \equiv < nl|\frac{-ie^2}{4} \int_{-\infty}^{\infty} d\tau \epsilon(\tau)e^{ih(x_1)\tau}\alpha_1^a D_{ab}(x_{12},\tau)\alpha_2^b e^{-ih(x_1)\tau}|n'l' > \qquad (2.17)$$

where the bra-ket notation implies integration over the electronic coordinates. Re-expressing the photon commutator (2.13)

$$D_{ab}(x,\tau) = \frac{c}{\pi}(\nabla_1^2\delta_{ab} + \nabla_1^a\nabla_2^b)\frac{1}{|x|} \int_{-\infty}^{\infty} \frac{dk}{k^2}e^{ikx}sin(ck\tau) \qquad (2.18)$$

and using,

$$\int_{-\infty}^{\infty} d\tau \epsilon(\tau)sin(ck\tau)e^{iW\tau}e^{-\eta|\tau|} = \frac{P}{ck+W} + \frac{P}{ck-W}$$

we get

$$\eta_{nl,n'l'} = \frac{-e^2}{4} < nl|\alpha_1^a \alpha_2^b \{(\nabla_1^2 \delta_{ab} + \nabla_1^a \nabla_2^b)r\left[\frac{2\hbar c}{w_{n,n'} r}\sin(\frac{w_{n,n'} r}{2\hbar c})\right]^2\}|n'l'>$$

$$w_{n,n'} \equiv E_n - E_{n'} \qquad r \equiv |x_{12}| \qquad (2.19)$$

where E_n is the energy eigenvalue of the Dirac orbital ϕ_n. We define the non relativistic limit by setting the quotient $(w_{n,n'} r/\hbar c) << 1$, (to be more precise, we mean that the expectation value of this operator is small). This limit is realized for low Z atoms, and for positive energy states where the electron velocity $v/c \approx (Z\alpha) << 1$. In this limit we can replace the term in brackets in (2.19) by unity and the matrix element becomes

$$\eta_{nl,n'l'} = < nl|\frac{-e^2}{4r}(\vec{\alpha}_1 \cdot \vec{\alpha}_2 + \vec{\alpha}_1 \cdot \hat{x}_{12}\vec{\alpha}_2 \cdot \hat{x}_{12})|n'l'>$$

which is the matrix element of the Breit operator divided by two. Therefore, in this limit, the matrix element of the non-local operator $v_B(x_{12})$ reduces to the matrix element of the local Breit operator. Because $v_B(x_{12})$ is always sandwiched between Dirac central field orbitals we can derive an alternative expression for $\eta_{nl,n'l'}$ that is obtained by integrating (2.19) by parts and then making use of the Dirac equation to get,

$$\eta_{nl,n'l'} = -\frac{e^2}{2} < nl|\frac{\vec{\alpha}_1 \cdot \vec{\alpha}_2}{r}\cos(\frac{w_{n,n'} r}{\hbar c}) - \frac{w_{l,l'}}{w_{n,n'}}\frac{1}{r}\left[1 - \cos(\frac{w_{n,n'} r}{\hbar c})\right]|n'l'> . \quad (2.20)$$

If we take the matrix element of both the Coulomb operator and the transverse non-local Breit operator we get

$$< nl|v_C + v_B|n'l'> =$$

$$e^2 < nl|\frac{1}{r} - \frac{\vec{\alpha}_1 \cdot \vec{\alpha}_2}{2r}[\cos(\frac{w_{n,n'} r}{\hbar c}) + \cos(\frac{w_{l,l'} r}{\hbar c})] +$$

$$\frac{w_{n,n'}}{w_{l,l'}}\frac{1}{2r}(1 - \cos(\frac{w_{l,l'} r}{\hbar c})) + \frac{w_{l,l'}}{w_{n,n'}}\frac{1}{2r}(1 - \cos(\frac{w_{n,n'} r}{\hbar c}))|n'l'> . \quad (2.21)$$

Consider the lowest order contribution to the energy from the operators $H_C + H_B$, it can be expressed in terms of the above matrix element and satisfies the energy on-shell condition, $w_{n,n'} + w_{l,l'} = 0$. We get

$$< nl|v_C + v_B|n'l'>_{on-shell} =$$

$$< nl|e^2\frac{(1 - \vec{\alpha}_1 \cdot \vec{\alpha}_2)}{r}\cos(\frac{w_{n,n'} r}{\hbar c})|n'l'> \qquad (2.22)$$

This is the matrix element of the two-body operator obtained in the covariant Feynman gauge. Therefore the lowest order, e^2, contribution to the energy is gauge invariant.

However, the two expressions are no longer equivalent for off-energy shell matrix elements that contribute to higher order corrections. In addition, if we resort to a Furry picture expansion in terms of Hartree-Fock orbitals the two electron-electron potentials are no longer gauge equivalent even to lowest order. This is due to the fact that the Hartree-Fock orbitals do not obey local current conservation.

There is another source of ambiguity in the two-body potentials due to the freedom of choice in decoupling the virtual photons. For example, instead of using the function $\epsilon(\tau)$ in deriving (2.9) we could have used $g(\tau) \equiv \beta\epsilon(\tau) - 2(1 - \beta)\theta(-\tau)$ where β is an arbitrary parameter. With this choice we obtain a new potential whose matrix elements differ from (2.20) by an additional term

$$\omega_{nl,n'l'} \equiv -\frac{ie^2}{2}(1 - \beta) < nl|\frac{1 + \vec{\alpha}_1 \cdot \vec{\alpha}_2}{r}sin(\frac{w_{n,n'}\tau}{\hbar c})|n'l' > . \qquad (2.23)$$

For on-shell matrix elements this term vanishes and terms proportional to β contribute only at higher order. One can show[9] that for $\beta \neq 1$ higher order potentials are singular. For this reason we use Schwinger's choice[10] $\beta = 1$.

III. THREE-ELECTRON POTENTIALS

A. Two-Photon Exchange

Following the procedure of Mittleman[2] we construct the three-electron potentials by going beyond the e^2 terms in the expansion (2.7). The next terms are of $O(e^3)$, these are off-diagonal in the photon number and contribute to the energy at the $O(e^6)$ level. Some of the e^4 terms are diagonal in the photon representation. They are given by,

$$H_{C-T} = -\frac{1}{2}[\sigma, [\sigma, H_C]],$$

and

$$H_{T-T} = -\frac{i}{8}[\sigma, [\sigma, [\sigma, H_I]]]. \qquad (3.1)$$

The subscripts C-T and T-T, refer to "Coulomb-transverse", and "transverse- transverse" parts respectively. This terminology reflects the fact that the first term describes processes where a virtual Coulomb photon, and transverse photon is exchanged; whereas in the latter process, two transverse photons are exchanged. Concentrating on the H_{C-T} terms first, we expand the commutator to get

$$H_{C-T} = -\frac{1}{8}\int_{-\infty}^{\infty} d\tau_1\epsilon(\tau_1)\int_{-\infty}^{\infty} d\tau_2\epsilon(\tau_2)J(1)[J(2), H_C][A(1), A(2)]e^{-\eta(|\tau_1|+|\tau_2|)} \quad (3.2)$$

where we have neglected all e^4 two and one-fermion terms, and all indices and coordinates are implicitly represented by the numerals one and two. The neglected one and two-fermion operators correspond to e^4 two and one-electron potentials. We ignore these terms and consider only the three-fermion operators. Expanding the fermion commutator, and discarding the two-body and one-body operators we get,

$$[J(x_2, \tau_2), H_C] = -e^3 \int dx_4 \psi^\dagger(x_2) e^{ih(2)\tau_2} \vec{\alpha}_2 e^{-ih(2)\tau_2} \psi^\dagger(x_4) v_C(x_{24}) \psi(x_4) \psi(x_2)$$

$$+ \psi^\dagger(x_2) \psi^\dagger(x_4) v_C(x_{24}) \psi(x_4) e^{ih(2)\tau_2} \vec{\alpha}_2 e^{-ih(2)\tau_2} \psi(x_2) \qquad (3.3)$$

where we have used (1.10), and (2.14). Inserting this expression into (3.2), we get,

$$H_{C-T} = \frac{1}{6} \int dx_1 \int dx_2 \int dx_3 : \psi^\dagger(x_1) \psi^\dagger(x_2) \psi^\dagger(x_3) S[\Delta_{C-T}(x_{123})] \psi(x_3) \psi(x_2) \psi(x_1) :$$

where,

$$\Delta_{C-T}(x_{123}) = -\frac{e^4}{8} \int_{-\infty}^{\infty} d\tau_1 \epsilon(\tau_1) \int_{-\infty}^{\infty} d\tau_2 \epsilon(\tau_2) \times$$

$$e^{ih(1)\tau_1} e^{ih(2)\tau_2} \alpha_1^a D^{ab}(1,2) \alpha_2^b e^{-ih(1)\tau_1} e^{-ih(2)\tau_2} v_C(x_{23}) e^{-\eta(|\tau_1|+|\tau_2|)} + H.c. \qquad (3.4)$$

and S symmetrizes the kernel with respect to particle coordinates. The matrix element of the kernel $\Delta_{C-T}(x_{123})$ is

$$V_{nlm,n'l'm'}^{C-T} \equiv < nlm|\Delta_{C-T}|n'l'm' > = \sum_s [\frac{(\eta_{nl,n's} - \eta_{ln,s'n'})}{w_{n,n'} + w_{l,s}}] C_{sm,l'm'} + H.c. \qquad (3.5)$$

where the matrix elements $\eta_{nl,n'l'}$ are defined in (2.19), $C_{sm,l'm'}$ is the matrix elements of the Coulomb operator,

$$C_{nl,n'l'} = < nl|v_C|n'l' >$$

and the sum $\sum_s$ represents the sum over all intermediate states. In deriving (3.5) we made use of the closure property of the Dirac wavefunctions and the identity

$$\int_{-\infty}^{\infty} d\tau_1 \epsilon(\tau_1) \int_{-\infty}^{\infty} d\tau_2 \epsilon(\tau_2) e^{iw_A \tau_1} e^{iw_B \tau_2} sin(ck(\tau_1 - \tau_2)) e^{-\eta(|\tau_1|+|\tau_2|)} =$$

$$4ick(w_A - w_B) \frac{P}{(ck + w_A)(ck - w_A)(ck + w_B)(ck - w_B)}. \qquad (3.6)$$

We now turn our attention toward the transverse-transverse term. Expanding the multiple commutators we get,

$$-\frac{i}{64}\int_{-\infty}^{\infty}\int_{-\infty}^{\infty}\int_{-\infty}^{\infty}d\tau_1\epsilon(\tau_1)d\tau_2\epsilon(\tau_2)d\tau_3\epsilon(\tau_3)e^{-\eta(|\tau_1|+|\tau_2|+|\tau_3|)}$$

$$[J(1)A(1),[J(2)A(2),[J(3)A(3),J(4)A(4)]]] \tag{3.7}$$

where all spatial indices and integrations are implicit, and index 4 does not contain the time component. We are interested in the terms that give us the three-fermion operators. After some commutator algebra we get,

$$H_{T-T} = -\frac{i}{64}\int_{-\infty}^{\infty}\int_{-\infty}^{\infty}\int_{-\infty}^{\infty}d\tau_1\epsilon(\tau_1)d\tau_2\epsilon(\tau_2)d\tau_3\epsilon(\tau_3)e^{-\eta(|\tau_1|+|\tau_2|+|\tau_3|)}\times$$

$$(3J(1)J(2)[J(3),J(4)] + J(3)J(4)[J(1),J(2)])D(2,3)D(1,4). \tag{3.8}$$

Evaluating the current-current commutators and repeating the above procedure we get,

$$H_{T-T} =$$

$$\frac{1}{6}\int dx_1 \int dx_2 \int dx_3 : \psi^\dagger(x_1)\psi^\dagger(x_2)\psi^\dagger(x_3)S[\Delta_{T-T}(x_{123})]\psi(x_3)\psi(x_2)\psi(x_1): \tag{3.9}$$

where,

$$\Delta_{T-T}(x_{123}) = -i\frac{e^4}{64}\int_{-\infty}^{\infty}\int_{-\infty}^{\infty}\int_{-\infty}^{\infty}d\tau_1\epsilon(\tau_1)d\tau_2\epsilon(\tau_2)d\tau_3\epsilon(\tau_3)e^{-\eta(|\tau_1|+|\tau_2|+|\tau_3|)}\times$$

$$(e^{ih(1)\tau_1}e^{ih(2)\tau_2}\alpha_1^a D_{ab}(x_{12},\tau_1-\tau_2)\alpha_2^b e^{-h(1)\tau_1}e^{-h(2)\tau_2})\times$$

$$(3e^{ih(3)\tau_3}\alpha_2^c D_{cd}(x_{23},\tau_3)\alpha_3^d e^{-ih(3)\tau_3} + e^{ih(3)\tau_3}\alpha_2^c D_{cd}(x_{23},\tau_3)\alpha_3^d e^{-h(3)\tau_3}) + H.c. \tag{3.10}$$

Expressing $\Delta_{T-T}(x_{123})$ in the occupation number representation we get,

$$V_{nlm,n'l'm'}^{T-T} \equiv < nlm|\Delta_{T-T}|n'l'm' > =$$

$$\frac{1}{2}\sum_s(\frac{\eta_{nl,n's} - \eta_{ln,sn'}}{w_{nn'}+w_{ls}})(3\eta_{ms,m'l'} + \eta_{sm,l'm'}) + H.c. \tag{3.11}$$

B. Virtual pair exchange

We must also include the terms that arise when electrons exchange pairs. We recognize that the two-body operator,

$$H_2 = H_C + H_B =$$

$$\frac{1}{2}\int dx_1 \int dx_2 \psi^\dagger(x_1)\psi^\dagger(x_2)v(x_{12})\psi(x_2)\psi(x_1) \tag{3.12}$$

422

where

$$v(x_{12}) = v_C(x_{12}) + v_B(x_{12})$$

can be decomposed into a set of operators that are characterized by their commutation relations with the electron (positron) number operator. Using (1.13) we note that H_2 consists of terms that:

i) Scatter two electrons,

ii) Scatter one electron, create a pair, or destroy a pair,

iii) Two pair terms, and positron scattering terms.

The no pair ansatz includes processes of the first type, however the effects induced by (ii) must also be included if an accurate e^4 configuration space equation is desired. We can realize the terms generated by (ii) if we introduce a decoupling transformation that diagonalizes the type (ii) terms with respect to the electron number to $O(e^4)$. The term we wish to diagonalize is

$$H_2' = \frac{1}{2}\int dx_1 \int dx_2 \psi_e^\dagger(x_1)\psi_e^\dagger(x_2)v(x_{12})\psi_e(x_2)\psi_p(x_1)$$

$$+\psi_p^\dagger(x_1)\psi_e^\dagger(x_2)v(x_{12})\psi_e(x_2)\psi_e(x_1) \tag{3.13}$$

where we have defined the field operators $\psi_e^\dagger(x)$, $\psi_e(x)$ which create, and destroy electrons respectively. $\psi_p^\dagger(x)$, $\psi_p(x)$ create and destroy positrons respectively. The Hamiltonian now has the form,

$$H' = H_0 + H_2 + H_2' + O(e^4).$$

where H_2 includes only the electron scattering terms. We employ the decoupling transformation (3.15) except that we now replace $H_I(\tau)$ in the generator σ with $H_2'(\tau)$. After carrying out the identical steps given previously we get,

$$H'' = e^{i\sigma}He^{-i\sigma} = H_0 + H_2 + \frac{i}{2}[\sigma, H_2'] + ... \tag{3.14}$$

where

$$\sigma = -\frac{1}{2}\int_{-\infty}^{\infty} H_2'(\tau)e^{-\eta|\tau|}. \tag{3.15}$$

When evaluating the commutator, we are interested only in the terms that commute with the electron number operator (the neglected terms appear at higher order). The resulting terms are three, two and one-electron operators. We are interested only in the three-electron operators

$$H_p \equiv \frac{i}{2}[\sigma, H_2'] = -\frac{i}{4} \int_{-\infty}^{\infty} d\tau \epsilon(\tau)[H_2'(\tau), H_2] =$$

$$-\frac{i}{4} \int_{-\infty}^{\infty} d\tau \epsilon(\tau) \int dx_1 \int dx_2 \int dx_3 : \psi_e^\dagger(x_1)\psi_e^\dagger(x_2)\psi_e^\dagger(x_3)$$

$$e^{-ih(1)\tau} e^{-ih(2)\tau} v(x_{12}) e^{ih(1)\tau} e^{ih(2)\tau} \Lambda_-(2) v(x_{23}) \psi_e(x_3)\psi_e(x_2)\psi_e(x_1) : +H.c. \quad (3.16)$$

where $\Lambda_- \equiv <0|\psi_p\psi_p^\dagger|0>$ is a negative energy projection operator. The matrix element of (3.16) is

$$V_{nlm,n'l'm'}^p \equiv <nlm|H_p|n'l'm'> = \sum_s \frac{\overset{-}{<nl|v|n's>}}{(w_{nn'} + w_{ls})} < sm|v|l'm' > \quad (3.17)$$

where $v = v_C + v_B$. Using the definitions for $\eta_{nl,n'l'}$ and $C_{nl,n'l'}$ we get

$$V_{nlm,n'l'm'}^p = \sum_s \frac{\overset{-}{(C_{nl,n's} + \eta_{nl,n's} + \eta_{ln,sn'})}}{(w_{nn'} + w_{ls})} (C_{sm,l'm'} + \eta_{sm,l'm'} + \eta_{ms,m'l'}) \quad (3.18)$$

the sum in (3.18) is over all negative energy states. Adding H_{C-T}, $H_{T-T'}$ and H_P the total three-electron interaction Hamiltonian is given by,

$$H_3 = \frac{1}{6} \int dx_1 \int dx_2 \int dx_3 : \psi_e^\dagger(x_1)\psi_e^\dagger(x_2)\psi_e^\dagger(x_3) \times$$

$$S\left(\Delta_{CT}(x_{123}) + \Delta_{T-T}(x_{123}) + \Delta_P(x_{123})\right) \psi_e(x_3)\psi_e(x_2)\psi_e(x_1) : \quad (3.19)$$

where, $\Delta_P(x_{123})$ is the virtual pair contribution, (3.16), to the three-body potential. In configuration space (3.19) manifests as a non-local three-electron potential.[2]

IV. NONRELATIVISTIC LIMIT OF THREE-ELECTRON POTENTIALS

Sixty years ago, H. Primakoff and T. Holstein derived an expression[4] for the three-electron potential in light atoms using both classical and quantal methods. They noted that, in the non-relativistic Hamiltonian, three-electron irreducible diagrams (in the time-ordered sense) are generated when the $\vec{A}^2$ term acts on an electron line followed by the subsequent action of the $\vec{p} \cdot \vec{A}$ term on two other distinct electron lines. The total energy depends on the instantaneous coordinates of the three electrons and is given by the potential[4]

$$H_{P-H} = \frac{e^4}{8m^3c^4} \sum_{i,j,k} \frac{1}{|r_{ij}||r_{jk}|} \times$$

$$\left\{ \vec{p}_i \cdot \vec{p}_k + (\vec{p}_i \cdot \hat{r}_{jk})(\vec{p}_k \cdot \hat{r}_{jk}) + (\vec{p}_i \cdot \hat{r}_{ij})(\vec{p}_k \cdot \hat{r}_{ij}) + (\vec{p}_k \cdot \hat{r}_{jk})(\hat{r}_{ij} \cdot \hat{r}_{jk})(\hat{r}_{ij} \cdot \vec{p}_i) \right\} \quad (4.1)$$

where $\vec{p}_i$ is the canonical momentum of the i'th electron. A more general expression was given later by Chanmugan and Schweber[11]. Mittleman[2] derived expression (3.19) valid for all values of nuclear charge Z. Below we recover the Primakoff-Holstein potential as a non-relativistic limit of expression 3.19. Knowing that the non-relativistic $\vec{A}^2$ term arises from the class of relativistic diagrams where virtual positrons lines are sandwiched between two electron-photon vertices we consider only those terms in 3.19 that involve diagrams containing negative energy intermediate states. In addition, we consider only those terms that exchange transverse photons. In the occupation represention these terms are contained in (3.17) and (3.11). Adding these two, including only the subset of diagrams of interest and invoking the on-shell condition, $w_{nn'} + w_{ll'} + w_{mm'} = 0$, we get

$$< nlm|H_{NR}|n'l'm' > = \sum_s^{(-)} 4\frac{\eta_{nl,n's}\eta_{ms,m'l'}}{w_{nn'} + w_{ls}}. \quad (4.2)$$

Note that the terms in (3.17) and (3.11) that depend on the frequency of the intermediate states cancel in the sum. Because of this cancellation the nonrelativistic limit is straightforward. We replace the matrix elements, $\eta_{ab,a'b'}$, by the matrix elements of the local Breit operator, and the energy denominator by $2mc^2$, to get

$$< nlm|H_{NR}|n'l'm' > = \sum_s^{(-)} \frac{< nl|v_B|n's >< sm|v_B|l'm' >}{2mc^2}. \quad (4.3)$$

Imposing the closure property of the Dirac-Coulomb states, $\sum_s^{(-)} |s>< s| = 1 + O(Z\alpha)$, neglecting terms to $O(Z\alpha)$ and using the anticommutation relations $\{\alpha^a, \alpha^b\}_+ = 2\delta_{ab}$ for the Dirac matrices we get,

$$< nlm|H_{NR}|nlm > = < nlm|H_{P-H}|n'l'm' > \quad (4.4)$$

where we substituted $\vec{\alpha}_i$ by $\vec{p}_i/mc$.

V. CALCULATIONS

Below we calculate the contribution of the three-electron potential (3.19) to the binding energy of, high-Z, lithium-like ions. The energy shift due to the three-electron interaction is (3.19) is,

$$\Delta E = < \Psi|H_3|\Psi > \quad (5.1)$$

where $|\Psi>$ is the unperturbed ground state of the ion. For high Z ions, screening effects are expected to be small and the use of the Dirac-Coulomb orbitals for the unperturbed states is justified. Expanding (5.1) gives the expression

$$\Delta E = \sum_P (-1)^P U_{abc,[abc]} \tag{5.2}$$

where $U_{abc,[abc]}$ is a symmetrized (with respect to particle interchange) matrix element of the kernel $\Delta_{C-T} + \Delta_{T-T} + \Delta_P$ in the occupation representation. The sum over P represent all permutations of the quantum numbers in the bracket $[abc]$. They represent the quantum numbers of the Dirac-Coulomb functions, $|(n,j,m,\kappa)>$, where n is the principal quantum number, j is the total angular momentum, m is the azimuthal quantum number, and κ is the Dirac quantum number. $P = 0$ for an even permutation and $P = 1$ for an odd one. The shift is then expressed in terms of the matrix elements $\eta_{nl,n'l'}$ and $C_{nl,n'l'}$. The matrix elements are simplified to expressions containing only products of the integrals.

$$g_{nl,n'l'} = < nl|\frac{1}{r}cos(\frac{w_{nn'}r}{hc})|n'l'>$$

$$h_{nl,n'l'} = < nl|\frac{\vec{\alpha}_1 \cdot \vec{\alpha}_2}{r}cos(\frac{w_{nn'}r}{hc})|n'l'> . \tag{5.3}$$

Selection rules limit $s_{1/2}$ and $d_{3/2}$ orbitals to contribute in the sum over intermediate states, the latter states give a small contribution and can be neglected. Because all matrix elements contain $s_{1/2}$ spinors only we further simplify (5.3) by carrying out the angular integrations to get,

$$g_{ab,a'b'} = 2X^{L=0}(m_a m_{a'}; m_b m_{b'})R^C(aa';bb')$$

$$h_{ab,a'b'} = \frac{2}{3}X^{L=0}(m_a m_{a'}; m_b m_{b'})R^B_-(aa';bb') - \frac{4}{3}X^{L=1}(m_a m_{a'}; m_b m_{b'})R^B_+(aa';bb') \tag{5.4}$$

where

$$X^L(m_a m_{a'}; m_b m_{b'}) = \sum_q (-1)^{1+q-m_a-m_b} \begin{pmatrix} \frac{1}{2} & L & \frac{1}{2} \\ -m_a & q & m_{a'} \end{pmatrix} \begin{pmatrix} \frac{1}{2} & L & \frac{1}{2} \\ -m_b & q & m_{b'} \end{pmatrix} . \tag{5.5}$$

The expressions in brackets are 3-j symbols and the radial integrals are defined by

$$R^C(aa';bb') = -\int_0^\infty dr_1 \int_0^\infty dr_2 \, \omega_{aa'} J_0(\omega_{aa'}r_<)\eta_0(\omega_{aa'}r_>)$$

$$\times [P_a(r_1)P_{a'}(r_2) + Q_a(r_1)Q_{a'}(r_2)][P_b(r_1)P_{b'}(r_2) + Q_b(r_1)Q_{b'}(r_2)]$$

$$R_{\pm}^{B}(aa';bb') = -\int_0^{\infty} dr_1 \int_0^{\infty} dr_2\, \omega_{aa'} J_1(\omega_{aa'}r_<)\eta_1(\omega_{aa'}r_>)$$

$$\times [P_a(r_1)Q_{a'}(r_2) \pm P_a(r_1)Q_{a'}(r_2)][P_b(r_1)Q_{b'}(r_2) \pm P_b(r_1)Q_{b'}(r_2)] \qquad (5.6)$$

where $P(r)$ and $Q(r)$ are the "large" and "small" components of the radial Dirac-Coulomb functions, and J_n, η_n are the regular and irregular spherical Bessel functions of order n respectively. At this stage, the shift is reduced to a quadrature problem and the remainder of the calculation is done numerically. The results are tabulated below[12]. We also tabulate the partial contributions coming from specific intermediate states.

We have also calculated (not shown) the total three-body energy. This energy is the sum of all three-body correlation diagrams originating from the two-electron potential, given in section II, and the intrinsic three-electron potential contribution tabulated below. We found that the correlation contribution is usually much larger than the three-electron potential contribution. However, we found that a class of correlation diagrams cancel with some of the three-body potential diagrams. The cancellation is such that the total three-body energy is independent of terms that depend on the frequency of the intermediate states, i.e there are no $cos(w_{n,s}r/\hbar c)$ terms when $w_{n,s}$ depends on energy differences with the intermediate states $\{s\}$. This cancellation greatly simplifies the numerical work.

Table I. - Contribution of the three-body potential to the binding energy (eV) of high-Z lithium-like ions. Powers of ten are in brackets.

Z	Negative Continuum	Positive Continuum	Bound States	Total (eV)
137	6.61(-1)	2.43(-1)	3.44(-1)	1.25(0)
130	1.59(-1)	4.33(-2)	-2.45(-3)	2.00(-1)
118	5.58(-2)	1.03(-2)	-3.18(-2)	3.43(-2)
110	3.13(-2)	3.64(-3)	-2.92(-2)	5.76(-3)
100	1.58(-2)	3.27(-4)	-2.20(-2)	-5.96(-3)
90	8.05(-3)	-6.37(-4)	-1.53(-2)	-7.89(-3)
80	4.08(-3)	-6.53(-4)	-9.81(-3)	-6.38(-3)

ACKNOWLEDGEMENTS

I would like to thank M. Mittleman for the many helpful discussions. I would also like to thank the organizers of this program, W. Johnson, P. Mohr and J. Sucher, for providing such a stimulating environment. This work was supported in part by DOE, Fundamental Interactions Branch, Division of Chemical Sciences, Office of Basic Energy Research.

REFERENCES

[1] G.E. Brown and D.G. Ravenhall, Proc. Roy. Soc. A208, 552 (1951).

[2] M.H. Mittleman, Phys. Rev. A 4, 893 (1971); Phys. Rev. A 5, 2395 (1972).

[3] J. Sucher (See his contribution in this volume).

[4] H. Primakoff and T. Holstein, Phys. Rev. 55, 1218 (1939).

[5] L.N. Labzovskii, Soviet Physics JETP 34, 749 (1972).

[6] J.L. Friar, Comments Nucl. Part. Phys. 11, 51 (1983); J.L. Friar, B.F. Gibson, G.L. Payne, Ann. Rev. Nucl. Part. Sci. 34, 403 (1984).

[7] E. E. Salpeter, Phys. Rev. 87, 328 (1952).

[8] M.H. Mittleman, Phys. Rev. A 24, 1167 (1981).

[9] B. Zygelman and M.H. Mittleman, J. Phys. B 19, 1891 (1986).

[10] J. Schwinger, Phys. Rev. 74, 1439 (1948).

[11] G. Chanmugan, and S. Schweber, Phys. Rev. A 1, 1369 (1970).

[12] The results presented in Table I. correct previous results given in Ref. 9 by a factor of six. The error was caused by the use of an incorrect normalization factor in the coding.

THREE-ELECTRON ATOMS FROM A
BETHE-SALPETER EQUATION APPROACH

Marvin H. Mittleman
The City College of New York, New York, NY 10031

ABSTRACT

The two-electron Bethe-Salpeter equation is reviewed. The transition to a Hamiltonian form is discussed, and the criterion that the first-order energy shift vanish is used to determine the electron-electron potential that appears in the Hamiltonian. A similar program is carried through for the three-electron case. The fundamental idea of Feldman and Fulton [Nuc. Phys. **B195**, 61 (1982)] is used for the transition to the Hamiltonian form, although some changes are necessary. The optimum two-electron interaction is the same as that in the two-electron case. It is necessary to introduce a static three-electron interaction, which then can be determined from the vanishing of the first-order energy shift. It is extremely complicated and unlikely to prove useful in the near future.

Published in Phys. Rev. A **39**, 1 (1989).

AN ALTERNATIVE APPROACH TO THE APPLICATION OF QED TO MANY–ELECTRON ATOMS

T. Fulton

The Johns Hopkins University, Baltimore, MD 21218

Abstract: The one and two–electron Green's function approach to atoms is outlined.

"Let a hundred flowers bloom.
Let a hundred thoughts contend."
Chairman Mao.

In the discussion which follows, I will call the "majority approach" the one taken, for example, by I. Lindgren and J. Sucher in these ITP seminar notes. Our alternative approach, which I will henceforth term the "minority approach" is represented by the work of my colleague, G. Feldman, our very able graduate student, S. S. Liaw, and myself. It uses the work of Fetter and Walecka ["Quantum Theory of Many Particle Systems", McGraw–Hill, New York, 1971], and of Csanak, Taylor and Yaris [Adv. Atom. Molec. Phys. 7, 287 (1971)] as take–off points. These original treatments are restricted to the non–relativistic (NR) limit for atoms. In our initial extensions of it, starting in 1981, we also considered only the NR case. However, recently we have dealt with the relativistic (R) case, as will the rest of this discussion.

Please note the appended list of selected references to our recent work. They constitute various stages in the progress of our program. I cannot possibly deal in any detail with all of this material here. I will therefore use this opportunity to "advertise" what we have done: I will allude to some salient features of our approach and discuss some of the tricks we found useful in moving our work forward. I will also attempt to compare and contrast this "minority approach" with the "majority approach".

The starting point for both these approaches is the full QED Hamiltonian density,

$$\mathcal{H}_{QED} = \mathcal{H}^{(0)}_{Lepton} + \mathcal{H}^{(0)}_{e.m.} + \mathcal{H}_{int} , \qquad (1)$$

with a fixed external nuclear Coulomb potential of a nucleus of charge $-Ze$. $\mathcal{H}_{QED}$ contains the Heisenberg field operators $\psi^\dagger(x)$, $\psi(x)$ and $A_\mu(x)$ in the usual way. The majority then would consider the M–lepton propagator (where eM is the total lepton charge, and $M \leq Z$, with the inequality holding for positive ions),

$$G^0_M(1,3,\ldots;2,4,\ldots,2M) = (-i)^M \left\langle \begin{matrix} 0 \\ 0 \end{matrix} \right| T[\psi(1)\psi(3) \ldots \psi(2M-1)\psi^\dagger(2) \ldots \psi^\dagger(2M)] \left| \begin{matrix} 0 \\ 0 \end{matrix} \right\rangle . \qquad (2)$$

The upper index denotes the fact that we are taking a vacuum expectation value. The lower index is the ground state energy of the vacuum, zero. The indices in parentheses denote space–time coordinates, e.g.

$$2 \equiv (r_2, t_2) .$$

All possible time orderings must be taken. In zeroth order, only Coulomb interactions are considered between electrons, but other electromagnetic interactions, including closed loops, can be treated. The homogeneous equation, generated from the residues of the particular time–ordering $t_\psi > t_{\psi^\dagger}$ (which corresponds to the inhomogeneous equation for the M–lepton propagator) is the M–lepton Bethe–Salpeter (BS) equation. The zeroth order BS equation is the modified Breit equation, with the electron–electron Coulomb potential, V_{ee}, sandwiched between two positive energy and two negative energy projection operators.

We in the minority also deal with propagators. However, in contrast to the majority, we consider the one– and two–lepton propagators (2–point and 4–point functions) and the one– and two–lepton e.m. vertices (3–point and 5–point functions). Rather than defining these propagators in terms of vacuum expectation values, we take N–particle expectation values, where N describes a non–degenerate state (for example a closed shell or subshell). In this report, I will mostly discuss the one–lepton propagator $G_1^N(1;2)$, where

$$G_1^N(1;2) = -i\left\langle {N \atop 0} \right| T[\psi(1)\psi^\dagger(2)] \left| {N \atop 0} \right\rangle . \tag{3}$$

There are two possible time orderings in this propagator. As we shall shortly see, the time ordering $t_1 > t_2$ describes atoms with non–degenerate cores plus one electron (corresponding in equation (2) to M=N+1). We term these the one–particle (1P) cases. These cases include, e.g.,the alkali atom isoelectronic sequences. The time ordering $t_1 < t_2$ describes atoms with non–degenerate cores minus one electron, or the one–hole (1H) cases (M=N–1). In short, we say that G_1^N refers to the 1P/1H system. In an analogous way, the two–lepton propagator,

$$G_2^N(1,3;2,4) = \left\langle {N \atop 0} \right| T[\psi(1)\psi^\dagger(2)\psi(3)\psi^\dagger(4)] \left| {N \atop 0} \right\rangle , \tag{4}$$

involves 4! time orderings. The particular time orderings $t_1,t_3 > t_2,t_4$ and $t_1,t_3 < t_2,t_4$ correspond to M=N+2 (2P) and M=N–2 (2H), while $t_1,t_2 > t_3,t_4$ and $t_1,t_2 < t_3,t_4$ correspond to M=N (PH), where N is again a non–degenerate core.

It is important to note that G_M^0, G_1^N and G_2^N are fully equivalent, exact formulations of the many–electron atomic problem. The seeming disadvantage of G_1^N and G_2^N is that they can be used to describe atoms with specific numbers of electrons, while G_M^0 can describe atoms with any number of electrons. As mentioned before, the 1P cases in G_1^N describe, among other atoms, the alkali atom isoelectronic sequences. The 2P cases of G_2^N include Be and C isoelectronic sequences, and the HP cases of G_2^N all of the noble gases. All of the 1P/1H, 2P/2H and PH cases would cover about 80% of atomic species, and so the restriction of the N core electrons to describe non–degenerate states is not a severe one. One could increase the number of atoms covered by such an approach by considering G_3^N, G_4^N, etc., but the formalism for these cases would get so cumbersome that one may as well use G_M^0 in the first place.

Why not deal with G_M^0, and the M–particle BS equation all the time?

1. $\underline{G_1^N \text{ and } G_2^N \text{ have natural approximations which are both useful and}}$ $\underline{\text{familiar in atomic physics.}}$

a. G_1^N: Restrictions to single V_{ee} kernels in the Green's function equations (the so–called "ladder approximation") lead to the Dirac–Fock (DF) equation, if G_1^N is the starting point. There are no projection operators sandwiching V_{ee}. In contrast, the DF equations, with G_M^0 as starting point, are obtained by a variational principle, and V_{ee} is sandwiched between projection operators.[1,5,6] (See also point 3 below).

b. G_2^N: The ladder approximation for G_2^N in the PH case is the random phase approximation[1,5] (particle–hole BS equation), and in the 2P/2H case is an He–like BS equation.[8,9]

2. <u>One can implement gauge invariance (GI) at every level of approximation.</u> Of course, GI is automatic if one deals with exact wave functions, but, failing that, GI is a good error–catching machine in devising various models.

a. $G_1^N{\to}DF$, in association with the 3–point irreducible vertex function $\Lambda(1;2|3)$, where 3 is the space–time point of the photon attachment, leads to the description of one–electron transitions, e.g. in alkali atoms, in terms of non–local currents.[1,2,5] Correlation corrections can also be included in this GI formalism.

b. $G_2^N{\to}HP$ (RPA) leads to the GI description of one–electron transitions in, e.g., noble gases. Correlation corrections can once again be included in this GI formalism.[1,5]

c. $G_2^N{\to}2P$, in association with the 5–point irreducible vertex function $\Lambda(1,3;2,4|5)$, where 5 is the photon attachment point, leads to one– and two–electron excitations of, e.g., Be and C.[8] In the ladder approximation, the somewhat intractable 5–point function simplifies so that only a product of the 3–point function and the inverse of G_1^N appears in the transition matrix elements.

3. <u>One can make systematic approximations, and know at each stage what terms one has left out, and one can also avoid double–counting.</u> (In contrast, one has a "single shot" in a variational approach.) This is particularly important in dealing with radiative corrections. In such corrections, our treatment makes it apparent, for example, that some positron states are in finite correlation corrections, while others have to be lumped with radiative corrections, which need to be renormalized.[6,7]

4. <u>One can gain insight into the physical effects involved in radiative corrections by using perturbation theory for G_1^N.</u>[7] Our starting point is the zeroth order Furry picture. Though this approach is not numerically useful, it leads to a clear indication of how screening appears in the formalism, and also results in initially unexpected (but, after the fact, obvious) contributions which yield the dominant Lamb shift effects in other than s states. The structures are so transparent that the results for the numerically interesting DF case can be conjectured easily[7]. We are presently engaged in attempting to prove our conjectures.

5. <u>Numerical implementation of our formalism (including correlation effects) appears to be feasible.</u>[3,8,9] The formulation in terms of G_1^N, G_2^N, etc. appears to be rather convenient for this purpose. For example, the absence of projection operators in the DF formalism arising from G_1 simplifies computations. (Of course, projection operators do appear in the 2P/2H and PH cases arising from G_2, but they can be handled quite simply in this context.)

Our work so far has focused almost entirely on formal developments. We have made only a start at numerical work, principally in order to test the feasibility of our approach:

A. W. R. Johnson and I have shown[3] that non–local effective currents in the radiative E1 transitions of alkali atoms in the dipole approximation lead to the equality of the length and velocity forms of oscillator strengths.

B. Our student, S. S. Liaw, has obtained some preliminary results for energy levels in the 2P/2H formalism.[8,9] In particular, for the (2s, 2s) double ionization energy of

432

neutral Be, he gets a value[8] of −1.0124 au, which differs from the experimental value by 0.04%. Liaw has also treated the ground state of neutral He, as the simplest test of our formalism and of the computational techniques[8,9]. The computed value for this state differs from experiment by 0.2%. Both of these results were obtained by going beyond the ladder approximation. We also included contributions (in perturbation theory beyond the ladder approximation) due to kernels associated with irreducible Feynman diagrams involving two V_{ee}. Although these results are by no means bad, they are clearly not yet competitive with the corresponding values in the published literature, obtained by other methods. We didn't expect them to be. One should probably include at least the next higher order irreducible Feynman diagrams in the kernels.

The B–spline approach, pioneered by Johnson, Sapirstein and collaborators, has been used in these computations. We hope to calculate numerical values for radiative corrections in the near future, using our minority approach. However, we first want to gain more confidence in and experience, with these computational techniques.

I previously discussed our approach at the 1986 workshop at ICTP, Trieste[4]. In what follows, I will emphasize some aspects of our work since that time. However, at the risk of being repetitious, I will backtrack a little and discuss some details of the kinds of information we can obtain from G_1^N. I will conclude these notes with a brief summary of salient aspects of the radiative corrections for many–electron atoms in perturbation theory, the nature of the physical effects which arise, and our conjectures for the form of such corrections in the DF approximation.

I will first focus on the energy eigenvalues of many–electron atoms. I will outline first how the same information on this point can be gotten as well from G_1^N as from G_M^0. Consider, as an illustration, the time ordering $t_1 > t_2$:

$$G_1^N(1;2)\Big|_{t_1>t_2} = -i \left\langle\begin{matrix}N\\0\end{matrix}\right| \psi(1) \quad\uparrow\quad \psi^\dagger(2)\left|\begin{matrix}N\\0\end{matrix}\right\rangle \qquad (5)$$

$$\sum_j \left|\begin{matrix}N+1\\j\end{matrix}\right\rangle\left\langle\begin{matrix}N+1\\j\end{matrix}\right| ;$$

we inserted a complete set of states in the middle of the expression, in which Σ is a sum over discrete and an integral over continuous states, of total energy E_j^{N+1}.

Let us define

$$f_j^N(1) \equiv \left\langle\begin{matrix}N\\0\end{matrix}\right|\psi(1)\left|\begin{matrix}N+1\\j\end{matrix}\right\rangle; \quad f_j(1,0) \equiv f_j(1) . \qquad (6)$$

The standard form of the time–translation is

$$\psi(r,t) = e^{iHt}\psi(r,0)e^{-iHt} , \qquad (7)$$

where H is the full Hamiltonian, e.g. for eigenstates $\left|\begin{matrix}N+1\\j\end{matrix}\right\rangle$,

$$H\left|\begin{matrix}N+1\\j\end{matrix}\right\rangle = E_j^{N+1}\left|\begin{matrix}N+1\\j\end{matrix}\right\rangle . \qquad (8)$$

Eq. (5) thus becomes

$$G_1^N(1,2)\Big|_{t_1>t_2} = -if_j^N(1)f_j^{\dagger N}(2)e^{i(E_o^N - E_j^{N+1})(t_1-t_2)}\Big|_{t_1>t_2}. \tag{9}$$

Let us define $t_1-t_2 \equiv t$, and carry out a time Fourier–transform, adding the infinitesimal iη, $\eta \to 0^+$ to the energy difference, to make the integral converge at t=∞. We obtain

$$G_\omega^N(1,2)\Big|_{\text{val-excit}} \equiv \int_0^\infty e^{i\omega t} G_1^N(1,2)\Big|_{t_1>t_2} dt = \sum_j \frac{f_j^N(1) \; f_j^{\dagger N}(2)}{\omega - e_j^N + i\eta}, \tag{10}$$

where e_j^N is defined by

$$e_j^N \equiv E_j^{N+1} - E_0^N. \tag{11}$$

A similar analysis can be carried out for the other time ordering.

The formal result we have just obtained is exact. In zeroth order perturbation theory (with V_{ee} neglected), $e_j^N \to \epsilon_j$. The superscript N has disappeared, because, in this approximation, an electron doesn't know about any of the other electrons. The DF theory is a quasi–particle theory of single dressed electrons. These electrons know about the electrons in the core, so that $e_j^N \to \left[\epsilon_j^N\right]^{DF}$ in this approximation.

Let us compare and contrast this analysis with the majority point of view. Take the time ordering $t_\psi > t_{\psi\dagger}$, set M=N+1, N$\geq$2, and we have

$$G_{N+1}^0\Big|_{t_\psi > t_\psi} = (-i)^{N+1} \Big\langle\begin{smallmatrix}0\\0\end{smallmatrix}\Big| \underbrace{\psi \dots \psi}_{} \qquad \underbrace{\psi^\dagger \dots \psi^\dagger}_{} \Big|\begin{smallmatrix}0\\0\end{smallmatrix}\Big\rangle. \tag{12}$$
$$\uparrow$$
$$\sum_j \Big|\begin{smallmatrix}N+1\\j\end{smallmatrix}\Big\rangle \Big\langle\begin{smallmatrix}N+1\\j\end{smallmatrix}\Big|$$

Since there are more than two time coordinates in this case, the analog of f, the N+1 particle BS amplitude, will have (N−1)$\geq$1 relative–time dependences. However, if we define the c.m. times

$$T \equiv \frac{1}{N+1} \times \text{sum of } t_\psi\text{'s} \quad \text{and} \quad T' \equiv \frac{1}{N+1} \times \text{sum of } t_{\psi\dagger}\text{'s}, \tag{13}$$

and Fourier transform with respect to T−T', the energy denominator becomes

$$\omega - \Omega_j^{N+1} + i\eta. \tag{14}$$

Ω_j^{N+1} above is defined by

$$\Omega_j^{N+1} \equiv E_j^{N+1} - E_0^0 = E_j^{N+1}, \tag{15}$$

and E_0^0, the lepton vacuum energy, vanishes.

The (N+1)–lepton Green's functions satisfy inhomogeneous integral equations. The corresponding homogeneous equations for the BS amplitude are, as usual, obtained by taking the residues at the appropriate poles.

We generate equations for e_j^N if we follow the minority way, and for E_j^{N+1}, if we follow the majority approach. However, if we are interested in single particle excitations, say between states j and k, where j≠k, we have formally the same exact energy difference, using both G_{N+1}^{0} and G_{1}^{N}:

$$e_{j}^{N} - e_{k}^{N} = \Omega_{j}^{N+1} - \Omega_{k}^{N+1} = E_{j}^{N+1} - E_{k}^{N+1} . \qquad (16)$$

e_{j}^{N}, with Z=N+1, directly gives the ionization energy for the jth state of a neutral atom. Correspondingly, and less directly, it is given by $\Omega_{j}^{N+1} - \Omega_{o}^{N}$ in the majority formalism. Clearly, in approximate treatments within this approach, core relaxation effects come into play.

Let me next focus on our most recent work.[6,7] The rigorous parts of references 6 and 7 are for perturbation theory, but we formulate our result in terms suggestive of the way in which they can be generalized to the DF approximation. We make some conjectures in these papers about these generalized forms. We are still concerned with G_{1}^{N}, and the energy levels generated from its use, including correlation terms and radiative corrections. To save writing, I will suppress the subscript 1 on G_{1}^{N}, and will define g^{N} as the zeroth order 1–lepton propagator, i.e. $G_{1}^{N} \equiv G^{N} \to g^{N}$. Suppressing the space–coordinates and writing g^{N} in Fourier transform space, we have

$$g_{\omega}^{N} = \sum_{\ell} \frac{|\ell\rangle\langle\ell|}{\omega - \epsilon_{\ell} - i\eta} + \sum_{a=1}^{N'} \frac{|a\rangle\langle a|}{\omega - \epsilon_{a} - i\eta} + \sum_{j=N+1}^{\infty} \frac{|j\rangle\langle j|}{\omega - \epsilon_{j} + i\eta} . \qquad (17)$$

The propagator g_{ω}^{N} can be represented in Feynman diagrams. We will use a double line to indicate it.

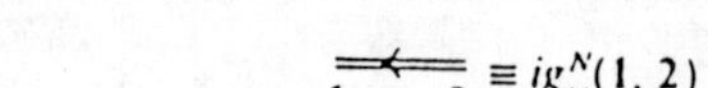

$$\overrightarrow{\underset{1 \quad \omega \quad 2}{\rule{3cm}{0.4pt}}} \equiv i g_{\omega}^{N}(1, 2)$$

Fig. 1. Feynman propagator for g_{ω}^{N}.

Above, we evaluated in some detail a term corresponding to the last one of Eq. (17) for the exact one–lepton propagator. It represents the contribution of valence and excited states, both discrete and continuous. The second term in (17) is due to the discrete core states, and the first term is the positron continuum contribution. The spectrum and contour in the complex ω–plane is given (not to scale) in Fig. 3 below. It is easy to show, using the same arguments as were employed in obtaining Eq. (17), that the vacuum Green's function, g_{ω}^{0} (which we will call g_{ω} for short) is given by

$$g_{\omega} = \sum_{\ell} \dots + \sum_{a=1}^{N} \frac{|a\rangle\langle a|}{\omega - \epsilon_{a} + i\eta} + \sum_{j=N+1}^{\infty} \dots . \qquad (18)$$

Notice the sign change in the infinitesimal imaginary part in the energy denominator of what were previously core states. There is no longer any core.

We will represent g_{ω} by the Feynman diagram element given in Fig. 2.

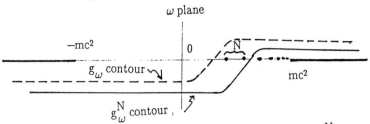

$$\overrightarrow{1 \quad \omega \quad 2} \equiv i g_{\omega}(1, 2);$$

Fig. 2. Feynman propagator for g_{ω}.

The ω–plane contour for g_{ω} is also indicated in Fig. 3.

Fig. 3. The spectrum and ω–plane contour corresponding to g_{ω}^{N} and g_{ω}.

The spectrum for g_{ω} is identical with that for g_{ω}^{N}. Only the two ω–plane contours, corresponding to g_{ω} and g_{ω}^{N}, differ.

We are now ready to tackle the essential problem with radiative corrections: the isolation of ultraviolet infinities, which have to be eliminated through renormalization. We carry out this process in three stages:

         1. Start with g^{N}.
         2. Go $g^{N} \rightarrow g$, thereby isolating the u.v. infinities in the latter.
         3. Go $g \rightarrow g_{free}$, where g_{free} is the Green's function, which is the vacuum expectation value in the absence of the nuclear potential. This is the conventional method used to achieve renormalization in bound state problems, for example in the H Lamb shift.

Stage 1. is easily achieved. We write

$$g^{N} = (g^{N} - g) + g , \tag{19}$$

and observe that

$$(g^{N} - g) = \sum_{a=1}^{N} |a\rangle\langle a| \left\{ \frac{1}{\omega - \epsilon_a - i\eta} - \frac{1}{\omega - \epsilon_a + i\eta} \right\} = \sum_{a=1}^{N} |a\rangle\langle a| \left\{ 2\pi i \delta(\omega - \epsilon_a) \right\} . \tag{20}$$

The u.v. singularities are now going to arise from expressions in which g appears, since these same expressions, with $(g^{N} - g)$ replacing g^{N}, consist only of a finite sum of individually finite contributions.

We will illustrate this point, but first have to invent some new symbols.

$$(g^{N}-g) \equiv \overset{(a)}{\longleftarrow}\!\!\!\vdash\!\!\!\longleftarrow \quad ; \quad \overset{1 \qquad a}{\longleftarrow\!\!\!\vdash} \equiv \langle 1|a\rangle \equiv \varphi_a(1) ; \quad \overset{a \qquad 2}{\vdash\!\!\!\longleftarrow} \equiv \langle a|2\rangle \equiv \varphi_a^{\dagger}(2)$$

Fig. 4. Graphical representation of $(g^{N}-g)$ of Eq. (20).

The letter "a" in the figure is a dummy variable of summation, and (a) denotes the finite sum over the core states.

Let us now graphically consider the lowest order self—energy contributions, those involving one virtual photon. The Feynman propagator of the photon will be represented as in Fig. 5.

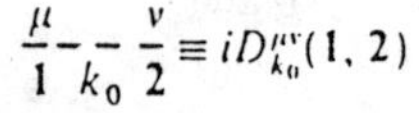

$$\frac{\mu}{1} - \frac{\nu}{k_0} \frac{\nu}{2} \equiv iD_{k_0}^{\mu\nu}(1, 2)$$

Fig. 5. Feynman propagator of the photon.

We will keep the same topologies, but "open up" the $(g^N - g)$ parts of diagrams, to emphasize the finite character (easily confirmed when the corresponding algebraic expressions are written down) of the tree—diagrams which result.

The "tadpole", or vacuum polarization term is given by Fig. 6 below.

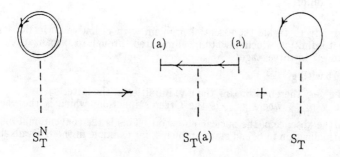

Fig. 6. Isolation of the usual vacuum polarization term in the "tadpole" diagram.

In the figure above, S denotes self—energy. (We don't use the more standard Σ to denote self—energy, because there are factors of γ_0 appearing in S which are not in Σ, due to the use of $\psi^\dagger$ instead of $\bar\psi$ in our theory. This form is more standard, and also more convenient to use in atomic calculations.) T stands for "tadpole". S_T^N is our starting point, and contains g^N's. S_T is the usual vacuum polarization (Uehling term) in the presence of the unscreened nucleus. (Only g's, which are vacuum expectation values, appear in it.) $S_T(a)$ is the remaining finite contribution.

In a similar way, the diagram for the non—vacuum—polarization part of the self energy (we will call this Lamb shift, in short) is given in Fig. 7.

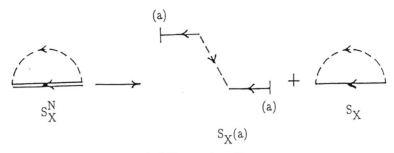

Fig. 7. Lamb shift term in self energy.

The symbol X stands for "exchange". $S_T(a)$ and $S_X(a)$ are leading terms in the perturbation iteration of the equations for the Green's function in the DF approximation, with $S_T(a)$ arising from the Hartree—, and $S_X(a)$ from the exchange—contribution. S_X is the usual leading Lamb shift contribution for an unscreened nucleus.

We carry out a similar process only for a single typical two—virtual—photon term. We label it $S_X(\Gamma)$, because it represents the insertion of a vertex into a lowest order exchange term. The strategy we follow is to use Eq. (19) and "open up" as few lepton lines as possible. That is, we stop with g^N's, as soon as we have a diagram corresponding to a finite contribution. In the present case, the ultimate finite contributions are both tree and box diagrams. We arbitrarily start with "opening" the middle g^N. (A reminder that (a), (b), (c) indicate summation over dummy indices. These summations are used only to identify parts of the g^N's involved.). This first stage is represented in Fig. 8.

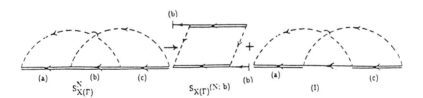

Fig. 8.

$S_{X(\Gamma)}(N;b)$ is finite (with N signaling that g^N's appear in the diagram), and is thus in an acceptable form. (I) is an arbitrary label for the term which still contains g^N's, and is infinite. We have to do further work on it.

Next, we use Eq. 19 again in (I), and "open up" the propagator labeled (a):

438

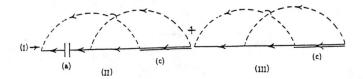

Fig. 9.

Both of the terms we generate still contain g^N's, and are infinite.

We have to carry out the process for one final time, for each of the figures labeled (II) and (III):

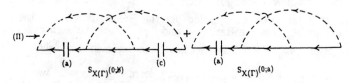

Fig. 10.

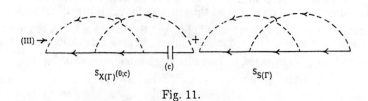

Fig. 11.

The symbol "0" in the labels of Figs. 10 and 11 above denotes that g rather than g^N appears. We no longer have any g^N's left. The remaining divergent terms can now be treated in the standard QED way.

As a final step, we "unfold", or "open up" the (g^N-g) parts of the diagrams, and collect the resulting final diagrams in Figs. 12–14:

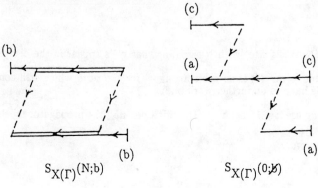

Fig. 12. Correlation terms (finite).

The diagrams of Fig. 12 contain some contributions from positrons.

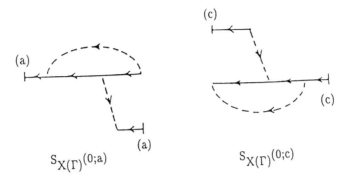

$$S_{X(\Gamma)}{}^{(0;a)} \qquad S_{X(\Gamma)}{}^{(0;c)}$$

Fig. 13. Vertex terms (infinite).

These terms of Fig. 13 contain one exchange and one radiative correction photon, and contribute eventually to the exchange part of the screening effect.

Finally, we have the second order Lamb shift contribution:

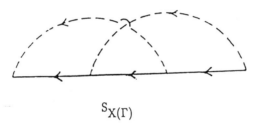

$$S_{X(\Gamma)}$$

Fig. 14. Second order Lamb shift (infinite).

There are two radiative correction photons in the diagram of Fig. 14. It leads to a contribution one order α higher than the leading Lamb shift, and we shall neglect it.

None of the terms above appear in the second order contributions in the expansion of the DF propagator.

The two–virtual–photon diagram I illustrated is one of 10 such diagrams. This single term went to 5 terms, when we isolated the infinite terms. The complete set of 10 two–photon diagrams go to a total of 50 diagrams. 40 of these 50 diagrams (all those which exclude the neglected higher order Lamb shift and Uehling contributions) are listed in ref. 6. Besides correlation terms, the other finite terms which appear also include the second order contributions in the expansion of the DF propagator and finite corrections due to the Breit interaction and retardation.

A conjecture for generalizing our result to the DF case immediately suggests itself.[6] The DF propagator, G_ω^{DF}, has the same pole structure as the zeroth order g^N, except that the energies and wave functions are now those for the DF case ($\epsilon_j \to [\epsilon_j^N]_{DF}$, etc. for ϵ_a and ϵ_l, $|a\rangle \to |a\rangle_{DF}$, etc. for $|l\rangle$ and $|j\rangle$). We can write the analog of Eqs. (19) and (20) as

440

$$(G_\omega^N)^{DF} = 2\pi i \sum_{a=1}^{N} |a\rangle^{DF}\ {}^{DF}\langle a| + R_\omega^{DF} .$$

(21)

We use a symbol R_ω^{DF}, rather than G_ω in Eq. (21), to emphasize that this term is not a vacuum expectation value of two time–ordered lepton operators (in contrast with the case for perturbation theory). Thus, it also cannot be represented by a Feynman diagram. Indeed, we can take Eq. (21) as the definition of R_ω^{DF}. It is a well–defined function, of the form of Eq. (18), but contains DF energies and wave functions.

If we try to assume that R_ω^{DF} is a propagator, and insert it in self energy terms in the conventional way, the resulting, erroneously labeled, S^{DF}, which is generated with its use, cannot be systematically renormalized. The S^{DF}'s so generated are also not GI. It is, further, not clear that we don't double count, or, alternatively, miss some of the terms leading to Breit and retardation corrections. A fall–back position is to restrict the photon propagators to be those yielding the Coulomb interaction, neglect R_ω^{DF} in the lowest order self energy kernel, and further neglect all positron terms in correlations. This set of approximations is sufficient to generate all the leading order effects, including the DF approximation, and the leading correlation corrections to it, which also survive in the non–relativistic limit.

In order to do better, we have to work harder. Let us return initially to perturbation theory, and consider the 12 infinite terms, 2 of which were illustrated in Fig. 13. This set of 12 terms can be grouped in 4 GI subsets, each containing 3 terms. All involve a singe V_{ee} exchange between electrons, and a closed loop. (Some of these loops are parts of vertices, some are self–energies.)

We have succeeded in separately evaluating each of these subsets of 3 terms,[7] by a trick of making the exchange V_{ee} mimic the external nuclear Coulomb field. We then use an extension of the Erickson–Yennie technique for the Lamb shift, to complete the evaluation of our terms. To save effort, we restrict our calculations to the explicit evaluation of the finite log terms ($\ell n(Z\alpha)^2$ and Bethe–log) in ref. 7. These terms provide the largest part of the radiative corrections in any case, and may be sufficient for the accuracies required in atomic calculations, for atoms of sufficiently low Z. (We have nothing to say about high Z atoms for the present.) Our perturbation theory results can be recombined into a very suggestive form, which is particularly simple for the $\ell n(Z\alpha)^2$ terms. (We call these terms leading log (LL) terms in ref. 7.) It is sufficient to evaluate all of these terms in the non–relativistic limit, just as in the standard Lamb shift calculations for low Z.

We define a "local Lamb shift operator", ℓ, where

$$\ell \equiv \frac{\alpha}{3\pi m^2}\ell n(Z\alpha)^2\nabla^2 V_{Nuc} ,$$

(22)

and where V_{Nuc} is the nuclear Coulomb potential. We will indicate ℓ diagrammatically by ∇. All of the perturbation theory results for this LL term, to the order we have considered, can be represented by an integral over a non–local function $\langle 1|L|2\rangle$ in coordinate space (reminder: $1\equiv r_1$, etc.). For state $|n\rangle$, evaluated through first order in V_{ee}, the contribution of all these LL radiative corrections is

$$(\Delta \mathcal{E}_n)_{LL} = \int d^3 1 d^3 2\langle n|1\rangle\langle 1|L|2\rangle\langle 2|n\rangle .$$

(23)

The function $\langle 1|L|2\rangle$ satisfies an integral equation,[7] given by Fig. 15:

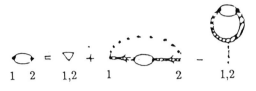

$$\underset{1 \quad 2}{\bigcirc} \equiv <1|L|2>$$

Fig. 15. Integral equation for $\langle 1|L|2\rangle$.

The hatched lines in the figure represent non–relativistic lepton propagators, accurate to first order in V_{ee}. The dotted lines represent just V_{ee}, the e–e Coulomb potential. The non–local terms (the 2nd and 3rd diagrams on the right of Fig. 15) have significant, possibly dominant, contributions for other than s states. They are, at first glance, unexpected results of our analysis (but see below for a reason for their appearance).

The simple structure of Fig. 15 almost forces the conjecture of a generalization to the non–relativistic DF (i.e., the Hartree–Fock – HF) case. It seems natural to let $|n\rangle \rightarrow |n\rangle^{HF}$, and to replace the hatched lepton propagator by $\left[G^N\right]^{HF}$. The Z in LL can remain, since the log–dependence makes it unnecessary to take shielding into account here, and replace Z by some $Z_{effective}$.

More fundamentally, the structure of the integral equation represented by Fig. 15 is very similar to that of the integral equation we derived earlier for the non–local effective current needed to describe one–electron radiative transitions in the DF approximation[1,2,5]. This similarity brings home the fact that GI plays an important role in radiative corrections, analogous to its role in radiative transitions in the DF approximation. This role has not as yet been exploited by our conjectured extension to the HF approximation. (Note, however, that GI is a property of the perturbation theory.)

My co–workers and I are currently engaged in developing an integral equation for the GI radiative corrections in the HF approximation from first principles, using the previous results for the non–local currents in the radiative transitions, and our perturbation theory results[7] for the radiative corrections, as a guide. The approach seems promising. A side benefit may well be that we will be able to take into account finite corrections due to the Breit interaction and to retardation effects in a way which, on the one hand, prevents double–counting, and, on the other hand, avoids the omission of terms.

Selected recent minority view references.

1. G. Feldman & T. Fulton, Ann. Phys. 152, 376 (1984).
2. T. Fulton, Nucl. Phys. A443, 77 (1985).
3. T. Fulton & W. R. Johnson, Phys. Rev. A34, 1686 (1986).
4. T. Fulton, Physica Scripta 36, 441 (1987).
5. G. Feldman & T. Fulton, Ann. Phys. 172, 40 (1986).
6. G. Feldman & T. Fulton, Ann. Phys. 179, 20 (1987).
7. G. Feldman & T. Fulton, Ann. Phys. 231, (1988).
8. S.S. Liaw, G. Feldman & T. Fulton, Phys. Rev. A38, #12 (Dec. 15, 1988).
9. S. S. Liaw, dissertation, Johns Hopkins (1988).

DIVERSE TOPICS

THE COULOMB GREEN'S FUNCTION

Michael Lieber
University of Arkansas, Fayetteville, Arkansas 72701

ABSTRACT

It is something of a miracle that the nonrelativistic Schrodinger equation with a Coulomb potential can be solved for the wavefunction in exact analytic form. Even more miraculous is the result of Schwinger which enables the Green's function to be solved in closed form, for this is in effect, an infinite sum of wavefunction products. In the relativistic case too the wavefunction can be found in closed form, but as yet no such result for the Green's function has been found. This lecture provides a brief overview of the situation with an emphasis on the "hidden symmetry" which underlies the nonrelativistic problem and its degenerate form which carries over to the relativistic case.

I. Introduction

The existence of degeneracies peculiar to the inverse square law force, and their explanation in terms of an additional conserved vector, was apparently discovered by Laplace nearly two centuries ago.[1]   Since that time, this fact has been repeatedly forgotten and rediscovered numerous times. It was the application to the Coulomb Green's function by Schwinger[2], published in 1964, that has (probably) cemented this knowledge in the consciousness of physicists permanently.

The Green's function (or propagator, or resolvent kernel), which we define as $(E-H)^{-1}$, plays a central role in many areas of atomic physics and quantum field theory. In the nonrelativistic case it is crucial in particular to atomic collision theory. Unfortunately, while the Green's function itself is well known, the development of the associated scattering theory is still in its infancy, and most applications, especially high energy calculations, rely upon methodology carried over from nuclear physics and developed for short-range potentials, and thus have questionable validity.

The relativistic Coulomb Green's function (CGF) is much less well developed, despite its importance in atomic structure calculations of the type that are being discussed in this workshop.

446

II. Nonrelativistic Coulomb Problem

A. Symmetries and Degeneracies

Laplace showed that in addition to the angular momentum vector $\vec{L}$, and the total energy E, there was an additional conserved vector in the case of the inverse square law force:

$$\vec{A} = \vec{v} \times \vec{L} - k\,\hat{r}/r$$

This vector, which is nowadays called the Runge-Lenz vector for reasons of historical amnesia, has a magnitude proportional to the orbital eccentricity and is directed along the principal axis of the planetary ellipse. Its conservation therefore represents well-known properties of the Kepler orbits: non-precessing ellipses.

Shortly after the invention of matrix mechanics, in 1926, Pauli[3] gave an elegant solution to the hydrogen atom problem using the hermitian matrix form of the Laplace vector:

$$\vec{A} = (\vec{p} \times \vec{L} - \vec{L} \times \vec{p})/2m - k\,\hat{r}/r \tag{1}$$

Here $k = Z\alpha$, m = reduced mass, and we will use units with $\hbar = c = 1$.

In that same year Schrodinger developed the wave mechanical formalism, and matrix techniques, including Pauli's paper, were put on the shelf. It was, however, noted that the hydrogen energy levels possessed an "accidental degeneracy" -- the energy was independent of the angular quantum number, a property not shared by other potentials.

With hamiltonian operator

$$H = p^2/2m - k/r$$

we have the following commutation relations:

$$[H, \vec{L}] = 0 \tag{2}$$
$$[H, \vec{A}] = 0 \tag{3}$$
$$[L_i, L_j] = i\,\varepsilon_{ijk} L_k \tag{4}$$
$$[L_i, A_j] = i\,\varepsilon_{ijk} A_k \tag{5}$$
$$[A_i, A_j] = -i\,\varepsilon_{ijk} L_k\,(2/m)\,H. \tag{6}$$

Equations (2) and (3) represent the conservation of $\vec{L}$ and $\vec{A}$. Eqs. (4) and (5) describe the vector nature of $\vec{L}$ and $\vec{A}$; that is, under ordinary rotations, $\vec{L}$ and $\vec{A}$ transform the same way. In particular, Eq. (4) exhibits the three components of $\vec{L}$ in their as members of the Lie algebra so(3), i.e. as generators of three-dimensional rotations.

In addition, we have the following relations:

$$\vec{L}\cdot\vec{A} = \vec{A}\cdot\vec{L} = 0 \tag{7}$$
$$A^2 = k^2 + H(L^2 + 1)/2m. \tag{8}$$

Equation (7) simply says that A lies in the plane of the orbit, while (8) is essential for deducing Bohr's formula for the energy levels.

Hulthen, in a 1933 paper little noticed at the time[4], observed that if one restricted oneself to a subspace of the Hilbert space corresponding to an eigenvalue E of H (E · 0) so that the H on the right side of (6) could be treated as a constant, then the six operators contained in L and A had communtation relations which closed among themselves. In a footnote to Hulthen's paper, O. Klein observed that by defining new operators:

$$\vec{J} = (\vec{L} + \vec{M})/2 \text{ and } \vec{K} = (\vec{L} - \vec{M})/2 \tag{9}$$

where

$$\vec{M} = (-m/2E)^{1/2}\vec{A}$$

the commutation relations (4) - (6) became simply

$$[J_i, J_j] = i\,\varepsilon_{ijk}\,J_k \tag{10}$$
$$[K_i, K_j] = i\,\varepsilon_{ijk}\,K_k \tag{11}$$
$$[J_i, K_j] = 0 \tag{12}$$

Thus $\vec{J}$ and $\vec{K}$ behave as independent angular momenta. Klein identified the resulting Lie algebra as so(3) x so(3) = so(4). It should be noted that representations of the corresponding Lie group (SO(4) are thus labelled by two quantum numbers j and k, eigenvalues of the two Casimir operators J^2 and K^2: $J^2 \Rightarrow j(j+1)$, $K^2 \Rightarrow k(k+1)$, where j and k take the values 0, 1/2, 1, 3/2, 2, ... as usual. However, because of the restrictions (7), which implies $J^2 = K^2$, only j = k representations are realized in the

Coulomb problems. It is then only two lines of algebra to go from Eq. (8) to the energy eigenvalues

$$E = -(Z\alpha)^2 / 2m (2j + 1)^2 \tag{13}$$

B. Wave functions

In 1935 V. Fock[5], apparently completely unaware of Hulthen's paper (in the same journal -- Zeitschrift!) gave an explicit construction of the Coulomb wave functions which exhibited the SO(4) symmetry. There were two essential tricks: the Schrodinger equation had to be presented in the momentum space representation, and then the three dimensional p-space had to be stereographically projected onto the surface of the unit sphere in a four dimensional Euclidean space. The Schrodinger equation then appears as the integral equation for the 4-dimensional spherical harmonics. By this method, Fock rederived the momentum-space wave functions originally obtained by Podolsky and Pauling[6] by Fourier transformation of the configuration space wave functions.

Rather than give the details here, which are almost identical to those used by Schwinger in his Green's function paper, we will turn to the latter subject, after remarking that , in the long period between Fock and Schwinger there were two other noteworthy contributions: in 1936 Bargmann[7] related the Fock and the Pauli discussions with the separation of variables in parabolic coordinates, and in 1957 Alliluev[8] generalized Fock's method to N-dimensions -- the N-dimensional Coulomb problem is solved by the spherical harmonics in N + 1 dimensions!

C. The Green's Function

In momentum space the Coulomb potential $-Z\alpha / r$ becomes an integral operator:

$$\langle \vec{p}|V|\vec{p}'\rangle = -Z\alpha / 2\pi^2 (\vec{p} - \vec{p}')^2 \tag{14}$$

Thus the equation for the CGF, $(E - H)G = 1$, can be written out as

$$(E - p^2 / 2m) \; G(\vec{p}, \vec{p}'|E) + \frac{Z\alpha}{2\pi^2} \int d^3p'' \frac{G(\vec{p}'',\vec{p}'|E)}{(\vec{p} - \vec{p}'')^2} = \delta(\vec{p} - \vec{p}') \tag{15}$$

This is defined for all E in the complex plane, apart from the points in the Coulomb spectrum, which is in fact defined as the set of points where E - H is not invertible: the infinity of discrete eigenvalues having E = 0 as an accumulation point and the continuum cut from E = 0 to infinity. (The coincidence of this accumulation point with the branch point is the source of some of the pathologies that lend the Coulomb potential its charm.).

We restrict ourselves for the time being to E < 0 and define $p_o = (-2mE)^{1/2}$, so that the first term in Eq. (15) can be written $- (p_0^2 + p^2) \, G(\vec{p}, \vec{p}\,' \, |E) \, /2m$. We now make the stereographic projection onto the unit sphere in 4-space. We define a unit 4-vector $\xi = (\xi_o, \vec{\xi})$ whose "spatial" part is proportional to $\vec{p}$ (see Fig. 1):

$$\vec{\xi} = 2p_o\vec{p}/\lambda(p) \, , \quad \xi_o = (p_0^2 - p^2)\,/\lambda(p) \tag{16}$$

where we have made the abbreviation

$$\lambda(p) = p_0^2 + p^2 \tag{17}$$

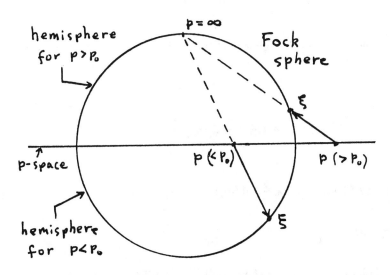

Fig. 1.   The Fock stereographic projection.

450

It will be useful to introduce spherical coordinates (the angle α should not be confused with $\alpha = 1/137$):

$$\begin{aligned}
\xi_0 &= \cos\alpha, \\
\xi_1 &= \sin\alpha \cos\theta, \\
\xi_2 &= \sin\alpha \sin\theta \cos\phi, \\
\xi_3 &= \sin\alpha \sin\theta \sin\phi.
\end{aligned} \tag{18}$$

The element of surface "area" is

$$d\Omega = \sin^2\alpha d\alpha \, \sin\theta d\theta \, d\phi = (8po^3/\lambda^3)d^3p \tag{19}$$

or, conversely,

$$(1/po^3) \, (dp) = d\Omega/(1 + \xi_0)^3 \tag{20}$$

Defining a 4-vector ξ' corresponding to momentum $\vec{P}'$ (but still the same p_0), we note that

$$(\xi - \xi')^2 = (4p_0^2/\lambda\lambda') \, (\vec{p} - \vec{p}')^2, \tag{21}$$

with λ' defined for $\vec{P}'$ in analogy with (17). Making the further definitions

$$\nu = Z \, \alpha \, m/p_0, \tag{22}$$

$$D(\Omega,\Omega') = 1/4 \, \pi^2(\xi - \xi')^2, \tag{23}$$

and

$$\Gamma(\Omega,\Omega') = -(\lambda^2\lambda'^2/ \, 16mp_0^3) \, G(\vec{p},\vec{p}'), \tag{24}$$

we can rewrite (15)

$$\Gamma(\Omega,\Omega') - 2\nu \int D(\Omega,\Omega'')\Gamma(\Omega'', \Omega')d\Omega'' = \delta(\Omega,\Omega'), \tag{25}$$

where the points on the Fock sphere are denoted Ω, Ω' and the δ function is given, according to (20) by

$$\delta(\Omega,\Omega') = (\lambda^3/8p_0{}^3)\ \delta(\vec{p} - \vec{p}').\tag{26}$$

We now introduce spherical harmonics

$$Y_{n\lambda m}(\Omega) = Z_{n\lambda}(\alpha)\ Y_{\lambda m}\ (\theta,\phi),\tag{27}$$

where $Y_{\lambda m}(\theta,\phi)$ is an ordinary spherical harmonic and n, λ, m are integers (do not confuse the integer λ with that of Eq. (17): $n \geq 1$, $0 \leq \lambda \leq n-1$, $|m| \leq \lambda$. The representation (27) makes explicit the three-dimensional rotation group 0(3) which acts on $\vec{p}$ (and $\vec{\xi}$, which is parallel to $\vec{p}$). The functions $Z_{n\lambda}(\alpha)$ are

$$Z_{n\lambda}(\alpha) = N_{n\lambda}\ (\sin\alpha)^\lambda\ C_{n-\lambda-1}{}^{\lambda+1}(\cos\alpha),\tag{28}$$

where $C_{n-\lambda-1}{}^{\lambda+1}$ is a Gegenbauer polynomial, defined by

$$\frac{1}{(1-2tx+t^2)^{\lambda+1}} = \sum_{j=0}^{\infty} t^j C_j^{\lambda+1}(x)\tag{29}$$

and the normalization constant is

$$N_{n\lambda} = \left[\frac{n(n-\lambda-1)!}{(n+\lambda)!}\frac{1}{K_\lambda}\right]^{1/2}\tag{30a}$$

where

$$K_\lambda = \pi/2^{2\lambda+1}(\lambda!)^2.\tag{30b}$$

(The phases have been chosen to make $N_{n\lambda}$ real and positive; this is not the usual choice, nor from the group-theoretic point of view the most natural choice, but we find it convenient to avoid the factor i^λ.) The functions $Z_{n\lambda}(\alpha)$ satisfy

$$\int_0^\pi Z_{n\lambda}(\alpha)Z_{n'\lambda}(\alpha)\ \sin^2\alpha d\alpha = \delta_{nn'},\tag{31}$$

which makes the $Y_{n\lambda m}(\Omega)$ a complete orthonormal set on the sphere.

The expansion

$$D(\Omega,\Omega') = \sum_{n\lambda m} \frac{1}{2n} Y_{n\lambda m}(\Omega) Y_{n\lambda m}(\Omega')^* \tag{32}$$

immediately gives the solution to (14):

$$\Gamma(\Omega,\Omega') = \sum_{n\lambda m} \frac{Y_{n\lambda m}\Omega) Y_{n\lambda m}(\Omega')^*}{1-\nu/n} \; ; \tag{33}$$

the poles at $\nu = n$ give the spectrum via (22)

$$E_n = -p_n{}^2/2m = -(Z\alpha)^2 m/2n^2 \tag{34}$$

and the residues in the E plane are the wave function

$$\psi_{n\lambda m}(p) = \frac{4p_n{}^{5/2}}{\lambda_n{}^2} Z_{n\lambda}\left(2\tan^{-1}\left(\frac{p}{p_n}\right)\right) Y_{\lambda m}(\theta,\varphi), \tag{35}$$

writing p_n whenever we mean p_0 corresponding to the energy level E_n, and λ_n correspondingly.

Equation (35) reveals the momentum space Coulomb wave function as precisely the surface spherical harmonics of the sphere in four dimensions, and are identical with Fock's result. It is worth noting that they are also essentially the same as the momentum representation version of the "Sturmian" wave functions popularized by Rotenberg, et. al.[9] The expansion of Eq. (33) seems to have no continuum part, yet is complete. That is because of our initial assumption of negative energy. Had we begun with positive E, we would have extracted continuum wave functions as 'hyperboloidal harmonics' appropriate to the non-compact group SO(1,3) (the Lorentz group -- but having nothing to do with relativity).

We have not yet obtained an expression for $G(\vec{p}, \vec{p}'|E)$. To do this we must generalize the expansion given in Eq. (32) to points off the surface of the sphere. From Eq. (23), which shows that D is really just the Green's function for the Poisson equation in four dimensions, we can immediately write

$$D(\xi - \xi') = \sum_{n=1} \frac{\rho_<^{n-1}}{\rho_>^{n+1}} \frac{1}{2n} \sum_{\lambda m} Y_{n\lambda m}(\Omega) \, Y_{n\lambda m}(\Omega')* \tag{36}$$

where the letter ρ stands for the length of the 4-vector ξ. A little algebra then gives us the expansion

$$\frac{1}{2\pi^2} \frac{1}{(1-\rho)^2 + \rho(\xi-\xi')^2} = \sum_{n=1}^{\infty} \frac{\rho^{n-1}}{n} \sum_{\lambda n} Y_{n\lambda m}(\Omega) \, Y_{n\lambda m}(\Omega')* \tag{37}$$

Use of the identity $(1-v/n)^{-1} = 1 + v/n + v^2/n(n-v)$ and the integral representation

$$\frac{1}{n-v} = \int_o^1 d\rho \, \rho^{-v} \rho^{n-1}$$

(which requires that $v < 1$), gives us

$$\Gamma(\Omega,\Omega') = \delta(\Omega - \Omega') + \frac{v}{2\pi^2} \frac{1}{(\xi - \xi')^2} + \frac{v^2}{2\pi^2} \int_o^1 \frac{d\rho \, \rho^{-v}}{(1-\rho)^2 + \rho(\xi-\xi')^2} \tag{38}$$

[The restrictions to $v < 1$ can be removed by replacing the real integral by a contour integral

$$\int_o^1 d\rho \, \rho^{-v} \, (\ ) = \frac{i \, e^{i\pi v}}{2 \sin \pi v} \int_c d\rho \, \rho^{-v} \, (\ ). \tag{39}$$

The contour C is shown in Fig. 2]. This equation, projected back onto ordinary momentum space becomes:

$$G(\vec{r}, \vec{r}'E) = \frac{\delta(\vec{r} - \vec{r}')}{E - T} - \frac{Z\alpha}{2\pi^2} \frac{1}{E - T} \left\{ \frac{1}{(\vec{\rho} - \vec{\rho}')^2} + i\eta \int \frac{d\rho\rho^{-i\eta}}{(\vec{\rho} - \vec{\rho}')^2 \, \rho - \frac{m}{2E}(E-T)(1-\rho)^2} \right\} \frac{1}{E - T} \tag{40}$$

where $T = p^2/2m$, $\eta = -iv = Z\alpha m/(2 m E)^{1/2}$ ($E > 0$).

454

Schwinger also give two alternative representations of the CGF, obtained from Eq. (40) by partial integrations. It is worth remarking that prior to Schwinger's paper Wichmann and Woo[10] had obtained a double-integral representation of the CGF, and Hostler[11] had obtained a configuration space representation, which he then Fourier transformed into momentum space. Indeed, Okubo and Feldman[12] had also obtained a one-parameter integral representation in 1960, but it was little noticed at the time, being buried in a much longer paper on the Bethe-Salpeter equation.

The one-parameter integral representations have proved to be most useful. Nevertheless they do not represent what I would call a "closed form" result. The contour integral in the above representations can, however, after decomposing the integrand into partial fractions, be seen to represent the difference of two well-known Gaussian hypergeometric functions of the form $_2F_1$ (1, 1-v 2-v; z) (Ford[13]).

Just as for any potential, the CGF does not exist for real positive energies. But the long-range behavior of the Coulomb potential gives rise to additional pathologies as the energy approaches the cut from above or below. This particular limit is of great interest in scattering theory, because the T-matrix, which is simply related to the physical scattering amplitude f (θ) is usually extracted from the Green's function via the

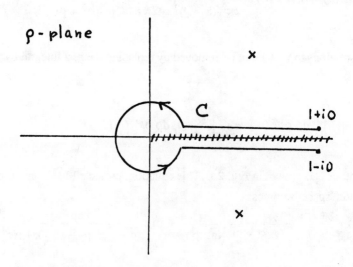

Fig. 2. The contour C in the complex ρ - plane.
C must avoid two poles located at
u $\pm$ i (1 - u2)1/2, where -1 <u <1,
with u = 1 - (ξ - ξ') 2 /2.

identity $T = V + V G V$. In the case of the Coulomb potential, however, the limit onto
the physical energy shell does not exist either in the strong (operator) or weak (matrix
element) senses. Schwinger shows that the usual (Rutherford times Coulomb phase)
expression for $f(\theta)$ can be extracted from the CGF only by first removing singular
factors before taking the on-shell limit. These factors represent the distorted plane wave
propagation of the particle before and after the collision ("creeping phase").
Nevertheless, most calculations in atomic collisions are based upon Born-series type
approximations using the short-range potential relation for the T-matrix, and it is not
surprising that singularities have been encountered, as well as poor agreement with
experiment, although there have also been successes .

III. Relativistic Coulomb Problem

A. Dirac-Coulomb Equation: Symmetries
Unlike the nonrelativistic CGF, there is at present no known exact closed-form
CGF for either the Dirac or Klein-Gordon equations. This appears to be a result of the
lower degree of symmetry possessed by the corresponding equation. In what follows
we will discuss only the Dirac case. I should mention that, even more than in the
nonrelativistic equation, individual authors differ in their notational conventions, and I
may not be fully consistent in these notes. ["Confusion cannot be eliminated -- only
reduced" (H. J. Lipkin)].

The Dirac CGF (or Feynman propagator) is defined, as above by
$(E - H_D)^{-1}$, and exists for all complex E except on the physical spectrum of H_D,
which consists of cuts from $-\infty$ to $-m$ and from $+m$ to $+\infty$, plus an infinite sequence of
bound state poles below $+m$ (or above $-m$ in the repulsive case, $Z\alpha < 0$). Here H_D is
the conventional Dirac-Coulomb Hamiltonian :

$$H_D = \vec{\alpha} \cdot \vec{p} + \beta m + V \tag{41}$$

As is well known[14], the operator $\vec{J} = \vec{L} + \vec{\sigma}/2$ commutes with H_D, representing
conservation of total angular momentum, but $\vec{\sigma}$ and $\vec{L}$ individually do not commute and
are not conserved. Thus eigenstates of (41) can be labelled by the eigenvalues of J^2
and J_z. Although it is not obvious, Dirac found an additional conserved quantity, the

456

operator $K = \beta(\vec{\sigma} \cdot \vec{L} + 1)$. This operator commutes with H_D and has the additional properties:

$$K^2 = J^2 + 1/4 \tag{42}$$

and

$$L^2 = K(K - \beta). \tag{43}$$

From (42), we see that the eigenvalues of K are $\pm (j + 1/2)$, i.e. all nonzero positive and negative integers. K also commutes with the radial part of $\vec{\alpha}$, $\alpha_r = \vec{\alpha} \cdot \vec{r}/r$. As we shall discuss in the next section, there is yet another conserved (pseudo-) scalar operator, discovered in 1950 by M.H. Johnson and B. A. Lippmann[15], and published by them only in an abstract! The almost totally unknown Johnson-Lippmann operator M has almost never been put to practical use. Unfortunately, it anti-commutes with K, so that states cannot generally be simultaneously eigenstates of K and M, and the eigenstates of K are almost universally used. (M maps a state with eigenvalue k of K into one of eigenvalue -k, without changing the energy eigenvalue; the eigenstates of M are thus the sums of these two K-eigenstates, with M-eigenvalue b = 1, or their differences, with b = 0). Nevertheless, M is somehow a remnant of the nonrelativistic Runge-Lenz vector. Biedenharn[16] has shown that M can be interpreted as a kind of helicity operator -- in the nonrelativistic limit, M is proportional to $\vec{\sigma} \cdot \vec{A}$.

While there are no other symmetries of the Dirac-Coulomb problem it is interesting to note that Barut[17], who extended the SO(4) invariance group of the nonrelativistic problem to the SO(4,2) "noninvariance" group (or "dynamical" group) which contains ladder operators connecting states of different n, has also succeeded in finding an SO(4,2) dynamical group for the relativistic case.

B. Wave Functions and Degeneracy

The most elegant treatment of the Dirac-Coulomb wave functions, in my opinion, is given incidentally in a 1958 Phys. Rev. paper by Martin and Glauber[18]on radiative electron capture. They first solve the second-order Dirac equation and then project out the correct solutions to the first-order equation. To solve the second order equation they introduce ("at the suggestion of K.A. Johnson") an operator

$$\Lambda = -K\beta - iZ\alpha\, \alpha_r. \tag{44}$$

This is NOT the Johnson-Lippmann operator referred to above, but has frequently been confused with it because of the attribution to the wrong Johnson! The K.A. Johnson operator satisfies

$$\Lambda^2 = K^2 - (Z\alpha)^2 \tag{45}$$

and

$$\Lambda(\Lambda+1) = L^2 - (Z\alpha)^2 - i\, Z\alpha\, \alpha_r \tag{46}$$

Using this operator the second order Dirac equation can be written:

$$[E^2 - m^2 + \frac{2EZ\alpha}{r} + \frac{1}{r^2}\frac{\partial^2}{\partial r^2}r^2 - \frac{\Lambda(\Lambda+1)}{r^2}]\,\psi(r) = 0 \tag{47}$$

It is very much like the nonrelativistic equation in form. By at first taking $\psi(r)$ to be an eigenstate of Λ, with a nonintegral eigenvalue

$$l = \pm[(j + 1/2)^2 - (Z\alpha)^2)]^{1/2}, \tag{48}$$

we get an equation virtually identical with the nonrelativistic Schrodinger equation. Unfortunately, does not commute with H_D, so the resulting wavefunction is not a Dirac eigenfunction. The latter are then easily projected out.

The Johnson-Lippmann operator is given in terms of Λ by:

$$M = \gamma_5 (\Lambda + K\, H_D/m) \tag{49}$$

Since the energy eigenvalues of the Dirac-Coulomb equation are independent of the sign of the K-eigenvalue, there is a two-fold "accidental" degenaracy in the problem. Since M maps one of the degenerate eigenstates onto its partner, it is clear that M is the symmetry operator which explains this degeneracy, just as the Laplace vector $\vec{A}$ explains the degeneracy of the nonrelativistic energy levels.

C. Relativistic Coulomb Green's Function

The Dirac CGF in configuration space was first derived in 1956 by Wichmann and

458

Kroll[19] and by Brown and Schaefer[20]. The aforementioned paper of Martin and Glauber contains an elegant solution to the problem of finding the exact Dirac CGF, in configuration space. Their solution is in the form of an infinite sum over eigenvalues of Eq. (47). (Actually, they first find the CGF in this form for the second order Dirac equation, and then project out the first order CGF). It is an elegant form because there is no continuum integral, the spectrum of $\Lambda (\Lambda +1)$ being discrete, and is analogous to a partial wave expansion.

In general, the standard wave functions for the Dirac-Coulomb problem have two terms of hypergeometric structure. Since the CGF expansions contain products of these wave functions, the radial parts of the CGF expansion have four terms, each a product of a hypergeometric function and a Whittaker function. Biedenharn[16] found a new representation of the wave function, in which Λ and M are diagonal (and so H_D is presumably not), and where each wave function consists of only one term. Using a modification of Biedenharn's wave functions, Wong and Yeh[21] claim to have found a simpler representation of the Dirac CGF, each radial part having only one term. Unfortunately, the style of the paper is very opaque. Other references to the Dirac CGF are given (Refs. 22-25).

REFERENCES

1. H. Goldstein, Am. J. Phys. 43 , 737 (1975). It may even predate Laplace -- see the "note aded in proof" to H. Goldstein, *ibid* ., 1123 (1976). Good reviews may be found in M. Bander andC. Itzykson, Rev. Mod. Phys. 38 , 330 (1966) and M. Engelfield, Group Theory and the Coulomb Problem (Wiley, N.Y., 1972).

2. J. Schwinger, J. Math. Phys. 5, 1606 (1964). The results were given in his lectures over a decade earlier.

3. W. Pauli, Zetischr. f. Physik 36, 336 (1926). (English translation in Sources of Quantum Mechanics, B.L. van der Waerden (Ed.) (Dover, N.Y., 1968).

4. L. Hulthen, Zeitschr. f. Physik 86, 21 (1933).

5. V.A. Fock, Zeitschr. f. Physik 98, 145 (1935).

6. B. Podolsky and L. Pauling, Phys. Rev. 34, 109 (1929).

7. V. Bargmann, Zeitschr. f. Physik, 99 576 (1936).

8. S.P. Alliluev, Zh. Eksp. Teor. Fiz. 33, 200 (1957) (English translation in Sov. Phys. -JEPT 6, 156 (1958).

9.   M. Rotenberg, Adv. in At. Mol. Phys. $\underline{6}$ (1970).

10.   E. Wichmann and C.H. Woo, J. Math. Phys. $\underline{2}$, 171 (1964).

11.   L.C. Hostler, J. Math. Phys. $\underline{5}$, 591, 1235 (1964). See also L. Hostler and R.H. Pratt, Phys. Rev. Lett. $\underline{10}$, 469 (1963). These authors also give the relativistic CGF for the Dirac equation in the Furry approximation.

12.   S. Okubo and d. Feldman, Phys. Rev. $\underline{117}$, 292 (1960).

13.   W.F. Ford, J. Math. Phys. $\underline{7}$, 626 (1966).

14.   G. Baym, Lectures on Quantum Mechanics (Benjamin, N.Y., 1969) or V.B. Berestetshii, E.M. Lifschitz and L.P. Pitaevskii, Quantum Electrodynamics (Pergamon, N.Y., 1982). These are the only two textbooks that include a discussion of the degeneracy-lifting operator M.

15.   M.H. Johnson and B.A. Lippmann, Phys. Rev. $\underline{78}$. 329(A) (1950).

16.   L.C. Biedenharn, Phys. Rev. $\underline{126}$, 845 (1962).

17.   A.O. Barut, Lectures on Dynamical Groups (Univ. of Canterbury, Christchurch, NZ, 1972).

18.   P.C. Martin and R.J. Glauber, Phys. Rev. $\underline{109}$, 1307 (1958).

19.   E. Wichmann and N.M. Kroll, Phys. Rev. $\underline{101}$ 843 (1956).

20.   G.E. Brown and G.W. Schaefer, Proc. Roy. Soc. London, Ser. A, $\underline{233}$, 527 (1956).

21.   M.F. Wong and E.H.Y. Yeh, J. Math. Phys. $\underline{26}$, 1701 (1985).

22.   L.C. Hostler, J. Math. Phys. $\underline{23}$, 2366 (1982); *ibid* . $\underline{24}$, 2366 (1983); *ibid* . $\underline{28}$, 2984 (1987).

23.   V.N. Baier and A.I. Mil'shtein and V.M. Strakhovenko, Phys. Lett. $\underline{90A}$, 447 (1982).

24.   B.A. Zon, N>L. Manakov, and L.P. Rapoport, Sov. J. Nucl. Phys. $\underline{15}$, 282 (1972); N.L. Manakov, L.P. Rapoport, and S.A. Zapryagin, Phys. Lett. $\underline{43A}$, 139 (1973) and J. Phys. B $\underline{7}$, 1076 (1984).

25.   Ya. I. Granovskii and V.I. Nechet, Theor. Mat. Fiz. $\underline{18}$, 262   (Engl. version, 185) (1974).

SOME ATOMIC EFFECTS IN β DECAY, WITH ONE APPLICATION TO ASTROPHYSICS AND ONE TO ELEMENTARY PARTICLE PHYSICS

Z. Chen and L. Spruch

Department of Physics, New York University, New York, NY 10003

ABSTRACT

This article is intended to be an introduction to some atomic effects on β decay in astrophysics and in elementary particle physics. The emphasis is on the development of physical insights rather than on rigor. Two limiting cases are of special interest. The first is low-energy high-Z β decay, for which atomic effects can be large, and the second is high-energy low-Z superallowed transitions, for which atomic effects are small but important because both theory and experiment are accurate to about 1 part in a thousand. The discussion of the low-energy high-Z domain includes a qualitative description of the β decay of ^{187}Re and of its application to studies of the age of our solar system and of our galaxy. The *extremely* large bound-state β-decay rate of highly ionized ^{187}Re, present in certain astrophysical environment, is of particular interest. We give a simplified derivation, applicable to any β decay in this domain, for both allowed and forbidden continuum decay, of overlap effects; these originate in the fact that the ground state wave functions of the initial atom and the final ion are different. The effects on high-energy low-Z β decay of screening by atomic electrons, a subject of considerable interest in connection with superallowed transitions and electro-weak theory, has only very recently been clarified. Since a conceptual understanding of some of the problems which have arisen is much simpler in the context of a non-relativistic static model, our considerations include such a model.

I. INTRODUCTION

Atomic effects in β decay include bound-state β decay, screening effects, overlap and exchange effects, and some secondary atomic effects. We will give a brief qualitative review of some of theses effects, with one application to astrophysics and one to elementary particle physics. The two applications, in Secs. II and III, respectively, can be read independently.

Normally, in β-decay studies, one simply ignores the atomic electrons, and rightly so; atomic effect are usually negligible. We will consider two types of situation for which the effect of the atomic electrons may have to be taken into account.

1) Atomic effects can be important for the low-energy β decay of a nucleus with high Z. (To make the discussion more concrete, we will here be concerned with the emission of an electron (e^-) and an electron anti-neutrino ($\bar{\nu}_e$) pair; much but not all of the discussion can be carried over, with minor alterations, to the emission of a positron (e^+) and an electron neutrino (ν_e) pair.) Since the magnitude $B(Z)$ of the total binding energy of the atomic electrons increases with increasing charge, there is available to the e^- - $\bar{\nu}_e$ pair an amount of energy $\Delta B(Z) = B(Z+1) - B(Z)$ over and above the energy ΔE_{nuc} associated with the difference in nuclear energies of the initial and final states. The additional energy $\Delta B(Z)$ is of course available in any β-decay, but $\Delta B(Z)$ will be negligible unless Z is large, and it will be negligible even if Z is large for high-energy β decays. Since the binding energy of the inner orbitals also increases with Z, *bound-state* β decay may be important for highly ionized atoms with large Z. For the β decay of neutral atoms, it is also expected that the effects due to the interaction between the β particle and the atomic electrons will increase with the number of atomic electrons.

2) Atomic effects may also be of interest for superallowed transitions, transitions in which the initial and final nuclear states would have identical spatial and spin wave functions if (charge independent) nuclear forces were the only forces present; in fact the states differ because of the presence of (relatively weak) Coulomb forces. Such nuclei exist only for low Z, and since the e^+ and the ν_e are emitted with reasonably large energies, atomic effects will be very small. The atomic effects may nevertheless be of interest, because of the extraordinary accuracy of both theory and experiment for these decays. Corrections of order one part in one thousand are currently of interest, and, should theoretical and experimental results improve further, even smaller atomic effects will be of interest. It should be noted that superallowed transitions play a significant role in studies of the electro-weak theory.

We will consider the two types of situation in turn.

II. LOW-ENERGY HIGH-Z β DECAY

A. The β decay of atomic ^{187}Re: its half-life and the possibility of bound-state decay

We will later consider low-energy β decay in a more general context, with emphasis on atomic effects on the spectrum and half-life. We will here consider one particular such decay, that of ^{187}Re. One decay mode of ^{187}Re is the continuum decay mode

$$^{187}Re \rightarrow {}^{187}Os^+ + e^- + \bar{\nu}_e.$$

We note that $Z = 75$ for ^{187}Re and $Z = 76$ for ^{187}Os$^+$, and that the maximum kinetic energy E_0 of the e^- is roughly 3 keV. (A more precise value is 2.64 keV.) For later purposes, it must be emphasized that E_0, which represents the sum of the difference between the nuclear energies of the initial and final states *and* the difference between the energies of the initial atomic state and the final ionic ground state, is determined experimentally.

From Thomas-Fermi theory the binding energy $B(Z)$ of an atom is, using an empirical constant 16,

$$B(Z) \approx 16Z^{7/3} \quad \text{eV},$$

so that the difference in binding energies of atoms differing by one unit of charge is roughly

$$\Delta B(Z) = (7/3)16Z^{4/3} \quad \text{eV},$$

or about 12 keV for ^{187}Os relative to ^{187}Re. A more accurate estimate, based on a relativistic Hartree-Fock calculation,[1] gives $\Delta B(Z) \approx 15$ keV. Since $\Delta B(Z)$ is larger than E_0, it follows that a neutral ^{187}Re atom can decay only because of atomic effects: the rest energy of a ^{187}Re nucleus is *less* than the sum of the rest energies of a ^{187}Os nucleus and of an e^- (by about $15 - 3$ or 12 keV). An isolated ^{187}Re nucleus cannot undergo continuum decay! With such a low value of E_0, ^{187}Re might then be expected to have a reasonable probability of undergoing bound-state β decay, with a ^{187}Re atom decaying to a *neutral* ^{187}Os atom (roughly speaking, with the β particle ending up in an unoccupied orbital), with the emission of a $\bar{\nu}_e$. Indeed, for some time it was believed that this bound-state β decay might be important. Thus, two different experimental methods were used to determine the half-life $T_{\frac{1}{2}}$ for the decay. The first was simply a laboratory measurement of the rate of emergence of β particles from a given amount of ^{187}Re. The second was based on geochemical studies of the present-day ratios of ^{187}Re, ^{187}Os, and ^{186}Os in meteorites and in samples from the crust of the Earth. (We will elaborate on the second method shortly.) Various measurements by the two methods often gave very different estimates of $T_{\frac{1}{2}}$. One possible explanation was that some of the β particles did not emerge but were captured in a bound state, and were therefore not detected in the laboratory measurement. The situation seems recently to have been clarified. On the theoretical side, two calculations, the first a relativistic modified Thomas-Fermi model calculation[2] and the second a relativistic Hartree-Fock calculation[1] both concluded that the probability of bound-state decay was only of the order of one per cent. (As of now, there is no experimental evidence whatever for bound-state decay of any nucleus. There have been suggestions for using storage rings to study such decays.[3]) On the experimental side,[4] a liquid initially containing ^{187}Re but essentially no ^{187}Os was studied four years later for its ^{187}Os content. This measurement gave the value $T_{\frac{1}{2}} \approx 43$ Gyr. (Note that in 4 years about 1 in 10^{10} of the ^{187}Re atoms will have decayed, so that the initial number of ^{187}Os atoms must be much less than one part in 10^{10} of the ^{187}Re atoms. Note too that this measurement determines the total decay rate; the ^{187}Os atoms present after 4 years represent the result of both continuum *and* bound-state decays.)

B. The β-decay of atomic ^{187}Re, and the determination of the age of the solar system and of our galaxy

We return to the second method for determining $T_{\frac{1}{2}}$ for atomic ^{187}Re.[5] We make the following assumptions:

      1) The solar system was formed at a rather well-defined time T.

      2) The various isotopes were uniformly mixed, and the ratio of the isotopes $^{187}Os(T)$ and $^{186}Os(T)$ at the time T was the same in all meteorites and in the Earth samples; this does *not* assume that the temperatures, pressures, and densities were the same throughout, but merely ignores the relatively small mass difference between the two Os isotopes. Re and Os being chemically different, the ratio of ^{187}Re to ^{186}Os *can* vary from meteorite to meteorite to sample of the Earth's crust.

      3) There was no nucleosynthesis within the solar system once the system was formed.

Noting that ^{186}Os is stable and is not the end-product of any decay process, we have

$$^{186}Os(T) = {}^{187}Os(t), \tag{1}$$

where t represents the present time. By conservation, we also have

$$^{187}Os(T) + {}^{187}Re(T) = {}^{187}Os(t) + {}^{187}Re(t). \tag{2}$$

Finally, with the decay rate $\lambda = \ln 2/T_{\frac{1}{2}}$, we have

$$^{187}Re(T) = e^{+\lambda(t-T)}\,{}^{187}Re(t). \tag{3}$$

Eqs. (1),(2), and (3) give

$$\frac{^{187}Os(t)}{^{186}Os(t)} = \left[e^{\lambda(t-T)} - 1\right]\frac{^{187}Re(t)}{^{186}Os(t)} + \left(\frac{^{187}Os(T)}{^{186}Os(T)}\right). \tag{4}$$

Note that Eq.(4) contains only *ratios* of isotope abundances; absolute values are not required. With the quantity in square brackets as the slope and the quantity in large parentheses as the intercept, Eq. (4) represents a straight line with $^{187}Os(t)/^{186}Os(t)$ the abscissa and $^{187}Re(t)/{}^{186}Os(t)$ the ordinate. In fact, a plot using present ratios of these isotopes for a number of meteorites and samples from the Earth's crust falls very nicely on a straight line, strong evidence for the validity of the three assumptions listed above. The slope of the straight line determines $\lambda(t-T)$, the intercept gives the ratio of the Os isotopes at the time of formation of the solar system, and the slight deviation of the data from a straight line gives a measure of the length of time it took the solar system to form; the length of time is found to be about 0.1 or 0.2 Gyr. With $\lambda(t - T)$ known, an independent determination of $(t - T)$, the age of the solar system, fixes λ and therefore $T_{\frac{1}{2}}$. Conversely, a laboratory measurement of $T_{\frac{1}{2}}$ and therefore of λ fixes $t - T$; one thereby finds the age of the solar system to be 4.55 ± 0.03 Gyr. The value of the intercept is found to be 0.805.

The above discussion should be of interest to those not familiar with the astrophysical literature, but in fact the importance of studies of Re and Os does not lie so much in connection with the age of the solar system, for that age has been determined in many other ways and is believed to be very well known. Rather the Re/Os studies are of interest because of what one learns about the age of the galaxy, an age which is far from well known. We will not go into this latter question in any detail because there exist a number of reviews on the subject[6,7,8,9] but will simply indicate how one goes about estimating the galactic age.

Thus we have used a laboratory measurement of λ and of isotope ratios at the present time t to determine not only the age $t - T$ of the solar system but (going *backward* in time) also the isotope ratios at the time T of formation of the solar system. Note that at this stage one knows $t - T$ but neither T nor t separately. Indeed, the origin of time has not yet been defined. One then makes the assumption that no nucleosynthesis took place from the time of the big bang to the time of formation of our galaxy. Choosing the latter time to be the origin of time, one then makes a number of assumptions about the rates of production of ^{187}Re, ^{187}Os, and ^{186}Os once our galaxy formed, and, now going *forward* in time, T is then the time at which one achieves the known ratio of isotopes at the time of formation of the solar system.[6]–[9] Estimates of T obtained this way range from 11 Gyr to 18 Gyr, in rough agreement with estimates obtained using other chronometric pairs and using studies of globular clusters.

C. The bound-state β decay of a bare ^{187}Re nucleus

In reality the production rates of the relevant Re and Os isotopes are not very well known, and whatever success the method described above has had is due to the fact that the estimate of T is not too sensitive to the values chosen for the parameters which appear in the somewhat arbitrary forms taken to represent the different production rates. (For example, T can be expressed in a form which involves only isotope ratios, and therefore only production ratios.) Physicists not in astrophysics tend to be unnerved by the free-wheeling nature of making such assumptions and may be reminded of Mark Twain's analysis of exactly how much the Mississippi river will stick out over the Gulf of Mexico in time, ending with "There is something fascinating about science. One gets such wholesale returns of conjecture out of such trifling investment of fact." The free-wheeling aspect of certain parts of astrophysics is in fact one of its charms, but of course it has its dangers. In the present context one of the dangers is that there may be periods in which ^{187}Re is in an environment in which it is completely or almost completely ionized, in which case its half-life, as we will now discuss, may be reduced by a factor of order 10^9! The probabilities of various degrees of ionization are of course determined, for specified temperature and density of electrons, by the Saha equation. Let us concentrate on the limiting case of complete ionization, for which, as noted above, continuum β decay is not possible. Bound-state β decay, however, *is* possible. The analysis of this process involves questions of energetics, of selection rules, and of the ratio of large and small components of Dirac wave functions at the nuclear radius. We will simply outline the analysis, noting that this is a one-electron problem, so that, apart from nuclear matrix elements, one can obtain almost arbitrary accuracy.

With regard to energetics, we remark that for the β decay of a bare ^{187}Re nucleus the orbitals available for capture include not only orbitals such as the $6p_{\frac{3}{2}}$ orbital with an energy of the order of a few eV, as in the case of bound-state decay by a neutral ^{187}Re atom, but the $1s_{\frac{1}{2}}$ orbital with an energy of about 86 keV and, ignoring fine structure, 2s and 2p orbitals with energies of about 20 keV. For capture to the $1s_{\frac{1}{2}}$ state, the energy available to the $\bar{\nu}_e$ will then be $-12+86$ or 74 keV, the 12 keV representing the energy, noted above, needed for continuum β-decay to be possible. For capture to an $n = 2$ state, the $\bar{\nu}_e$ energy will be $-12 + 20$ or 8 keV. Capture to states with $n \geq 3$ will not be energetically possible.

The selection rules follow from the properties of the nuclear ground states of ^{187}Re and ^{187}Os, with angular momenta and parities $5/2^+$ and $1/2^-$, respectively. The dominant decay mode will therefore be one for which the spin of the e^-, the spin of the $\bar{\nu}_e$, and one unit of orbital angular moment carried off by the e^- or the $\bar{\nu}_e$, will be parallel, representing two units of angular momentum carried off and a change of parity. For a fixed very low energy, the finite mass e^- has a much larger maximum momentum than the massless $\bar{\nu}_e$, and the e^- can therefore much more readily carry off the unit of orbital angular momentum.

Finally, we note that the square of the ratio of small to large components at the nuclear radius R is of order $(\alpha Z)^2$ for $j = l + \frac{1}{2}$ states, with $\alpha = e^2/\hbar c$, but $(Z\alpha)^2/(ZR/na_0)$ for $j = l - \frac{1}{2}$ states with principal quantum number n.

The above remarks provide at least some qualitative understanding of the relative decay rates, which are as follows:

$1s_{1/2}$	$2s_{1/2}$	$2p_{1/2}$	$2p_{3/2}$	$1s^*_{1/2}$
10^4	1	10^{-1}	10^2	10^9

In studying relative decay rates, the decay rate for capture to the 2s state can be set equal to 1, a convenient choice for that decay rate happens to be roughly equal to the continuum decay rate of the neutral ^{187}Re atom. We begin with the first four entries. The entry $1s^*_{1/2}$ will be discussed shortly.

For capture to the $1s_{1/2}$ or $2s_{1/2}$ state, the $\bar{\nu}_e$ must carry off the unit of orbital angular momentum, and the $1s_{1/2}$ state has much the greater decay rate because of the much greater energy available to the $\bar{\nu}_e$, and because $|\psi_{1s}(r)|^2$ within the nucleus, where the nucleon decays, is much larger than $|\psi_{2s}(r)|^2$ within the nucleus. In the comparison of the $n = 2$ states, energy considerations do not enter, for the energies are the same. The $2p_{3/2}$ state has the fastest of the $n = 2$ decay rates since the $\bar{\nu}_e$ need not carry off any orbital angular momentum. The $2p_{1/2}$ state has the slowest decay rate because it is the *small* component of the $2p_{1/2}$ state which is relevant and because the $\bar{\nu}_e$ must also be in a p state — two units of angular momentum must be carried off — and p-state wave functions are small at the nuclear radius.

By far the most interesting result is the decay to the $1s^*_{1/2}$ state. As noted by Takahashi et al.,[10] there is a $3/2^-$ excited state of ^{187}Os at an energy 9.8 keV above the ground state. The bound-state β decay to this state is energetically possible and requires only *one* unit of total angular momentum to be carried off by the e^--$\bar{\nu}_e$ pair. We can ignore hyperfine effects, and the e^- wave functions for the $1s_{1/2}$ and $1s^*_{1/2}$ states are therefore identical, the presence of the star being simply a reminder that the ^{187}Os nucleus is in an excited state. We can then have conservation of angular momentum by allowing the e^- to be captured into (the small component of) the $1s^*_{1/2}$ state and the $\bar{\nu}_e$ to be in a continuum $s_{1/2}$ state, with the total angular momentum of the e^--$\bar{\nu}_e$ pair equal to unity. We thereby avoid having any p states, having conserved parity by capture to the small component of the $1s^*_{1/2}$ state; (in fact that component is not really very small, the square of the ratio of the small to large component being of order $(\alpha Z)^2$, which is about 0.3).

Note that the ratios of the $1s_{1/2}$ and the $n = 2$ decay rates are independent of nuclear matrix elements, all decays involving the same nuclear matrix element. In arriving at the relative decay rate for the $1s^*_{1/2}$ state it was assumed that the matrix elements for capture to this state and of capture to the ground state were roughly equal.

D. Overlap effects for the continuum β decay of a neutral atom

We are here concerned with the overlap effects[11] for a neutral atom with Z protons and Z electrons whose nucleus undergoes continuum β decay, with the emission of an e^- and a $\bar{\nu}_e$. We first consider the case for which E_0 is much larger than the binding energy of a K-shell electron, so that the time required for a β particle to pass through the K-shell of the atom is much less than the orbital period of the K-shell electron. We can then use the sudden approximation in which the probability for a neutral atom with initial state $|i>$ to end up in a final ionic eigenstate $|n>$ with $Z + 1$ protons and Z electrons is $|<n|i>|^2$. If the ionic state $|n>$ has an excitation energy

$\Delta\varepsilon_n$ relative to the ground state of that ion, the maximum kinetic energy of the β particle for this particular channel will be $E_0 - \Delta\varepsilon_n$. The total decay rate $\lambda(E_0)$ can then be written as the weighted sum of the probabilities of ending up in particular states,

$$\lambda(E_0) = \sum_n |<n|i>|^2 \int_0^{E_0-\Delta\varepsilon_n} P(E_0 - \Delta\varepsilon_n, E)dE.$$

$$= \sum_n |<n|i>|^2 \lambda_0(E_0 - \Delta\varepsilon_n), \qquad (5)$$

where $P(E_0, E)$ is the β spectrum, for a β particle of energy E, that would exist if there were no overlap effects and if the end-point energy were E_0, and

$$\lambda_0(E_0) = \int_0^{E_0} P(E_0, E)dE$$

gives the decay rate if one ignores the overlap effect. Generally the average excitation energy

$$<\Delta\varepsilon> \equiv \sum_n |<n|i>|^2 \Delta\varepsilon_n$$

of the atom in the sudden approximation is much smaller than E_0 even for low energy β decay. (Using the Hellmann-Feynman theorem and the Thomas-Fermi approximation, $<\Delta\varepsilon>$ is approximated by[12]

$$<\Delta\varepsilon> \approx 30Z^{1/3} \ eV \ .)$$

Thus we can expand $\lambda_0(E_0 - \Delta\varepsilon_n)$ in Eq. (5) about E_0 and use the closure relation to obtain

$$\lambda(E_0) \approx \lambda_0(E_0) - \lambda_0'(E_0) <\Delta\varepsilon>, \qquad (6)$$

where the prime denotes an energy derivative. The decay rate can be rewritten in the simple form

$$\lambda(E_0) \approx \lambda_0(E_0- <\Delta\varepsilon>) = \int_0^{E_0-<\Delta\varepsilon>} P(E_0- <\Delta\varepsilon>, E)dE$$

$$\approx \int_0^{E_0} P(E_0- <\Delta\varepsilon>, E)dE; \qquad (7)$$

in the last step we used the fact that $P(E_0, E_0) = 0$. It should be noted, as emphasized by Bahcall,[11] that the end-point kinetic energy in the β spectrum, determined experimentally, is E_0, not $E_0- <\Delta\varepsilon>$, and that the result in Eq. (7) is *not* to be interpreted as simply a decrease in the energy available to the β particle. In fact, the atom could not end up in an excited state were it not for the imperfect overlap between the initial atomic and final ionic ground-state wave functions.

From the nature of the proof, the result in Eq. (7) is valid for a forbidden as well as for an allowed decay; the form of $P(E_0, E)$ will of course depend upon the nature of the decay. (The proof is perhaps somewhat simpler than that of Bahcall [11], and the result represents a slight extension of his result, which was for the allowed case.) The overlap effect decreases the decay rate.

We now consider the other limit in which E_0 is very much *less* than the K-shell binding energy. We may then use the adiabatic approximation, so that the probability that the daughter ion ends up in its ground state is close to unity, giving an overlap effect close to zero. It is to be noted that the overlap effect approaches zero when the end-point energy is very much smaller *or* very much larger than the K-shell binding energy. The overlap effect seems to be important only when the end-point energy is comparable to the K-shell binding energy, and for this case no simple approximation scheme exists. The self-consistent field approximation may be used in this case, and for ^{187}Re the overlap integral between the initial and final atomic ground states was calculated to be about 0.87.[13,1]

E. Exchange effects for the bound-state β decay of a neutral atom

There also exist overlap effects for bound-state β decay, for the same physical reasons discussed in subsection D. However for low-energy high-Z *bound*-state β decay there exists an effect more important than overlap, namely, exchange.[14] We will comment briefly on the exchange effect.

During the β transition, an atomic electron may be "excited" to an outer orbital, leaving a vacant inner orbital, with the β particle captured into that vacant state. (The "excitation" from an initial atomic state to an outer atomic state need not have been caused by an interaction with the emerging β particle. The nuclear charge seen by the electrons changes by one unit by virtue of the β transition, and this change, and the Pauli principle, can also cause the excitation.) In the relativistic Hartree-Fock approximation, the exchange effect was found to increase the bound-state β decay rate by a factor of roughly 2 for a neutral ^{187}Re atom,[1] by about 13% for a ^{187}Re ion with 18 electrons, and by about 6% for a ^{187}Re ion with 10 electrons[9]. The effect decreases with a decrease in the number of atomic electrons, as is to be expected since fewer exchange orbitals are involved.

The exchange effect for bound-state β decay is more important than for continuum β decay, because exchange integrals involving two bound states are of greater magnitude than those involving one bound state and one continuum state.

III. ATOMIC EFFECTS IN SUPERALLOWED TRANSITIONS

A. Some background

Superallowed transitions are low-Z high-energy β transitions, precisely the domain for which one might expect atomic effects to be least significant.

The background material sketched in this subsection[15] is intended to provide a motivation for studying that domain, but is in no way required for an understanding of that study as presented in subsection B.

There are eight isotopic triplets ranging from mass number $A = 14$ (C, N, and O) to $A = 54$ (Mn, Co and Ni), with the central element having $N = Z = \frac{1}{2}A$, where A is of the form $4n + 2$ with n an integer, for which both the half-life t and the maximum total energy $W_0 = E_0 + m_e c^2$ given up to the positron in the 0^+ to 0^+ allowed Fermi transition between two of the three elements of the triplet have been measured with great accuracy. For these triplets the Coulomb energy is rather small compared to the total binding energy — Coulomb energies become relatively large only for large Z, and for large Z there are no isotopic triplets. The spatial and spin wave functions of each element of the triplet are therefore very similar, and the Fermi matrix element $\int 1$ is expected to be quite close to $\sqrt{2}$. If so, the ft values of the various decays, with

$$f \equiv C \times \int_0^{W_0} pW(W_0 - W)^2 F_+(Z, W)dW, \qquad (8)$$

where W is the *total* energy of the e_+ and p is the associated momentum, $F_+(Z, W)$ is the (relativistic) Fermi function for positrons, and C is a constant, should be almost the same, and indeed they are.

Superallowed transitions had always served as a corner-stone in β decay, but, now that β decay has been incorporated into a larger scheme, "almost the same" is not nearly good enough. Thus, the value close to $\sqrt{2}$ for the Fermi matrix element follows from conserved vector current theory, on much less model-dependent grounds than those commented on above, and, further, β decay plays a role in determining some of the matrix elements of the Kobayashi-Maskawa-Cabibbo (KMC) matrix $\mathcal{V}$ which connects "weak eigenstates" d', s', b' with mass eigenstates d, s, b. For our purposes, all that is essential to keep in mind is that $\mathcal{V}$ is supposed to be unitary, and that the use of the simple form of f given by Eq. (8) leads to a violation of unitarity. (Some of the matrix elements of $\mathcal{V}$ are poorly known, but if the sum of the squares of two elements of the top row of $\mathcal{V}$ exceeds unity, as they do by about 4 percent for f of Eq. (8), $\mathcal{V}$ is surely not unitary.) Corrections to f have been made. Thus, $F_+(Z, W)$ is appropriate to a point nucleus, and one must correct for the finite size of the nucleus. Further, radiative corrections are large, about 4 percent.[15,16] When these and other corrections are made, one finds, with $\mathcal{F}$ the corrected value of f, that the $\mathcal{F}t$ values are constant to within about one part in one thousand. (The sum of the squares of the elements of the first row of $\mathcal{V}$ is then unity to about the same accuracy.) At that level of accuracy, and for such a basic problem in physics, it seems well worth while to consider the (clearly small) atomic physics corrections to f for superallowed transitions.

B. Electron screening corrections to $0^+ \to 0^+$ β decay

Electron screening corrections to β decay were first considered more than 50 years ago by Rose,[17] at a time when the original Fermi theory and experiment were far from agreement, as a possible explanation of the discrepancy. It turned out that the experiments had been wrong, and that there was agreement between theory and the new experimental results, but the experimental errors were much larger than the screening corrections, and those corrections were therefore neglected until recently, when questions such as the validity of the conserved vector current theory and the unitarity of the KMC matrix $\mathcal{V}$ arose. Rose had considered the effect of a static screening potential $V_{sc}(r)$ generated by the atomic electrons, and, within that model, his results were confirmed by analytic[18] and numerical studies[19] of model screening potentials, the corrections to his results being negligible; bounds on screening effects for a large class of screening potentials were also determined.[20] It was later claimed that if one allowed for possible (real and virtual) excitation of the atomic electrons, there were significant (Z-dependent) corrections to Rose's static results, so much so that the $\mathcal{V}$ was not unitary. More careful analysis, however, showed this not to be the case; allowance for electron excitation gave negligible corrections to the static results.[21]

That allowance for excitation of the atomic electrons has negligible consequences on the spectrum is a believable result, since the velocity of the e^+ for superallowed transition is of order c, so that the time it takes the e^+ to pass through any atomic shell is much less than the orbital period of the electrons in that shell, and the atomic electrons are therefore effectively stationary.[21,22] (That the correction to the static approximation is proportional to the (very small) characteristic excitation energy[22] is also not surprising.)

Let t_0 be the time it takes the e^+ to pass through the K shell and T_K be the orbital period of a K-shell e^-. The statement that $t_0 \ll T_K$ is equivalent to the statement that $\lambda \ll 2\pi a_0/Z$, where λ is the reduced wavelength of the e^+ and a_0/Z is the K-shell radius. We will make the slightly stronger assumption that $\lambda \ll a_0/Z$. This latter inequality very strongly suggests that scattering by the nuclear Coulomb potential Ze^2/r, which then takes place largely within a distance λ of the nucleus, occurs well before scattering by the (stationary) atomic electrons.[23] This in turn suggests that to a high level of accuracy nuclear Coulomb-effects are much the same, that is, are represented by $F_+(Z, W)$, whether or not screening is taken into consideration.[23,24] (We will shortly discuss this point more quantitatively).

We begin with three observations, that the potential seen by the e^+ at the origin is $-V_0$, where

$$V_0 \equiv V_{sc}(0) = < \sum_{i=1}^{Z} \frac{e^2}{r_i} >,$$

where here and below $<>$ represents the expectation value with respect to the initial atomic ground state wave function, that $V_{sc}(r)$ varies very little over the (small) sphere of radius λ within which the e$^+$ is created so that $V_{sc}(r)$ can be replaced by $-V_0$ in that region, and, finally, that when the charge of the nucleus changes from Z to $Z - 1$ as the e$^+$ is created, the potential energy of the atomic electrons changes from

$$\left\langle -\sum_{i=1}^{Z} \frac{Ze^2}{r_i} \right\rangle \text{ to } \left\langle -\sum_{i=1}^{Z} \frac{(Z-1)e^2}{r_i} \right\rangle,$$

that is, decrease in magnitude by

$$\left\langle \sum_{i=1}^{Z} \frac{e^2}{r_i} \right\rangle, \text{ or } V_0.$$

(The e$^-$-e$^-$ contribution to the potential energy does not change.) Thus V_0 plays a number of roles.

The effect of $V_{sc}(r)$ is to introduce a factor

$$F_{sc}(V_0, W) = \Psi_{sc}^+(0)\Psi_{sc}(0) \tag{9}$$

into the spectrum, where $\Psi_{sc}(\vec{r})$ is a solution of the Dirac equation with a potential $V_{sc}(r)$ and has unit amplitude at very large r. (The notation and form of Eq. (9) are suggested by the relation

$$F_{\pm}(Z, W) = \Psi_c^+(0)\Psi_c(0)$$

for the Fermi functions, with Ψ_c Dirac Coulomb functions of unit amplitude at large r.) That the correction factor to the spectrum in the absence of a potential is determined by the wave function at the origin for a unit amplitude wave incident on that potential follows formally from the fact that the e$^+$-ν_e pair is created at the origin, (or more generally, to allow for forbidden decay, within the nucleus.) More physically, V_{sc} has two effects. Firstly, V_{sc} changes the energy available to the e$^+$-ν_e pair; this is accounted for by using the experimental value of W_0. Secondly, V_{sc} causes the normalization of $\Psi(0)$ to be different from unity since, for our choice of the density of states of the e$^+$ on emergence, $\Psi(\vec{r})$ must be normalized to unity asymptotically.

The subject of screening effects is one which has been beset by errors, some of which will be noted shortly. A number of these errors are unrelated to the relativistic aspects of the problem. Thus, since the algebra is much simpler while the physics is much the same, it will be convenient, even though superallowed transitions require a relativistic analysis, to consider a non-relativistic model. For similar reasons we make the static approximation. Finally, having partially justified the assumption that the effects of the nuclear Coulomb potential can be "factored out", we begin by considering the effects of $V_{sc}(r)$ and Ze^2/r separately, and later provide a more complete justification. Thus, we consider the determination of F_{sc} under the assumption that the positron moves with non-relativistic(NR) speeds in the static potential $V_{sc}(r)$. (A rigorous relativistic treatment which includes atomic excitation has been published.)[21,22] We then have

$$F_{sc}^{NR}(V_0, W) = |\psi(0)|^2,$$ (10)

where $\psi(\vec{r})$ is a solution of the Schroedinger equation with a potential $V_{sc}(r)$ for an incident wave of unit amplitude. The determination of $\psi(0)$ is relatively trivial because i) $V_{sc}(r)$ is very weak and a calculation good to lowest order in V_{sc} is therefor adequate, ii) the β transition of interest is allowed, and iii) $V_{sc}(r)$ is a smooth function of r which is roughly constant inside the K shell. For an incident plane wave $\psi_0(\vec{r}) = \exp(i\vec{k} \cdot \vec{r})$, $\psi(\vec{r})$ is given by

$$\psi(\vec{r}) = \psi_0(\vec{r}) + \int G_0(\vec{r}, \vec{r}')V_{sc}(r')\psi_0(\vec{r}')d^3r',$$
$$+ \int\int G_0(\vec{r}, \vec{r}')V_{sc}(r')G_0(\vec{r}', \vec{r}'')V_{sc}(r'')\psi_0(\vec{r}'')d^3r'd^3r'' + \cdots (11)$$

where $G_0(\vec{r}, \vec{r}')$ is the free Green's function given by

$$G_0(\vec{r}, \vec{r}') = -\frac{2m}{4\pi\hbar^2}\frac{e^{ik|\vec{r}-\vec{r}'|}}{|\vec{r} - \vec{r}'|}.$$ (12)

Since we need evaluate $\psi(\vec{r})$ only within the nucleus, we can, by ii), choose $\vec{r} = 0$. Since $V_{sc}(r)$ and

$$G_0(0, \vec{r}') = -\frac{2m}{4\pi\hbar^2}\frac{e^{ikr'}}{r'}.$$

are spherically symmetric, we can use the $l = 0$ term of the expansion of the Green's function $G_0(\vec{r}, \vec{r}')$. Eq. (11) then reduces to

$$\psi(0) \approx 1 - \frac{1}{k}\int_0^\infty I(r)\tilde{V}_{sc}(r)R(r)dr$$
$$+ \frac{1}{k^2}\int_0^\infty\int_0^\infty I(r)\tilde{V}_{sc}(r)\tilde{V}_{sc}(r')R(r_<)I(r_>)R(r')drdr',$$
$$\equiv \psi_0(0) + \psi_1(0) + \psi_2(0),$$ (13)

where

$$R(r)/(kr) = \sin(kr)/(kr) \text{ and } I(r)/(kr) = e^{ikr}/(kr)$$ (14)

are the regular and outgoing irregular solutions of the $l = 0$ free Schroedinger equation, appropriately normalized, and $\tilde{V}_{sc}(r) \equiv \left(2m/\hbar^2\right)V_{sc}(r)$.

We retained $\psi_2(0)$ in Eq. (13) because, as we will see, the contribution of the $\text{Re}\psi_2(0)$ to $|\psi(0)|^2$ is of the same order as the contribution of $\psi_1(0)$; note that only the real components of $\psi_1(0)$ and $\psi_2(0)$ interfere with the dominant term, the $\psi_0(0) = 1$ term. Thus, we have

$$|\psi(0)|^2 \approx 1 + 2\text{Re}\psi_1(0) + |\text{Im}\psi_1(0)|^2 + 2\text{Re}\psi_2(0).$$ (15)

Initially, without justification, we drop the last two terms in Eq. (15). By
Eqs. (10) and (13) we then have

$$F_{sc}^{NR}(V_0, W) \approx 1 + 2\mathrm{Re}\psi_1(0).$$
$$= 1 - \frac{1}{k}\int_0^\infty \tilde{V}_{sc}(r)\sin(2kr)dr. \qquad (16)$$

Because of the factor $\sin(2kr)$, only the region $r \le 1/(2k)$ is significant,
and by iii), $\tilde{V}_{sc}(r)$ can therefore be replaced by $\tilde{V}_{sc}(0) = -2mV_0/\hbar^2$, with a
convergence factor in the integrand understood. We thereby arrive at

$$F_{sc}^{NR}(V_0, W) \approx 1 + \frac{m}{\hbar^2 k^2}V_0. \qquad (17)$$

Rose's result, to the first order in V_0, which can be obtained from the relativistic analog of the above procedure, is

$$F_{sc}(V_0, W) = 1 + \left[\left(W/(p^2 c^2)\right) + (1/W)\right]V_0; \qquad (18)$$

F_{sc} of Eq. (18) reduces to F_{sc}^{NR} of Eq. (17) in the non-relativistic limit. (V_0
can be estimated in a variety of ways, including the Thomas-Fermi model
or Schwinger's corrected Thomas-Fermi model.[25,26] An approximation to
V_0 is

$$V_0 \approx 2\frac{e^2}{a_0}Z^{4/3}, \qquad (19)$$

where a_0 is Bohr radius. The correction term in Eq. (18) ranges from 0.1
to 0.2% for superallowed transitions.[22])

That the omission of the last two terms in Eq. (15), the approach
which was used initially by some authors, gives the correct non-relativistic
result is somewhat accidental. Thus, it was later pointed out[27] that the
$|\mathrm{Im}\psi_1(0)|^2$ term is of the same order as the $2\mathrm{Re}\psi_1(0)$ term. Still later it
was found that the $2\mathrm{Re}\psi_2(0)$ term is also of the same order, and, further,
that these last two terms cancel to the accuracy under consideration.[23]
This cancellation, between the third and forth terms of Eq. (15) in our
non-relativistic approximation, follows easily. By Eqs. (13) and (14) we
have

$$|\mathrm{Im}\psi_1(0)|^2 = \left|\frac{1}{k}\int_0^\infty \sin^2(kr)\tilde{V}_{sc}(r)dr\right|^2, \qquad (20)$$

and

$$2\mathrm{Re}\psi_2(0) = \frac{2}{k^2}\mathrm{Re}\left[\frac{1}{2}\left(\int_0^\infty R(r)I(r)\tilde{V}_{sc}(r)dr\right)^2 + \int_0^\infty dr I^2(r)\tilde{V}_{sc}(r)\int_0^r dr' R^2(r')\tilde{V}_{sc}(r')dr'\right]. \qquad (21)$$

With $R(r)$ and $I(r)$ defined by Eq. (14), one finds that

$$2\mathrm{Re}\psi_2(0) = a + b + c,$$

where

$$a \equiv \frac{1}{k^2} \left(\frac{1}{2} \int_0^\infty \sin(2kr) \tilde{V}_{sc}(r) dr \right)^2,$$

$$b \equiv -\frac{1}{k^2} \left(\int_0^\infty \sin^2(kr) \tilde{V}_{sc}(r) dr \right)^2,$$

$$c \equiv \frac{2}{k^2} \int_0^\infty \left[\tilde{V}_{sc}(r) \cos(2kr) \int_0^r dr' \tilde{V}_{sc}(r') \sin^2(kr') \right] dr.$$

Thus the b component of $2\mathrm{Re}\psi_2(0)$ exactly cancels $|\mathrm{Im}\psi_1(0)|^2$. The a and c terms are of higher order; thus, since k is much larger than $1/r_0$, with r_0 a characteristic length, of order a_0/Z, we have, with $\tilde{V}_0 \equiv \tilde{V}_{sc}(0)$,

$$a \approx \frac{1}{k^2} \left(\frac{1}{4k} \tilde{V}_0 \right)^2$$

and

$$b \approx -\frac{1}{k^2} \left(\frac{1}{2} \tilde{V}_0 r_0 \right)^2,$$

so that

$$\left| \frac{a}{b} \right| \approx \left[\frac{1}{2kr_0} \right]^2 \ll 1;$$

c is of the same order as a.

One also finds that

$$[\mathrm{Im}\psi_1(0)]^2 \approx \left(\frac{1}{2k} \tilde{V}_0 r_0 \right)^2 \tag{22}$$

and that

$$\mathrm{Re}\psi_1(0) \approx -\tilde{V}_0/(4k^2). \tag{23}$$

It follows from Eqs. (19), (22) and (23) that the ratio of $[\mathrm{Im}\psi_1(0)]^2$ to $2|\mathrm{Re}\psi_1(0)|$ is of order $2/Z^{2/3}$, which is of order unity for Z not too large.

Combined with the nuclear Coulomb correction, the spectrum would be expected to be given by

$$C F_+(Z, W) F_{sc}(V_0, W) p W (W_0 - W)^2 dW, \tag{24}$$

where, with only partial justification, we have assumed that Ze^2/r and $V_{sc}(r)$ act independently. That can not be rigorously true, but we will now argue that interference effects are negligible. Thus, since the nuclear Coulomb potential manifests itself very largely in the region $r \lesssim \lambda$, whatever interference there is will occur very largely in that region, where $V_{sc}(r)$ can be approximated by $-V_0$. It is then natural to replace $F_+(Z, W) F_{sc}(V_0, W)$ by $F_+(Z, W') F_{sc}(V_0, W)$ with $W' \equiv W + V_0$, but this makes a negligible difference; using an explicit expression for F_+, one finds,[22] to first order in V_0, and using $Ze^2/\hbar v \ll 1$, that

$$\frac{F_+(Z, W')}{F_+(Z, W)} = 1 + \alpha Z \pi m^2 V_0 / p^3,$$

which is unity to the accuracy being considered.

A more convincing proof of the effective independence of Ze^2/r and V_{sc} can readily be made. Thus, in the non-relativistic domain, and in the static potential approximation, one would start with Eq. (11), but with ψ_0 and G_0 replaced by their Coulombic analogs, or, since we will set $r = 0$, with ψ_0 equal to, and G_0 proportional to, the $l = 0$ regular and outgoing irregular Coulomb functions, $R_c(r')$ and $I_c(r')$, respectively. We can therefore factor out the regular coulomb function at the origin, but Coulomb effects remain in the Coulomb Green's function. However, retaining only the first order correction, the dominant contribution to the integral in Eq. (11) can be assumed to come from values of r' large enough to allow the replacements of R_c and I_c by their asymptotic forms. The slowly varying logarithmic phase factor η_c can also be dropped in this approximation; one is approximating $\sin^2(kr + \eta_c)$ by $1/2$. This is the assumption made earlier, that screening effects occur well outside the region where the Ze^2/r plays its role, but in the present context one can more readily estimate the order of the error being made. The extension of Eq. (11), for ψ_0 and G_0 Coulombic, from potential scattering to dynamic scattering, allows one to show[24] that the "independence" of Ze^2/r and $V_{sc}(r)$ remains a very good approximation in this more general context.

There exist similar screening corrections for $e^+ - \nu_e$ or $e^- - \bar{\nu}_e$ angular correlations, and in asymmetry effects (the angular distribution of the e^- or e^+ with respect to a uniform external magnetic field). The asymmetry measurements of the mirror nuclei can be used to determine the axial-vector coupling constant C_A and the Cabibbo angle,[28] and hence can provide additional experimental checks on the standard model and the partially conserved axial-vector current hypothesis. The experimental accuracy in these measurements is not comparable to that of the spectra for superallowed transitions, but it is nevertheless of interest to obtain theoretical estimates of the atomic effects for these cases. Apart from the usual over-all factor of $F_+(Z.W)$, Ze^2/r has hardly any effect on either the angular correlation or the asymmetry effect. On the other hand, we have checked[29] in relativistic perturbation theory that the effect of $V_{sc}(r)$, to order V_0, is to demand that in the expressions in the absence of $V_{sc}(r)$, the velocity v be replaced by v', the velocity associated with W'.

We would be remiss if we did not take at least some notice of the very elegant (if tersely presented) derivation by Rose[17] of the screening effect, in the static approximation, on the spectrum for an allowed transition. Rose used the fact that the potential within the nucleus (of radius R) could be well approximated by $U \equiv (Ze^2/r) \pm V_0$, with the $+$ for e^- and the $-$ for e^+. It follows for any wave equation in which the energy and potential enter only in the combination $W - U$, as they do for the Dirac and Schroedinger equations, that the wave equation for a point within the nucleus is the same for the potential U and the energy W as for the potential Ze^2/r and the energy $W \pm V_0$. He then showed that

$$Cn(W)|\psi_U(W,R)|^2 = Cn(W \pm V_0)|\psi_c(W \pm V_0, R)|^2, \qquad (25)$$

where C is the constant in Eq. (24), $n(W)$ is the number of states per unit energy at energy W, and ψ_c and ψ_U are the solutions for the potentials Ze^2/r and U, respectively. [Eq. (25) remains valid if, with $\phi = \psi_c$ or ψ_U, we replace $|\phi|^2 = \phi^+\phi$ by $\phi^+G\phi$, where G is an arbitrary energy-independent Hermitian operator.] The effect of screening is therefore obtained very simply, by replacing W_0 by $W_0' \equiv W_0 \pm V_0$, W by W', and p by the momentum p' associated with W', in the expression for the spectrum. We thereby arrive at the spectrum

$$CF_\pm(Z, W')p'W'(W_0 - W)^2 dW,$$

in agreement with the result in Eq.(18) to the order considered. [Note that the neutrino factor $(W_0 - W)^2$ in the unscreened case is unchanged under $W_0 \to W_0'$ and $W \to W'$, consistent with the fact that a neutrino is not influenced by $V_{sc}(r)$.] It may seem remarkable that the screening effect can be obtained without the wave equation to be used having been specified, but, of course, the screened spectrum is expressed in terms of the unscreened spectrum, and the latter is obtained with the use of the Dirac equation.

It does not seem possible to generalize Rose's proof to allow for excitation of the atomic electrons. It might be possible to generalize his proof, again in the static approximation, to include screening effects on angular correlation and asymmetry since, as noted above, screening effects for those cases have been shown by other methods to follow from the expressions for the unscreened case by the replacement of v by v'.

References

[1] Z. Chen, L. Rosenberg, and L. Spruch, Phys. Rev. A **35**, 1981(1987).

[2] R. D. Williams, W. A. Fowler, and S. E. Koonin, Astrophys. J. **281**, 363(1984).

[3] K. Takahashi, R. N. Boyd, G. J. Mathews, and K. Yokoi, Phys. Rev. C**36**, 1522(1987).

[4] M. Lindner, D. A. Leich, R. J. Borg, G. P. Russ, J. M. Bazan, D. S. Simons, and A. R. Date, Nature **320**, 246(1986).

[5] J. M. Luck and C. J. Allegre, Nature **302**, 130(1983).

[6] D. D. Clayton, Astrophys. J. **139**, 637(1964). This is the original paper on the subject.

[7] See, for example, D. D. Clayton, *Principles of Stellar Evolution and Nucleosynthesis*(University of Chicago Press, Chicago, 1983), Chap. 7.

[8] K. Yokoi, K. Takahashi, and M. Arnould, Astron. Astrophys. **117**, 65(1983).

[9] Z. Chen, L. Rosenberg, and L. Spruch, submitted to *Advances in Atomic, Molecular and Optical Physics*, 1989. This paper was written with an atomic physicist as the reader in mind.

[10] K. Takahashi and K. Yokoi, Nucl. Phys. A**404**, 578(1983).

[11] J. N. Bahcall, Phys. Rev. **129**, 2683(1963).

[12] Similar estimates can be found in Ref. [11].

[13] For the bound-state β decay of atomic ^{187}Re, one has $E_0 \approx 3$ keV and a K-shell binding energy $E_K \approx 86$keV. With $E_0/E_K \approx 1/30$, one might expect the adiabatic approximation to be quite good. However, one might want to compare t_0 with T_K rather than compare E_0 with E_K, where t_0 is the time required for the β particle to pass through the K-shell and T_K is the orbital period of the K-shell electron, and one finds $t_0/T_K \approx 1$. In fact, one might want to use as the comparison time not T_K but $T_K/2\pi$, the time required for the K-shell electron to rotate through one radian, and $t_0/(T_K/2\pi)$ is greater than 1. For a careful formal discussion of the domain of validity of the adiabatic approximation see, for example, Messiah, *Quantum Mechanics* (North-Holland Publishing Company, Amsterdam, 1965), p. 750.

[14] As was the case for overlap, the first thorough analysis of the exchange effect was carried out by Bahcall. See Ref. [11] and the references therein to the earliest papers on the subject.

478

[15] For a more serious discussion, see, for example, A. Sirlin, Nucl. Phys. B(Proc. Suppl) **3**, 417(1988).

[16] A. Sirlin and R. Zucchini, Phys. Lett. **57**, 1994(1986).

[17] M. E. Rose, Phys. Rev. **49**, 727(1936).

[18] L. Durand, Phys. Rev. **135**, B310(1964).

[19] W. Bühring, Z. Phys. **A312**, 11(1983); Nucl. Phys. **A430**, 1(1984).

[20] L. S. Brown, Phys. Rev. **135**, B314(1964).

[21] L. Durand and J. L. Lopez, Phys. Lett. **B198**, 249(1987).

[22] J. L. Lopez and L. Durand, Phys. Rev. **C37**, 535(1988).

[23] E. G. Drukarev and M. I. Strikman, Zh. Eksp. Teor. Fiz. **91**, 1160(1986) [Sov. Phys. — JEPT **64**, 686(1986)].

[24] R. D. Williams and S. E. Koonin, Phys. Rev. **C27**, 1815(1983).

[25] J. Schwinger, Phys. Rev. A **22**, 1827(1980); Phys. Rev. A **24**, 2353(1981).

[26] The decomposition of Schwinger's form for the total energy into components — kinetic, electron-electron, and electron-nuclear — is completely trivial [e.g. see, Z. Chen and L. Spruch, Phys. Rev. A **35**, 4035(1987)], as is the evaluation of the electron potential at the origion.

[27] J. Arafune and T. Watanabe, Phys. Rev. C **34**, 336(1986).

[28] J. D. Garnett, E. D. Commins, K. T. Lesko, E. B. Norman, Phys. Rev. Lett. **60**, 499(1988).

[29] L. Spruch and Z. Chen, to be submitted for publication.

THEORETICAL ATTEMPTS TO EXPLAIN THE GSI POSITRON LINES

B. Müller, J. Reinhardt, O. Graf, S. Graf, C. Ionescu, S. Schramm, M. Grabiak,
W. Greiner

Institut für Theoretische Physik,
Johann Wolfgang Goethe - Universität, Postfach 111 932,
D-6000 Frankfurt am Main, Federal Republic of Germany

A. Schäfer, G. Soff

Gesellschaft für Schwerionenforschung, Postfach 110 552,
D-6100 Darmstadt, Federal Republic of Germany

INTRODUCTION

Slow collisions of two heavy atoms or ions offer a unique laboratory for the study of the motion of electrons, and the behaviour of the QED vacuum, in very strong electric fields. It is for this reason that the theoretical and experimental aspects of such collisions have been studied very intensely during the last two decades. Early on, theory focussed on the change of the vacuum state from a neutral to a charged realization [1] at a critical nuclear charge $Z_c = 173$, and the associated spontaneous emission of positrons of well-defined energy. Later, the view broadened into the quest of gaining an understanding of dynamic phenomena caused by the time-dependence of the collision process, e.g. dynamically induced pair production, K-shell ionization, and delta-electron production.

In recent years two further developments have attracted widespread interest: the "atomic clock" phenomenon, and the narrow electron-positron coincidence lines detected at GSI. The first of these, which allows for a fairly precise, model-independent determination of nuclear reaction times, is by now well established. The narrow $e^+ - e^-$ lines, on the other hand, have been a surprising experimental discovery that is still controversial and unexplained. After a (theorist's) discussion of the present status of the experiments, we will present the pro-s and (mostly) con-s of various mechanisms that have been suggested as eplanation of the data. Much of this discussion will be concerned with the hypothesis that the lines are the result of the $e^+ - e^-$ decay of new neutral particles.

NARROW LINES IN POSITRON SPECTRA

We summarize the experimental results [2-10] for the line structures in positron singles spectra:

- Lines have been observed for a large variety of collision systems, ranging from $Z_u = 163$ (Th+Ta) up to $Z_u = 188$ (U+Cm) and involving nuclei with widely different structure.

- The line positions appear to fall into several groups between 250 and 400 keV; their width is about 70 keV, if all positron emission angles are covered. This value corresponds to the Doppler width of a sharp line emitted by a source moving with centre-of-mass velocity.

- A number of lines are common to different collision systems and to both experiments (ORANGE-collaboration [6] and EPOS-collaboration [7]).

- Nuclear pair conversion processes ($A^* \rightarrow Ae^-e^+$) appear to be excluded from γ-ray and electron spectra, linewidth, and A-independence.

Finally, it should also be mentioned that the line intensity may depend very sensitively, almost erratically, on some yet unknown parameter, possibly the beam energy or the target quality. This circumstance has been a constant source of worry for the experimental groups, and it may well convey some important clue toward the origin of the positron peaks.

Correlated Electron-Positron Lines

The A- and Z-invariance of the line energies strongly hint at a common source that in itself is not related to the nuclei or the strong electric field, although the strong Coulomb field might play a role in the production of this source. Since no Z-dependence is seen, the most natural candidate for such a source would be some (neutral) object that moves with the velocity of the centre of mass and eventually decays into a positron and a single other particle. A two-body decay must be invoked to explain the narrow linewidth. Could the second decay product simply be a second electron, i.e. could it be that one sees the pair decay of a neutral particle, $X^0 \rightarrow e^+e^-$, with a mass somewhat below 2 MeV?

When the EPOS collaboration, inspired by this hypothesis, added an electron counter opposite of the positron detector to their device in order to investigate this question, they indeed found a correlated line structure in the electron spectrum at precisely the same energy of 380 keV in the U+Th system [7]. Fig. 1 shows the published results, which contain a comparison of the measured spectra with Monte Carlo simulations of the decay of a slowly moving neutral particle and of the pair decay of one of the nuclei. An event required the coincident detection of an electron, a positron, and two elastically scattered nuclei. Part (a) shows the positron spectrum in coincidence with an electron in the energy range 340-420 keV; part (b) shows the electron spectrum in coincidence with a positron in that energy range; while (c) shows the sum energy spectrum under the condition that electron and positron energy are the same within the limits provided by Doppler broadening from c.m. motion. Finally, part (d) shows the difference energy spectrum for all events where the sum of electron and positron energy is 760 ± 40 keV. Parts (e) - (h) show the same spectra, but for adjacent energy windows. Evidently a clear peak is visible in any of the top four spectra, but in none of the four reference spectra. The peak intensity is just what would be expected if every positron in the singles peak is accompanied by an electron of the same energy.

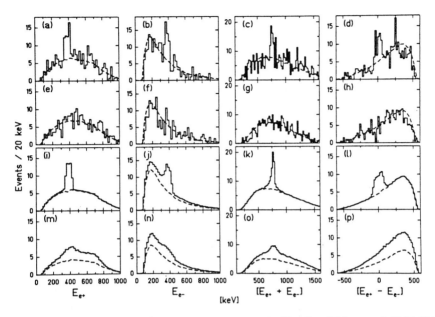

Figure 1: Spectra taken from e^+e^- coincidence events in U+Th collisions at 5.83 MeV/u (EPOS data), compared with computer simulations of pair decay of a neutral particle and of a nucleus.

The computer simulations reproduced in the two bottom rows are quite conclusive: Pair decay of a slowly moving neutral particle X^0 of mass 1.78 MeV, parts (i) - (l), nicely fits all the spectra of the top row, whereas nuclear pair decay, parts (m) - (p), does not. In particular, the appearence of a peak in the difference spectrum (last column) cannot be understood except for a two-body decay. That nuclear pair decay does not even correctly reproduce the narrow sum energy line is due to the fact that the nuclear pair is not emitted back-to-back, so that the linear Doppler shift from the nuclear motion does not cancel. A second argument against nuclear E0-pair decay comes from the coincidence yield, which saturates the positron singles line if back-to-back emission is assumed. For the $(1 + \cos \theta_{ee})$ distribution of nuclear E0-pairs the observed intensity of e^+e^- coincidences would be too high by a large factor [7].

When the same system was remeasured in 1986, at 5.82-5.87 MeV/u beam energy, again narrow sum energy lines were found, but now at different energies [9] (see Fig. 2). The top two subfigures show the sum and difference spectra obtained requiring a prompt coincidence between the two leptons. A narrow line now emerges at the sum energy $E_1 = 809 \pm 8$ keV with a width $\Gamma_1 = 41 \pm 5$ keV. The two lower subfigures show similar spectra, now taken with the condition of a slightly delayed coincidence, with the positron arriving about 5 ns after the electron. Such a delay would not necessarily indicate that the positron was emitted after the electron, but could simply mean that the positron followed a longer trajectory before it arrived at the detector. This would occur if the positron is emitted at a large angle with respect to the axis of the solenoidal transport system, being forced to move on a large helical orbit. Fig. 2 shows that an even narrower sum peak appears for such events, but now at the sum energy $E_2 = 608 \pm 8$ keV and with width $\Gamma_2 = 23 \pm 3$ keV. In both cases there is also a broad peak in the

482

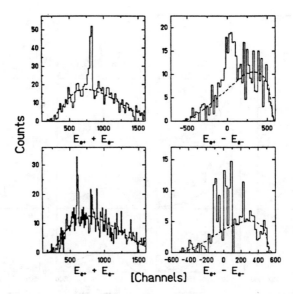

Figure 2: Sum and difference energy e^+e^- spectra in U+Th collisions at 5.82-5.87 MeV/u (EPOS data), for prompt (top row) and delayed (bottom row) coincidences .

difference energy spectrum, again around equal energies of electron and positron.

For the next run, in 1987, the EPOS group modified their lepton detectors so that electrons and positrons emitted forward from, or in front of, the target could be distinguished from those emitted backwards. The two parts of the detectors will be denoted by "F" and "B". The aim was to find out whether the leptons were, indeed, emitted back to back as required by the particle decay hypothesis, or with uncorrelated angles, or even preferentially in the same direction as in nuclear pair decay. This could be done by comparing the combinations (FB+BF) and (FF+BB), where the first letter indicates the part of the electron counter, and the second letter that of the positron counter. Tests with a radioactive ^{90}Sr source, which emits pairs from a 1.77 MeV transition in ^{90}Zr, proved that the discrimination could be achieved if the pair is produced within a few mm from the target.

The measurements were this time performed with the collision system U+Ta for reasons of target quality, and a broad range of beam energies around the Coulomb barrier was investigated. In the energy range 5.93-6.16 MeV/u a sum energy peak was found for delayed positrons at $E_3 = 748 \pm 8$ keV with width $\Gamma_3 = 26 \pm 5$ keV. In the higher energy range 6.24-6.38 MeV/u two peaks were observed, which appeared to be the same as those discovered in 1986 in the U+Th system. The peaks could again be isolated by selecting prompt or delayed coincidences. All the measured positions and widths of peaks in the sum energy spectrum are listed in Table 1.

The two lines observed at the higher beam energy, i.e. the lines at 620 and 805 keV, were only visible in the spectrum from the detector combination (FB+BF), indicating a back-to-back correlation of the lepton pair. This was more or less the expected result, because trajectory simulations for the lines seen in the 1986 run with U+Th had yielded that the relative angle of emission had to be at least 150° to explain the narrow width of

Table 1: Sum energies and widths of correlated e^+e^- lines observed at GSI.

System	Expt.	Beam energy (MeV/u)	condition	Energy (keV)	width (keV)
U + Th	EPOS	5.82-5.87	delayed	608 ± 8	23 ± 3
U + Ta	EPOS	6.24-6.38	delayed	620 ± 8	20 ± 3
U + Th	EPOS	5.83		760 ± 20	80 ± 20
U + Ta	EPOS	5.93-6.16	delayed	748 ± 8	26 ± 5
U + Th	EPOS	5.82-5.87	prompt	809 ± 8	41 ± 5
U + Ta	EPOS	6.24-6.38	prompt	805 ± 8	30 ± 4
U + U	ORANGE	5.9	(180 ± 18)°	815 ± 8	40 ± 15
U + Pb	ORANGE	5.9	(180 ± 18)°	802 ± 8	32 ± 15

the sum energy line. However, the 748 keV line seen at the lower beam energies showed up preferentially in the FF detector. Moreover, the difference energy spectrum in this case does not show a peak at equal energies of electron and positron. Note that 748 keV is precisely the sum energy associated with the 1.77 MeV transition in ^{90}Zr discussed above in connection with the radioactive source test. The long lifetime of the ^{90}Zr state (62 ns) allows to rule this explanation out on the basis of timing measurements. Also, no line was seen at 1.77 MeV in the γ-ray spectrum, but a transition of multipolarity E0 has not been ruled out.

Intrigued by this discovery, the ORANGE group has added a second orange-type magnetic spectrometer to their set-up, permitting simultaneous measurement of the electron (forward) and positron (backward) spectrum. Since both lepton detectors are subdivided into six azimuthal angular ranges of 60° ("pagoda" counters), spectra for specific angular correlation, e.g. 60° or 180°, can be selected. In 1987 a first run with U+U at 5.9 MeV/u was carried out [10]. The sum energy spectra of e^+-e^- coincidence events obtained in this run is shown in Fig. 3. The energies of the most prominent line detected in the 180° correlated spectrum coincides almost exactly with one of the two lines that appeared to be back-to-back in the latest EPOS measurement: 815 keV. As the figure shows, this line structure appears to be absent in the spectrum taken for an angular correlation in the range 40° − 170°, confirming the EPOS results. A third line appears at about 630 keV sum energy, but no line was seen that could be identified with the third EPOS energy (748 keV). The 810 keV line was also observed in the system U+Pb.

Recently, also the TORI group has measured e^+-e^- coincidences in the U+Th system at 5.85 MeV/u with kinematical conditions corresponding to a large relative angle between electron and positron (70° - 180°) or to a small opening angle (0° - 110°), respectively. As listed in Table 1, two line structures were found in the backward correlated events at roughly the same energies as in the 1986 data of the EPOS group. However, structures were also observed in the spectrum of forward correlated events, and no firm conclusion about the opening angle distribution has been reached so far [11].

We conclude the section by summarizing the experimental results for the correlated

484

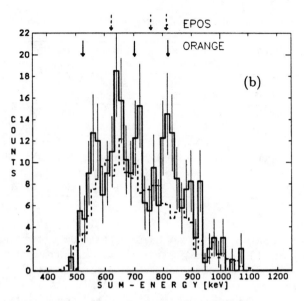

Figure 3: Sum energy e^+e^- spectra measured in U+U collisions at 5.9 MeV/u with the double-orange spectrometer. Solid line: $180° \pm 18°$ correlations; dotted line: correlations with $40° - 170°$ opening angle (scaled).

line structures in electron-positron coincidence spectra:

- Lines at 620 and 810 keV sum energy have been observed by the EPOS and the ORANGE collaboration. A third line at 750 keV was only seen by the EPOS group. The peaks seem to occur at the same positions in various systems, e.g. U+Ta ($Z_u = 165$), U+Th ($Z_u = 182$) and U+U ($Z_u = 184$).

- The width of the sum energy peaks lie in the range 20-40 keV; they are much narrower than the positron singles peaks. The source must move slowly ($\beta_s \leq 0.05$); if it is not at rest, the data imply an opening angle close to 180° between the two leptons.

- The 620 and 810 keV lines appear to be caused by back-to-back emission. The 750 keV line in U+Ta could be forward correlated.

- The difference-energy spectra exhibit a broad peak near zero energy, indicating that the lepton pair is not produced inside the strong Coulomb field, or in the vicinity of a third body. Again, the 750 keV line in U+Ta appears to be an exception to this rule.

- The e^+e^- coincidence line intensity exhausts the full strength of the positron singles line, when the experimental efficiency is taken into account ($d\sigma/d\Omega \approx 10\mu b/sr$). [The line intensity appears to be smaller in the data taken by the double-orange spectrometer.]

- The different spectra obtained for prompt and delayed coincidences hint at non-isotropic emission of the lepton pair. This may also explain the lack of intensity exhibited by the double-orange data.

Thus almost everything appears to be compatible with the assumption that one observes the pair decay of at least three neutral particle states in the mass range between 1 and 2 MeV. These states must have a lifetime of more than 10^{-19}s (because of the narrow linewidth) and less than about 10^{-9}s (because the vertex of the lepton pair is within 1 cm of the target). Surely, a few pieces of data do not really fit into this picture, e.g. the characteristics of the 748 keV line observed in U+Ta. But it is by no means clear at the present time, whether these features provide conclusive evidence against the particle hypothesis, especially if one is not dealing with simple, elementary particle states.

Nevertheless, the very idea that a whole family of neutral particle states in the MeV mass range should have remained undetected through more than 50 years of nuclear physics research is hard to accept for the conservative mind. Most physicists, when first confronted with the GSI data, have therefore tried to explain the data in terms of known nuclear or atomic physics. As mentioned before, nuclear pair decay would be the most natural explanation. In fact, it is rather easy to invent some scenario that would give rise to nuclear pairs of the correct sum energy, e.g. due to target impurities, deexcitation of fission products, neutron activation of the target frame, and so on. However, none of those scenarios has yet stood up against a detailed comparison with the experiments. As long as this is so, nuclear pair decay cannot be considered as a viable alternative to the particle hypothesis. Similar remarks apply to attempts to explain the GSI peaks in terms of atomic physics. None of the ideas that were studied quantitatively have been successful, even if they were based on plausible, but unfounded, ad hoc assumptions. We will discuss some of these attempts later, after a review of the neutral particle models.

THE CASE AGAINST NEW ELEMENTARY PARTICLES

In this chapter we will review the experimental arguments against the existence of neutral *elementary* particles with mass below 2 MeV that can decay into an e^+-e^- pair. We will see that precision experiments set severe limits on the permissible coupling strength of such particles to the electron-positron field, but do not rule out the full range of lifetimes relevant to the GSI peaks. Beyond that, dedicated searches have now eliminated the possibility of an axion in the relevant mass range. Beam dump experiments have closed the remaining lifetime gap for weakly interacting point-like particles, but have left room for extended particle states. Bhabha scattering below 2 MeV centre-of-mass energy, which is the unique model-independent probe for the hypothetical particle states, has just begun to set new limits but is still far from the sensitivity required to reject the particle hypothesis once and for all.

Limits on Light Neutral Bosons from Precision Experiments

Even when the hypothesis of a new neutral particle was first seriously discussed [12,13], it was recognized that the precision experiments of quantum electrodynamics provide stringent limits on the coupling of such light particles to the electron-positron field and to the electromagnetic field. The strength of this argument lies in the fact that any particle X^0 which decays into an e^+-e^- pair must couple to the electron-positron field. At least in the low-energy limit, the coupling can be expressed by an effective interaction of the form:

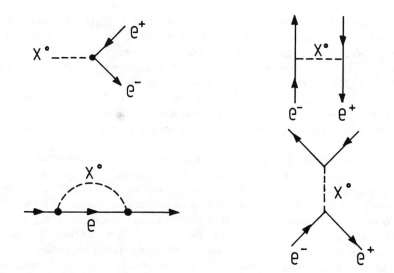

Figure 4: Feynman diagrams for (a) pair decay of X^0, (b) contribution to the electron anomalous magnetic moment, (c,d) positronium hyperfine splitting.

$$L_X = g_i(\bar{\psi}\Gamma_i\psi)\phi, \tag{1}$$

where ψ denotes the electron-positron field, ϕ the X^0 field, and Γ_i with $i = $ S,P,V,A stands for the vertex operator associated with the various possible values of spin and parity of the X^0 particle. Given the interaction Lagrangian (1) one can calculate the lifetime of the X^0 particle against pair decay as well as the contributions to QED processes by virtual exchange of an X^0. The most sensitive of these is the anomalous magnetic moment, because of the high experimental accuracy [14]:

$$a_e^{exp} = (g_e - 2)/2 = (1159652193 \pm 4) \times 10^{-12},$$

and its excellent agreement with the value predicted by QED [15]:

$$a_e^{QED} = (1159652230 \pm 52 \pm 41) \times 10^{-12}.$$

where the first error is due to the uncertainty in the electromagnetic coupling constant $\alpha = 0.00729735308 \pm 33$, and the second error contains the uncertainties in the evaluation of the third and fourth order Feynman diagrams. A reasonable estimate for the 95% confidence limit is $\Delta a_e < 2 \times 10^{-10}$.

As illustrated in Fig. 4, the contribution of a hypothetical X^0 particle to the value of a_e involves two vertices between an electron or positron and an X^0, and are thus proportional to the effective coupling constant $\alpha_{Xe} = g_i^2/4\pi$. The same applies to the decay rate τ_X^{-1}, which involves the square of an amplitude with a single vertex, and to the contribution of an X^0 particle to the hyperfine splitting of the positronium ground states. (The last process does not yield a limit rivalling that derived from a_e in accuracy, but permits to eliminate possible cancellations between contributions from different particles to the anomalous magnetic moment [16,17].)

Table 2: Limits on the coupling constant, lifetime, and pair decay width of neutral bosons with mass $M_X = 1.8$ MeV derived from the anomalous magnetic moment of the electron.

Particle type	Spin J^π	Vertex Γ_i	Max. coupling $\alpha_{Xe} = g_i^2/4\pi$	Min. lifetime τ_X (s)	Max. width Γ_X^{ee}(meV)
Scalar (S)	0^+	1	7×10^{-9}	2×10^{-13}	3.0
Pseudoscalar (P)	0^-	$i\gamma_5$	1×10^{-8}	1×10^{-13}	6.8
Vector (V)	1^-	γ_μ	3×10^{-8}	4×10^{-14}	16
Axial vector (A)	1^+	$\gamma_\mu\gamma_5$	5×10^{-9}	5×10^{-13}	1.4

The limits derived from these considerations [16] on the X^0-coupling constant and its lifetime are listed in Table 2. Particles with lifetime $\tau_X > 10^{-13}$s cannot be ruled out by this argument. Considering that the experimental conditions only require a lifetime below about 1 ns, there remains an unexplored range of four orders of magnitude in τ_X.

Similar upper limits can be derived for the coupling of an X^0 boson to other known particles [16]. That for the coupling to the muon, $\alpha_{X\mu}$, is about one order of magnitude weaker because of the lower accuracy in the value of a_μ. A limit on the product of the coupling constants to the electron and to nucleons is obtained from the Lamb shift in hydrogen and from the K-shell binding energy in heavy elements, one finds $\alpha_{Xe}\alpha_{XN} < 10^{-14}$. For scalar particles an extremely stringent bound on the coupling to nucleons can be derived from low-energy neutron scattering: $\alpha_{XN} < 10^{-9}$ [18]. Special bounds for vector (gauge) bosons were considered by Zee [19]. Finally, measurements of nuclear Delbrück scattering yield an upper limit on the coupling of a spinless X^0 boson to the electromagnetic field through an effective interaction of the type

$$L_{X\gamma\gamma} = g_S(E^2 - H^2)\phi_X \quad \text{(scalar)}$$
$$L_{X\gamma\gamma} = g_P(\vec{E} \cdot \vec{H})\phi_X \quad \text{(pseudoscalar)}. \tag{2}$$

The limits are: $g_S < 0.02$ GeV^{-1} and $g_P < 0.5$ GeV^{-1}. They provide lower limits for the lifetime against decay into two photons: $\tau_{\gamma\gamma}(X^0) > 6 \times 10^{-11}$s for a scalar particle and $\tau_{\gamma\gamma}(X^0) > 4 \times 10^{-13}$s for a pseudoscalar particle [20].

Inadequacy of Perturbative Production Mechanisms

One consequence of these results is that the particle hypothesis cannot be rejected off-hand. On the other hand, the condition that the coupling constant between the hypothetical X^0 boson and the particles involved in the heavy ion collision, i.e. electrons and nucleons, must be very small creates severe problems for any attempt to explain the measured production cross section of about 100 μb by a perturbative interaction of the type shown in eq. (1) [12,13,16]. Also the cross section for production by the strong electromagnetic fields present in the heavy ion collision falls short by several orders of magnitude, if it is based on the Lagrangian (2) or similar perturbative interactions.

A second serious difficulty with the interactions (1) and (2) is that they favour the production of particles with high momenta due to phase space enhancement. For collisions with nuclei moving on Rutherford trajectories the calculated spectra typically are very broad, peaking at velocities $\beta_X > 0.5$. As discussed in the previous chapter, the experiments would require an average particle velocity $\beta_X < 0.05$!

Both these problems could, in principle, be circumvented by the assumption that a very long-lived, excited giant compound nucleus is formed [21,22], but only at the price of violating other boundary conditions set by the experimental data, e.g. the absence of a much larger peak in the positron spectrum caused by spontaneous pair production [23]. A further counterargument is that the emission of slow particles is tremendously suppressed by phase space factors [24]. One might also consider the possibility that the X^0 particles are somehow slowed down after production, but this cannot be achieved with the interactions discussed above.

Axion Searches

At first the axion, i.e. the light pseudoscalar Goldstone boson associated with the breaking of the Peccei-Quinn symmetry required to inforce time-reversal invariance in quantum chromodynamics, seemed like a plausible candidate for the suspected X^0 boson. The interest in an axion was revived when it was realized that there was, indeed, a gap left by previous axion search experiments for a short-lived axion in the mass range around 1 MeV [25]. However, new experimental studies of J/Ψ and Υ decays [26,27,28] quickly ruled out the standard axion.

The reason, why the heavy quarkonium states provide the best test for the Peccei-Quinn axion, is that it couples to all quarks according to their mass, except for a parameter x that determines the ratio of the coupling constants to quarks with weak isospin $+1/2$ (u,c,t) and $-1/2$ (d,s,b). It is possible to modify the axion model in such a way that the axion coupling to the light quarks (u,d) is essentially independent of the coupling to heavy quarks [29,30]. However, even in these variant axion models the coupling constants to the light quarks, g_u and g_d, remain predictable numbers, if the axion mass is known. In terms of the quark masses m_u, m_d, the pion mass and decay constant m_π, f_π, and a basis symmetry breaking scale $f \leq 250$ GeV, on has:

$$g_u = \frac{m_u x}{f}, \qquad g_d = \frac{m_d}{x f}, \qquad m_a = \frac{m_\pi f_\pi \sqrt{2 m_u m_d}}{2 f (m_u + m_d)} \left(x + \frac{1}{x} \right), \qquad (3)$$

and a similar coupling to the electron, g_e, which is proportional to the electron mass m_e. The existence of a variant axion with mass above 1 MeV can therefore be ruled out by experiments on $e^+ e^-$-decays of excited hadronic states involving light quarks, e.g. pion decays and decays of excited nuclear states. There are also strong limits from electromagnetic decays of strange hadrons, such as $\Sigma^+ \to p e^+ e^-$ and $K^+ \to \pi^+ e^+ e^-$ [31,32,33,34], and from neutral pion decay [35].

The rare radiative pion decay ($\pi^+ \to e^+ \nu e^+ e^-$) was studied at SIN in order to search for a decay branch ($\pi^+ \to e^+ \nu \phi$) where the axion ϕ decays later into an electron-positron pair [36]. No such events were found, limiting the branching ratio of this decay mode to less than 10^{-10}. In order to rigorously eliminate all variant axions one must turn to nuclear decays. Hallin et al. [37] analyzed a new experimental study of the pair decay of the 9.17 MeV 2^+-state in ^{14}N [38] and older experiments on ^{10}B and ^{6}Li, and showed that these combined results ruled out all axions that decay into an electron-positron pair. Similar analyses by Bardeen et al. [39] and by Krauss and Zeller [17] reached the same conclusion. (However, a variant axion below 1 MeV is not yet completely ruled out, see ref. [40].)

Further experiments on the 3.59 MeV 2^+-state in ^{10}B [41], the 3.68 MeV $\frac{3}{2}^-$-state in ^{13}C [42], the 15.1 MeV state in ^{12}C [43], the 18.15 and 17.6 MeV 1^+-states in ^{8}Be [46], and the 1.115 MeV state in ^{65}Cu [45] have also not shown any indication for the presence of a short-lived, pair-decaying axion. Searches for a light, pair-decaying scalar or vector particle emitted in the decay of the 6.05 MeV 0^+-state in ^{16}O [43,46] have also been negative, yielding an upper limit of 2×10^{-9} for the coupling constant α_{XN} of such a particle to nucleons. Of course, no useful limits are provided for particles that interact only with leptons, not with quarks, except electromagnetically. (An experiment that does not rely on the coupling of the axion to quarks was suggested in ref. [47].)

Beam Dump experiments

Beam dump experiments, in particular those with a high-energy electron beam, are an excellent source of rather model-independent bounds on the properties of hypothetical light neutral particles. The idea behind these experiments is rather simple: When an electron enters a solid piece of material, e.g. a lead block, it is slowed down and deflected by collisions with the target electrons and nuclei. In particular, the electron can be scattered off mass shell in the Coulomb field of a target nucleus, and return to the mass shell by emission of a real photon. This process is called bremsstrahlung. Of course, instead of radiating a photon, the electron can get rid of its excess energy by emitting some other light neutral particle X^0, if any exists. Except for effects from the particle mass and spin, the expected cross section is given by the cross section for photon radiation, multiplied by the ratio of the coupling constant of the emitted particle to the electron and the electromagnetic coupling constant:

$$\frac{d\sigma_X}{d\Omega dE} = \frac{\alpha_{Xe}}{\alpha} \frac{d\sigma_\gamma}{d\Omega dE}. \tag{4}$$

An upper limit for the measured cross section for the X^0-particle cross section hence yields an upper limit for the coupling constant α_{Xe}. A simple formula for the bremsstrahlung cross section has been given by Tsai [48] (see also ref. [49]).

However, for every beam dump experiment there is not only a lower bound for the range of excluded coupling constants but also on upper bound, for the following reason. The lifetime of the hypothetical particle against pair decay is inversely proportional to the α_{Xe}. For sufficiently large values of the coupling constant almost all produced particles therefore decay inside the beam dump, and the e^+e^- pair produced in the decay is absorbed in the target. It is clear that a good value for this other limit requires a short beam dump, whereas high cross-section and low background require a thick target.

Hence, the result of a beam dump experiment is a region of excluded values of α_{Xe}, i.e. the coupling cannot be in the range $\alpha_{Xe}^{min} < \alpha_{Xe} < \alpha_{Xe}^{max}$. In the analysis one assumes that the neutral particles interact so weakly that they pass essentially undisturbed through the target. For elementary particles this assumption is not critical: for the SLAC experiment discussed below it has been shown that even a nuclear absorption cross section as large as 50 mb per nucleon (this is more than the total nucleon-nucleon

Table 3: Excluded ranges of the coupling constant α_{Xe} of a pseudoscalar particle of mass 1.8 MeV derived from beam dump experiments.

Experiment	Beam	Target	α_{Xe}^{min}	α_{Xe}^{max}
Konaka et al. (KEK) [50]	e^- (2.5 GeV)	W + Fe(2m)	10^{-14}	4×10^{-8}
Davier et al. (Orsay) [51]	e^- (1.5 GeV)	W (10cm)	10^{-11}	10^{-8}
Riordan et al. (SLAC) [52]	e^- (9.0 GeV)	W (10-12cm)	10^{-12}	10^{-7}
Bechis et al. [53]	e^- (45 MeV)		10^{-13}	10^{-10}
Brown et al. (FNAL) [54]	p (800 GeV)	Cu (5.5m)	10^{-10}	10^{-7}

cross section at high energy!) would not seriously affect the limits derived from the data.

Three such beam dump experiments with high-energy electrons have been performed recently, at KEK [50], Orsay [51], and at SLAC [52], and there exists a relevant older experiment at much lower energy [53]. The conditions and results of these experiments are listed in Table 3, where the excluded ranges of the coupling constant are given for pseudoscalar particles of mass 1.8 MeV. For a scalar particle the bounds would be similar, but for spin-one particles about one order of magnitude better lower limits would be obtained. Also listed in Table 3 is a proton beam dump experiment performed at Fermilab. Due to the production of secondary electrons and positrons in the target, a limit is obtained also for the coupling to electrons.

Together, the experiments exclude the range of coupling constants α_{Xe} between 10^{-14} and 10^{-7}, corresponding to lifetimes against pair decay in the range $10^{-14}s < \tau_X < 10^{-7}s$. When combined with the bounds derived from the electron anomalous magnetic moment a_e and by experimental conditions, the beam dump results conclusively rule out any elementary neutral particle as source of the GSI e^+e^- events. To wit, the bound from a_e gave $\alpha_{Xe} < 10^{-8}$ (see Table 2), and the experimental conditions required a mean lifetime below $10^{-9}s$, or $\alpha_{Xe} > 10^{-12}$. The remaining interval is covered by the beam dump results.

Does this rule out the particle scenario altogether? As we shall see in a moment, the answer is, no! That there is still a "loop-hole" for neutral particles left was revealed by a recent analysis by A. Schäfer who calculated the bremsstrahlung production cross section for extended particles [55]. He showed that a finite form factor can invalidate the experimental bounds, if the emitted particle has a radius of more than about 100 fm (10^{-11}cm). Let us try to understand why this is so.

In lowest order of perturbation theory the radiation emitted by a Coulomb scattered electron is given by the sum of the two Feynman graphs depicted in Fig. 5. In order for the electron to be able to emit an X^0-boson, e.g., in the left-hand diagram it must be off mass shell. More precisely, its invariant mass must exceed the sum of the rest masses of the two particles in the final state:

$$(p + k)^2 = (p^0)^2 - (\vec{p} + \vec{k})^2 \approx m_e^2 - 2\vec{p} \cdot \vec{k} > (m_e + m_X)^2. \tag{5}$$

This means that the longitudinal momentum of the virtual photon must be larger than a minimal value $|k_{\parallel}| > m_X(m_e + m_X/2)/E$, where E is the incident energy of the

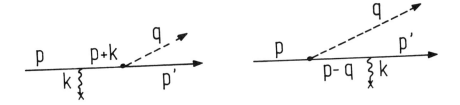

Figure 5: Feynman diagrams for (a) radiation of a neutral boson by an electron scattering in the Coulomb field of a nucleus.

electron. Since the Coulomb scattering cross section falls as $|\vec{k}|^{-4}$, the bremsstrahlung cross section is dominated by events where the invariant mass $(p+k)^2$ just barely exceeds the critical threshold. In other words: as seen from the rest frame of the emitted boson, the electron is slow before and after the emission process. The width of the momentum distribution of the electron in the X^0 rest frame, measured by the Lorentz invariant variable $\bar{p} = [(p + k) \cdot q]/m_X$, is of order m_X.

We are now in a position to discuss the effect of a finite form factor of the X^0-boson on the bremsstrahlung cross section. According to our considerations, the characteristic de Broglie wavelength of the electron in the rest frame of the X^0-boson is given by $\bar{\lambda}/2\pi = \langle \bar{p}^{-1} \rangle \approx m_X^{-1}$. If this is much larger than the size R_X of the radiated particle, the emission will remain unaffected, otherwise the emission of an X^0 will be suppressed. The amount of suppression is expressed by the form factor $G_X(\bar{p}R_X)$, which has the limits $G_X(0) = 1$ and $G_X(x) \ll 1$ for $x \gg 1$. The bremsstrahlung cross section will, therefore, be strongly suppressed if $R_X \gg m_X^{-1} \approx 100$ fm.

This qualitative consideration is confirmed by the results of the complete calculation [55], which are represented in Fig. 6. The cross section is here expressed in terms of the dimensionless strength function $F(x)$, defined by

$$\frac{d\sigma_X}{dx} = \frac{2(Z\alpha)^2 \alpha_X}{m_e^2} \cdot F(x), \tag{6}$$

where $x = |\vec{q}|/|\vec{p}|$ is the fraction of the initial electron momentum carried away by the X^0-boson. The X^0 form factor has been assumed to be of monopole form

$$G_X(\bar{p}R_X) = \frac{1}{1 + (\bar{p}R_X)^2}. \tag{7}$$

In Fig. 6 the value of $F(x = \frac{1}{2})$ is plotted as function of R_X for emission of a pseudoscalar or vector boson. Clearly, for $R_X > 1000$ fm the bremsstrahlung cross section is so much reduced that the limits derived from the beam dump experiments discussed above are no longer relevant.

The curious fact that F initially rises rapidly is explained as follows. The contributions from the two diagrams in Fig. 5 almost cancel for point particles, because the exchange of the order of the vertex operators γ^0 and γ^μ or γ_5 changes the sign of the scattering amplitude, leaving the magnitude almost unchanged. This cancellation is gradually alleviated in the presence of a form factor and the cross section rises, until

Pseudoscalar Bremsstrahlung

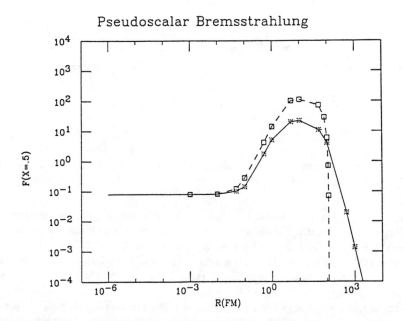

Figure 6: Bremsstrahlung strength function $F(x)$ at $x = \frac{1}{2}$ versus particle size. Two different models for the formfactor have been investigated.

the cut-off due to the finite size sets in. A consequence of this effect is that the beam dump limits would actually become considerably sharper, not weaker, for an X^0 boson with a radius in the range between 1 and 10 fm! (This remark bears upon the "micro-positronium" states discussed in the next section.)

MODELS OF NEW EXTENDED NEUTRAL PARTICLES

General Considerations

The postulate of new neutral particles with finite size, or substructure, can simultaneously solve several general difficulties of any explanation of the GSI data in terms of particle decay. These are:

- The fact that several line structures have been seen is naturally explained as the decay of internally excited states of the same particle.

- The small velocity of the pair-decaying source may be explained in two ways: either as a high-momentum cut-off due to the X^0 form factor, if $R_X > 20m_X^{-1} \approx$ 2000 fm; or by production of the X^0-boson in a bound state around both nuclei.

- A composite particle with electrically charged constituents could be efficiently produced by some non-perturbative mechanism that requires the presence of strong Coulomb fields.

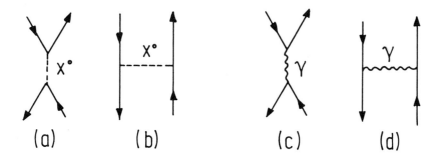

Figure 7: Diagrams for electromagnetic decay modes of a neutral boson: (a) two-photon decay, (b) three-photon decay, (c) pair decay ($J^\pi = 1^-$), (d) pair decay (other J^π).

- As already argued in the previous section, a general bonus is that all experimental limits are rendered irrelevant for a sufficiently large radius R_X, with the exception of those derived from resonant Bhabha scattering.

Moreover, a general conclusion can be drawn with respect to the competition between two-photon and pair decay. Unless the particle is a bound state of electron-positron pairs, or has a fundamental coupling to the electron field (as the axion would!), the photon decay dominates for all states except those with spin one and negative parity. The argument goes as follows: One may first rule out pair decay due to the weak neutral current, because this would yield a lifetime much longer than 1 ns. For electromagnetic decays, two-photon decay and pair decay via a virtual photon (diagrams (a) and (c) in Fig. 7) are both of order α^2, three-photon decay (diagram (b)) is of order α^3 and pair decay via two virtual photons (diagram (d)) is of order α^4. Decay into two photons is possible for all particles except those with spin one, for these three photons are needed in the final state. On the other hand, pair decay via a single virtual photon requires that the particle carries the quantum numbers of the photon, i.e. $J^\pi = 1^-$. Thus pair decay dominates for a vector boson, three-photon decay is the main decay mode for an axial vector boson, and all other particles would predominantly decay into two photons. (Again it must be emphasized that this conclusion does not apply if the particle is already composed of one or several electron-positron pairs, which may be emitted without interaction with an intermediate virtual photon.)

Two general routes can be taken by the theorist who wants to construct a model of extended particles in the mass range between 1 and 2 MeV:

- One can speculate that there exists an undiscovered, "hidden" sector of low-energy phenomena within the framework of the standard model of particle physics, i.e. within the $SU(3) \times SU(2) \times U(1)$ gauge theory. This might be a non-perturbative, strongly coupled phase of quantum electrodynamics, low-energy phenomena associated with the Higgs sector of the Glashow-Salam-Weinberg model, or some unknown long-range properties of QCD. It has even been speculated that the standard electromagnetic interaction between charged particles with spin behaves quite differently at short distances than normally assumed in perturbation theory.

- One can invoke new interactions which, for some reason that remains to be explained, do not normally show up in experiments. Examples are many-body forces between electrons and positrons that do not contribute in positronium, or new light fermions that are confined by equally new, medium ranged interaction.

Both roads have been extensively explored during the past three years, overall with little success. One must be aware that any attempt to fit a scheme of new low-energy composite particles into the standard model faces awesome obstacles, viz. the wealth of experimental data and precision measurements accumulated over fifty or more years.

Micro-positronium and the Virial Theorem

The idea that leptons can form (quasi-)bound states on a distance scale of about 10^{-13} cm due to the interaction of their magnetic moments has been entertained by Barut and his collaborators for many years [56]. They were first associated with the GSI data by C.Y. Wong and Becker.[57] on the basis of a two-body equation that treated the magnetic moments of the electron and positron as those of classical, spinning point particles. This is clearly inappropriate because it does not take into account the "Zitterbewegung" of relativistic leptons. Similar problems exist in Barut's approach, where the binding force originates with the anomalous magnetic moment which, as a radiative effect, is distributed over a region of the size of the Compton wavelength (386 fm) around the bare electron, according to perturbative QED [58]. Arguments that the form factor could shrink to dimensions of 1 fm due to nonperturbative effects in the micro-positronium bound state [59] have never been substantiated.

Moreover, it was shown by Geiger et al. [60] that Barut's equation fails to reproduce the spectrum of (normal) positronium due to an incorrect treatment of retardation effects. It therefore certainly cannot be relied upon to describe the highly relativistic, hypothetical micro-positronium states. Also, such states could not even be found as solutions of Barut's equation for a highly localized magnetic form factor [60].

Recently, Wong and collaborators have made an attempt to derive strongly localized, quasi-bound micropositronium states in the framework of the quasi-potential approach to the Bethe-Salpeter equation. In principle, although not in practice, the two-body problem in QED can be treated exactly by this approach, and the spectrum of positronium can be described correctly.

However, a very general argument on the basis of the relativistic virial theorem has been presented by Grabiak et al. [62] against the existence within QED of quasi-bound states that are mainly composed of a single electron-positron pair. Because it is quite illuminating, let us investigate the basic idea behind this argument in some detail. We proceed in a number of steps:

(a) Let us start with an electron or positron interacting with a static potential $V(\vec{r})$. The motion of the particle is then described by the Dirac Hamiltonian

$$H_D = -i\vec{\alpha} \cdot \vec{\nabla} + \beta m_e + V(\vec{r}). \tag{8}$$

For a normalizable eigenstate Ψ_0 with energy E_0 one has

$$\langle \Psi_0 | i\vec{\alpha} \cdot \vec{\nabla} + \vec{r} \cdot \vec{\nabla} V | \Psi_0 \rangle = \langle \Psi_0 | [\vec{r} \cdot \vec{\nabla}, H_D] | \Psi_0 \rangle = (E_0 - E_0) \langle \Psi_0 | \vec{r} \cdot \vec{\nabla} | \Psi_0 \rangle = 0. \tag{9}$$

For a Coulomb potential $V(r) = \pm e^2/r$ one finds $\vec{r} \cdot \vec{\nabla} V = -V(r)$, and therefore

$$0 = \langle \Psi_0 | i\vec{\alpha} \cdot \vec{\nabla} - V(r) | \Psi_0 \rangle = \langle \Psi_0 | - H_D + \beta m_e | \Psi_0 \rangle = -E_0 + m_e \langle \Psi_0 | \beta | \Psi_0 \rangle. \quad (10)$$

Now it is easy to see that the expectation value of the Dirac matrix β is always less than unity. One thus obtains the general inequality $|E_0| < m_e$, i.e. that all normalizable eigenstates must lie below the continuum threshold.

(b) In order to see how this relation can be generalized to quasi-bound states, i.e. resonance states, one must study the derivation of the virial theorem by means of scale transformations. This is implicit in the above equations, because the operator $\vec{r} \cdot \vec{\nabla}$ is just the generator of such transformations. However, whereas a normalizable eigenstate does not vary under infinitesimal changes of scale, a continuum solution does so. It is possible to show that the magnitude of this variation is inversely proportional to the energy derivative of the continuum phase shift, and hence proportional to the resonance width in the case of a quasi-bound state. The relation derived above is therefore generalized to resonance states in the form

$$|E_{res}| < m_e + g\Gamma_{res}, \quad (11)$$

where g is a numerical factor of order one.

(c) The argument can be extended to arbitrary electromagnetic fields generated by point particles, where the interaction Hamiltonian is $V(x) = e\gamma^0\gamma^\mu A_\mu$. The electromagnetic potential of a point charge e satisfies the wave equation

$$\partial^\nu \partial_\nu A^\mu = e \int d\tau u^\mu(\tau)\delta^4[x - z(\tau)], \quad (12)$$

where $z^\mu(\tau)$ is the world line of the point charge and u^μ its velocity four-vector. The relativistic scaling operator $(x \cdot \partial) = x^\nu\partial_\nu$ yields the following relations:

$$\begin{aligned}
(x \cdot \partial)\partial^\nu\partial_\nu A^\mu &= \partial^\nu\partial_\nu[(x \cdot \partial)A^\mu] - 2\partial^\nu\partial_\nu A^\mu \\
(x \cdot \partial)\delta^4(x) &= -4\delta^4(x) \\
(x \cdot \partial)x^\mu &= x^\mu,
\end{aligned} \quad (13)$$

and therefore one finds that $(x \cdot \partial)A^\mu$ also satisfies a wave equation like eq. (12), but with the opposite sign of the charge, i.e. with $(-e)$. Hence, $(x \cdot \partial)A^\mu = -A^\mu$, and the remainder of the argument given under (a) and (b) goes through in the same way. For N pairs, i.e. $2N$ particles, the energy bound becomes:

$$|E^{res}_{(e^+e^-)^N}| < 2Nm_e + g\Gamma_{res}. \quad (14)$$

(d) The core of this argument based on the virial theorem is that the scale in electrodynamics is set solely by the particle mass, here the electron rest mass m_e. This statement must be somewhat modified in the context of *quantum* electrodynamics, because here a second scale, the four-momentum cut-off Λ, has to be introduced in the renormalization procedure. A detailed analysis shows that the cut-off scale can be absorbed in the renormalized mass of the electron, with the exception of correction terms due to radiative processes, which are of order α. The final bound on the energy of a resonance due to a quasibound state of N electron-positron pairs then takes the following form:

$$|E^{res}_{(e^+e^-)^N}| < 2N(1 + c\alpha)m_e + g\Gamma_{res}, \quad (15)$$

where c is a numerical constant of order one, but depending (logarithmically) on the spatial extension of the quasibound state.

From these considerations one can conclude that *a narrow resonance that is mainly composed of a single e^+e^--pair cannot occur at 1.6 or 1.8 MeV*. This eliminates the idea originally conceived by Wong and Becker to explain the GSI events as the result of the decay of magnetically bound positronium states.

Poly-positronium

The argument presented in the previous section cannot be used to rule out states that are predominantly built up from two or more e^+e^--pairs. However, no mechanism is known within the framework of QED for the strong binding required to bring such states far below the threshold of at least $4m_e$. This constitutes the main objection against recent claims that strongly bound states of two electron-positron pairs ("quadronium") hold the key to the solution of the GSI positron puzzle [63]. An equation of Bethe-Salpeter type has been derived for this system [64], but no indication for a strongly bound state has been found.

(The $(e^+e^-)^2$ system is known to have a very weakly bound state with binding energy of a few eV [65], which has the structure of an ordinary positronium molecule. The analogous system $(e^+e^+e^-)$ was originally proposed as explanation for the GSI events by Wong [66] when only the positron singles peaks were known. However, the branching ratios for the interesting two-body decay modes $(e^+e^+e^-) \rightarrow e^+\gamma$ and $(e^+e^-)^2 \rightarrow e^+e^-$ are minute [67,68], eliminating any such possibility for reasons of intensity. An attempt by Wong to circumvent this difficulty [69] turned out to be based on an incorrect interpretation of the Dirac equation [70].)

Thus, strongly bound $(e^+e^-)^n$ states, the so-called "poly-positronium" states, appear to require the assumption that some new, non-QED force exists between electrons and positrons [71]. On this basis, a rather satisfactory phenomenological explanation of the GSI events could be constructed, if the poly-positronium system would have a size of several 100 fm. The widths for decay into a single e^+e^--pair or into two photons would then be roughly in agreement with the bounds derived from QED. The states would be expected to be produced in the heavy ion collision by the action of the strong electric fields with a cross section and kinematic charcteristics similar to that of the QED pairs[71,72]. (The behaviour of micro-positronium in strong electric fields was studied in refs. [73,74]. The latter of these seems to suffer from an incorrect treatment of gauge invariance.)

Is the required new interaction between electrons and positrons compatible with our knowledge of e^+e^- physics? E.g., one might postulate the existence of a short range attractive many-body force that does not act between a single e^+e^--pair, thus avoiding problems in electron-positron scattering at high energy and in the normal positronium system that is well described by QED. The question was systematically studied by Ionescu et al. [75], who considered the limits set by spectroscopic data from heavy atoms on nonlinear interactions of the form

$$L_{int} = \lambda(\bar{\psi}\psi)^n, \tag{16}$$

where n is some integer greater than one. Such forces would contribute measurably to the K-shell binding energy in heavy atoms, if the effective coupling constant λ is too large. The following limits were obtained in this way: $\lambda(n = 2) < 5 \times 10^{-4}$ and $\lambda(n = 3) < 2 \times 10^{-3}$. On the other hand, the values of λ required to support a poly-positronium bound state are at least $8(n = 2)$ or $130(n = 3)$, respectively [75]. Higher exponents n or even nonpolynomial interactions do not give more favourable results. Thus, *poly-positronium states based on a new e^+e^--interaction of type (16) can be excluded.*

Abnormal QED Vacuum

In spite of the fact that it is not tenable for the reason explained above, the idea that the GSI events could be attributed to the decay of tightly bound e^+e^- states has many attractive aspects. If a new interaction among electrons and positrons is incompatible with atomic physics, and if the standard forces of QED do not appear to support such states, one may ask whether the QED force between electrons and positrons could not be modified by the action of very strong electric fields such as those present in heavy ion collisions.

The first considerations in the context of pure QED were concerned with the possible existence of collective modes in the electron-positron vacuum in intense external fields [76]. A high vacuum polarization density could indeed support plasma oscillations of virtual e^+e^--pairs with a plasma frequency between 1.5 and 2 MeV, which would predominantly decay into an electron-positron pair. However, if such collective excitations exist at all, they appear to be much too broad to be associated with the GSI events [76]. Furthermore, the strong fields are present only for a very brief time, and it is not at all clear how these modes could survive sufficiently long to give rise to narrow e^+e^- lines, simply on the basis of the uncertainty relation $\Delta E \cdot \Delta t \geq \hbar$.

A more radical approach is based on the hypothesis, first put forward by Celenza et al. [77,78], that QED may possess an alternative vacuum state resembling in its properties the normal vacuum of QCD (quantum chromodynamics), and that this new vacuum may be formed in heavy ion collisions. Indeed, it has been known for some time that the U(1) lattice gauge theory has a strong coupling phase, which confines electric charges due to the formation of a condensate of magnetic monopoles. If a similar phase transition also occurs in the (non-compact) continuum gauge theory with dynamical fermions, i.e. in the QED of real physics, then the GSI events might be explained crudely as follows [77,79,80] (see Fig. 8): A region of the confining, strongly coupled phase of QED (shaded in the figure) is formed in the vicinity of the colliding heavy ions due to the action of their strong electric fields. An electron-positron pair may be captured in this region, where it becomes permanently bound because the force between the two leptons is now of a long-ranged, confining nature. The meson-like structure survives for some time after the collision of the nuclei and finally decays into a free e^+e^--pair with simultaneous disappearance of the "bag" of abnormal QED vacuum.

If one is willing to follow this speculation for the moment being, the multiplicity of e^+e^- peaks observed at GSI may be naturally interpreted as decays of various "abnormal QED mesons". Depending on the specific details of the slightly different models, the energy splittings between the various GSI peaks can be interpreted as mass splittings due to internal (radial or rotational) excitations or different spin couplings of the QED mesons. Typical values required to fit the data are $\kappa = (166 \text{ keV})^2$ for the slope of the

498

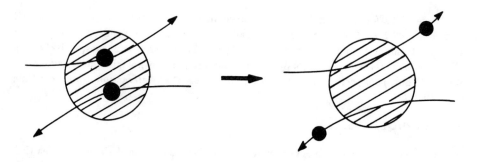

Figure 8: Schematic illustration of the "abnormal phase" interpretation of the GSI events. See text for detailed explanation.

linear confinement potential [80], or $B = (250\ \text{keV})^4$ for the QED bag constant [81], while the size of these states lies in the range between 1000 and 3000 fm.

As alluring as the speculation of abnormal, confined $e^+ e^-$ states may be, one must be aware that it relies on an unproven hypothesis, viz. the existence of a second, strongly coupled QED vacuum state. It is now quite clear that QED has a phase in which chiral symmetry is broken, if the intrinsic coupling constant α is sufficiently large (probably for $\alpha > \alpha_c = \pi/3$) [82,83]. However, we all know that in the real world we have $\alpha \approx 1/137 \ll \alpha_c$, which is why the normal, chirally symmetric QED vacuum state is realized. The question is then, whether the strong electric fields present in a heavy ion collision can somehow catalyze a transition to the abnormal vacuum state, i.e. effectively increase the electromagnetic coupling constant beyond the critical value.

This question was recently studied by Dagotto and Wyld [84], who investigated lattice QED in the presence of an electric background field. They found that the critical coupling constant α_c actually *increases* if an external electric field is present. This behaviour is not really surprising, because it has an analogue in superconductivity. There the superconducting phase, which exhibits confinement of magnetic fields (the Meissner effect), is based on the condensation of Cooper pairs, i.e. of electric charges. This condensate is perturbed, and finally disrupted, by a strong external magnetic field. Similarly, the magnetic monopole condensate of strongly coupled QED is weakened by an external electric field, and a higher value of α is required to keep it in existence. Dagotto and Wyld found that the chirally broken, abnormal phase altogether ceases to exist beyond a certain strength of the electric background field, where it goes over to the charged QED vacuum already studied in Chapter 1 in connection with supercritical nuclear charge Z. (This charged vacuum is also present in the lattice gauge theory.) The phase diagram of QED in the presence of an external charge Z as function of the inverse (!) coupling constant α^{-1} is shown schematically in Fig. 9. Thus, at present, the results of lattice gauge theory do not at all support the hypothesis of the formation of abnormal QED meson states in heavy ion collisions.

The question whether strong external electric fields may facilitate the transition into a new, abnormal phase of QED, was studied by Peccei et al. from a more phenomeno-

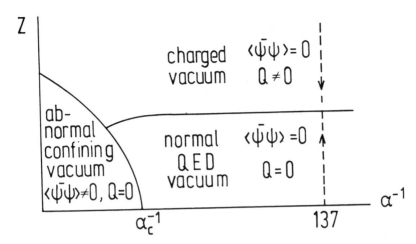

Figure 9: Schematic representation of the phase diagram of QED. In heavy ion collisions the diagram is explored along the dashed line.

logical point of view [85]. Their analysis of nonlinear effects in QED showed that these are generally suppressed by a factor of order α, compared to possible naive expectations relying on the large value of $Z\alpha$. Another interesting, non-perturbative approach to QED in strong background fields indicated the possible presence of thresholds in the production of virtual e^+e^--pairs as function of the external field strength [86]. However, this treatment suffers from severe analytic approximations whose validity in the real physical situation is difficult to assess.

In conclusion, let us discuss some other problematic aspects of this speculative explanation of the GSI peaks:

- Would the region of space filled with abnormal QED vacuum and the e^+e^--pair confined in it detach itself from the heavy ions after the collision and somehow lead a "life of its own"? This would require that the abnormal QED phase can be metastable even in the absence of a strong background field. No clear answer to this question has been provided.

- The precise mechanism by which the QED meson states would decay into a free electron-positron pair has not been described in the literature. Would the decay occur still in the presence of the strong background field, or in a more or less field-free region of space after the QED meson has left the interaction region? It appears that the latter alternative must be realized, because a background field would act as a third body that can absorb momentum, thereby destroying the two-body characteristics of the e^+e^- decay. This, as we remember, was the basis for the particle hypothesis in the first place.

- Would the QED meson states show up as resonances in Bhabha scattering? It has been argued that they would not, because the strong electric field is instrumental in their creation [80]. However, if the decay can occur in a field-free region, the creation should also be possible in the absence of an external field, although the

500

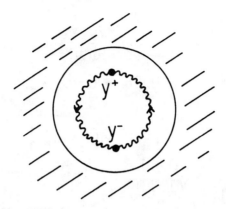

Figure 10: Bag containing two charged fermionic constituents y$^+$ and y$^-$, as schematic model for an extended X^0-boson.

cross section for this process might be very small. Thus, the states should be visible in e$^+$-e$^-$ scattering unless time reversal invariance is broken, for which there is no indication.

A Schematic Model of New Extended Particles

In view of the fact that none of the specific models discussed above appears to be a good candidate for the particle interpretation of the GSI events, it may be worthwhile to study a schematic model of extended particles that has no direct connection with known physics. Such a model, which resembles those studied in the context of an abnormal QED vacuum state [77,79,80,81] was recently suggested by Schramm et al. [87]. Here one postulates the existence of a *new* kind of electrically charged light fermions, called Y$^+$ and Y$^-$, that interact among each other by an also *new*, confining interaction. This interaction is *not* assumed to be of electrodynamic origin, although it might be described by a strongly coupled U(1) gauge group, and the Y-particles are not assumed to have any relation to electrons.

Following the experience with QCD, the bound states of a y$^+$y$^-$-pair were described in terms of the MIT-bag model but, of course, with a value B_X for the bag constant that has no relation at all with the bag constant of QCD. (It will turn out that the motion of the Y-constituents in the bag is non-relativistic for the preferred values of the parameters; a confining potential model would, therefore, furnish a more appropriate description. Such a model yields only minor quantitative modifications which are not essential in the present context.) Denoting the constituent mass by m_y, the total energy of a Y-bag of radius R_X, representing the X^0-boson, is:

$$M_X = \frac{4\pi}{3} R_X^3 B_X + 2\frac{\xi(m_y R_X)}{R_X}, \tag{17}$$

where ξ is the dimensionless eigenvalue of the Dirac equation with bag boundary condition. ξ depends on $m_y R_X$; for $m_y = 0$ the lowest eigenvalue is $\xi = 2.04$. The expression (17) must be minimized with respect to the bag radius R_X, which is thus determined

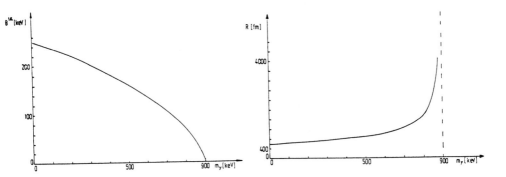

Figure 11: Bag constant (left axis) and bag radius (right axis) as function of the constituent mass for $M_X = 1.8$ MeV.

as function of the two parameters m_y and B_X of the model. If the total bag mass M_X is fixed, e.g. to the energy of one of the e^+e^- lines seen at GSI, a one parameter family of solutions is obtained. The solutions are shown in Fig. 11 versus m_y as independent parameter. Taking $M_X = 1.8$ MeV, one finds $B_X^{1/4} \approx 250$ keV for $m_y = 0$, and the bag radius is about 600 fm. For larger values of m_y one finds that B_X decreases while R_X increases, which is desirable as we shall see. (Of course, m_y must remain smaller than $\frac{1}{2} M_X = 900$ keV.) A reasonable solution is, e.g., $m_y = 880$ keV with $R_X = 4200$ fm [87].

In order to be compatible with the GSI data, the model must satisfy several minimal requirements:

- The spectrum of excited states must be dense, but not too dense. In our schematic bag model, excited states are separated by $\Delta E \approx n\pi^2 / R_X^2 m_y$, i.e. about 40 keV for $R_X = 4200$ fm. This is not incompatible with the splitting between the lines observed at GSI.

- The size of the particles should be sufficiently large to evade the limits posed by the beam-dump experiments, i.e. R_X should not be smaller than about 1000 fm.

- The lifetime against electromagnetic decay must be longer than about 10^{-13}s, because of the bounds from QED and from Bhabha scattering. Since the decay rates are proportional to R_X^{-3}, large radii are preferred.

- An efficient production mechanism requires that the strong electric fields present in the heavy ion collision can strongly interact with the overall neutral X^0 boson. This interaction is proportional to the electric polarizability of the particle, which grows as the volume, R_X^3.

The rates for decay into an e^+e^--pair or into two photons according to the Feynman diagrams in Fig. 7 are of order

$$\Gamma(y^+y^- \to e^+e^-), \Gamma(y^+y^- \to \gamma\gamma) \sim \left(\frac{\alpha}{M_X}\right)^2 |\psi_{yy}(0)|^2 \sim \frac{\alpha^2 M_X}{(M_X R_X)^3}, \qquad (18)$$

502

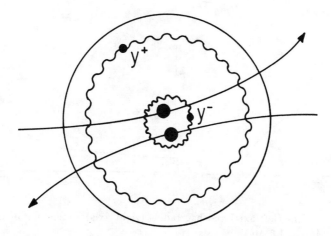

Figure 12: Schematic illustration of the strong binding of a composite X^0-boson in the field of two heavy ions.

where ψ_{yy} is the relative wavefunction of the constituents. For $M_X = 1.8$ MeV and $R_X = 4200$ fm one finds lifetimes between 10^{-14}s and 10^{-13}s, but the precise value will depend on details of the relative wavefunction. Even larger radii are preferable from this point of view.

The interaction energy between our composite X^0-boson and the electric field of the heavy ions cannot be calculated perturbatively, because the available field energy is much larger than the particle rest mass. To wit, the total available field energy inside the bag volume is

$$E_{int} \sim (Z\alpha)^2 \int_{R_N}^{R_X} \frac{1}{r^4} r^2 dr \approx \frac{(Z\alpha)^2}{R_N} \approx 20 \text{ MeV}, \qquad (19)$$

where $R_N \approx 15$ fm is the radius of the nuclear charge distribution. Schramm et al. [87] calculated the interaction energy by solving the Dirac equation for the $y^\pm$ with the two nuclei at the centre of the bag. This leads to a large reduction of the effective mass of the bag state in the presence of the two heavy ions, e.g. two uranium nuclei, from 1.8 MeV down to a mere 46 keV in the case of the special parameter set ($m_y = 880$ keV).

The reason for this effect is just the same as that for the strong binding of K-shell electrons in the field of two heavy nuclei, which was discussed in detail in the first section of this report. The negatively charged constituent y^- becomes bound in a tight orbit around the two nuclei, while the positively charged constituent y^+ is repelled toward the edge of the bag, as illustrated in Fig. 12. The net result of this rearrangement is a stronly attractive interaction between the X^0-boson and the two nuclei.

This has two major advantages: First, the lowering of the effective mass due to binding will increase the production cross section tremendously. This was already pointed out earlier, where it was estimated that the electromagnetic production cross section could approach the desired value of about 100 μb, if the effective mass is reduced below 300 keV. Another way to estimate the production cross section in a heavy ion collision is to neglect the presence of the bag and to calculate the total cross section for production of y^+y^--pairs in the same way as for e^+e^--pairs in section 1. In this way one obtains similar values.

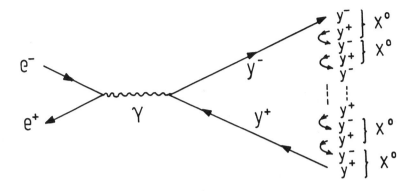

Figure 13: Formation of a jet of y^+y^--pairs in electron-positron collisions at high energy.

Secondly, the production of the X^0-boson occurs in a bound state that moves with the centre of mass of the heavy ions. When the nuclei separate, the binding is strongly reduced, and the X^0-boson may be released almost at rest in the centre of mass, as desired according to the experimental results. Unfortunately, for the specific set of parameters discussed here in detail the bag remains bound even to the individual nuclei [87]. A certain fraction of X^0-bosons may be stripped off by the action of the collision dynamics, but these will not be slowly moving in the c.m. frame. The particles that remain bound would decay in the presence of a third body, hence its decay products (e^+ and e^-) would not exhibit the desired two-body characteristics. This difficulty has not yet been satisfactorily resolved.

Is this schematic model compatible with what we know about QED, atomic and high energy physics? Although we have constructed the model in such a way that the e^+e^--width of any single particle state does not violate the bounds imposed by the precision experiments, the answer is not simple, because there exists an infinite number of excited states that can contribute to certain physical processes. Let us discuss two specific examples. In high energy e^+-e^- collisions a pair of constituents y^+y^- could be created with high relative momentum. Because there is a confining force between the y^+ and y^- one would expect that a force string extends between the two particles that finally breaks into a multitude of y^+y^--pairs that eventually form X^0-bosons, just as the glue string between a high energy quark-antiquark pair fragments into hadrons (see Fig. 13).

However, if one estimates the breaking rate of the y^+-y^- string according to the standard Schwinger formula [88]

$$\frac{dP}{d^4x} = \frac{\kappa^2}{4\pi^3} \sum_{n=1}^{\infty} \frac{1}{n^2} \exp(-\frac{n\pi m_y^2}{\kappa}), \tag{20}$$

where $\kappa = (32\alpha_y B_X/3)^{1/2}$ is the string constant, one encounters a surprise. For m_y = 880 keV one has $B_X^{1/4} \approx 25$ keV and $\kappa \approx (45 \text{ keV})^2 = 1$ MeV/Å(with $\alpha_y = 1$). Multiplying the expression (20) by the volume $V = E_{yy}/B_X$ when the string is fully extended, one finds that the breaking rate is negligible by all comparisons:

504

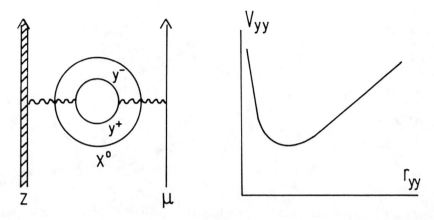

Figure 14: Left: Contribution of virtual X^0-boson exchange to vacuum polarization. Right: y-y potential with short-range repulsion and long-range confining attraction.

$$\frac{dP}{dt} = \frac{8\alpha_y}{3\pi^3} E_{yy} \exp(-\frac{\pi m_y^2}{\kappa}) = \frac{1}{12} E_{yy} \exp(-1200). \tag{21}$$

Clearly, formation of a high energy y^+y^- jet does not occur. But then what are the observable final states predicted by this model? At the moment we simply do not know the answer to this question.

The second kind of process, where the large number of internally excited states becomes relevant, is the exchange of virtual X^0-bosons, e.g. in the vacuum polarization correction to energy levels in muonic atoms (see Fig. 14). The interesting muonic states have an orbital radius of 100 fm or less, i.e. they are much smaller than the X^0-boson states in our schematic model. It appears therefore justified to neglect the confinement force between the virtual y^+y^--pair, so that the total contribution of all excited X^0-boson states to the vacuum polarization can be estimated by the Feynman diagram shown in the left-hand part of Fig. 14 with free $y^\pm$ propagators. Because m_y is not much different from the electron mass m_e, this diagram yields a contribution of almost the same size as the standard electronic vacuum polarization, and thus is in flagrant contradiction to experiment. Is there a way out of this difficulty? One mechanism that could possibly save our model is a short-range repulsive force between the y^+y^--pair, so that the total y-y potential looks as illustrated in the right-hand part of Fig. 14. If the repulsion is sufficiently strong it would probably reduce the contribution of the virtual y^+y^--loop. However, one should not be too optimistic about the prospects of saving the schematic extended particle model with the help of such "tricks".

In conclusion, it seems fair to say that no convincing candidate for a model of extended neutral particles exists at present that could provide an explanation for the e^+e^- peaks observed at GSI. The speculations about an "abnormal", confining phase of QED have not been supported by results from lattice gauge theory. A schematic model of composite X^0-bosons runs into serious difficulties when the contribution to vacuum polarization in muonic atoms is computed. If the GSI events are really caused by the two-body decay of a neutral particle, its nature remains a mystery.

NON-PARTICLE MODELS

General Considerations

The difficulty of accomodating new neutral particles in the framework of established physics and the lack of success of all attempts to find their trace in other phenomena beside the GSI electron-positron peaks has motivated the search for explanations of the GSI events that do not involve new particles. Because these models must involve phenomena which almost look like particle decay, but not quite so, it is useful to remind oneself of what constitutes a particle and what not. A "particle" is a quantum mechanical state which has a definite value of energy, momentum, invariant mass, spin, and parity, or expressed in somewhat more formal terms, which transforms under an irreducible representation of the Poincaré group. In the eyes of an experimentalist, any system that decays strictly back-to-back into an electron-positron pair of equal energy in some reference frame is a particle, and everything else is not a particle. It is with phenomena of the latter kind that we are concerned in this section.

Broadly speaking, three types of models not involving new particles have been proposed: (1) Atomic physics models, invoking some complex process that occurs during the collision of two heavy ions; (2) models which are based on some specific nonlinear property of QED in strong fields; and (3) models which exploit some scheme involving pair conversion processes of nuclear excited states. Some aspects of these models have been briefly touched upon in the previous sections, but we will here discuss all non-particle models in connection.

Models involving Atomic Physics

The ·possibility that complex atomic excitation mechanisms can produce structures in the electron and positron spectra observed in heavy ion collisions was emphasized by Lichten et al. [89,91]. Indeed, it is well known that the matrix elements $\langle \varphi_k | \partial/\partial t | \varphi_j \rangle$ in the coupled channel equations can have pronounced structures at some internuclear distance R_0, if one goes beyond the monopole approximation. A typical example is provided by the radial coupling between the $2p_{1/2}\sigma$ and the $2p_{3/2}\sigma$ states, which has a large, broad peak at $R_0 \approx 500$ fm. Assuming an analogous coupling also to states of the negative energy continuum, structures emerged in the positron spectrum [89]. However, a dynamical calculation by Reinhardt et al. [90] showed that this phenomenon is absent if one uses realistic single-particle matrix elements in the coupled channel equations.

It was argued by Lichten that this absence of dynamic structures might be an artefact of the single-particle approximation, and that many-body correlation effects might possibly yield highly structured matrix elements at some particular value of R [91]. No indication was given as to the specific nature of these correlations, and it was not explained why they could be as narrow as the line structures seen in the experiment.

A specific mechanism for narrow structures was considered by de Reus et al. [92], who took up an old suggestion that sudden rearrangements could occur in the ionic charge clouds when the inner electron shells begin to penetrate each other [93]. In principle, a sudden change in the screening potential could lead to rather sharp structures in *all* matrix elements in the coupled equations, at a characteristic two-centre distance $R_0 \approx 2000$ fm. The contribution to the pair creation amplitude from this structure

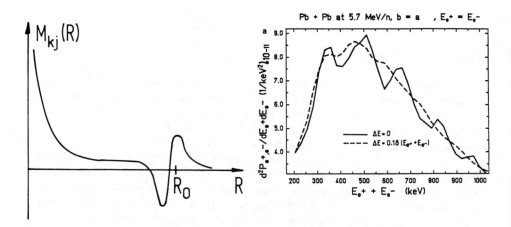

Figure 15: Left: Assumed structure in the matrix elements for dynamical pair production, attributed to sudden change in the screening potential. Right: Small structures in the sum-energy spectrum disappear after averaging over experimental cuts.

would interfere with the main contribution created at very small distances $R < 50$ fm by the strong electric fields. Depending on the value of the energy of the electron or positron this interference would be constructive or destructive.

Further study of this model revealed that such a general structure in the dynamic matrix elements would ruin the good agreement of the calculations with experimental data on the impact parameter dependence of K-shell ionization [94]. This problem could be avoided, if the structure in the matrix elements at R_0 was assumed to be of a particular nature, i.e. extremely narrow ($\Delta R \approx 50$ fm) and oscillating with vanishing total integral. Then the structure does not contribute to excitation processes with low average energy transfer, such as ionization, but it is still actively contributing to pair creation. Unfortunately, the effect of the structures was much reduced, even in the singles spectra [94]. The effect essentially disappeared in the sum-energy coincidence spectrum, when the same wedge-shaped cut was taken as in the EPOS experiment, as shown in the right-hand part of Fig. 15. Besides the fact that the model does not seem to work, it is not clear at present whether the assumed sudden rearrangements actually exist.

Several authors have claimed that the continuum matrix elements of the two-centre Dirac equation could contain resonance structures even in the adiabatic approximation, because charged particle may be captured along the molecular axis between the two nuclei [107-110]. In one space dimension this is indeed correct, but the phenomenon does not seem to occur for the real three-dimensional two-centre potential. A phase shift analysis of the positron continuum of the two-centre Dirac equation did not reveal any resonance structure except the one caused by the supercritical bound state [99]. But even if such resonances would occur, their specific location would vary with nuclear distance R, so that it is hard to understand how they could result in narrow spectral structures.

CONCLUSIONS AND FUTURE PROSPECTS

Before discussing some possible future prospects of the study of QED of strong fields with heavy ions, let us briefly summarize the results of the previous sections:

- The narrow positron peaks and the even narrower e^+e^- coincidence lines appear to be well established experimentally. The different experiments give a consistent picture overall, but a few aspects, e.g. the beam energy dependence, require further investigation. It has also not been clarified, precisely under what conditions the various lines appear and disappear, and why.

- From an experimental point of view the three most important questions that remain are:

    1. Are the electron and positron really emitted back to back, i.e. do the coincident pairs have a definite invariant mass?
    2. What is the lowest nuclear charge Z for which the lines can be observed?
    3. Is nuclear pair decay really excluded?

- On the side of theory, there is still some room left for the "new particle" interpretation, but it is shrinking rapidly. Elementary particles have been conclusively ruled out, and the remaining alternatives are rather exotic. It is hard to believe that any of them could be true.

- Other explanations have also been unsuccessful, and the "principle of the most conservative assumption" is the only serious argument presently in favour of nuclear pair decay.

- Bhabha scattering is the unique method to look for neutral particles without specific model-dependent assumptions, but probably the required sensitivity can only be reached with an e^+-e^- collider. It is not yet clear whether such an experiment is feasible.

Finally, it may be good to come back to the original aim of the QED experiments with heavy ions, i.e. detecting spontaneous pair creation in supercritical electric fields! Two routes are open to attack this problem. First, a distinct line is predicted to appear in the positron spectrum for sufficiently long nuclear time delay. Such long reaction times have not yet been observed for systems with $Z_u > 170$, but they are known to occur for lighter collision systems. Maybe they can be found and put to use in the positron experiments with improved methods of selecting nuclear final states, e.g. collisions with large mass transfer, large negative Q-value, etc.

The second route becomes viable with the advent of fully stripped beams of very heavy ions, now at Berkeley, and in the near future at GSI. Bringing the K-vacancies into the collision increases the positron yield almost by two orders of magnitude [100]. Since this is true for spontaneously emitted positrons as well as for the dynamic background, the use of fully stripped beams alone is not enough. However, one could make use of the fact that there is no background in the electron spectrum due to ionized atomic electrons if the projectile and target nuclei are both fully stripped. Each emitted electron is then associated with a partner positron. It was recently demonstrated by O. Graf et al. [101]

508

that, under such conditions, delayed collisions would give a unique signal for spontaneous pair creation due to a change in the angular correlation between electron and positron that occurs only in supercritical systems: For nuclear delay times of 2×10^{-21}s the normal forward correlation turns into a backward correlation.

Of course, the feasibility of such experiments depends on many things, e.g. the magnitude of the background from nuclear pair decay, but the situation does not appear hopeless. With further studies of the yet unexplained narrow e^+e^- coincidence lines, and with the prospects of colliding fully stripped uranium beams, the future promises to remain exciting.

References

[1] W. Greiner, B. Müller, and J. Rafelski, "Quantum Electrodynamics of Strong Fields", Springer, Berlin-Heidelberg (1985).

[2] J. Schweppe, A. Gruppe, K. Bethge, H. Bokemeyer, T. Cowan, H. Folger, J.S. Greenberg, H. Grein, S. Ito, R. Schule, D. Schwalm, K.E. Stiebing, N. Trautmann, P. Vincent, and M. Waldschmidt, Phys. Rev. Lett. 51:2261 (1983).

[3] M. Clemente, E. Berdermann, P. Kienle, H. Tsertos, W. Wagner, C. Kozhuharov, F. Bosch, and W. Koenig, Phys. Lett. B137:41 (1984).

[4] T. Cowan, H. Backe, M. Begemann, K. Bethge, H. Bokemeyer, H. Folger, J.S. Greenberg, H. Grein, A. Gruppe, Y. Kido, M. Klüver, D. Schwalm, J. Schweppe, K.E. Stiebing, N. Trautmann, and P. Vincent, Phys. Rev. Lett. 54:1761 (1985).

[5] H. Tsertos, F. Bosch, P. Kienle, W. Koenig, C. Kozhuharov, E. Berdermann, S. Huchler, and W. Wagner, Z. Phys. A326:235 (1987).

[6] W. Koenig, F. Bosch, P. Kienle, C. Kozhuharov, H. Tsertos, E. Berdermann, S. Huchler, and W. Wagner, Z. Phys. A328:129 (1987).

[7] T. Cowan, H. Backe, K. Bethge, H. Bokemeyer, H. Folger, J.S. Greenberg, K. Sakaguchi, D. Schwalm, J. Schweppe, K.E. Stiebing, P. Vincent, Phys. Rev. Lett. 56:444 (1986).

[8] "Physics of Strong Fields", W. Greiner, ed., NATO Advanced Study Institute Series B, vol. 153, Plenum, New York (1986).

[9] T.E. Cowan and J.S. Greenberg, Narrow correlated positron-electron peaks from superheavy collision systems, in: ref. [8], p. 111.

[10] E. Berdermann, F. Bosch, P. Kienle, W. Koenig, C. Kozhuharov, H. Tsertos, S. Schuhbeck, S. Huchler, J. Kemmer, and A. Schröter, Nucl. Phys. A488:683c (1988)

[11] B. Blank, E. Bozek, E. Ditzel, H. Friedemann, H. Jäger, E. Kankeleit, G. Klotz-Engmann, M. Krämer, V. Lips, C. Müntz, H. Oeschler, A. Piechaczek, M. Rhein, I. Schall, Sha Yin, and C. Wille, in: "GSI Scientific Report 1987", p. 175, Report GSI-88-1 (1988).

[12] A. Schäfer, J. Reinhardt, B. Müller, W. Greiner, and G. Soff, J. Phys. G11:L69 (1985).

[13] A.B. Balantekin, C. Bottcher, M. Strayer, and S.J. Lee, Phys. Rev. Lett. 55:461 (1985).

[14] R.S. Van Dyck, P.B. Schwinberg, and H.G. Dehmelt, in: "Atomic Physics 9", p. 53, ed. R.S. Van Dyck and E.N. Fortson, World Scientific, Singapore (1984).

[15] P. Mohr, in: ref. [8], p.17. [The value of α was replaced by the new recommended value of the 1986 fit of fundamental physical constants, see Physics Today, August 1987.]

[16] J. Reinhardt, A. Schäfer, B. Müller, and W. Greiner, Phys. Rev. C33:194 (1986).

[17] L.M. Krauss and M. Zeller, Phys. Rev. D34:3385 (1986).

[18] R. Barbieri and T.E.O. Ericson, Phys. Lett. B57:270 (1975); U.E. Schröder, Mod. Phys. Lett. A1:157 (1986).

[19] A. Zee, Phys. Lett. B172:377 (1986).

[20] A. Schäfer, J. Reinhardt, W. Greiner, and B. Müller, Mod. Phys. Lett. A1:1 (1986).

[21] A. Chodos and L.C.R. Wijewardhana, Phys. Rev. Lett. 56:302 (1986).

[22] D. Carrier, A. Chodos, and L.C.R. Wijewardhana, Phys. Rev. D34:1332 (1986).

[23] B. Müller and J. Reinhardt, Phys. Rev. Lett. 56:2108 (1986).

[24] A. Schäfer, J. Reinhardt, B. Müller, and W. Greiner, Z. Phys. A324:243 (1986).

[25] N.C. Mukhopadhyay and A. Zehnder, Phys. Rev. Lett. 56:206 (1986).

[26] G. Mageras, P. Franzini, P.M. Tuts, S. Youssef, T. Zhao, J. Lee-Franzini, and R.D. Schamberger, Phys. Rev. Lett. 56:2672 (1986).

[27] T. Bowcock et al. (CLEO collaboration), Phys. Rev. Lett. 56:2676 (1986).

[28] H. Albrecht et al. (ARGUS collaboration), Phys. Lett. B179:403 (1986).

[29] R.D. Peccei, T.T. Wu, and Y. Yanagida, Phys. Lett. B172:435 (1986).

[30] L.M. Krauss and F. Wilczek, Phys. Lett. B173:189 (1986).

[31] E. Ma, Phys. Rev. D34:293 (1986);

[32] M. Suzuki, Phys. Lett. B175:364 (1986).

[33] C.M. Hofman, Phys. Rev. D34:217 (1986).

510

[34] N.J. Baker, H.A. Gordon, D.M. Lazarus, V.A. Polychronakos, P. Rehak, M.J. Tannenbaum, J. Egger, W.D. Herold, H. Kaspar, V. Chaloupka, E.A. Jagel, H.J. Lubatti, C. Alliegro, C. Campagnari, P.S. Cooper, N.J. Hadley, A.M. Lee, and M.E. Zeller, Phys. Rev. Lett. 59:2828 (1987).

[35] E. Massó, Phys. Lett. B181:388 (1986).

[36] R. Eichler, L. Felawka, N. Kraus, C. Niebuhr, H.K. Walter, S. Egli, R. Engfer, Ch. Grab, E.A. Hermes, H.S. Pruys, A. van der Schaaf, W. Bertl, N. Lordong, U. Bellgardt, G. Otter, T. Kozlowski, and J. Martino, Phys. Lett. B175:101 (1986).

[37] A.L. Hallin, F.P. CalaPrice, R.W. Dunford, and A.B. McDonald, Phys. Rev. Lett. 57:2105 (1986).

[38] M.J. Savage, R.D. McKeown, B.W. Filippone, and L.W. Mitchell, Phys. Rev. Lett. 57:178 (1986).

[39] W.A. Bardeen, R.D. Peccei, and T. Yanagida, Nucl. Phys. B279:401 (1987).

[40] L.M. Krauss and D.J. Nash, Phys. Lett. B202:560 (1988).

[41] F.W.N. deBoer, K. Abrahams, A. Balanda, H. Bokemeyer, R. van Dantzig, J.F.W. Jansen, B. Kotlinski, M.J.A. de Voigt, and J. van Klinken, Phys. Lett. B180:4 (1986).

[42] C.V.K. Baba, D. Indumathi, A. Roy, and S.C. Vaidya, Phys. Lett. B180:406 (1986).

[43] C.V.M. Datar, S. Fortier, S. Gales, E Hourani, H. Langevin, J.M. Maison, and C.P. Massolo, Phys. Rev. C37:250 (1988).

[44] F.W.N. deBoer, J. Deutsch, J. Lehmann, R. Prieels, and J. Steyaert, J. Phys. G14:L131 (1988).

[45] F.T. Avignone III, C. Baktash, W.C. Barker, F.P. CalaPrice, R.W. Dunford, W.C. Haxton, D. Kahana, R.T. Konzes, H.S. Miley, D.M. Moltz, Phys. Rev. D37:618 (1988).

[46] M.J. Savage, B.W. Filippone, and L.W. Mitchell, Phys. Rev. D37:1134 (1988).

[47] S.J. Brodsky, E. Mottola, I.J. Muzinich, and M. Soldate, Phys. Rev. Lett. 56:1763 (1986).

[48] Y.S. Tsai, Phys. Rev. D34:1326 (1986).

[49] H.A. Olsen, Phys. Rev. D36:959 (1987).

[50] A. Konaka, K. Imai, H. Kobayashi, A. Masaike, K. Miyake, T. Nakamura, N. Nagamine, N. Sasao, A. Enomoto, Y. Kukushima, E. Kikutani, H. Koiso, H. Matsumoto, K. Nakahara, S. Ohsawa, T. Taniguchi, I. Sato, and J. Urakawa, Phys. Rev. Lett. 57:659 (1986).

[51] M. Davier, J. Jeanjean, and H. Nguyen Ngoc, Phys. Lett. B180:295 (1986).

[52] E.M. Riordan, M.W. Krasny, K. Lang, P. de Barbaro, A. Bodek, S. Dasu, N. Varelas, X. Wang, R. Arnold, D. Benton, P.Bosted, L. Clogher, A. Lung, S. Rock, Z. Szalata, B.W. Filippone, R.C. Walker, J.D. Bjorken, M. Crisler, A. Para, J. Lambert, J. Button-Shafer, B. Debebe, M. Frodyma, R.S. Hicks, G.A. Peterson, and R. Gearhart, Phys. Rev. Lett. 59:755 (1987).

[53] D.J. Bechis, T.W. Dombeck, R.W. Ellsworth, E.V. Sager, P.H. Steinberg, L.J. Tieg, J.K. Joh, and R.L. Weitz, Phys. Rev. Lett. 42:1511 (1979).

[54] C.N. Brown, W.E. Cooper, D.A. Finley,A.M. Jonckheere, H. Jostlein, D.M. Kaplan, L.M. Lederman, S.R. Smith, K.B. Luk, R. Gray, R.E. Plaag, J.P. Rutherford, P.B. Straub, K.K. Young, Y. Hemmi, K. Imai, K. Miyake, Y. Sasao, N. Tamura, T. Yoshida, A. Maki, J.A. Crittenden, Y.B. Hsiung, M.R. Adams, H.D. Glass, D.E. Jaffe, R.L. McCarthy, J.R. Hubbard, and Ph. Mangeot, Phys. Rev. Lett. 57:2101 (1986).

[55] A. Schäfer, Phys. Lett. B211:207 (1988)

[56] A.O. Barut and J. Kraus, Phys. Lett. B59:175 (1975);

[57] C.Y. Wong and R.L. Becker, Phys. Lett. B182:251 (1986).

[58] B. Lautrup, Phys. Lett. B 62:103 (1976).

[59] A.O. Barut and J. Kraus, Phys. Rev. D16:161 (1977).

[60] K. Geiger, J. Reinhardt, B. Müller, and W. Greiner, Z. Phys. A329:77 (1988).

[61] C.Y. Wong, in: "Windsurfing the Fermi Sea", Vol.2, T.T.S. Kuo and J. Speth, Eds., Elsevier, Amsterdam (1987), p. 296.

[62] M. Grabiak, B. Müller, and W. Greiner, Ann. Phys. 185:284 (1988)

[63] J.J. Griffin, Quadronium: Rosetta stone for the e^+e^- puzzle, preprint PP#88-245, Maryland (1988); J.J. Griffin, Quadronium: Key to the e^+e^- puzzle, preprint, GSI (1988).

[64] S.K. Kim, B. Müller, and W. Greiner, Mod. Phys. Lett. in print (1988).

[65] E.A. Hylleraas and A. Ore, Phys. Rev. 71:493 (1947); Y.K. Ho, Phys. Rev. A33:3584 (1986).

[66] C.Y. Wong, Polyelectron P^{++-} production in heavy ion collisions, preprint, Oak Ridge (1985).

[67] K.G. Lynn, D.N.Lowy, and I.K. Mackenzie, J. Phys. C13:919 (1980).

[68] M.C. Chu and V. Pönisch, Phys. Rev. C33:2222 (1986).

[69] C.Y. Wong, Phys. Rev. Lett. 56:1047 (1986).

[70] H. Lipkin, (Comment), Phys. Rev. Lett. 58:425 (1987).

[71] B. Müller, J. Reinhardt, W. Greiner, and A. Schäfer, J. Phys. G12:L109 (1986); ibid., 12:477 (Erratum).

[72] J.M. Bang, J.M. Hansteen, and L. Kocbach, J. Phys. G13:L281 (1987).

[73] D.H. Jakubassa-Amundsen, Phys. Lett. A120:407 (1986).

[74] C.Y. Wong, Condensation of (e^+e^-) due to short-range non-central attractive forces, preprint, Oak Ridge (1988). C.Y. Wong, Interaction of a neutral composite particle with a strong Coulomb field, preprint, Oak Ridge and Tokyo (1988).

[75] D.C. Ionescu, J. Reinhardt, B. Müller, W. Greiner, and G. Soff, J. Phys. G14:L143 (1988).

[76] R.H.Lemmer and W. Greiner, Phys. Lett. B162:247 (1985).

[77] L.S. Celenza, V.K. Mishra, C.M. Shakin, and K.F. Liu, Phys. Rev. Lett. 57:55 (1986).

[78] L.S. Celenza, C.R. Ji, and C.M. Shakin, Phys. Rev. D36:2144 (1987).

[79] D.G. Caldi and A. Chodos, Phys. Rev. D36:2876 (1987).

[80] Y.J. Ng and Y. Kikuchi, Phys. Rev. D36:2880 (1987).

[81] C.W. Wong, Phys. Rev. D37:3206 (1988).

[82] J. Kogut, E. Dagotto, and A. Kocić, Phys. Rev. Lett. 60:772 (1988).

[83] P. Fomin, V. Gusynin, V. Miransky, and Yu. Sitenko, Riv. Nuovo Cim. 6:1 (1983); V. Miransky, Nuovo Cim. A90:149 (1985).

[84] E. Dagotto and H.W. Wyld, Phys. Lett. B205:73 (1988).

[85] R.D. Peccei, J. Solà, and C. Wetterich, Phys. Rev. D37:2492 (1988).

[86] H.M. Fried and H.T. Cho, On strongly coupled (QED)$_4$ and the emission of e^+e^- pairs in heavy ion collisions, preprint HET#631.

[87] S. Schramm, B. Müller, J. Reinhardt, and W. Greiner, Mod. Phys. Lett. A3:783 (1988).

[88] N.K. Glendenning and T. Matsui, Phys. Rev. D28:2890 (1983).

[89] W. Lichten and A. Robatino, Phys. Rev. Lett. 54:781 (1985).

[90] J. Reinhardt, B. Müller, and W. Greiner, Phys. Rev. Lett. 55:134 (1985).

[91] W. Lichten, in: "Relativistic and QED Effects in Heavy Atoms", AIP Conf. Proceed. 136:319 (1985). W. Lichten, On an alternative to the X^0 particle, preprint, Yale (1986)..

[92] T. de Reus, G. Soff, O. Graf, and W. Greiner, J. Phys. G12:L303 (1986).

[93] R.K. Smith, B. Müller, W. Greiner, J.S. Greenberg, and C.K. Davis, Phys. Rev. Lett. 34:117 (1975).

[94] T. de Reus, U. Müller-Nehler, G. Soff, O. Graf, B. Müller, W. Greiner, Z. Phys. D8:305 (1988).

[95] Yu.N. Demkov and S.Yu. Ovchinnikov, Pis'ma ZhETF 46:14 (1987).

[96] M. O'Connor, Continuum structure in strong electromagnetic fields, M. Sc. thesis, Cape Town, unpublished (1987).

[97] L.G.Ixaru and D. Pantea, Periodicity effects in the two-center Dirac equation, preprint FT-323-1987, Bucharest.

[98] L. G. Ixaru, e^+e^- pair resonant production in heavy-ion collisions: A theoretical scenario, preprint FT-329-1988, Bucharest.

[99] K. Rumrich,W. Greiner, G. Soff, K.H. Wietschorke, and P. Schlüter, Continuum states of the two-centre Dirac equation, J. Phys. B (1989) in print.

[100] U. Müller, T. de Reus, J. Reinhardt, B. Müller, W. Greiner, and G. Soff, Phys. Rev. A37:1149 (1988).

[101] O. Graf, J. Reinhardt, B. Müller, W. Greiner, and G. Soff, Angular correlations of coincident electron-positron pairs produced in heavy ion collisions with nuclear time delay, preprint GSI-88-22, Darmstadt (1988).

AIP Conference Proceedings

		L.C. Number	ISBN
No. 1	Feedback and Dynamic Control of Plasmas – 1970	70-141596	0-88318-100-2
No. 2	Particles and Fields – 1971 (Rochester)	71-184662	0-88318-101-0
No. 3	Thermal Expansion – 1971 (Corning)	72-76970	0-88318-102-9
No. 4	Superconductivity in d- and f-Band Metals (Rochester, 1971)	74-18879	0-88318-103-7
No. 5	Magnetism and Magnetic Materials – 1971 (2 parts) (Chicago)	59-2468	0-88318-104-5
No. 6	Particle Physics (Irvine, 1971)	72-81239	0-88318-105-3
No. 7	Exploring the History of Nuclear Physics – 1972	72-81883	0-88318-106-1
No. 8	Experimental Meson Spectroscopy –1972	72-88226	0-88318-107-X
No. 9	Cyclotrons – 1972 (Vancouver)	72-92798	0-88318-108-8
No. 10	Magnetism and Magnetic Materials – 1972	72-623469	0-88318-109-6
No. 11	Transport Phenomena – 1973 (Brown University Conference)	73-80682	0-88318-110-X
No. 12	Experiments on High Energy Particle Collisions – 1973 (Vanderbilt Conference)	73-81705	0-88318-111–8
No. 13	π-π Scattering – 1973 (Tallahassee Conference)	73-81704	0-88318-112-6
No. 14	Particles and Fields – 1973 (APS/DPF Berkeley)	73-91923	0-88318-113-4
No. 15	High Energy Collisions – 1973 (Stony Brook)	73-92324	0-88318-114-2
No. 16	Causality and Physical Theories (Wayne State University, 1973)	73-93420	0-88318-115-0
No. 17	Thermal Expansion – 1973 (Lake of the Ozarks)	73-94415	0-88318-116-9
No. 18	Magnetism and Magnetic Materials – 1973 (2 parts) (Boston)	59-2468	0-88318-117-7
No. 19	Physics and the Energy Problem – 1974 (APS Chicago)	73-94416	0-88318-118-5
No. 20	Tetrahedrally Bonded Amorphous Semiconductors (Yorktown Heights, 1974)	74-80145	0-88318-119-3
No. 21	Experimental Meson Spectroscopy – 1974 (Boston)	74-82628	0-88318-120-7
No. 22	Neutrinos – 1974 (Philadelphia)	74-82413	0-88318-121-5
No. 23	Particles and Fields – 1974 (APS/DPF Williamsburg)	74-27575	0-88318-122-3
No. 24	Magnetism and Magnetic Materials – 1974 (20th Annual Conference, San Francisco)	75-2647	0-88318-123-1
No. 25	Efficient Use of Energy (The APS Studies on the Technical Aspects of the More Efficient Use of Energy)	75-18227	0-88318-124-X